The Urea Cycle

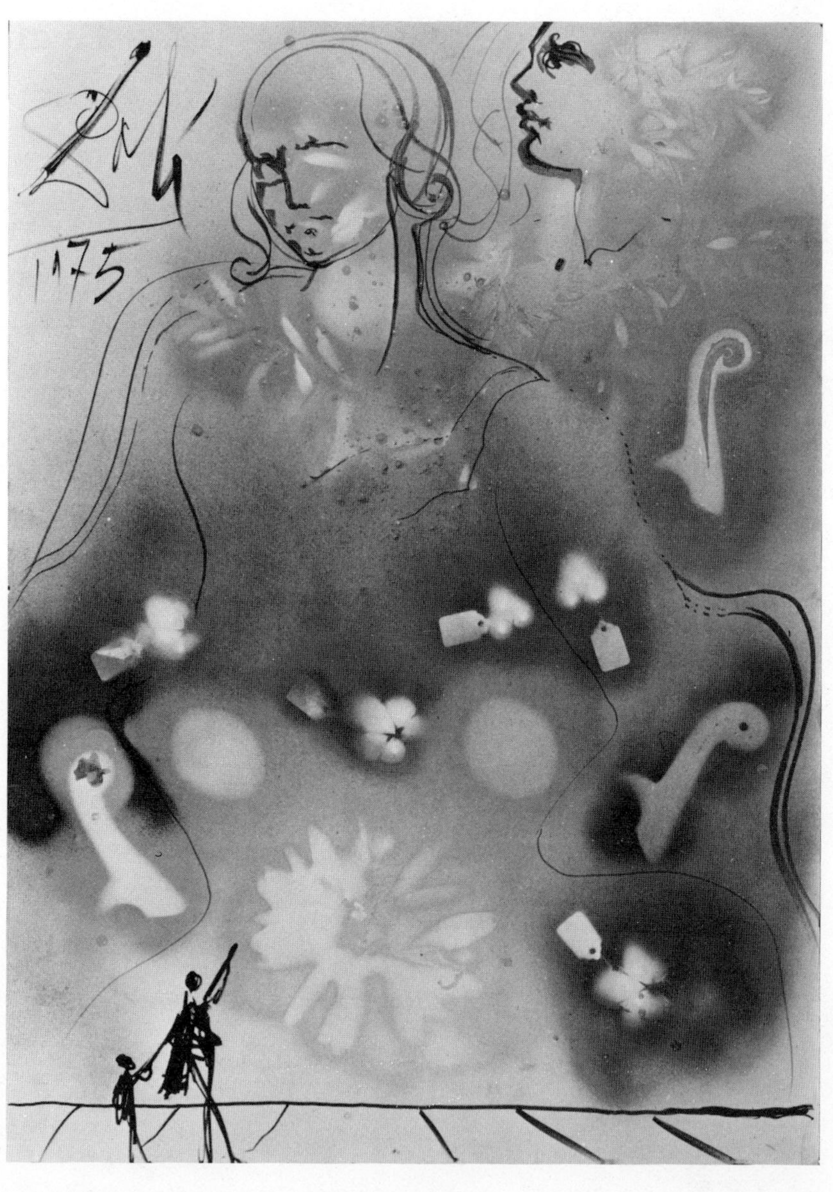

The Urea Cycle

Santiago Grisolia
The University of Kansas Medical Center,
Kansas City, Kansas

Rafael Báguena
The University of Valencia,
Valencia, Spain

Federico Mayor
The Autonomous University of Madrid,
Madrid, Spain

A Wiley-Interscience Publication

John Wiley & Sons

New York / London / Sydney / Toronto

Library of Congress Cataloging in Publication Data:

Main entry under title:

The Urea cycle.

"A Wiley-Interscience publication."
Proceedings of a symposium held at the University of Valencia, Valencia, Spain, 1975.
Includes bibliographical references and index.
1. Urea—Congresses. 2. Biological chemistry—Congresses. 3. Krebs, Sir Hans Adolf. I. Grisolia, Santiago, 1923- II. Báguena, Rafael. III. Mayor, Federico.
QP801.U7U73 599'.01'3 76-7382
ISBN 0-471-32791-3

Printed in the United States of America

10 9 8 7 6 5 4 3 2 1

Conference Participants

Hugo Aebi, Ph.D., Professor
Medizinisch-Chemisches Institut, Bern (Switzerland)

Maria Jose Báguena, Medical Student
Valencia (Spain)

Rafael Báguena Candela, M.D., Professor
Rector Magnífico de la Universidad de Valencia, Valencia (Spain)

Samuel P. Bessman, M.D., Professor
Departments of Pharmacology and Pediatrics, School of Medicine, University of Southern California, Los Angeles, California (U.S.A.)

John T. Brosnan, Ph.D., Assistant Professor
Department of Biochemistry, Memorial University of Newfoundland, St. John's, Newfoundland (Canada)

Jadwiga Bryla, Ph.D., Assistant Professor
Institute of Biochemistry, Warsaw (Poland)

José Carreras Barnés, M.D., Associate Professor
Departamento de Fisiología y Bioquímica, Facultad de Medicina de la Universidad de Barcelona, Barcelona (Spain)

Francesco Cedrangolo, M.D., Professor
Instituto di Chimica Biologica, I Facoltà di Medicina e Chirurgia, Naples (Italy)

Amparo Chabás, Ph.D., Research Associate
Instituto Provincial de Bioquímica Clínica, Universidad Autónoma de Barcelona, Barcelona (Spain)

José Chabás López, Ph.D., Technical Director
Boehringer Mannheim Gmbh., Barcelona (Spain)

Official Photograph of the Symposium in Honor of Sir Hans Krebs Front row (left to right): J. Grisolia, A. Chabás, P. González, L. Raijman, P. Lund, M. Marshall, M. Stubbs-Spry, J. Bryla, V. Shih, and B. Feijoo. Second row: F. Vivanco, R. Mills, J. B. Chappell, M. Tatibana, R. Crist, S. Bessman, J. P. Colombo, R. Báguena, S. Grisolia, M. E. Jones, H. Krebs, S. Ratner, P. Cohen, F. Mayor, J. Kennedy, J. Viña, J. Chabas, M. Walser, V. Rubio, M. Grassl, M. J. Báguena, and J. T. Brosnan. Third row: F. Cedrangolo, G. Illiano, E. de la Morena, K. R. F. Elliott, G. Sóberon, J. I. Fernandez-Alonso, J. Mendelson, and C. Frieden. Fourth row: J. Carreras, G. Ramponi, N. Raiha, H. Aebi, A. Prieto, P. Manao, R. Hems, and G. Sainsbury. Not shown: A. Ventura, G. Forteza, and S. de Miguel.

J. B. Chappell, Ph.D., Professor
Department of Biochemistry, Medical School, University of Bristol, Bristol (England)

Philip P. Cohen, Ph.D., M.D., Professor
Department of Physiological Chemistry, Medical School, University of Wisconsin, Madison, Wisconsin (U. S. A.)

Jean Pierre Colombo, Ph.D., Privatdozent
Chemisches Zentrallaboratorium, Bern (Switzerland)

Robert D. Crist, M.D., Ph.D., Associate Professor
Obstetrics-Gynecology Laboratory, University of Kansas Medical Center, Kansas City, Kansas (U. S. A.)

Keith R. F. Elliott, Ph.D.
Biological Laboratory, The University Canterbury, Kent (England)

Blanca Feijóo, Ph.D., Research Associate
Departamento de Bioquímica, Facultad de Farmacia, Universidad Complutense de Madrid, Madrid (Spain)

Gerónimo Forteza Bover, M.D., Professor
Instituto de Investigaciones Citológicas de la Caja de Ahorros y Monte de Piedad de Valencia, Valencia (Spain)

Carl Frieden, Ph.D., Professor
Department of Biological Chemistry, School of Medicine, Washington University, St. Louis, Missouri (U. S. A.)

Pilar González, Ph.D., Assistant Professor
Departamento de Bioquímica, Facultad de Farmacia, Universidad Complutense de Madrid, Madrid (Spain)

Marianne Grassl, Ph.D.
Boehringer Mannheim Gmbh., Biochemica Werk Tutzing, Tutzing, Obb (Germany)

James Grisolia, B.S.
Yale University, New Haven, Connecticut (U. S. A.)

Santiago Grisolia, M.D., Ph.D., Distinguished Professor
Department of Biochemistry, The University of Kansas Medical Center, Kansas City, Kansas (U.S.A.)

Reginald Hems, B.S.
Metabolic Research Laboratory, Nuffield Department of Clinical Medicine, Radcliffe Infirmary, Oxford (England)

Gennaro Illiano, Ph.D., Associate Professor
Instituto di Chimica Biologica, I Facolta di Medicina e Chirurgia, Naples (Italy)

Mary Ellen Jones, Ph.D., Professor
Department of Biochemistry, School of Medicine, University of Southern California, Los Angeles, California (U. S. A.)

James Kennedy, M.D., Associate Professor
Department of Medicine, University of Kansas Medical Center, Veterans Administration Hospital, Kansas City, Missouri (U.S.A.)

Hans A. Krebs, M.D., Professor
Metabolic Research Laboratory, Nuffield Department of Clinical Medicine, Radcliffe Infirmary, Oxford (England)

José M. López Piñero, M.D., Professor
Cátedra de Historia de la Medicina y Documentación Médica, Facultad de Medicina, Universidad de Valencia, Valencia (Spain)

Patricia Lund, Ph.D.
Metabolic Research Laboratory, Nuffield Department of Clinical Medicine, Radcliffe Infirmary, Oxford (England)

Margaret Marshall, Ph.D., Research Associate
Department of Physiological Chemistry, Medical School, University of Wisconsin, Madison, Wisconsin (U. S. A.)

Federico Mayor Zaragoza, Ph.D., Professor
Departamento de Bioquímica, Facultad de Ciencias, Universidad Autónoma de Madrid, Madrid (Spain)

Judith Mendelson, B.S., Chemist
Department of Biochemistry, The University of Kansas Medical Center, Kansas City, Kansas (U. S. A.)

Sandalio de Miguel, M.D.
Valencia (Spain)

Russell C. Mills, Ph.D.
Associate Vice-Chancellor. The University of Kansas Medical Center, Kansas City, Kansas (U. S. A.)

Enrique de la Morena, M.D.
Departamento de Bioquímica, Fundación Jiménez Díaz, Madrid (Spain)

Eduardo Primo Yúfera, Ph.D., Professor
Presidente del Consejo Superior de Investigaciones Científicas, Madrid (Spain)

Pedro Puig Parellada, Ph.D.
P.E.Y.V.A. Laboratories, Barcelona (Spain)

Niels Räihä, M.D., Professor
Department of Obstetrics and Gynecology, Helsinki University Central Hospital, Helsinki (Finland)

Luisa Raijman, Ph.D., Assistant Professor
Department of Biochemistry, School of Medicine, University of Southern California, Los Angeles, California (U. S. A.)

Giampietro Ramponi, Ph.D., Professor
Instituto di Chimica Biologica, Università di Firenze, Florence (Italy)

Sarah Ratner, Ph.D., Member Emeritus
Department of Biochemistry, The Public Health Research Institute of the City of New York, New York (U. S. A.)

Gillian Sainsbury, B.S.
Metabolic Research Laboratory, Nuffield Department of Clinical Medicine, Radcliffe Infirmary, Oxford (England)

Vivian E. Shih, M.D., Assistant Professor
Department of Neurology, The Massachusetts General Hospital, Boston, Massachusetts (U. S. A.)

Guillermo Soberón, M.D., Ph.D.
Rector de la Universidad Nacional de México, Ciudad de México (Mexico)

Marion Stubbs, B.S.
Metabolic Research Laboratory. Nuffield Department of Clinical Medicine, Radcliffe Infirmary, Oxford (England)

Masamiti Tatibana, M.D., Professor
Department of Biochemistry, School of Medicine, Chiba University, Chiba (Japan)

Antonio Ventura Cervera, M.D.
Valencia (Spain)

José Viña, Medical Student
Valencia (Spain)

José Viña Giner, M.D., Professor
Decano de la Facultad de Medicina, Universidad de Valencia, Valencia (Spain)

Francisco Vivanco, M.D., Professor, Head of Department of Metabolism and Nutrition, Director of Research
Fundación Jiménez Díaz, Madrid (Spain)

Mackenzie Walser, M.D., Professor
Department of Pharmacology and Experimental Therapeutics, School of Medicine, The Johns Hopkins University, Baltimore, Maryland (U. S. A.)

Additional Contributors

C. Bachmann, Ph.D.
Department of Clinical Chemistry, Inselspital, University of Bern, Bern (Switzerland)

M. L. Batshaw, Ph.D.
Department of Pharmacology and Experimental Therapeutics, School of Medicine, The Johns Hopkins University, Baltimore, Maryland (U. S. A.)

E. B. Brown, Ph.D., Associate Vice-Chancellor
University of Kansas Medical Center, Kansas City, Kansas (U. S. A.)

S. W. Brusilow, M.D., Professor
Department of Pediatrics, School of Medicine, The Johns Hopkins University, Baltimore, Maryland (U. S. A.)

T. Caldes, Ph.D., Research Associate
Departamento de Bioquimica, Facultad de Farmacia, Universidad Complutense de Madrid, Madrid (Spain)

Dennis Diederich, M.D., Associate Professor
Department of Medicine, University of Kansas Medical Center, Kansas City, Kansas (U. S. A.)

F. Lavinha, Ph.D.
Department of Neurochemistry, Born-Bunge Foundation, Berchem-Antwerp (Belgium)

A. Lowenthal, Ph.D.
Department of Neurochemistry, Born-Bunge Foundation, Berchem-Antwerp (Belgium)

Masataka Mori, Ph.D., Research Associate
Department of Biochemistry, School of Medicine, Chiba University, Chiba (Japan)

C. Thomas Nuzum, M.D., Assistant Professor
Department of Medicine, University of North Carolina School of Medicine, Chapel Hill, North Carolina (U. S. A.)

Nandita Pal, Ph.D. Research Associate
Department of Pharmacology, School of Medicine, University of Southern California, Los Angeles, California (U. S. A.)

Rafael Palacios, Ph.D., Research Associate
Universidad Nacional de Mexico, Ciudad de Mexico (Mexico)

Vicente Rubio, M.D., Research Associate
Department of Biochemistry, University of Kansas Medical Center, Kansas City, Kansas (U. S. A.)

Katsuya Shigesada, Ph.D., Research Associate
Department of Biochemistry, School of Medicine, Chiba University, Chiba (Japan)

Philip J. Snodgrass, M.D., Professor
Department of Medicine, Indiana School of Medicine, Indianapolis, Indiana (U. S. A.)

H. G. Terheggen, Ph.D.
Department of Pediatrics, Kinder-Krankenhaus der Stadt Koln, Cologne (Germany)

E. Ventura, Ph.D., Research Associate
Departamento de Bioquimica, Facultad de Farmacia, Universidad Complutense de Madrid, Madrid (Spain)

Malcolm Watford, Ph.D.
Metabolic Research Laboratory, Nuffield Department of Clinical Medicine, Radcliffe Infirmary, Oxford (England)

Preface

Urea has the distinction of being the first organic substance produced in the laboratory, and its synthesis marked the transition from inorganic to organic chemistry. Moreover it is the substance that suggested the concept of the first metabolic cycle and, as such, led to the development of modern biochemistry.

We feel that the elucidation and extensive basic knowledge of the urea cycle have reached the point at which clinical application, which has already been initiated, can be markedly extended. Although a great deal of information is available, as in some excellent reviews, there has been no single publication in which the historical, chemical, and physiopathological relationships of the discovery and present status, as well as the future outlook, have been considered together.

During a visit to the University of Kansas, Sir Hans Krebs commented on the importance of recording the development of biochemistry because memories are short and often biased, and the history of medicine and science has, for the most part, been written secondhand. Indeed he suggested that we write about the more modern aspects of the biochemistry of the urea cycle since he himself had just completed a comprehensive historical article. This led to further discussions, and it seemed that it would be not only quite proper to pay homage to Sir Hans, but also of great interest to bring together his urea cycle "family," for (with notable exceptions) much of the confirmation and basic understanding of the urea cycle has been made by his scientific sons, grandsons, and even great-grandsons! We therefore sought a convenient and pleasant location where we could discuss these matters, and to that end, largely through the interest and enthusiasm of Antonio Ventura and Sandalio de Miguel, we selected the University of Valencia.

It seemed to us that it would be quite fitting and proper, and in line with ideas of Sir Hans, to incorporate these proceedings into a book and to devote whatever royalties (which have been generously donated by the participants) the book might yield to create a modest fellowship for a medical student in honor of the late Jose Garcia Blanco, Professor of Biochemistry at the Medical School of the University of Valencia, who a few years earlier had also been a student of Thannhauser in the laboratory in which Sir Hans first discovered the urea cycle. Duplicate records of the proceedings of this meeting, tapes with the papers, discussions, and correspondence, and so forth will be given to the History of Medicine Departments of the Universities of Kansas and Valencia.

Although there are and have been many great scientists, few will remain as widely known as Lavoisier, Newton, and Einstein. However Krebs and his cycles, as is proper, will remain eminent in the history of science. It is a pleasure to acknowledge that we have had the good fortune to have enlisted the imagination of the great painter Salvador Dali who, with his characteristic ability, recognized the historical importance of this event and, at our suggestion, commemorated this great scientific event by a painting that incorporates the chemistry of the urea cycle and its relation to health and disease. The unveiling of this painting at the end of the meeting provided a golden finale to a most exciting symposium. Through the courtesy of Dali, this painting is reproduced as the frontispiece of this book, and numbered posters have been made for the Symposium participants. The original will be deposited at the Figueras Museum.

It is appropriate to indicate that Krebs very early realized the intrinsic importance of the urea cycle as well as many of its physiological implications. At that time he visualized enormous difficulties in arriving at the enzyme level, as shown by a quotation from one of his early masterpiece reviews*:

"These reactions take place only in the intact liver cell. Both the chemical mechanisms used in the cell to bring about these reactions and the physical mechanisms for transmitting the energy required for the urea synthesis are unknown, and there is yet little hope of progress along these lines. Our lack of knowledge of the chemical nature of enzymes makes impossible an understanding of the chemical mechanism even of simple enzyme reactions, and even more remote must be the possibility of penetrating into the complex system of urea synthesis."

On the other hand it must be remembered that the enormous developments in enzyme chemistry and intermediary metabolism, which no one could visualize at the time the urea cycle was discovered, prompted an immediate interest in other metabolic areas and led to the rapid clarification of the bulk of metabolism. Indeed, it was so successful that in about 10 years, about 1946 to 1956, the main metabolic pathways were uncovered. Intermediary metabolism became less fashionable, and in the last 15 years it attracted a smaller percentage of biochemists.

It is said that while at lunch a few years ago, someone made the comment to Sir Hans that both metabolism and enzyme chemistry at this level were

* *Ergeb. Enzymforsch.*, **3**, 247 (1934).

essentially passé. Sir Hans did not comment immediately, but at the end of the meal said, "But it will come back." It will be evident to the reader from the presentations included here that the time to come back to metabolism and to one of the hardest and perhaps most important questions of modern biochemistry—to relate what goes on in the test tube to what happens in the living cell—has already arrived, again initiated largely by Sir Hans and his followers.

The absence of the names of several individuals intimately involved in the development of the urea cycle may have become apparent to the reader. This has not been a matter of omission, although of course we had decided to limit the conference to a manageable number of people; for a variety of reasons, most often prior commitments, some invited individuals were unable to attend.

We dedicate this book to Sir Hans in the names of all the participants, who speak for biochemists throughout the world, with our deepest gratitude and joy in having been part of this homage.

The Editors acknowledge with deep gratitude the financial support of the Ministry of Education and Science; the National Research Council; the University of Valencia; the Spanish Society of Internal Medicine; the Boehringer Foundation; Sandoz, S.A.E.; the Banco de Valencia; the Ministry of Foreign Affairs; and particularly that of the Caja de Ahorros y Monte de Piedad of Valencia, which also provided the facilities of the Institute of Cytology for one of our sessions. As indicated elsewhere, the other sessions were held in the Paraninfo at the University of Valencia.

Santiago Grisolia
Rafael Báguena
Federico Mayor

Kansas City, Kansas
Valencia, Spain
Madrid, Spain
February 1976

Contents

Historical Introduction

The Discovery of the Ornithine Cycle

Hans A. Krebs
Metabolic Research Laboratory,
Nuffield Department of Clinical Medicine,
Radcliffe Infirmary, Oxford, England

APPRENTICESHIP

I started to study urea synthesis in 1931, in my thirty-first year. At that time I was for the first time completely free to choose my own field of research. Until then I had served a long apprenticeship. This began with the study of medicine from 1919 to 1924, continued with a course in chemistry in 1925, and was completed by an assistantship in the laboratory of Otto Warburg in Berlin, lasting four years and three months, from 1926 to 1930. Early in 1930 I was faced with the question of whether to make a living in biochemistry or in clinical medicine. Prospects in biochemistry appeared to be bleak, for the number of openings was small, and biochemistry was not regarded as a "profession." As Warburg forcefully put it to me, "If you want to marry and buy a house and you ask a banker for a loan, he will enquire about your job. When you tell him that you are a biochemist, he is very likely to say 'no,' but if you tell him that you are a doctor, you will get what you need." And so I decided to return to clinical medicine, in the hope that I might be able to combine scientific research with clinical work. I was lucky to find a suitable position as an assistant at the Altona (today part of Hamburg) Municipal Hospital under Professor Leo Lichtwitz, who himself was keenly interested in the application of biochemistry to clinical problems. He was the author of an important book, *Klinische Chemie*, first published in 1913. This hospital, though of very high standing, was not connected with a university, which was a disadvantage. A year later, in 1931, I was offered an opportunity of moving to the University of Freiburg im Breisgau, where Professor S. J. Thannhauser accepted me as one of the "assistants" thanks to a recommendation from a personal friend of mine, Professor Klothilde Gollwitner-Meier, of Berlin, a distinguished and influential physician and cardiophysiologist.

NEW APPROACHES TO THE STUDY OF INTERMEDIARY METABOLISM

It was only at this stage that I became free in the choice of my research topic. When I left Warburg's laboratory, he agreed to support an application for research equipment—mainly the Warburg and the Van Slyke manometric apparatus—on the understanding that I would study a topic of his choice, namely, the proteolytic capacity of tumor tissue. Having demonstrated that tumor tissue is distinguished from most other tissues by a high glycolytic capacity, Warburg thought it useful to compare the proteolytic potential of animal tissues. The work (1) showed that malignant tumor tissue did not differ strikingly in this respect from normal tissues.

I should mention that I had in fact published several independent papers—mainly on cell permeability—before I joined Warburg's laboratory, but although I had been free in the choice of topic, inasmuch as I had not been told by a boss what to do or what not to do, I had been grossly handicapped by lack of training in research and especially by my ignorance of biochemical and physiological techniques. This imposed great limitations on my freedom in choosing a research topic. The experiences in Warburg's laboratory removed some of these limitations.

The basic idea behind my own choice of a problem was a simple one. In Warburg's laboratory I had seen the value of the tissue-slice technique in studying the respiration and glycolysis of animal tissues *in vitro* under controllable conditions. Before Warburg it had not been possible to study respiration and glycolysis satisfactorily *in vitro*, because the earlier tissue preparations—chopped and minced tissue—had lost most of the respiratory and glycolysing activity. Warburg (2, 3) had used the slice technique solely for measuring the overall rate of the degradative reactions which supply energy. My idea was to use slices for the study of other aspects of metabolic processes. I had in mind especially synthetic reactions, and I hoped that slices would also lend themselves to the study of the intermediary metabolism of respiration; I also hoped that a knowledge of normal processes would soon be of value to the study of diseases.

As a first subject I chose to study the synthesis of urea in the liver, because it appeared to be one of the simplest biosynthetic processes and one that occurs at a high rate.

A NEW SALINE SOLUTION AS A PLASMA SUBSTITUTE

Since all previous attempts to obtain a synthesis of urea in isolated liver preparations—other than the perfused liver—had been unsuccessful, I thought it necessary, as a preliminary, to design a medium which simulated blood plasma as closely as possible in respect to the inorganic constituents. This meant modifications of the type of saline solutions which Ringer (4), Locke (5), Tyrode (6), and Warburg (7) had introduced. The media of Ringer, Locke, and Tyrode were grossly deficient in the major plasma constituents—especially bicarbonate and CO_2. They were arrived at mainly on the basis of trial and error, in particular by testing for a medium which maintained the activity of the frog heart. Warburg included bicarbonate and CO_2, but his medium lacked magnesium, phosphate, and sulphate, and the concentration of other ions was not so close to the physiological range as is possible. The idea behind the design of the new saline was the conviction that the ionic concentrations in blood plasma (which are almost identical in all mammalian species and very similar in other vertebrates) are not accidental but have evolved to be optimally attuned to the functions of the various organs. At that time it was already well established that the function of many organs depends on the composition of the ionic environment. The new medium which included magnesium, phosphate, and sulfate at physiological concentrations has subsequently proved decisively superior to all earlier saline plasma substitutes, not only in biochemical, but also in physiological and pharmacological work. It is now widely used.

This medium still differs from plasma in respect to the colloid-osmotic pressure and the organic constituents of plasma such as glucose, fatty acids, amino acids, and hormones. The colloid-osmotic pressure can be adjusted by the addition of commercial bovine serum albumin, and other substances can easily be added; but attention must be paid to maintaining isotonicity.

Before testing the capacity of liver slices to synthesize urea, I thought it necessary to devise a quick, sensitive, and specific method for the determination of urea, because Warburg had conveyed to me the importance of carrying out large numbers of experiments when exploring unknown territories. Many years later, in 1964, Warburg expressed this philosophy in print in a prefatory chapter of the *Annual Reviews of Biochemistry* when he wrote, "A scientist must have the courage to attack the great unsolved problems of his time. Solutions have to be forced by carrying out innumerable experiments without much critical hesitation." So I adapted the urease method of Marshall (8) and Van Slyke (9) to manometry. A solution of urease was added at pH 5.0 to the solution to be examined and the CO_2 evolved was measured (10). This procedure permitted a dozen determinations to be carried out within less than 1 hour. Quantities of approximately 0.05 mg could be determined with an accuracy of a few per cent. Incidentally, the principle of the urease method, introduced by Marshall in 1913, was one of the earliest analytical procedures employing enzymes as analytical tools. Now there are many hundreds of enzymic analytical methods.

PLAN OF WORK

Preliminary tests with tissue slices showed that liver slices synthesize urea at rates expected from what was known for the intact body, and the rates were greatly raised by the addition of ammonium salts or certain amino acids. After this hopeful start I decided to measure systematically the rate of urea synthesis in the presence of a variety of precursors, hoping that the results may throw light on the chemical mechanism of urea synthesis. I had no preconceived idea and no hypothesis about this mechanism but I had in mind Warburg's work (11) on cell respiration in which he had studied the rate of respiration in the presence of cyanide, carbon monoxide, narcotics, and other substances and succeeded in drawing far-reaching conclusions about the properties of the enzymes of respiration. In this work I was joined by a medical student, Kurt Henseleit, who wished to do a piece of research for his M.D. thesis. He proved a very skilful experimenter who rapidly acquired the necessary laboratory techniques.

Questions which suggested themselves for investigation included the following:

1. Is ammonia an obligatory intermediate in the conversion of amino nitrogen to urea nitrogen? If it is, ammonia must yield urea at least as rapidly as amino acids. If it is not, the rate of urea formation from amino acids might be more rapid than the rates from ammonia. (It became known more than 20 years later that half of the urea nitrogen can be derived directly from amino acids, without passing through the stage of ammonia.)
2. How do the rates of urea formation from various amino acids compare?
3. Do substances which had been suspected to be intermediates, for example, cyanate, behave like intermediates in that they can be converted into urea?
4. Do pyrimidines yield urea directly or via ammonia?

THE FIRST CRUCIAL FINDING

So we measured the rate of urea synthesis under many different conditions, and these included the presence of mixtures of ammonium ions and amino acids (10). It was in the course of these experiments that we discovered the exceptionally high rates of urea synthesis when both ornithine and ammonium ions were present. The interpretation of this finding was not at once obvious. It took a full month to find the correct interpretation. At first we were skeptical about the correctness of the observations. Was the ornithine perhaps contaminated with arginine? The answer was no. Then it occurred to us that the effect of ornithine might be related to the presence of arginase in the liver, the enzyme which converts arginine into ornithine and urea, known since the work of Kossel and Dakin (12) in 1904:

$$\text{Arginine} + H_2O \rightarrow \text{ornithine} + \text{urea}$$

Arginase was known to occur in very high activity in the liver of those animals (mammals, amphibians, and certain reptiles) which can synthesize urea from ammonia and amino acids (13, 14). The coincidence of the exceptionally high activity of this enzyme and the occurrence of urea synthesis in the liver raised the suspicion of a connection, but for some weeks we were unable to visualize one, nor had this occurred to any previous investigator. The solution of the problem developed gradually as the ornithine effect was studied in detail.

THE ANALYSIS OF THE ORNITHINE EFFECT

In the first experiments which revealed the ornithine effect, the concentration of ornithine had been high, because it had been the intention to explore whether ornithine can act as a nitrogen donor. When lower ornithine concentrations were tested, the stimulating effect remained, and the final result of this aspect of the work was the discovery that 1 molecule of ornithine can bring about an extra formation of over 20 molecules of urea, provided that ammonia was present. Moreover the amino nitrogen content of the medium as measured by the Van Slyke technique did not decrease during the synthesis of urea, and the total urea nitrogen could be accounted for by the disappearance of ammonia. This established the fact that ornithine acts like a catalyst; it is not used up, and there is no simple stoichiometry between the amount of ornithine present and its effect on the rate of urea production.

When considering the mechanism of this catalytic action, I was guided by the idea that a catalyst must take part in the reaction and form intermediates. The reactions of the intermediates must eventually regenerate ornithine and form urea. Once these postulates had been formulated, it became obvious that arginine fulfilled the requirements of an expected intermediate. This meant that a formation of arginine from ornithine had to be postulated according to the overall equation

$$\text{Ornithine} + CO_2 + 2NH_4^+ \rightarrow \text{arginine}$$

SEARCH FOR INTERMEDIATES

It also was at once obvious that the synthesis of arginine from ornithine must involve more than one step, since four molecules—one ornithine, one CO_2, two NH_3—had to interact. So I began to search for possible intermediates between ornithine and arginine. Paper chemistry suggested that citrulline

$$COOH{\cdot}CH(NH_2){\cdot}CH_2{\cdot}CH_2{\cdot}CH_2{\cdot}NH{\cdot}CO{\cdot}NH_2$$

might play a role as an intermediate. This substance had just been identified independently by two biochemists in entirely different contexts. Wada (15) isolated it in 1930 from watermelons (Citrullus), and Ackermann (16) isolated it as a product of the bacterial degradation of arginine in 1931. I wrote to both and obtained a few milligrams from each, sufficient to do the decisive tests. These entirely fulfilled expectations: they demonstrated the rapid formation of urea in the presence of citrulline and ammonium salts in accordance with the following scheme:

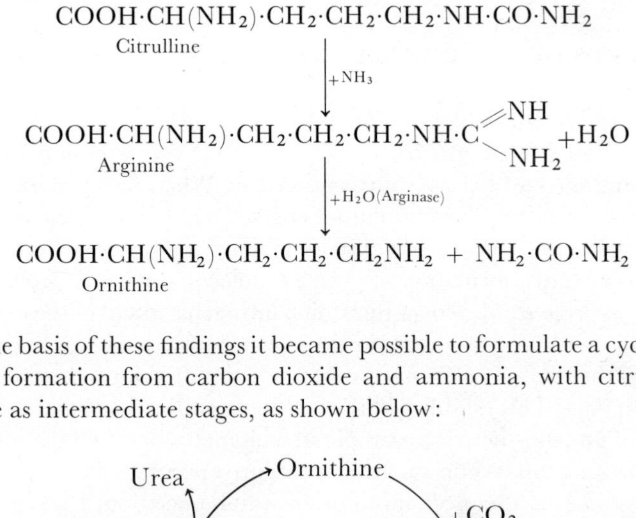

$$\underset{\text{Citrulline}}{COOH{\cdot}CH(NH_2){\cdot}CH_2{\cdot}CH_2{\cdot}CH_2{\cdot}NH{\cdot}CO{\cdot}NH_2}$$
$$\downarrow +NH_3$$
$$\underset{\text{Arginine}}{COOH{\cdot}CH(NH_2){\cdot}CH_2{\cdot}CH_2{\cdot}CH_2{\cdot}NH{\cdot}C}\begin{matrix}\diagup NH \\ \diagdown NH_2\end{matrix} + H_2O$$
$$\downarrow +H_2O\text{(Arginase)}$$
$$\underset{\text{Ornithine}}{COOH{\cdot}CH(NH_2){\cdot}CH_2{\cdot}CH_2{\cdot}CH_2NH_2} + NH_2{\cdot}CO{\cdot}NH_2$$

On the basis of these findings it became possible to formulate a cyclic process of urea formation from carbon dioxide and ammonia, with citrulline and arginine as intermediate stages, as shown below:

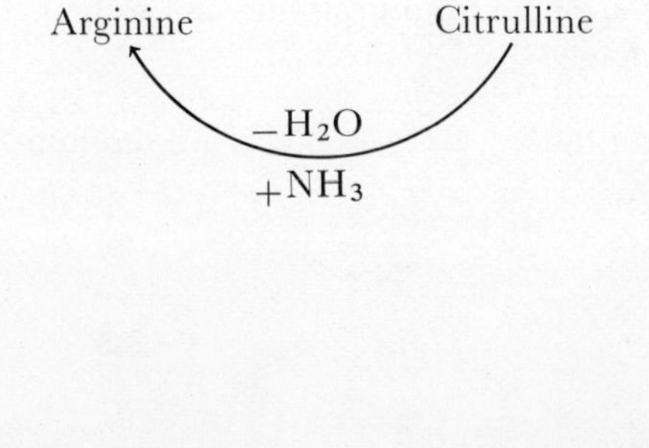

Urea ← $+H_2O$ — Arginine

Ornithine → $+CO_2$, $+NH_3$, $-H_2O$ → Citrulline

Citrulline → $-H_2O$, $+NH_3$ → Arginine

Arginine → $+H_2O$ → Ornithine (+ Urea)

THE ORNITHINE CYCLE AS A PATTERN OF METABOLIC ORGANIZATION

No cycle of this kind—which may be called "metabolic cycle"—had been known before, a cycle in which low molecular intermediates are formed cyclically. Entirely different kinds of cycles of course were familiar to biologists and chemists. There are the biological cycles like the menstrual cycle and the diurnal cycle. There is the life cycle of metamorphosing insects. There is the cell generation cycle. Closest to the metabolic cycles are the catalytic cycles of chemistry, for instance, the actions of heavy metals in catalyzing the knall-gas reaction. The combination of hydrogen and oxygen to form water is catalyzed by platinum or palladium, and this catalysis involves the intermediate formation of the hydride of the catalyst. Meyerhof and Cori spoke of a lactic acid cycle in muscle and liver. Lactate is produced from glycogen under anaerobic conditions during exercise and resynthesized aerobically to glycogen. None of these cycles is strictly analogous to the ornithine cycle, but subsequently a large number of metabolic cycles, exactly analogous to the ornithine cycle, have been discovered. Thus the ornithine cycle revealed a new pattern of the organization of metabolic processes.

LATER ELABORATIONS OF THE CYCLE

The experiments of 1931 and 1932 established the outlines of the pathway by which urea is synthesized in the liver. We carried out a few experiments on urea synthesis in ground-up liver, but all these experiments were negative. Retrospectively we now appreciate that the time was not yet ripe for attempts to unravel details of the enzymic mechanisms. The history of the development of the ornithine cycle illustrates the general experience that at any one time the solution of a problem can be advanced only to a limited extent. Soon seemingly impenetrable walls obstruct progress. After a time, however, advances in collateral fields overcome the barriers—sometimes by circumventing them, sometimes by demolishing piece by piece the barriers of ignorance. Where I had failed, Cohen and Hayano (17) succeeded 14 years later. By that time much had been learned in general about the preservation of complex metabolic processes in cell-free homogenates. Other contributors will discuss these developments in later sessions. But I would like to add some comments on a few points arising from the discovery of the ornithine cycle.

SOME REACTIONS OF THE SCIENTIFIC COMMUNITY

The response of fellow scientists was very typical. I have seen it later in many other situations. The great majority accepted the evidence presented in 1932 as convincing and commented enthusiastically. Warburg arranged a formal invitation for me by the Kaiser Wilhelm Gesellschaft, signed by Max Planck, the President, to speak on the subject in Berlin. Meyerhof invited me to lecture at Heidelberg. Hopkins (18) referred to it in his Presidential Address at the Royal Society in November 1932. Knoop, to whom I submitted the paper for publication in his capacity as an editor of the *Zeitschrift für physiologische Chemie*, commented in a very complimentary way and added that he felt really stupid that it had never occurred to him before that arginase might play a role in the synthesis of urea.

But there were adverse criticisms. In 1934 a Russian physiologist of Leningrad, E. S. London, argued that ornithine had no effect in his experiments on the isolated perfused dog liver (19). In 1942 Trowell, of the Cambridge Physiological Laboratory, reported that the perfused rat liver did not respond to ornithine (20). At the same time Bach and Williamson (21), of the Cambridge Biochemical Laboratory, claimed to have shown that dog liver can form urea from ammonia even when arginase is completely inhibited by high concentrations of ornithine, and they concluded that liver can synthesize urea without participation of arginase. As late as 1956 Bronk and Fisher (22) reported—from my own department at Oxford—that under certain conditions citrulline is less effective than ornithine in promoting urea formation from ammonia, and they concluded that citrulline cannot be an intermediate. The results of London and of Trowell were due to inadequate perfusion technique. The conclusions of Bach and Williamson were based on the wrong assumption that arginine and ornithine readily penetrate liver slices. In fact the rates of penetration are relatively slow. Although the observations of Bronk and Fisher were correct, their interpretation was mistaken. They expected the kinetics of an homogenous solution in the highly compartmented system of living cells. Permeability barriers, often unpredictable, can interfere and may cause deviations, usually of a minor quantitative kind, from a postulated kinetic behavior. At that time, in 1956, the importance of permeability barriers at the plasma membrane and mitochondrial membrane was not yet sufficiently appreciated. Experience shows that to postulate, as Bronk and Fisher did, a second mechanism on the basis of some kinetic discrepancies is far more rash and fanciful than to accept kinetic abnormalities on account of permeability barriers and other complications when dealing with very complex systems.

EVOLUTIONARY ASPECTS

An interesting development was the demonstration by Srb and Horowitz (23) in 1944 that the mold Neurospora can synthesize arginine via ornithine through the reactions of the ornithine cycle, with citrulline as an intermediate stage. Later many investigators showed that this is also true generally for microorganisms and for plants (24). Thus the detoxication of ammonia in vertebrates uses a pathway evolved already at very low levels of life for biosynthetic purposes.

Exactly analogous is the situation in birds in which uric acid is the main end product of nitrogen metabolism. The elucidation of the pathways of uric acid synthesis in birds and of the purine bases required by all organisms as constituents of nucleic acids led to the realization that hypoxanthine is the precursor of the excreted uric acid and that the two pathways are thus essentially the same. So again a general metabolic pathway, that of nucleic acid synthesis, occurring in lower organisms has been used in the course of evolution to serve as a detoxication mechanism for ammonia in higher organisms.

CLINICAL ASPECTS

It might well be asked why, as a clinician working in a department of clinical medicine, I investigated urea synthesis, a normal function of the liver. I was convinced at that time that studying intermediary metabolism in an academically orientated hospital was justified, because I felt that sooner or later information on intermediary metabolism would be relevant to clinical problems.

German clinical medicine emphasized—especially the writings of Ludolf Krehl (25)—that a proper understanding of the nature of disease and of rational therapy depends on intimate knowledge of normal organ function and of the disturbances of function caused by disease, a point of view first clearly expressed by Claude Bernard (26). Traditionally academic clinicians in Germany were expected to participate in the search for this kind of knowledge. In the long run my expectations of clinical relevance were not disappointed. Diseases are now known in which intermediates of the urea cycle accumulate in the body fluids and the urine as a result of inborn errors. Other contributors will go into details of these diseases.

CONCLUDING REMARKS ON THE HISTORY OF THE ORNITHINE CYCLE

If there is a lesson to be drawn from this historical account it is perhaps that the story brings home the importance to progress of new techniques, especially of techniques which make it possible to conduct a large number of experiments, and of studying a phenomenon under many different conditions, with a view to establishing factors which affect the rate of progress. It also illustrates the importance of following up an unexpected and puzzling observation arising in the course of the experiments. Luck, it is true, is necessary, but the more experiments are carried out, the greater is the probability of meeting with luck. The story also shows that adverse criticisms are liable to be raised on the grounds that either the observations are not confirmed or that some other observations do not fit in with the interpretation of the findings. Almost every major development in science meets with criticisms of this kind.

NOTE ON THE STOICHIOMETRIC FORMATION OF ASPARTATE AND CARBAMOYLPHOSPHATE

After this historical account I would like to discuss a new aspect with which I have been recently concerned. For some time I have been puzzled by a problem which, I believe, has not received adequate consideration in the past. When urea is synthesized, half the nitrogen passes through the stage of carbamoylphosphate and half through the stage of aspartate, as indicated by diagrams 1 and 2. Normally most of the nitrogen originates from amino acids and arises in the form of glutamate as the result of transamination of various amino acids with α-oxoglutarate. Some nitrogen arises primarily in the form of ammonia from the amide groups of glutamine and asparagine, from serine, threonine, histidine, from aminopurines, nucleotides, and a few other precursors. As glutamate may form either aspartate or ammonia (see diagram 1), there must be a mechanism by which stoichiometry between carbamoylphosphate and aspartate is achieved.

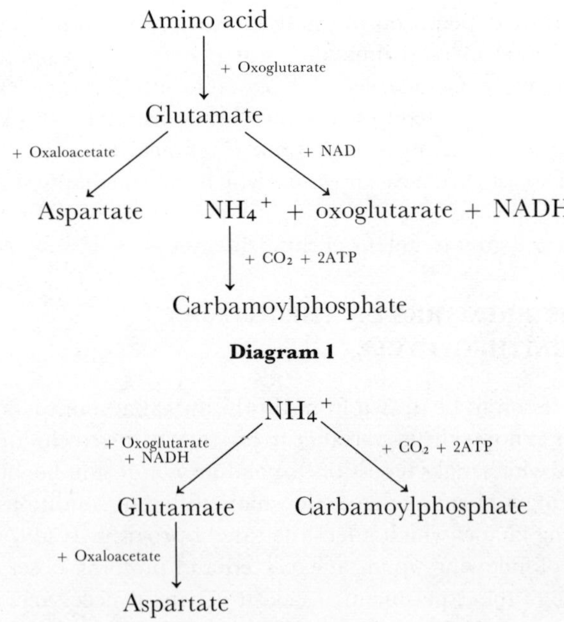

Diagram 1

Diagram 2

I can now offer an explanation which is based on the fact that the value of the mass action ratio of the aspartate aminotransferase system in the normal liver corresponds very nearly to the equilibrium constant. This has been noted in many experiments (27, 28), and it means that the given steady-state concentrations of glutamate, α-oxoglutarate, and oxaloacetate in the liver fix the concentration of aspartate in the region of 0.7 m*M*. Appreciable amounts of aspartate, it should be noted, can be formed in animal tissues only by the aminotransferase reaction (minor amounts can arise by proteolysis). Thus whenever aspartate is removed in the course of the synthesis of urea by the formation of argininosuccinate from citrulline and aspartate, the tendency to reestablish equilibrium automatically replaces the aspartate by the aminotransferase reaction and maintains it at a semiconstant concentration.

Any excess of ammonia formed either by the glutamate dehydrogenase reaction or from the other precursors then enters the carbamoylphosphate pathway leading to the formation of citrulline. Thus stoichiometry is achieved by the thermodynamic limitation of aspartate formation and the disposal of the excess nitrogen by the formation of carbamoylphosphate. The regulatory device is thus exceedingly simple. No regulatory mechanism controlling the carbamoylphosphate synthetase is required except that there must be a mechanism which prevents tissue ammonia from being removed completely. In the liver a steady-state concentration of about 0.7 m*M* ammonia is maintained, and it is not yet clear why carbamoylphosphate synthetase does not lower this concentration.

REFERENCES

1. Warburg, O. (1923). *Biochem. Z.* **142**, 317.
2. Warburg, O. (1908). *Z. Physiol. Chem.* **57**, 1.
3. Warburg, O. (1914). *Ergeb. Physiol.* **14**, 253.
4. Ringer, S. (1883), *J. Physiol.* **4**, 29, 222; (1886), *J. Physiol.* **7**, 291.
5. Locke, F. S. (1894), *Zb. Physiol.* **8**, 166; (1900), *Zb. Physiol.* **14**, 670; (1901), *Zb. Physiol.* **15**, 490.
6. Tyrode, M. J. (1910). *Arch. Int. Pharmacodyn.* **20**, 205.
7. Warburg, O. (1924). *Biochem. Z.* **152**, 51.
8. Marshall, E. K. (1913). *J. Biol. Chem.* **14**, 283; **15**, 487, 495.
9. Cullen, G. E., and van Slyke, D. D. (1914). *J. Biol. Chem.* **19**, 211.
10. Krebs, H. A., and Henseleit, K. (1932). *Z. Physiol. Chem.* **210**, 33.
11. Warburg, O. (1932). *Z. Angew. Chem.* **45**, 1.
12. Kossel, A., and Dakin, H. D. (1904). *Z. Physiol. Chem.* **41**, 321; **42**, 181.
13. Hunter, A., and Dauphinée, J. A. (1925). *Proc. Royal Soc. Ser. B.* **97**, 227.
14. Clementi, A. (1913), *Atti Reale Accad. Lincei* **23**, 612; (1915), *Atti Reale Accad. Lincei* **25**, 483; (1918), *Atti Reale Acad. Lincei* **27**, 299.

15. Wada, M. (1930). *Biochem. Z.* **224**, 420.

16. Ackermann, D. (1931). *Biochem. Z.* **203**, 66.

17. Cohen, P. P., and Hayano, M. (1946). *J. Biol. Chem.* **166**, 239.

18. Hopkins, F. G. (1932), *Nature* **130**, 869; (1933), *Proc. Royal Soc. Ser. B.* **122**, 175.

19. London, E. S., Alexandry, A. K., and Nedswedski, S. W. (1934). *Z. Physiol. Chem.* **227**, 229; (1934). *Z. Physiol. Chem.* **230,** 279.

20. Trowell, O. A. (1942). *J. Physiol.* **100**, 432.

21. Bach, S. J., and Williamson, S. (1942). *Nature* **150**, 575.

22. Bronk, J. R., and Fisher, R. B. (1956). *Biochem. J.* **64**, 118.

23. Srb, A. M., and Horowitz, N. H. (1944). *J. Biol. Chem.* **154**, 129.

24. Wiame, J. M. (1971). *Current Topics in Cell Regulations* **4**, 1.

25. Krehl, L. (1922). *Klinische Physiologie*, 12th ed., Verlag F.C.W. Vogel, Leipzig.

26. Bernard, C. (1957). *An Introduction to the Study of Experimental Medicine*, Dover, New York (originally published in 1865).

27. Williamson, D. H., Lopes-Vieira, O., and Walker, B. (1967). *Biochem. J.* **104**, 497–502.

28. Brosnan, J. T., and Williamson, D. H. (1974). *Biochem. J.* **138**, 453–462.

2

Alternate Paths Artifacts, and Confirmation at the Preenzyme Level

Santiago Grisolia
Department of Biochemistry
University of Kansas Medical Center
Kansas City, Kansas

A few years ago, Sir Hans, while visiting Kansas and having himself just finished an historical review, suggested that I should write some of the more modern developments regarding the urea cycle. In reality, this suggestion gave me the idea to enlist the help of many of you as main protagonists in compiling not only the past historical events but to put together the present state of urea synthesis. Therefore, my presentation will be restricted as indicated in the program to "Alternate Paths, Artifacts, and Confirmation at the Preenzyme Level," and I will, of course, present personal overviews rather than detailed experimental data.

In 1944, as a young instructor here in Valencia, I was assigned to compose an examination for my first group of medical students. I remember vividly having chosen among other questions the urea cycle which had been suggested by Krebs and Henseleit (1) in 1932. I would have been rather skeptical at that time if I had been told that I would spend approximately 25 years of my life working and thinking mostly on the urea cycle and related problems.

The notion that the Krebs urea cycle was not the correct path for the synthesis of urea appeared sporadically but persistently for many years after the initial postulation. For example, in an otherwise outstanding review, Borsook and Dubnoff (2) in 1943 begin the discussion on urea formation

This work was supported in part by Grant AM-01855 from the National Institutes of Health.

with the following: "There is an increasing expression of doubt regarding the ornithine cycle."

It should be said that the difficulties in accepting the urea cycle and, for that matter, the tricarboxylic acid cycle, for a number of years, were other than the common tendency to suspect and defer new ideas, albeit, of such novelty and elegance, mostly due to misunderstandings and technical difficulties. It is proper to remember that one of the most serious objections to both the tricarboxylic acid cycle and the urea cycle was due to the misinterpretation of the newer tool of isotopic labeling. We cannot and need not dwell in this instance on what is now a classical example in biochemistry. I am referring to the Ogston hypothesis, which clarified the problem in so far as the TCA cycle is concerned. However, although the incorporation of CO_2 into urea had been shown very early by Evans and Slotin (3) with ^{11}C and by Rittenberg and Waelsch (4) with ^{13}C, later work in trying to elucidate whether or not citrulline was truly an intermediate in the urea cycle led others, using labeled citrulline, to conclude that this material was not an intermediate. In reality, as Krebs has properly and clearly expressed in the historical review to which I just referred (5), most of the experiments from 1932 to about 1946 mainly strengthened the concept of the urea cycle but provided no additional knowledge or understanding of the cycle. The cycle more or less according to Krebs' conceptions from 1932 to 1946 is illustrated in Fig. 1.

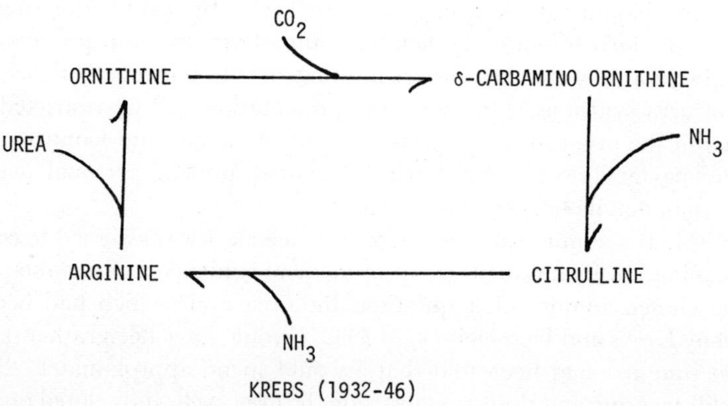

Figure 1

As illustrated in Fig. 2, the main findings supporting Krebs' theory were that the addition of CO_2 or ornithine and citrulline to slices of liver resulted in the incorporation of CO_2 into urea and in an increase in the amount of

1932-1946

A) WORK DONE WITH SLICES

B) MAIN FINDINGS IN SUPPORT: CO_2 INCORPORATION, CATALYTIC ROLE OF ORNITHINE & CITRULLINE; RATIO OF AMMONIA NITROGEN TO UREA ~ 1 WITH ORNITHINE, ~ 0.5 WITH CITRULLINE

C) MAIN OPPOSING FINDINGS:

1) RATIO OF AMMONIA NITROGEN TO UREA ~ 1 WITH CITRULLINE
2) FAILURE TO ISOLATE CITRULLINE FROM LIVER
3) POOR OR LOW ENHANCEMENT OF UREA SYNTHESIS FROM CITRULLINE; LACK OF INCORPORATION OF RADIOACTIVE CITRULLINE INTO UREA
4) RAPID INCORPORATION OF GLUTAMINE INTO UREA

Figure 2

urea beyond what could be explained on the basis of direct conversion into urea; thus the catalytic function of ornithine was demonstrated.

The figure also illustrates some of the apparent contradictory findings. These contradictory findings and other perfectly sound experiments were interpreted by Leuthardt, Bach, and other workers as an indication of the existence of a mechanism other than the ornithine cycle of Krebs. These criticisms we now know stemmed from misconceptions particularly in regard to permeability problems. For an admirable review of this subject see Krebs (6). It should be noted that the work of Gornall and Hunter (7) was outstanding in confirming the work of Krebs.

The concept invoked most often as an alternative, and indeed one which led eventually to the clarification of the urea cycle, was the participation of an amide donor. The rapid incorporation of isotopic ammonia into glutamine had been known since the work of Schoenheimer and coworkers and of Krebs. The fact that glutamine in the absence of ornithine could increase urea synthesis suggested another mechanism first proposed by Leuthardt (8). It should be made clear that Leuthardt's initial and subsequent modifications involved glutamine, as well as asparagine, as direct amino donors; however, Bach modified and extended this concept (9). The alternate concepts are illustrated in Fig. 3. At a later time Leuthardt and Glasson postulated that the synthesis of citrulline could result from a coupling reaction with oxalacetic amide (10). An extension of these and later ideas of Leuthardt, as we will see

1)

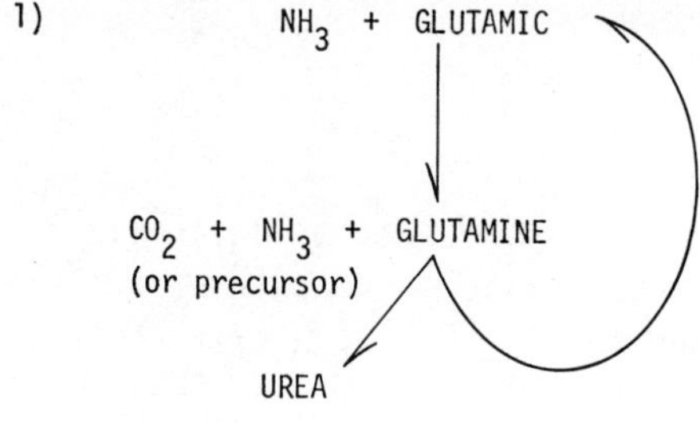

2) $CONH_2$ | R + NH_3 → UREA + R

Glutamine (or other amide)

Leuthardt; Bach (1938-48)

Figure 3

in a moment, had a profound influence on the eventual clarification of the mechanisms of the urea cycle. It should be noted that the participation of glutamine proved up to a certain point to be right, for we know today glutamine can serve as a specific donor for carbamyl phosphate synthesis although not via the urea path. And of course CO_2 and ammonia can form an intermediate.

Without a doubt the most important findings for the period 1946 to 1949, as indicated in Fig. 4, were the demonstration by Cohen and Hayano that urea synthesis could be shown in homogenates and, what is more important, that the activity leading to citrulline synthesis resided in a particular fraction which later on was identified as mitochondria, while the activity to carry on the incorporation or the transformation of citrulline into urea resided in the supernatant fraction (11). It is remarkable that with the relatively rudimentary knowledge of the time in regard to some basic principles, particularly regarding ATP synthesis and use* and including the soon to be made rediscovery of the mitochondrion as a biochemical entity as well as a cellular component, such good, solid, and rapid advances were made at such an early time.

On my arrival in Cohen's laboratory in 1947, I knew two techniques which were somewhat foreign to that laboratory, namely, how to work in the cold with enzymes and with unstable preparations, which I had learned

* It should be noted that John Speck showed the need for ATP for glutamine synthesis in 1947 just before my leaving Chicago and shortly before his premature death and that citrulline and argininosuccinate were the other two first examples of ATP use for biosynthetic reactions.

1946-1948

HOMOGENATES AND FRACTIONS THEREOF

A) SEPARATION OF STEPS ORNITHINE → CITRULLINE (WASHED RESIDUE)
CITRULLINE → UREA (SUPERNATANT)

B) ASPARTIC ACID AS A SPECIFIC AND DIRECT AMINO DONOR. CONCEPT OF INTERMEDIARY BETWEEN CITRULLINE AND ARGININE.

C) DEMONSTRATION OF EQUAL SPECIFIC ACTIVITY FOR $^{14}CO_2$ → ^{14}C-CITRULLINE → ^{14}C-UREA.

D) THEORETICAL OBJECTION TO ORNITHINE AS CARBAMINO ACCEPTOR LEADING TO THE DISCOVERY OF THE CATALYTIC ROLE OF CARBAMYLGLUTAMATE AND CONCEPT OF INTERMEDIARY BETWEEN ORNITHINE AND CITRULLINE.

E) DEMONSTRATION OF ATP PARTICIPATION IN BIOSYNTHESIS.

Figure 4

in Ochoa's laboratory, and a rudimentary knowledge of how to handle isotopes, which I had learned with Birgit Vennesland and Earl Evans in Chicago.

Phil Cohen gave me a number of problems to choose from; however, the main reason to go to work with Phil, and on the advice of Severo Ochoa, was because of my interest in CO_2 fixation. Since he had a logical system and Mika Hayano was leaving the laboratory and since Sarah Ratner had coincidently demonstrated in a most outstanding and elegant manner the specific role of aspartic acid as an amino donor to citrulline for arginine synthesis (12), I decided to concentrate (*a*) on the demonstration of whether or not citrulline was truly an intermediate in the urea cycle by the use of isotopes and by isolation methods and (*b*) on the mechanism of biosynthesis of citrulline. I must confess that very soon Phil Cohen managed to bias me in at least two ways which, however, became rather useful. One was his great affection, possibly second only to Sir Hans, for glutamic acid and his vigorous support of the term "transaminase" with which he had rebaptized the aminophorases of Braunstein.

I was of course aware of the outstanding review by Borsook and Dubnoff (2) stating that although the uptake of carbon dioxide as such had been proven by isotopic studies, it was not proved that CO_2 was added directly to the ornithine and that with ammonia could form citrulline.

It was then a relatively simple procedure to check the incorporation of $^{14}CO_2$ from the media into citrulline, to isolate the citrulline with the same specific activity as the initial $^{14}CO_2$ and demonstrate that this citrulline could be converted into urea with the same specific activity. It should be noted that before that time no isolation of citrulline, in metabolic context as such, had been carried out.

It should be noted, of course, that we are talking about times when one had to prepare his own ATP and when there was only one Geiger counter on the whole Wisconsin campus and not even at the Medical School. Therefore, while walking to the Agriculture Building and waiting for my counts or doing other routine work, I had time to start thinking about mechanism(s). Borsook's review suggested the possibility that the δ-carbamino ornithine postulated by Krebs might not be correct and suggested another explanation, and one which should favor the free energy relations, that is, that the carbon dioxide is first incorporated into a metabolite and that this substance, in giving up carbon dioxide for urea formation, might provide the free energy necessary; they pointed out very properly that the formation of urea from ammonia and carbon dioxide under physiological conditions entails a gain of nearly 14,000 cal of free energy per mole of urea regardless of the type of mechanism. In the fall of 1947 shortly after my arrival in Cohen's laboratory I made a trip back to Chicago and took along copies of papers of Leuthardt and Glasson (10) and of Leuthardt and Brunner (13), who had pointed out that carboxylation of the delta amino group of ornithine would be improbable in view of the high pK value of this group which is in the order of 10.8. I was, of course, as I have already indicated, very much enamored with glutamic acid. Then, since, as had been shown before by Cohen and Hayano, glutamic acid was best in supporting citrulline synthesis, it occurred to me that perhaps this amino acid would be the proper acceptor for CO_2, for the pK would be just about right for CO_2 fixation at physiological pH. I remember clearly how I read these papers on the train returning from Chicago. Again that night I dreamed that since the pK of ornithine was wrong, the glutamic acid could serve as a CO_2 acceptor. The next day I went to the laboratory and preincubated "washed residue" preparations with glutamic acid, which was by far the best supporter of citrulline synthesis in particular preparations under a variety of conditions (14). That day we obtained the first evidence for the presence of an intermediate in citrulline synthesis.

We tried to make the carbamino derivative of glutamic acid, and I remember the frustration which I had when I spent several days in trying to make this compound which had been suggested or described by Siegfried (15) in the early part of the century. We were lucky in that some of the organic chemists close to us were not available, but that Karl Paul Link, a rather close personal friend of Phil's and later on of mine, was, and that Doherty was working with him. He suggested making carbamyl glutamic acid, which proved to be a good guess and model, since it is a catalytic agent for the biosynthesis of citrulline. For a number of years, we worked with it, and a new puzzle developed, for although we had believed initially in a direct transcarbamylation to ornithine, isotopic experiments clearly demonstrated this not to be the case. Nevertheless, a fairly accurate scheme of the urea cycle was soon available, as shown in this reproduction (Fig. 5) of a "tentative scheme for urea cycle" presented at the First Phosphorus Metabolism Symposium at Johns Hopkins (16).

In retrospect, it is important to mention that Waelsch and later on Mehler wrote to me, as well as others, because of their interest in histidine and the possibility that formyl glutamic acid could in some way be related to carbamyl glutamic acid. This was indeed lucky, for although we had at the very beginning tried a sample of synthetic believed to be formyl glutamic acid, Mehler's product had a much different melting point. This led me to reexamine a sample of Mehler's formyl glutamic acid, and although of lower reactivity than carbamyl glutamic acid, it was active, and it seems that our first sample was not the correct material. This finding suggested checking

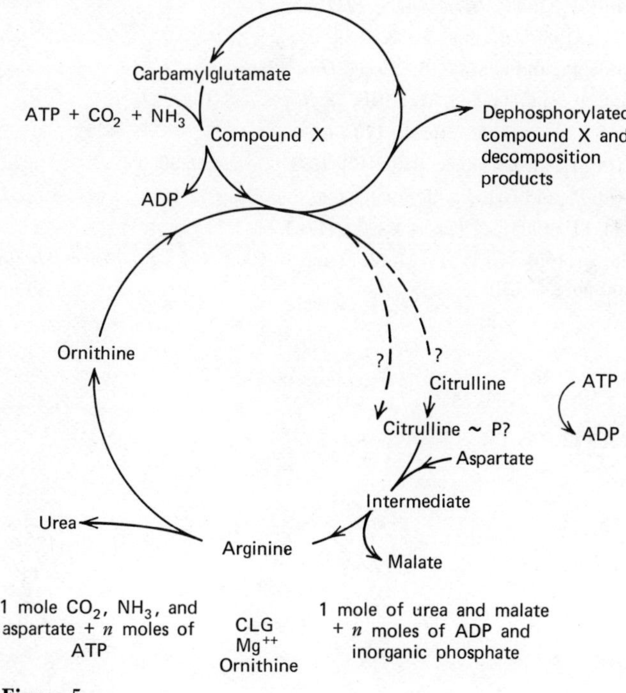

Figure 5

other *N*-acetyl derivatives, which naturally led to the discovery of *N*-acetyl-glutamic acid. In conclusion, I would like to emphasize this point, for as you know, *N*-acetyl-glutamic acid is an indispensable activator of citrulline synthesis at the enzyme level, and I fear that if it had not been for the curiosity of my friends Mehler and Waelsch, carbamyl glutamic acid might have remained a model, and the intimate mechanism of citrulline biosynthesis could have been delayed for much longer.

REFERENCES

1. Krebs, H. A., and Henseleit, K. (1932). *Z. Physiol. Chem.*, **210**, 33.

2. Borsook, H., and Dubnoff, J. W. (1943). in J. M. Luck and J. H. C. Smith, Eds., *Ann. Rev. of Biochem.*, **XII** Annual Reviews, Inc., Stanford, Calif.

3. Evans, E. A., Jr., and Slotin, L. (1940). *J. Biol. Chem.*, **136**, 805.

4. Rittenberg, D., and Waelsch, H. (1940). *J. Biol. Chem.*, **136**, 799.

5. Krebs, H. A. (1973). *Biochem. Ed.*, **1**, 19.

6. Krebs, H. A. (1952). in J. B. Sumner and K. Myrback, Eds., *The Enzymes*, **II**, Academic, New York.

7. Gornall, A. G., and Hunter, A. (1943). *J. Biol. Chem.*, **147**, 593.

8. Leuthardt, F. (1938). *Hoppe-Seyl. Z.*, **252**, 238.

9. Bach, S. J. (1939). *Biochem. J.*, **33**, 1833.

10. Leuthardt, F., and Glasson, B. (1944). *Helv. Physiol. Acta*, **2**, 549.

11. Cohen, P. P., and Hayano, M. (1946). *J. Biol. Chem.*, **166**, 239.

12. Ratner, S. (1947). *J. Biol. Chem.*, **170**, 761.

13. Leuthardt, F., and Brunner, R. (1947). *Helv. Chim. Acta*, **30**, 958.

14. Cohen, P. P., and Grisolia, S. (1948). *Fed. Proceedings*, **7** , 150.

15. Siegfried, M. (1905). *Z. Physiol. Chem.*, **44**, 85.

16. Grisolia, S. (1951). in W. D. McElroy and B. Glass, Eds., *Phosphorous Metabolism*, **I** Johns Hopkins, Baltimore, p. 619.

Comparative Aspects

3

Evolutionary and Comparative Aspects of Urea Biosynthesis

Philip P. Cohen
Department of Physiological Chemistry
University of Wisconsin
Madison, Wisconsin

This paper is not intended to be a comprehensive review of the evolutionary significance of urea biosynthesis but rather a retrospective on the paper published by Brown and Cohen (1) in 1960. On the basis of newly developed assay procedures, Brown and Cohen proposed a scheme relating evolutionary development to persistence or deletion of the enzymes involved in urea biosynthesis. Since that time, additional information has appeared in the literature which extends the enzyme assays to a greater variety of animal species, and new concepts have emerged regarding genetic aspects of evolution. The references which have been selected from a larger literature have been chosen on the basis that they are directly or significantly related to the purpose of this paper.

The biosynthesis of urea has evolved as a biochemical mechanism which exploited and concentrated in a single organ (the liver, in the case of ureotelic animals) enzyme activities which existed in nature long before urea was of paramount physiological usefulness. The capacity for urea biosynthesis emerged very early in organismic evolution but did not fully develop as a biochemical mechanism until problems of limited water supply and osmotic and excretory demands were imposed on certain organisms.

The studies reported in this paper from the author's laboratory were supported in part by Grant CA-03571 from the National Institutes of Health.

Huggins et al. (2) have suggested the following terminology to describe animals capable of converting ammonia plus carbon dioxide to urea:

Ureogenic: Animals with a full complement of the enzymes necessary for urea biosynthesis but which do not necessarily, or under all conditions, excrete urea as the major urinary nitrogenous end product.

Ureotelic: Ureogenic animals with a full complement of the enzymes necessary for urea biosynthesis and which excrete urea as the major nitrogenous urinary end product.

Ureosmotic: Ureogenic animals which form urea and retain it rather than excrete it to maintain osmotic balance.

Ammonia is toxic to animals in relatively dilute concentrations. Maintenance of low concentrations within an organism is no problem when there is an abundant water supply. Land animals have overcome the problem of a limited water supply by development of enzyme systems for the conversion of ammonia to the less toxic compounds, urea and uric acid. Marine mammals, turtles, and other animals which reinvaded the sea retained excretory patterns characteristic of their ancestors which first made successful adaptations to invade the land, probably from fresh water (3).

A two-volume, comprehensive review of nitrogen metabolism in different phyla of invertebrates and vertebrates has been published (4). In addition, several review papers on various aspects of urea biosynthesis are called to the reader's attention (5–9).

The existence of a functioning ornithine-urea cycle in primitive vertebrates appears to be unquestioned. On the basis of evidence available, a tentative map of the evolutionary development of the ornithine-urea cycle is shown in Fig. 1. This map is a composite of that previously published by Brown and Cohen (1) and Brown and Brown (10) and indicates that at some time primitive fishes, possibly the Placodermi, possessed a functioning ornithine-urea cycle. These primitive fishes were derived from ancestors which possessed some or all of the enzymes of the ornithine-urea cycle. In the ancestral forms the component enzymes of the cycle may not have all been present in a single organ in such a way as to permit the enzymes to function as an integrated cycle but rather were presumably functioning initially for the purpose of synthesizing arginine and phosphagens. From our present state of knowledge, it would appear that the extramitochondrial enzyme concerned with synthesis of carbamyl phosphate for pyrimidine biosynthesis (CPS II) emerged independently (see Fig. 2). As can be seen from Fig. 1, ureotelism was replaced by ammonotelism at the stage of emergence of the Actinopterygii, for the extant Holostei and Teleostei are for the most part ammonotelic and do not possess all the enzymes or, in most cases, very low levels of the urea cycle. Synthesis of urea from CO_2 and NH_3 has been demonstrated to take

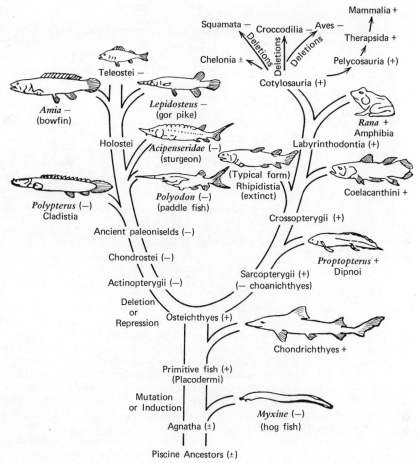

Figure 1. Evolution of vertebrates and the ornithine-urea cycle. Composite from Brown and Brown (10) and Brown and Cohen (1). Presence of cycle indicated by + ; absence by – ; presumed presence by (+); and presumed absence by (–).

place in Chondrichthyes (sharks), Dipnoi (lung fish), Coelacanthini, and Amphibia. At this point the Cotylosauria (stem reptiles) emerged with ureotelism, but reptiles (with the exception of Chelonia, the turtles) and birds, which followed, have lost their ureotelism and exploited uricotelism. The reptiles antecedent to Mammalia, namely, the Pelicosauria and Therapsida, are thought to have been ureotelic, because all mammals studied to date have a fully functioning urea cycle. Denton et al. (11) have presented evidence for ureotelism in the case of the adult echidna (*Tachyglossus aculeatus*) and platypus (*Ornithorhyncus*).

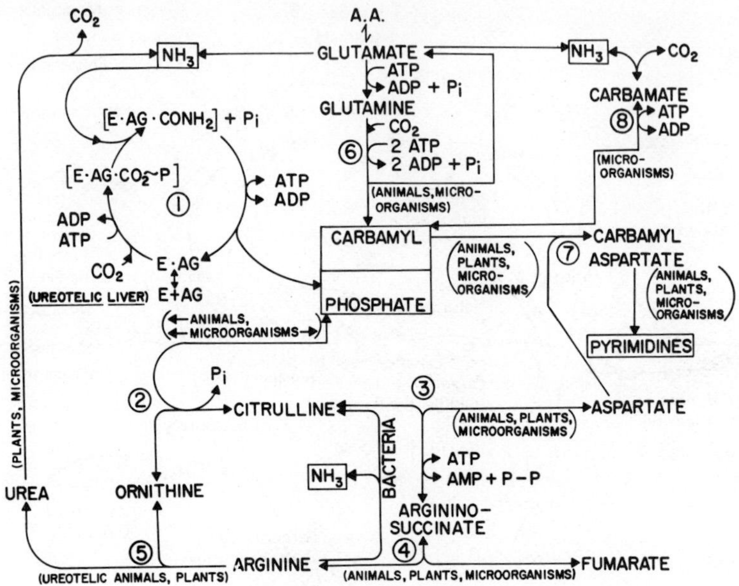

Figure 2. Scheme of reactions generating and using carbamyl phosphate. Enzymes catalyzing the reactions are (1) carbamyl phosphate synthetase-I; (2) ornithine transcarbamylase; (3) argininosuccinate synthetase; (4) argininosuccinase; (5) arginase [reactions (1) through (5) constitute the ornithine-urea cycle]; (6) carbamyl phosphate synthetase-II; (7) aspartate transcarbamylase; and (8) carbamate kinase. The two compartments of carbamyl phosphate shown are intended to represent the situation in liver of ureotelic animals. The lower compartment is the pool of carbamyl phosphate synthesized by the mitochondrial enzyme CPS I. This carbamyl phosphate is used for synthesis of citrulline by means of the mitochondrial enzyme, ornithine transcarbamylase. The upper compartment represents the carbamyl phosphate generated by the enzyme CPS II which is extramitochondrial in the liver of all animals. This carbamyl phosphate is used for the synthesis of carbamyl aspartate, a precursor of pyrimidines. The biosynthesis of arginine (and urea) from citrulline occurs extramitochondrially in liver of ureotelic animals. The enzyme carbamate kinase (E. C. 2.7.2.7) catalyzes a reversible reaction (in contrast to CPS I and II) and serves primarily as a pathway for generation of ATP from arginine and citrulline in certain microorganisms rather than for synthesis of carbamyl phosphate. The symbols used in the reaction scheme 1 are AG, *N*-acetylglutamate; E·AG, enzyme-*N*-acetylglutamate complex; [E·AG·CO_2~P], postulated intermediate enzyme complex; [E·AG·$CONH_2$], postulated intermediate enzyme complex; P_1, inorganic phosphate.

PRIMITIVE VERTEBRATES

The cyclostomes (lampreys and hagfish) are the only known surviving members of the class Agnatha which occupy a position near the base of vertebrate evolution (see Fig. 1). Read (12) has studied ammonia and urea production and excretion in the freshwater-adapted lamprey, *Entosphenus Tridentatus*, and has measured the enzymes of urea biosynthesis in fresh liver preparations. Although the animals proved to be essentially ammonotelic, liver was reported to contain carbamyl phosphate synthetase and arginase activities but not the other enzymes of urea biosynthesis. Arginase activity is commonly found in tissues of not only ureotelic animals but also of ammonotelic and even uricotelic animals (13). The carbamyl phosphate synthetase activity reported by Read (12) was not activated by *N*-acetylglutamate. Thus, it appears likely that the activity being measured was in fact extramitochondrial CPS (CPS II) concerned with pyrimidine biosynthesis, an enzyme which is widely distributed in animals, microorganisms, etc., and, as far as is known, is not involved in any way with urea biosynthesis. A remote possibility, yet to be investigated, is that mitochondrial CPS I in primitive fish does not require activation by *N*-acetylglutamate.

A teleost which represents an unusual case is that of the marine toadfish, *Opsanus tau*. Read (14) has reported that the liver of this fish has relatively high levels of all the enzymes of urea biosynthesis (see Table 1), yet excretes mainly ammonia and has a low blood urea level. This observation appears to be so unusual as to raise the question of proper evolutionary classification of this fish.

The relation of the environment to ureotelism is seen in the case of three lungfish with different estivating habits. The African lungfish, *Protopterus* (of the family Lepidosirenidae), estivates out of water in a dry burrow during the dry season and is enveloped by a thick, leathery cocoon (which develops after desiccation of a mucous excretion); the South American lungfish, *Lepidosiren* (of the family of Lepidosirenidae), estivates during the dry season in a burrow which is very moist and does not form a cocoon; the Australian lungfish, *Neoceradotus* (of the family Ceratodontidae), uses its lung only as an accessory breathing organ, does not estivate, and never leaves its aqueous environment. As can be seen from Table 1, the levels of activities of the urea-biosynthetic enzymes and the capacity for urea biosynthesis are consistent with the observed ammonotelism of *Neoceradotus* (15), the limited capacity for urea biosynthesis of *Lepidosiren* (16), and the significantly greater capacity for urea biosynthesis in the case of *Protopterus* (16–20). However, the studies of Janssens and Cohen (19) have demonstrated that *Protopterus* does not shift from ammonotelism when in the water to an apparent ureotelism during estivation because of a change in enzyme concentration but rather because, when actively swimming, *Protopterus* produces a large amount of metabolic ammonia which exceeds the limited capacity of the urea-biosynthetic enzymes for conversion to urea. On the other hand, during estivation relatively low levels of ammonia are produced, because of the low rate of metabolism, most of which can be converted to urea. Thus the description by Homer Smith (21) of a shift from ammonotelism during its aquatic state to ureotelism during estivation of *Protopterus* is not a primary regulatory event involving increased activity of the urea-biosynthetic enzymes, as is seen in the case of the transition of the ammonotelic tadpole to the ureotelic frog. In keeping with the classification offered by Huggins et al. (2), *Protopterus* represents a ureogenic animal which under conditions of dehydration exploits its limited capacity for urea biosynthesis to excrete a urine in which the major nitrogenous end product is urea.

Further evidence relating to the influence of the environment on urea biosynthesis is seen in the case of the elasmobranchs. Marine elasmobranchs are capable of synthesizing urea (22) and of maintaining high concentrations of urea in their body fluids to maintain osmotic balance with seawater (23). Freshwater elasmobranchs, on the other hand, have a low urea content (24). A study of the activities of the urea-biosynthesizing enzymes in the liver of *Potamotrygon*, a freshwater ray, revealed levels of one-half to one-twentieth of those found in liver of marine stingrays (*Dasyatis* and *Urolophus*), (25), (see Table 1).

The coelacanth, *Latimeria chalumnae*, has been reported to retain urea at a high concentration (comparable to that found in marine elasmobranchs) to maintain its serum osmolality at approximately that of seawater (26). Assays for arginase and ornithine transcarbamylase activities plus urea content of liver from the same specimen indicate that the coelacanth is ureogenic (10). Later studies on another specimen by Goldstein et al. (27) demonstrated the presence of all the enzymes for urea biosynthesis (see Table 1).

AMPHIBIANS

The amphibians occupy a unique position in the evolutionary tree in that they represent the transition forms between the terrestrial and aquatic vertebrates. Their uniqueness is further apparent from the striking mimicry of the evolutionary sequence which the larval forms show in their metamorphosis and transition from ammonotelism to ureotelism. The levels of carbamyl phosphate synthetase in liver of different amphibians and their stage of development are shown in Fig. 3. If one considers that the present limit of amphibian development is reached in the tailless, lung-breathing frogs and toads, a striking correlation is seen between the extent of ontogenic development of different amphibians and the level of carbamyl phosphate synthetase I. The correlation between larval characteristic and carbamyl phosphate synthetase levels (the progressive loss of larval characteristics is associated with a progressive increase in liver carbamyl phosphate synthetase) in the case of different amphibians and the identical correlation in the case of anuran metamorphosis provide the basis for interpreting the findings of studies on anuran metamorphosis in terms of evolutionary significance.

The relation of urea excretion to development of the enzymes of urea biosynthesis in the metamorphosing tadpole, *Rana catesbeiana*, is shown in

Table 1 Enzymes of Urea Biosynthesis in Liver of Fish

Order	Species	Cps	Oct	Ass	Asl	Arg	Habitat	Ref.
		Enzyme activity (μmoles/g liver/hour)						
Elasmobranchs								
	Dasyatis americana	6.5	14,360	21.6	—	34,880	Marine	(24)
	Urolophus jamaicensis	4.5	8,540	16.5	—	13,920	Marine	(24)
	Potamotrygon	0.36	1,600	9.4	—	4,310	Fresh water	(24)
	Dogfish (various species)	2.2–41.0	1,000	—	—	7,486	Marine	(2)
Dipnoi								
	Lepidosiren	4.0	623	4.5	—	15,340	Estivates; moist burrow	(15)
	Protopterus	31.2	1,675	6.6	56.8	34,800	Estivates; cocoon	(16)
	Neoceratodus	0.4	155	2.2	—	1,289	No estivation	(14)
Coelacanth								
	L. Chalumnae	4.8	8,240	10.0	—	30,000	Marine	(26)
Teleosts								
Order	Species							
Apodes	*Anguilla anguilli* (eel)	0.16	6	1.6	1.8	4,565	Catadromus	(2)
Clupeoida	*Clupea harengus* (herring)	0.23	1.7	Bld	1.2	664	Marine	(2)
Haplomi	*Esox lucius* (pike)	0.06	72	0.8	2.6	691	Fresh water	(2)
Salmonidae	*Salmo salar* (salmon)	0.03	4	0.6	0.7	1,117	Anadromus	(2)
	Salmo gairdnerii (trout)	0.13	1.4	1.7	1.1	1,326	Fresh water	(2)
Cyprinoidea	*Cyprinus carpo* (carp)	0.42	1	0.5	0.4	58	Fresh water	(2)
	Rutilus rutilus (roach)	0.11	1	0.3	0.8	329	Fresh water	(2)
	Opsanus tau (toadfish)	9.8	10,210	14.6	184	31,800	Marine	(12)
	Siluris glanis (catfish)	Bld	—	9.4	0.3	3,542	Fresh water	(2)
	Ictalurus punctatus (catfish)	3.0	220	Bld	—	5,500	Fresh water	(57)

Selected reports of assays of urea biosynthetic enzymes since 1960 (1).

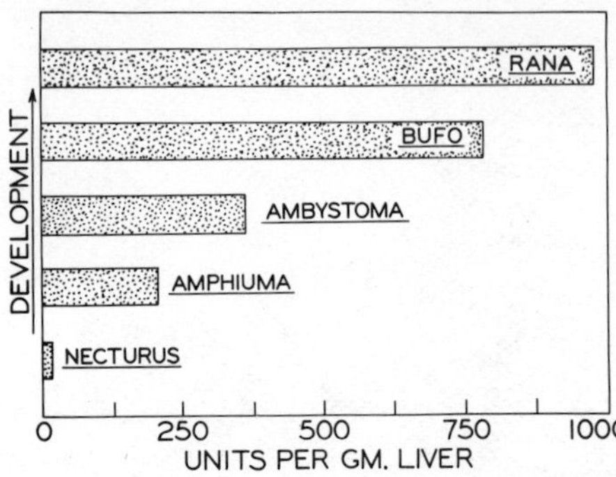

Figure 3. Relation of extent of ontogenic development of amphibians and the level of liver carbamyl phosphate synthetase I activity, See Cohen and Brown (5).

Fig. 4. The ratios of activities of a number of enzymes in liver of premetamorphic and metamorphic (natural and thyroxine-induced) tadpoles and adult frogs are shown in Table 2. In both natural and thyroxine-induced metamorphosis, the development of the enzymes of urea biosynthesis takes place as a "concerted" response (28), as shown in Fig. 5. The nature of the factors responsible for regulating the synthesis and levels of the two mitochondrial enzymes (carbamylphosphate synthetase I and ornithine transcarbamylase) and the three extramitochondrial enzymes (argininosuccinate synthetase, argininosuccinase and arginase) are as yet unknown.

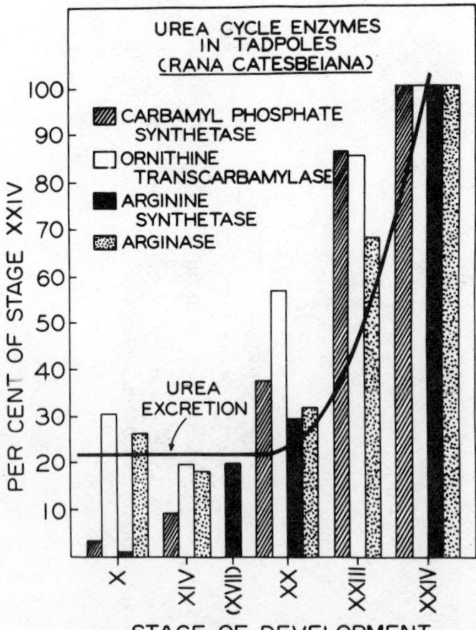

Figure 4. Urea excretion and development of enzymes involved in urea biosynthesis in the metamorphosing tadpole. Arginine synthetase represents the overall synthesis of arginine from citrulline plus aspartate and involves argininosuccinate synthetase (see reaction 3, Fig. 2) and argininosuccinase (see reaction 4, Fig. 2). (5)

Table 2 Relative Activities of Enzymes in Liver of Premetamorphic, Metamorphic, and Adult *Rana Catesbeiana* (8, 28)

Enzymes involved in	E.C. no.	Ratio of enzyme activities: Metamorphic : premetamorphic, Thyroxine	Metamorphic : premetamorphic, Natural	Adult : premetamorphic tadpole
	Urea biosynthesis			
Carbamyl phosphate synthetase I	(2.7.2.5)	8	15	30
Ornithine transcarbamylase	(2.1.3.3)	3.3	7.7	8
Argininosuccinate synthetase	(6.3.4.5)	2.4	3.2	35
Argininosuccinase	(4.3.2.1)			20
Arginase	(3.5.3.1)	2.3	3.6	30
	Dehydrogenation			
Glutamate dehydrogenase	(1.4.1.2)	6	6	10
Lactate dehydrogenase	(1.1.1.27)	0.6	0.6	0.4
Glucose-6-phosphate dehydrogenase	(1.1.1.49)	0.8	0.8	0.5
Malate dehydrogenase	(1.1.1.37)		1.2	1.4
	Amino acid activation			
20 amino acid-RNA ligases	(6.1.1)	1		2
	Transamination			
Glutamic-oxaloacetic transaminase	(2.6.1.1)	2	3	5
Glutamic-pyruvic transaminase	(2.6.1.2)	1	1	0.5
Tyrosine-α-ketoglutaric transaminase	(2.6.1.5)	0.5	0.2	0.2
Ornithine-α-ketoglutaric transaminase	(2.6.1.13)	0.6	0.7	0.5
	Nucleic acid metabolism			
Uridine kinase	(2.7.1.48)	1	1	1
Uridine phosphorylase	(2.4.2.3)	2.2	2.3	0.2
	Phosphate hydrolysis			
Pyrophosphatase	(3.6.1.1)	1.2		2.2

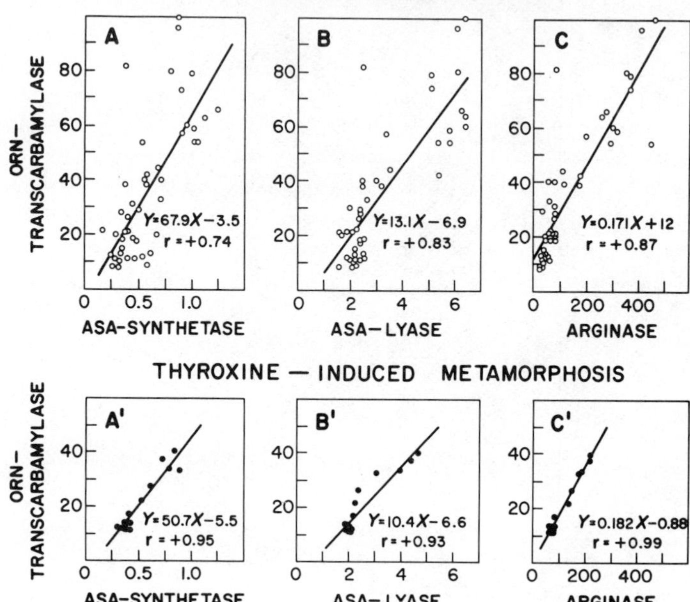

Figure 5. Concerted responses of the enzymes of urea biosynthesis during metamorphosis of *Rana catesbeiana*. The data for the tadpoles undergoing natural metamorphosis were plotted with the activity of ornithine transcarbamylase (ORN-TRANSCARBAMYLASE) as the ordinate against each of the other three enzymes on the abscissa (A, B, and C). The data for tadpoles undergoing thyroxine-stimulated metamorphosis were plotted similarly in A′, B′, and C′. ASA-, argininosuccinate. (28)

A different situation is seen in the case of *Xenopus laevis*, which maintains a wholly aquatic life and undergoes metamorphosis without changing from its larval ammonotelism (29). More recent studies indicate that when *Xenopus laevis* tadpoles are placed in moist peat or in 0.9% saline, urea biosynthesis increases along with an increase in the activity of carbamyl phosphate synthetase (20, 30). A similar change was observed by Balinsky et al. (31) in the case of tadpoles naturally estivating in mud near a dried-up pool. In further studies, Balinsky et al. (32) clearly demonstrated that *Xenopus laevis* tadpoles undergoing natural metamorphosis in water do not acquire ureotelism, primarily because of a failure of carbamyl phosphate synthetase and ornithine transcarbamylase activities to increase, although the other enzymes involved in urea biosynthesis were observed to increase.

In the case of *Rana cancrivora*, an amphibian capable of adapting to a wide range of salinity (33), it appears that in short-term exposures to a hypertonic environment (34, 35), there is a temporary decrease in urea biosynthesis.

However, with long-term exposures (4 weeks) urea biosynthesis is increased as a result of an increase in the activity of all the enzymes involved in urea biosynthesis (35).

Although it is clear that environmental factors such as salinity and estivation have an effect on the levels of one or more of the enzymes involved in urea biosynthesis, in the case of *Rana catesbeiana* at least, there is a clear case to be made for the relationship of an aquatic environment and ammonotelism in the tadpole and a terrestrial environment and ureotelism in the case of the adult frog. Although Wald (36) considered these changes to be consistent with the concept of recapitulation, Simpson (37) proposed that the concept of the "adaptive system" is more appropriate than that of recapitulation. However, it must be emphasized that, in the case of *Rana catesbeiana*, the tadpole makes its transition from ammonotelism to ureotelism in anticipation of a terrestrial environment rather than as an adaptation to it.

A selected list of assays for urea-biosynthesis enzymes in liver of amphibia and higher vertebrates which have appeared in the literature since the publication by Brown and Cohen (1) is given in Table 3.

Table 3 Enzymes of Urea Biosynthesis in Liver of Amphibia and Higher Vertebrates

	Enzyme activity (μmoles/g liver/hour)					
Animal	Cps	Oct	Ass	Asl	Arg	Ref.
Frog (*R. montezumae*)	225	18,000	90	—	80,000	(58)
Axolotl (*Ambystoma mexicanum*)						
non-metamorphosed	81	5,670	31	—	37,000–78,000	(58)
metamorphosed	225	8,100	41	—	52,200	(58)
Turtle (semiaquatic)						
Pseudemys scriptae	140	7,570	46	—	45,600	(58)
Kinosternon hirtipes	133	8,900	45	—	37,800	(58)
Turtle (terrestrial)						
Gopherus flavormarginatus	9	276	10	—	28,000	(58)
Gopherus berlandieri	27	900	18	—	29,600	(58)
Rat (*Rattus norvegicus*)	590	20,077	106	249	42,906	(59)
	340	14,500	182	310	35,900	(60)
African lion (*Felis Leo*)	68	14,880	105	—	39,480	(61)
Brown bear (*Ursidae*)						
(*euarctos americanus*)	345	16,820	196	347	11,732	(62)
Monkey (*macacca mulatta*)	256	6,600	121	248	122,000	(63)
Man (adult)	279	6,600	90	220	86,000	(63)

Selected reports of assays of urea biosynthetic enzymes in animals since 1960 (1).

INVERTEBRATES

Although the relationship between ureotelism and the emergence of animals from an aquatic to a terrestrial environment has stressed the vertebrate amphibians, a similar situation can be described for invertebrates. Evidence for the existence of a functioning ornithine-urea cycle has been reported by Bishop and Campbell (38) for the earthworm, *Lumbricus terrestris*, and by Campbell (39) for the land planarian, *Bipalium kewense*. Campbell (39) points out that, on the basis of the above, the ornithine-urea cycle can be considered to represent a pathway which has been continuous in its evolution from primitive cells to the vertebrates. Further, he states that it demonstrates the importance of this pathway in the invasion of land by animals and that, although the origin of the Metazoa is not established, primitive flatworms bear certain similarities to the hypothetical first metazoans, and thus the land planarian may represent the first successful invasion of land by animals.

Janssens and Bryant (40) have investigated four cestodes, a trematode, a nematode, and an acanthocephalan for enzymes involved in urea biosynthesis. No worm was found to have a full complement of the ornithine-urea cycle enzymes. These findings are considered by the authors to indicate that the synthesis of urea via the ornithine-urea cycle does not constitute a significant part of the nitrogen excretion of these helminths. The authors suggest that enzymes of the ornithine-urea cycle, where present, may contribute to metabolic processes other than urea biosynthesis in these animals.

On the other hand, Bishop and Campbell (38) report that the earthworm, *Lumbricus terrestris*, has all the enzymes needed for urea biosynthesis. The presence of these enzymes provides an explanation, according to these authors, for the ability of the earthworm to undergo a transition from ammonotelism to ureotelism with alternating periods of starvation and feeding (41).

In further studies on the enzymes of urea biosynthesis in invertebrates, Campbell and his colleagues observed that carbamyl phosphate synthesis in mitochondria from land snail (*Otala lactea* and *Helix aspersa*) hepatopancreas, earthworm (*Lumbricus terrestris*) gut tissue, and land planarian (*Bipalium kewense*) whole body is mediated, in part, by an enzyme that uses L-glutamine as a substrate and requires *N*-acetylglutamate as an activator (41, 43). In addition to the glutamine enzyme, mitochondria from the earthworm and land planarian also possess an enzyme that uses ammonia in the presence of *N*-acetylglutamate. Further studies will be required to determine whether these glutamine and *N*-acetylglutamate-dependent enzymes represent a primitive system involved in both pyrimidine as well as arginine biosynthesis or whether an independent enzyme system exists for pyrimidine biosynthesis which requires glutamine but not *N*-acetylglutamate. It is worth noting that, in *Neurospora*, there are two independent systems for synthesis of carbamyl phosphate, one concerned with arginine biosynthesis and the other with pyrimidine biosynthesis (44). Both enzymes use glutamine, and neither requires *N*-acetylglutamate for activity.

DISCUSSION

Although there is compelling evidence that the enzymes necessary for urea biosynthesis existed in primitive piscine ancestors, there is as yet no adequate molecular explanation for the evolutionary emergence of ammonotelism, ureotelism, and uricotelism in higher vertebrate forms. An hypothesis of "deletion" of the enzymes of urea biosynthesis was proposed by Brown and Cohen (1) as an explanation for the loss of ureotelism in some of the higher vertebrate forms (see Fig. 1). However, studies reported since this hypothesis was advanced indicate the presence at low levels of activity of some or, in some cases, of all of the enzymes necessary for urea biosynthesis. The persistence of enzyme activity, even though at low levels, must, of course, mean that a simple mutation with deletion of the gene coding for a particular enzyme involved in urea biosynthesis could not have occurred. However, there are several problems of a technical nature which need to be commented on.

Carbamyl phosphate synthetase I (the enzyme known to be present in mitochondria of liver and requiring *N*-acetylglutamate as an activator in higher ureotelic vertebrates) has been frequently selected as the enzyme of choice in assessing potential for urea biosynthesis. Since CPS I is the first enzyme in the sequence leading to the conversion of ammonia and carbon dioxide to urea, this is an understandable choice. However, it should be mentioned that there may be difficulties in assaying for this enzyme, particularly at low levels of activity. Although assay conditions established for the frog and the mammalian enzymes appear to be essentially the same (1), there is no assurance that this is the case for lower vertebrates. The extramitochondrial enzyme, CPS II, which is present in all tissues, uses glutamine in preference to ammonia as a substrate, does not require activation by *N*-acetylglutamate, and is involved in pyrimidine biosynthesis. In the case of a ureotelic animal, the activity of CPS II is of the order of 500 times lower than that of CPS I, and, further, the latter can be separated from the former by use of a mitochondrial fractionation. However, in the case of nonureogenic animals, great care and a great deal of study to establish optimal conditions are necessary before one can accept low levels of CPS activity as valid measures of mitochondrial CPS I activity. An example of the uncertainty of the validity of CPS I measurements is seen in the report by Watts and Watts (44), who reported greater activity in liver of dogfish with glutamine as a substrate than with ammonia. Assay conditions for CPS I activity in shark liver (*Squalus acanthius*) have been shown to be quite different from those used for assay of frog liver (46).

On the other hand, a comparative study of ornithine transcarbamylase prepared from liver of *R. Temporaria*, *Scylliorhinum caniculum*, and ox revealed no significant differences in kinetic properties but some differences in substrate binding properties (47, 48). The evolutionary significance of their findings is discussed by the authors (48).

The emergence of higher vertebrate forms which are primarily ammonotelic (such as the *Actinopterygii* and the *Teleosts*) were considered by Brown and Cohen (1) to have undergone a genetic deletion with the resultant loss of ureogenesis and ureotelism. On the other hand, the marine sharks, lungfish, coelecanths, and amphibians retained their ureogenic capacity. In the aquatic forms, ureogenesis was exploited in several species chiefly for osmotic purposes, and thus such animals might be considered to be ureosmotic. The stem reptiles, which emerged from primitive amphibian, are considered to have been ureogenic, but higher forms, such as reptiles and birds, exploited uricotelism when they became terrestrial and thus were considered by Brown and Cohen (1) to have also undergone a genetic deletion of the enzymes concerned with urea biosynthesis. The ureogenic capacity of the stem reptiles (*Cotylosauria*) persisted in the reptiles antecedent to the mammals with full exploitation of ureotelism as terrestrial habitats were established.

It is clear from studies to date that at the amphibian stage, the full complement of the enzymes involved in urea biosynthesis have emerged as a functioning cycle in one organ, the liver, with sequestration of CPS I and ornithine transcarbamylase in liver mitochondria and with an extramitochondrial location of argininosuccinate synthetase, argininosuccinase, and arginase. This pattern has persisted in higher forms, including man (see Table 3). Although detailed comparative studies have not yet been done for all the enzymes, it seems clear in the case of CPS I (as will be related in later presentations of this symposium) that the major physical, kinetic, immunochemical, and regulatory properties of the frog enzyme have been maintained in the mammalian enzymes studied to date.

The ability of a purified antibody, prepared by use of a purified frog liver mitochondrial carbamyl phosphate synthetase, to cross-react with liver preparations of all ureotelic animals examined and to inhibit enzyme activity practically completely at the equivalence zone (69) suggests a useful approach for study of the evolutionary development of this enzyme. The antibody neither cross-reacts with nor inhibits extramitochondrial carbamyl phosphate synthetase in the case of the rat (50) or the frog (51).

In addition to evolutionary development, amphibian metamorphosis provides a biological system for the study in biochemical terms of a number of currently challenging biological problems such as morphogenesis and differentiation, mitochondriagenesis, hormonal action, metabolic regulation, enzyme biosynthesis, and the roles of DNA and RNA in these processes.

Of particular interest from the standpoint of studying biochemical differentiation in the liver of the metamorphosing tadpole is the fact that the liver cells do not undergo cell division during metamorphosis (52). The experimental advantages of an animal tissue which is capable of responding to the hormone thyroxine to undergo biochemical differentiation without simultaneous cell division are obvious. Further, the ability to control the rate of the processes under study by means of raising or lowering the temperature and the dose of thyroxine in a free-swimming larval form provides obvious advantages over studies in mammalian fetuses.

In addition this system provides an opportunity for the study of the molecular events which simulate those which must have occurred in the evolutionary emergence of animals from an aquatic to a terrestrial environment.

The concept of genetic deletion as originally proposed by Brown and Cohen (4) was intended to be grossly descriptive rather than explanatory of an evolutionary process. In the light of newer knowledge about evolution at the molecular level, it is clear that more than a simple deletion of specific genes is involved.

Gene rearrangements and point mutations affecting regulatory genes are more likely events in adaptive evolution than are simple point mutations or deletions (53).

King and Wilson (54) have discussed the molecular basis for the evolution of organisms. They point out that regulatory mutations may be of at least two types: (1) point mutations which could affect regulatory genes by (*a*) nucleotide substitutions in a promoter or operator gene which would affect the production but not the amino acid sequence of proteins in that operon and (*b*) nucleotide substitutions in a structural gene coding for a regulatory protein such as a repressor, hormone, or receptor protein, with resulting amino acid substitutions which would alter the regulatory properties of the protein, and (2) the order of genes may change on a chromosome, owing to inversion, translocation, addition, or deletion of genes as well as fusion or fission of chromosomes.

Although the molecular mechanisms involved in gene rearrangements are not understood as yet, these gene rearrangements would be expected to have important effects on gene expression. Wilson and his co-workers have reported studies (54–56) which demonstrate that there is a parallel between the rate of gene rearrangement and the rate of anatomical evolution in the three major groups of vertebrates that have been studied in this respect, namely, birds, mammals and frogs.

SUMMARY REMARKS

Among the many unique features of the ornithine-urea cycle, the following stand out:

1. The cycle operates with the initial two enzymatic steps in liver mitochondria.
2. The remaining three enzymes are extramitochondrial.
3. The cycle operates under the influence of as yet unknown factors which regulate the levels of the mitochondrial and extramitochondrial enzymes in a concerted fashion.
4. The initiation of urea biosynthesis is dependent on the availability of a suitable concentration of *N*-acetyl-L-glutamate which is required as an activator of carbamyl phosphate synthetase.

These considerations prompt a number of questions which need to be answered if we are to understand not only the evolution but even the operation of the ornithine-urea cycle. The following are some of the questions which need to be answered:

1. What is the evolutionary relationship between CPS I and CPS II? Do they have subunits in common?
2. Does the requirement for *N*-acetylglutamate activation of CPS I represent an evolutionary modification of an enzyme originally serving a related but different function?
3. In view of the current views on the symbiotic origin of mitochondria in eucaryotic cells, what explanation can be offered for the location of CPS I and ornithine transcarbamylase in liver mitochondria?
4. What explanation can be offered for the absence of these enzymes in mitochondria of organs other than liver of ureogenic animals?
5. What mechanism(s) exist for the extramitochondrial synthesis of the precursor proteins of CPS I and ornithine transcarbamylase and their transport to liver mitochondria?
6. If the operation of ureogenesis in liver cells of different species is controlled by positive operators rather than suppressors, can these be isolated and identified? Can the absence of operators account for the deletion of urea biosynthesis in nonureogenic species? Can nonhepatic cells in ureogenic animals be induced to synthesize the enzymes of urea biosynthesis by introduction of these operators? If suppressors or repressors are involved, what

is the nature of these factors? Is thyroxine a "derepressor" in its role as an inducer of the enzymes of urea biosynthesis in the metamorphosing tadpole? Are repressors or operators affected by osmotic changes to account for responsiveness of the levels of the enzymes involved in urea biosynthesis in animals sensitive to osmotic changes?

REFERENCES

1. Brown, G. W., Jr., and Cohen, P. P. (1960). *Biochem. J.* **75**, 82–90.

2. Huggins, A. K., Skutsch, G., and Baldwin, E. (1969). *Comp. Biochem. Physiol.* **28**, 587–602.

3. Smith, H. W. (1961). From "Fish to Philosopher," The Natural History Library Anchor Books, Doubleday, Garden City, N.Y.

4. *Comparative Biochemistry of Nitrogen Metabolism*, Vol. 1, *The Invertebrates*, Vol. 2, *The Vertebrates*, J. W. Campbell, Ed., Academic, New York (1970).

5. Cohen, P. P., and Brown, G. W., Jr. (1960). In M. Florkin and H. S. Mason, Eds., *Comparative Biochemistry*, **2** Academic, pp. 161–244, New York.

6. Cohen, P. P., and Brown, G. W., Jr. (1963). In A. I. Oparin, Ed., *Evolutionary Biochemistry*, **III**, Macmillan (Pergamon), New York, pp. 129–138.

7. Cohen, P. P. (1964–65). *Harvey Lect.* **60**, 119–154.

8. Cohen, P. P. (1970). *Science* **168**, 533–543.

9. Ratner, S. (1973). Enzymes in Arginine and Urea Synthesis, in A. Meister, Ed., *Advances in Enzymology*, **39**, Wiley, New York, 1–90.

10. Brown, G. W., and Brown, S. G. (1967). *Science* **155**, 570–573.

11. Denton, D. A., Reich, M., and Hird, F. J. R. (1963). *Science* **139**, 1225.

12. Read, L. J. (1968). *Comp. Biochem. Physiol.* **26**, 455–466.

13. Brown, G. W., Jr. (1966). *Arch. Biochem. Biophys.* **114**, 184–194.

14. Read, L. J. (1971). *Comp. Biochem. Physiol.* **39B**, 409–413.

15. Goldstein, L., Janssens, P. A., and Forster, Roy P. (1967). *Science* **157**, 316–317.

16. Funkhouser, D., Goldstein, L., and Forster, R. P. (1972). *Comp. Biochem. Physiol.* **41A**, 439–443.

17. Janssens, P. A., and Cohen, P. P. (1966). *Science* **152**, 358–359.

18. Forster, R. P., and Goldstein, L. (1966). *Science* **153**, 1650–1652.

19. Janssens, P. A., and Cohen, P. P. (1968). *Comp. Biochem. Physiol.* **24**, 879–886.

20. Janssens, P. A., and Cohen, P. P. (1968). *Comp. Biochem. Physiol.* **24**, 887–898.

21. Smith, H. W. (1930). *J. Biol. Chem.* **88**, 97–130.

22. Schooler, J. M., Goldstein, L., Hartman, S.C., and Forster, R. P. (1966). *Comp. Biochem. Physiol.* **18**, 271–281.

23. Smith, H. W. (1936). *Quart. Rev. Biol.* **11**, 49–82.

24. Thorson, T. B., Cowan, C. M., and Watson, D. E. (1967). *Science*, **158**, 375–377.

25. Goldstein, L., and Forster, R. P. (1971). *Comp. Biochem. Physiol.* **39B**, 415–421.

26. Pickford, G. E., and Grant, F. B. (1967). *Science* **155**, 568–570.

27. Goldstein, L., Harley-DeWitt, S., and Forster, R. P. (1973). *Comp. Biochem. Physiol.* **44B**, 357–362.

28. Wixom, R. L., Reddy, M. K., and Cohen, P. P. (1972). *J. Biol. Chem.* **247**, 3684–3692.

29. Munro, A. F. (1953). *Biochem. J.* **54**, 29–36.

30. McBean, R. L., and Goldstein, L. (1970). *Amer. J. Physiol.* **219**, 1124–1130.

31. Balinsky, J. B., Choritz, E. L., Coe, C. G. L., and Van der Schans, G. S. (1967). *Comp. Biochem. Physiol.* **22**, 59–68.

32. Balinsky, J. B., Choritz, E. L., Coe, C. G. L., and Van der Schans, G. S. (1967). *Comp. Biochem. Physiol.* **22**, 53–57.

33. Gordon, M. S., Schmidt-Nielsen, K., and Kelley, H. M. (1961). *J. Exp. Biol.* **38**, 659–678.

34. Colley, L., Rowe, W. C., Huggins, A. K., Elliott, A. B., and Dicker, S. E. (1972). *Comp. Biochem. Physiol.* **41B**, 307–322.

35. Balinsky, J. B., Dicker, S. E., and Elliott, A. B. (1972). *Comp. Biochem. Physiol.* **43B**, 71–82.

36. Wald, G. (1963). In A. I. Oparin, Ed., *Evolutionary Biochemistry*, Macmillan (Pergamon), New York, pp. 12–51.

37. Simpson, G. G. (1964). *Science* **146**, 1535–1538.

38. Bishop, S. H., and Campbell, J. W. (1965). *Comp. Biochem. Physiol.* **15**, 51–71.

39. Campbell, J. W. (1965). *Nature* **208**, 1299–1301.

40. Janssens, P. A., and Bryant, C. (1969). *Comp. Biochem. Physiol.* **30**, 261–272.

41. Needham, A. E. (1957). *J. Exp. Biol.* **34**, 425–446.

42. Tramell, P. R., and Campbell, J. W. (1970). *J. Biol. Chem.* **245**, 6634–6641.

43. Tramell, P. R., and Campbell, J. W. (1971). *Comp. Biochem. Physiol.* **40B**, 395–406.

44. Davis, R. H. (1972). *Science* **178**, 835–840.

45. Watts, D. C., and Watts, R. L. (1966). *Comp. Biochem. Physiol.* **17**, 785–798.

46. Schooler, J. M. (1964). Ph.D. Thesis, University of Wisconsin.

47. Joseph, R. L., Baldwin, E., and Watts, D. C. (1963). *Biochem. J.* **87**, 409–416.

48. Joseph, R. L., Watts, D. C., and Baldwin, E. (1964). *Comp. Biochem. Physiol.* **11**, 119–129.

49. Marshall, M., and Cohen, P. P. (1961). *J. Biol. Chem.* **236**, 718–724.

50. Nakanishi, S., Ito, K., and Tatibana, M. (1968). *Biochem. Biophys. Res. Comm.* **33**, 774–781.

51. Kent, R. J., Strahler, J. R., and Cohen, P. P., unpublished studies.

52. Kaywin, L. (1936). *Anat. Record* **64**, 413–441.

53. Wilson, A. C., Sarich, V. M., and Maxson, L. R. (1974). *Proc. Natl. Acad. Sci.* **71**, 3028–3030.

54. King, M. C., and Wilson, A. C. (1975). *Science* **188**, 107–116.

55. Wilson, A. C., Maxson, L. R., and Sarich, V. M. (1974). *Proc. Natl. Acad. Sci.* **71**, 2843–2847.

56. Prager, E. M., and Wilson, A. C. (1975). *Proc. Natl. Acad. Sci.* **72**, 200–204.

57. Wilson, R. P. (1973). *Comp. Biochem. Physiol.* **46B**, 625–634.

58. Mora, J., Martuscelli, J., Ortiz-Pineda, J., and Soberón, G. (1965). *Biochem. J.* **96**, 28–35.

59. Charbonneau, R., Andrie, R., and Berlmquet, L. (1967). *Can. J. Biochem.* **45**, 1371–1374.

60. Schimke, R. F. (1962). *J. Biol. Chem.* **237**, 459–468.

Chemistry

4

Properties and Chemistry of Urea and Related Intermediates

James Kennedy
Veterans Administration Hospital
Project No. 4888–01)
Kansas City, Missouri
and Department of Medicine
University of Kansas Medical Center
Kansas City, Kansas

INTRODUCTION

The chemistry of each of the compounds involved in the synthesis of urea is discussed under its respective heading. Their chemical and physical properties, listed in Tables 1 and 2, are taken largely from ref. 34.

UREA

Formula: CH_4N_2O

$$H_2N-\overset{\overset{\displaystyle O}{\|}}{C}-NH_2$$

In the mid-eighteenth century a number of chemists had drawn attention to the presence of "a particular nitrogenous substance" in urine. In November 1773, F. M. Rouelle (1) extracted from urine a "substance savoneuse" which was rich in nitrogen and gave, on fermentation, carbonic acid and ammonia. At the turn of the eighteenth century, Fourcroy and Vauquelin (2) obtained a crystalline mass on treatment of fresh concentrated urine with nitric acid. They demonstrated that these crystals were a definite compound of urea and nitric acid and called the material in the urine, urée. Prout, in 1818, established the composition of naturally occurring urea and determined its correct empirical formula. It was first obtained in pure form by M. Proust (3) in 1820. Urea probably was synthesized first but not recognized by John Davy in 1812. He reported that phosgene (carbonyl chloride) reacted with dry ammonia to

Table 1 Properties of Urea and Related Intermediates

	Mol wt	Crystallization	Melting point	Solubility	Stability to storage
N-Acetyl-L-glutamic acid	189.2	Prisms from water	199		Stable
L-Arginine	174.2	Prisms + $2H_2O$ from water; anhydrous monoclinic plates from 60% ethanol	238 d	15.0 g in 100 ml H_2O at 21°	Stable
L-Argininosuccinic acid	290.3	Hygroscopic powder		Very soluble in water	One year or more if kept dry and at 0°; neut. solution stable several weeks at −18°
Carbamyl phosphate	141.0				Solid salt stable if kept dry and at 0°; aqueous solutions stable indefinitely if frozen; aqueous solutions at neutral pH have half-life of 50 minutes at 37°, 2 hours at 30° and 18 hours at 0°
L-Citrulline	175.2	White, long, thin prisms from ethanol-H_2O	220 d	Soluble in water; insoluble in ethanol	Stable
L-Ornithine	132.2	Crystals from ethanol–ethyl ether	227	Very soluble in H_2O, ethanol	Stable
Urea	60.1	Colorless, tetragonal prisms or needles	133	78 g in 100 ml H_2O at 5°; 119.3 g in 100 ml H_2O at 25°; 15.8 g in 100 ml ethanol at 20°	Stable

Table 2 Properties of Urea and Related Intermediates

	$[\alpha]_D^{25}$		pK_a	pI
N-Acetyl-L-glutamic acid	-16.6 (C = 2 in H_2O)			
	$+3.9$ (C = 2 in N NaOH)			
L-Arginine	$+12.5$ (C = 2 in H_2O)	(1)	2.17 (COOH)	10.76
	$+27.6$ (C = 2 in 5 N HCl)	(2)	9.04 (NH_3^+)	
		(3)	12.48 (guanidino)	
L-Argininosuccinic acid	$+16.4$ (C = 2.9 in H_2O)	(1)	1.62 (COOH)	3.2
	$+26.6$ (C = 2.9 in 0.5 N NaOH)	(2)	2.70 (COOH)	
		(3)	4.26 (COOH)	
		(4)	9.58 ($\alpha - NH_3^+$)	
		(5)	>12 (guanidino)	
Carbamyl phosphate		(1)	1.1	
		(2)	4.9	
L-Citrulline	$+4$ (C = 2 in H_2O)	(1)	2.43 (COOH)	5.92
	$+24.2$ (C = 2 in N HCl)	(2)	9.41 (NH_3^+)	
	$+10.8$ (C = 1.1 in 0.1 N NaOH)			
L-Ornithine	$+12.1$ (C = 2 in H_2O)	(1)	1.71 (COOH)	
	$+28.4$ (C = 2 in 5 N HCl)	(2)	8.69 ($\alpha - NH_3^+$)	
		(3)	10.76 ($\delta - NH_3^+$)	
Urea		(1)	0.18	

give a solid that, when treated with acetic acid, did not liberate carbon dioxide. However, he failed to identify the material with naturally occurring urea. Consequently, credit for the synthesis of urea has been given to Wöhler, who in 1828 recognized that the product obtained by boiling a solution of ammonium cyanate with water is identical with urea isolated from urine (4). Wöhler's experiment was a milestone in the history of chemistry in that it set the stage for eventual rejection of the theory of the "vital principle."

During this period, structural formulas were being increasingly used to represent the molecular arrangement of organic compounds. Based on the work of Dumas (5), Regnault (6), Hofmann (7), Basarov (8), and others, the well-known symmetrical "diamide" or carbamide formula was generally accepted as the proper representation of urea (formula **1**).

$$H_2N{-}C({=}O){-}NH_2 \quad \textbf{1}$$

$$HN{=}C\langle NH_3, O \rangle \text{ (three-membered ring)} \quad \textbf{2}$$

$$H_3\overset{+}{N}{-}C(O^-){=}NH \quad \textbf{3}$$

$$H_2\overset{+}{N}{=}C(O^-){-}NH_2 \quad \textbf{4}$$

$$H_2N{-}C(OH){=}NH \quad \textbf{5}$$

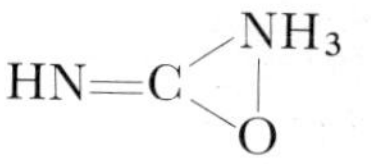

Doubts began to arise, however, in the latter part of the nineteenth century regarding the structure of the urea molecule. This was due largely to difficulties encountered in explaining certain decompositions of urea on the basis of the carbamide formula. Confusion concerning the structure of urea persisted into the twentieth century. In a series of papers beginning in 1913, Emil Werner (9) proposed a cyclic structure (formula **2**) for urea which gained wide acceptance. His views are summarized in a monograph (10) published in 1923. The cyclic structure is extremely improbable and eventually was discarded. Existing evidence is consistent with the carbamide representation (formula **1**). Formula **5**, the tautomer of **1**, has been assigned the common names, isourea and pseudourea. Formula **4** represents a resonance hybrid of **1** and seems considerably more probable than **3**.

Urea can be considered to be the diamide of carbonic acid, hence the designation, carbamide. The monoamide, carbamic acid, is unstable. Thus, the reactions of urea, most of which are well known, are those of an amide. In spite of the fact that urea is a very weak base ($pK_a = 0.18$), in concentrated nitric acid it forms a precipitate of urea nitrate. Since considerable resonance stabilization would result from addition of a proton to the oxygen rather than a nitrogen, the cation of urea is best represented as a resonance hybrid shown in formulas **6** and **7**. Nitrourea is produced by dissolving urea nitrate in concentrated sulfuric acid.

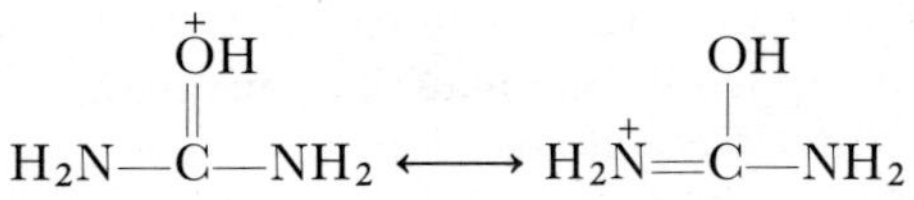

$$\underset{\mathbf{6}}{H_2N{-}\overset{\overset{+}{O}H}{\overset{\|}{C}}{-}NH_2} \longleftrightarrow \underset{\mathbf{7}}{H_2\overset{+}{N}{=}\overset{OH}{\overset{|}{C}}{-}NH_2}$$

Urea decomposes into ammonia and isocyanic acid when heated rapidly above its melting point (reaction 1). The isocyanic acid immediately polymerizes to approximately 70% its cyclic trimer, cyanuric acid, and 30% the linear polymer, cyamelide (reaction 2).

$$CO(NH_2)_2 \rightarrow NH_3 + HN{=}C{=}O \tag{1}$$

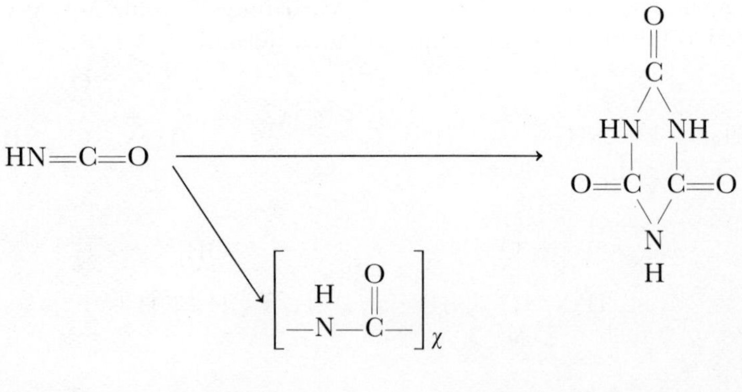

$$HN{=}C{=}O \longrightarrow \text{cyanuric acid (ring: } C(=O)\text{, HN, NH, } O{=}C\text{, } C{=}O\text{, NH)} \;+\; \left[{-}\overset{H}{N}{-}\overset{O}{\overset{\|}{C}}{-}\right]_x \tag{2}$$

Isocyanic acid is tautomeric with cyanic acid (reaction 3), and the former predominates.

$$\mathrm{HN{=}C{=}O} \rightleftharpoons \mathrm{N{\equiv}COH} \tag{3}$$

Hydrolysis by nucleophilic attack on the carbon atom of isocyanic acid yields carbamic acid, which decomposes at once to carbon dioxide and ammonia (reaction 4).

$$\mathrm{HN{=}C{=}O} + \mathrm{H_2O} \rightarrow \mathrm{H_2N{-}\overset{\overset{\displaystyle O}{\|}}{C}{-}OH} \rightarrow \mathrm{CO_2} + \mathrm{NH_3} \tag{4}$$

Isocyanic acid reacts with ammonia to form urea; thus it is likely that Wöhler's synthesis of urea proceeded in two stages. The first phase is dissociation of ammonium cyanate into isocyanic acid and ammonia. Then the products react by addition to form urea (reaction 5). Thiourea can be obtained in analogous fashion from ammonium thiocyanate.

$$\mathrm{NH_4N{=}C{=}O} \rightleftharpoons \mathrm{HN{=}C{=}O} + \mathrm{NH_3} \rightarrow \mathrm{CO(NH_2)_2} \tag{5}$$

Other reactions of interest include formation of biuret from urea and isocyanic acid, hydroxyurea from hydroxylamine and isocyanic acid, and urethans from isocyanic acid and alcohols. Ureides are formed by diacylation of urea. The ureide, barbituric acid, is obtained from malonic acid. Since it is an amide, urea undergoes the Hofmann rearrangement to form hydrazine, which in turn is oxidized by hypobromite to nitrogen and water.

Finally, it should be emphasized that in aqueous solution at high concentration, urea slowly decomposes to cyanate and ammonium ion. This is particularly relevant to studies of subunit structure of proteins by gel electrophoresis in buffers containing urea, in which the cyanate concentration should be kept as low as possible to prevent carbamylation of free amino groups of proteins. The equilibrium constant of the urea–ammonium cyanate equilibrium at 25° is of the order of 3.65×10^{-5} (11).

A number of analytical methods is available for quantitative estimation of urea. One method is dependent on reaction with xanthydrol (12). Other methods include measurement of ammonia (13–15) or CO_2 (15) liberated by the action of urease. Most colorimetric methods presently employed for the estimation of urea, citrulline, and related compounds are based on Archibald's modification (16) of Fearon's method (17), in which the ureido compound reacts with diacetyl monoxime to form a colorless condensation product, which in turn is oxidized to a yellow or red chromogen. Compounds with the general formula $R_1NH{-}CO{-}NHR_2$, where R_1 is hydrogen or a single aliphatic radical and R_2 is not an acyl radical, will react. The nature of the protochromogen has not been clearly established; triazine (18), glycouril (19), and diacetyl diureide (20) structures have been proposed. Major difficulties with the method are that it does not obey strictly Beer's law, and the chromogens formed are photosensitive. An amazing number of modifications of Archibald's method have been developed in an attempt to improve the sensitivity of the method and overcome the problems mentioned above. Among the compounds that have been employed in modifying the reaction are persulfate (21), arsenate (22), perchlorate (23), ferric salts (24, 25), a ferrous-ferric redox buffer (26), phenylanthranilic acid (27), semidine (24), thiosemicarbazide (28), and glucuronolactone (29). One of the most sensitive modifications, and one that is suitable for estimation of carbamyl aspartate as well, was developed by Prescott and Jones (30). α-Isonitrosopropiophenone (1-phenyl-1,2-propanedione-2-oxime) has been substituted for diacetyl monoxime for measurement of urea (31, 32). Moore and Kaufmann (33), using the diacetyl monoxime method, have developed a nomograph to aid in the simultaneous determination of urea and citrulline in biological fluids.

CARBAMYL PHOSPHATE

Formula: CH_4NO_5P

$$H_2N-\overset{\overset{\displaystyle O}{\|}}{C}-O-PO_3H_2$$

Carbamyl phosphate

The list of intermediates involved in urea biosynthesis was completed in 1955, when Jones et al. (35) synthesized carbamyl phosphate. The chemical synthesis of this compound was instrumental in clarifying the structure of the carbamyl donor required for the conversion of ornithine to citrulline. They succeeded in preparing carbamyl phosphate by reacting KH_2PO_4 with KCNO and isolated carbamyl phosphate as the dilithium salt by ethanol fractionation (35). A method for chemical synthesis of diammonium carbamyl phosphate and its conversion to the disodium salt was developed later by Metzenberg et al. (36).

From the time of its discovery, it was recognized that aqueous solutions of carbamyl phosphate are quite unstable (see Table 1). The mechanisms responsible have been studied in detail (37–39). Initial studies by Jones and Lipmann (37) revealed that decomposition of carbamyl phosphate varied with pH and was not simply a reversal of its chemical synthesis. In strongly acid solution, ammonia, CO_2, and orthophosphate are produced, but, in strongly alkaline solution, carbamyl phosphate yields cyanate and phosphate. Ammonia is not produced in the base-catalyzed reaction. However, over a range of intermediate pH values, that is, from pH 2 to 8, the rate of orthophosphate release remains constant, but the rate of ammonia formation rises with decreasing pH. Both Halmann et al. (38) and Allen and Jones (39) excluded the existence of a general acid or base catalysis of the decomposition of carbamyl phosphate at these intermediate pH values. The latter investigators, having revised the

pK_1 and pK_2 values for carbamyl phosphate to 1.1 and 4.9, respectively, proceeded to measure the rates of orthophosphate release over a pH range of 0.35 to 10.2. Two distinct rates were found between pH 2 and 8.5. The first-order rate constant for the monoanion, the predominant ionic species between pH 2 and 4, is 1.42×10^{-2} minute^{-1}. The rate of hydrolysis between pH 6.5 and 8.0, where the dianion exists alone, is 1.64×10^{-2} minute^{-1}. Furthermore, they were able to demonstrate that ^{18}O incorporation from $H_2{}^{18}O$ into orthophosphate was maximal at pH 3.0, a pH value at which the carbamyl phosphate monoanion accounts for 97% of the ionic species present. On either side of pH 3.0, ^{18}O incorporation rapidly decreased. For example, at pH 6, where the dianion predominates, ^{18}O incorporation into orthophosphate was less than 20% of maximum. Thus, it was concluded that cleavage of the monoanion occurs at the phosphorus-oxygen bond and that dianion (and possibly the neutral form) cleavage occurs at the carbon-oxygen bond.

$$H_2N{-}C({=}O){-}O{-}P({=}O)(OH){-}O^- \xrightarrow{\text{slow}} H_2N{-}C({=}O){-}OH + PO_2{-}O^- \quad (6)$$

$$PO_3^- \xrightarrow[\text{fast}]{H_2{}^{18}O} H_2PO_3{}^{18}O^- \qquad H_2N{-}C({=}O){-}OH \xrightarrow{\text{fast}} NH_3 + CO_2$$

$$H_2N{-}C({=}O){-}O{-}P({=}O)(O^-){-}O^- \xrightarrow{\text{slow}} H{-}N{=}C{=}O + O^-{-}P({=}O)(OH){-}O^- \quad (7)$$

$$H{-}N{=}C{=}O \xrightarrow{\text{slow}} NH_3 + CO_2 \qquad H{-}N{=}C{=}O \xrightarrow{\text{fast}} NCO^-$$

Reaction 6 illustrates cleavage of the monoanion to carbamic acid and monomeric metaphosphate. The latter reacts fast with water to form orthophosphate. The unstable carbamic acid rapidly decomposes to CO_2 and ammonia. In reaction 7, decomposition of the dianion by cleavage at the carbon-oxygen bond releases cyanic acid and phosphate dianion. The cyanic acid, depending on the pH of the solution, ionizes to the stable cyanate ion or goes to carbamic acid and thence to CO_2 and ammonia.

Despite its instability, it has been possible to measure the amount of carbamyl phosphate in hepatic tissue. Raijman (40) found that rat liver contains approximately 0.1 μmole/g wet weight. On the other hand, it has not been found in blood. Herzfeld et al. (41), using a procedure that would have detected carbamyl phosphate at a concentration as low as $5 \times 10^{-6}\ M$, were unable to detect it in freshly collected rabbit plasma.

Carbamyl phosphate can be estimated enzymatically by measuring colorimetrically the amount of citrulline formed in the presence of excess ornithine and ornithine transcarbamylase (39). Two chemical methods for estimation of carbamyl phosphate are differential phosphate analysis (42) and quantitative conversion of carbamyl phosphate to urea (39).

ACETYLGLUTAMIC ACID

Formula: $C_7H_{11}NO_5$

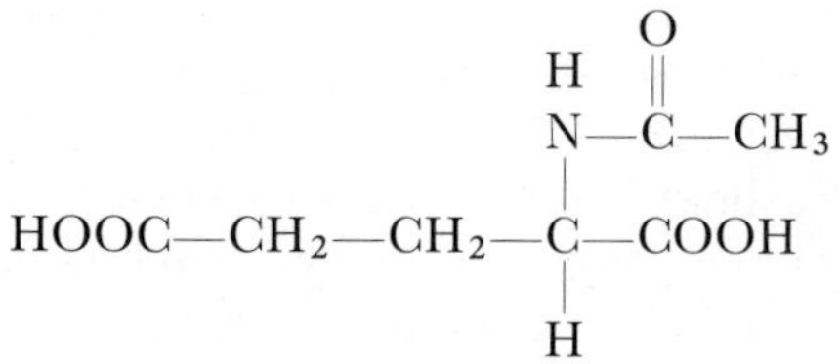

```
                           H   O
                               ‖
                           N—C—CH3
                           |
HOOC—CH2—CH2—C—COOH
                           |
                           H
```

N-Acetyl-L-glutamic acid

In 1952 Grisolia and Cohen (43) reported that synthesis of citrulline by preparations from rat liver required the addition of an appropriate derivative of glutamic acid. This requirement was met by a number of synthetic glutamyl derivatives, including *N*-acetyl-, *N*-carbamyl-, *N*-formyl-, *N*-chloroacetyl-, and *N*-propionyl-L-glutamic acids (44). The high activity of acetylglutamate and its isolation from mammalian liver (45) suggest that this particular glutamate derivative is the natural cofactor for carbamyl phosphate synthesis in mitochondria. Shigesada and Tatibana (46) have estimated acetylglutamate levels in rat and mouse liver to be approximately 30 nmole/g of tissue. A level of 10 nmole/g of small intestine was found. The compound could not be detected in heart, kidney, or spleen. In liver it appears to be localized to mitchondria at a concentration estimated to be in the range of 10^{-4} M.

Chemical synthesis of acetylglutamate involves direct acetylation of glutamate by acetic anhydride in alkaline solution (47). Quantitative determination of acetylglutamate is carried out enzymatically by measuring its stimulatory effect on mitochondrial carbamyl phosphate synthesis (45, 46).

Biological synthesis of acetylglutamate and its role as a cofactor in enzymatic synthesis of carbamyl phosphate are discussed in later sections of this symposium.

ORNITHINE

Formula: $C_5H_{12}N_2O_2$

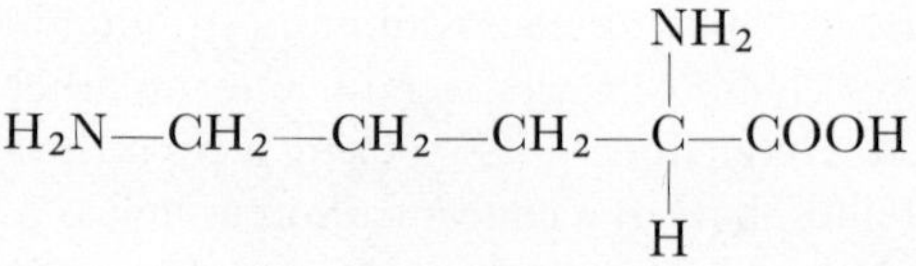

```
                              NH2
                              |
H2N—CH2—CH2—CH2—C—COOH
                              |
                              H
```

L-Ornithine

Conclusive evidence for the existence of ornithine in proteins is lacking, although it is well known that it may arise from arginine during protein hydrolysis. However, the L-amino acid, as well as its D-isomer, has been found in peptide linkage in the tyrocidine (48, 49) and gramicidin (50, 51) antibiotics. A number of ornithine-containing lipids has been isolated from bacteria. In at least two such instances, hydroxy fatty acids are amide-linked to the α-amino group of ornithine (52, 53).

Ornithine occurs in the free state in plants (54), animal tissues, and body fluids (55, 56). In a recent paper (40), Raijman reported the ornithine content of rat liver to be in the range of 0.11 to 0.19 μmole/g of tissue. Ornithuric acids, formed by acylation of ornithine by coenzyme A–activated benzoate, have been found in the excreta of fowl (57). δ-*N*-Acetyl-ornithine has been isolated from plants (58, 59) and has been shown to be a competitive inhibitor of argininosuccinase (60).

CITRULLINE

Formula: $C_6H_{13}N_3O_3$

$$H_2N-\overset{\overset{\displaystyle O}{\|}}{C}-\underset{\displaystyle H}{N}-CH_2-CH_2-CH_2-\underset{\underset{\displaystyle H}{|}}{\overset{\overset{\displaystyle NH_2}{|}}{C}}-COOH$$

L-Citrulline

The amino acid, citrulline, was first isolated in 1914 from watermelon by Koga and Odake (61). However, their work was lost sight of, and citrulline was reisolated almost simultaneously by Wada (62) and Ackermann (63) in 1931. The discovery of the urea cycle in 1932 by Krebs and Henseleit (64) focused the attention and interest of investigators on this amino acid.

Citrulline, as well as ornithine, may arise from arginine during protein hydrolysis; therefore, numerous early reports of citrulline-containing proteins should be viewed with some reservation. However, there is at least one carefully documented occurrence of this amino acid in peptide linkage in proteins. Rogers and associates (65, 66) isolated citrulline from the medullary protein of porcupine quills and from the root-sheath protein of guinea-pig hair follicles. They have proposed that citrulline arises by desimidation of arginine residues after incorporation of the latter in peptide linkage. Free citrulline has been found in a variety of animal tissues (55), plasma (67, 68), and urine (56). Curiously, the citrulline content of sweat is markedly increased compared with other body fluids (69).

Citrulline can be prepared enzymatically by the action of bacterial arginine desimidase (70). A simpler and preferred method is chemical synthesis from ornithine and urea (71). An ornithine copper chelate is prepared in neutral or alkaline solution to mask the α-amino group without interfering with reactions of the δ-amino group. Addition of urea, followed by heating, results in formation of copper citrullinate, from which the free amino acid can be obtained by decomposing the chelate with hydrogen sulfide. The free citrulline is crystallized from a water-ethanol solution. A modification of this procedure in which radioactive KCNO is substituted for urea has been described for preparation of labeled citrulline (72).

Most colorimetric methods employed for the estimation of citrulline represent modifications of Fearon's method (17) using diacetyl monoxime, and these are discussed in the section on urea. Another method for estimation of citrulline involves separation on columns of strongly basic anion exchange resins followed by treatment with ninhydrin reagent (56, 73).

ARGININOSUCCINIC ACID

Formula: $C_{10}H_{18}N_4O_6$

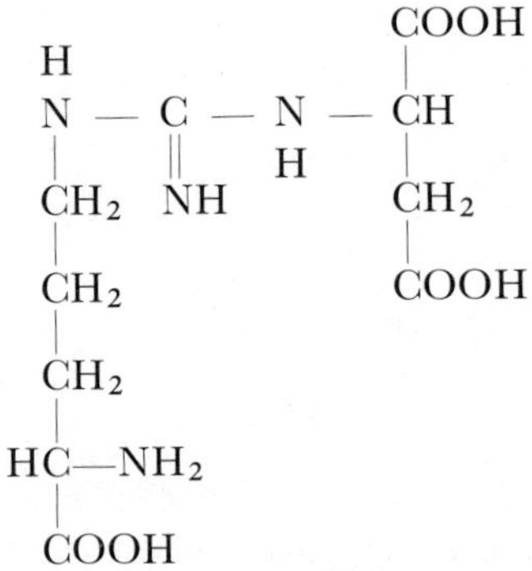

```
                                COOH
  H                              |
  N  —  C  —  N  —  CH
  |         ||       H         |
  CH2   NH                CH2
  |                              |
  CH2                        COOH
  |
  CH2
  |
HC—NH2
  |
  COOH
```

L-Argininosuccinic acid

In 1941 Borsook and Dubnoff (74) suggested that a hypothetical intermediate compound of citrulline and aspartate or glutamate is formed during the enzymatic conversion of citrulline to arginine. Subsequently, Ratner and associates (75) demonstrated the existence of such an intermediate. It was isolated as the barium salt after enzymatic synthesis both from citrulline, aspartate, and ATP and from arginine and fumarate (76). They characterized the compound and assigned to it the name argininosuccinic acid. Argininosuccinate, in aqueous solution, is relatively unstable in that it undergoes reversible ring closure to form two different anhydrides (76, 77). Both have been obtained in crystalline form (76, 78). The chemical and physical properties of the two anhydrides are quite different (78), and neither can serve as a substrate for argininosuccinate synthetase (79) or for argininosuccinase (76). The structures of the anhydrides (shown below) have been determined from their chemical properties (78) and from nuclear magnetic resonance data (80).

Argininosuccinate and anhydride I readily undergo interconversion. Increases in temperature and acidity favor conversion of argininosuccinate to

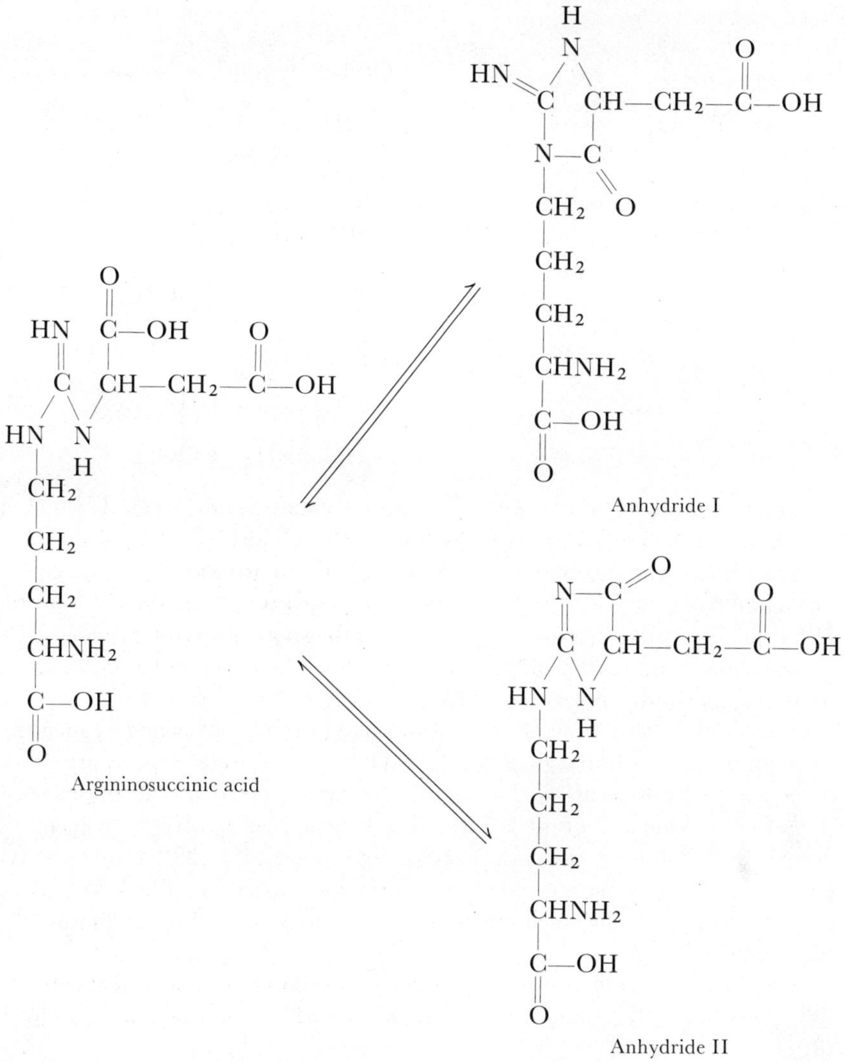

Argininosuccinic acid

Anhydride I

Anhydride II

the anhydride. In general, argininosuccinate is the dominant form at pH 7 and above, and anhydride I predominates at acid pH. Anhydride I, but not argininosuccinate or anhydride II, is susceptible to mild oxidative cleavage in alkaline solution, resulting in the release of arginine, pyruvate, and CO_2 (78). Oxidation at the α-carbon of the aspartate moiety causes formation of a "pseudobase" which then hydrolyzes at the anhydride linkage and rearranges to arginine and oxaloacetic acid (reaction 8). Pyruvate and CO_2 arise from β-decarboxylation of oxaloacetate.

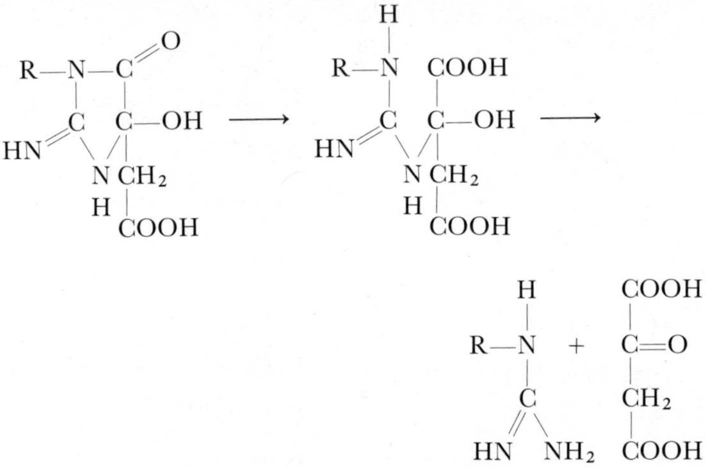

```
                                  H
                                  |
R—N—C=O                      R—N   COOH
  |   |                        |    |
  C   C—OH        ——→          C    C—OH        ——→
HN=  \ /|                    HN=  \ /|
   N  CH2                        N  CH2
   H  |                          H  |
      COOH                          COOH

                     H        COOH
                     |         |
                   R—N    +   C=O
                     |         |
                     C        CH2                    (8)
                    // \       |
                  HN   NH2    COOH
```

The magnitude of conversion of argininosuccinate to anhydride II is much less than that to anhydride I. Compared with anhydride I, anhydride II is more stable to heat and changes in pH, more resistant to oxidation, has a lower dissociation constant for the guanidino anhydride group, and only anhydride II has an absorption maximum (215 nm) in the short ultraviolet region (81).

Argininosuccinate is stable for 1 year or more if kept dry and at 0°. Neutral solutions are stable for several weeks at – 18°.

Quantitative estimation of argininosuccinate can be performed by measuring arginine (76) or fumarate (82) released by enzymatic cleavage by argininosuccinase or by measuring the quantity of urea formed in the presence of the latter enzyme and arginase (76). Another method involves conversion to anhydride I, followed by oxidative cleavage to arginine (83). Conventional ion-exchange chromatographic procedures have been modified to permit estimation of argininosuccinate or its anhydrides in biological fluids (77, 78, 84, 85).

Naturally occurring argininosuccinate has been found in pea-meal extracts (86), Jack bean (87), planaria (88), a land snail (89), and rat serum and liver (81).

ARGININE

Formula: $C_6H_{14}N_4O_2$

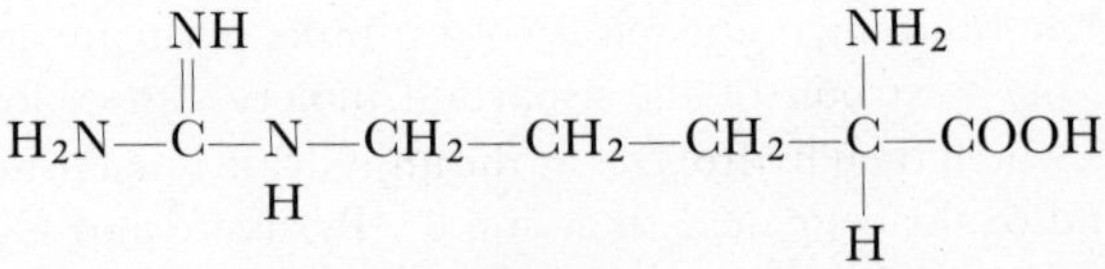

```
      NH                               NH2
      ||                                |
H2N—C—N—CH2—CH2—CH2—C—COOH
        H                               |
                                        H
```

L-Arginine

Arginine was isolated by Schulze and Steiger (90) in 1886 from etiolated lupine seedlings. It was first obtained from an acid hydrolysate of a protein by Hedin (91) in 1895. Free crystalline arginine was prepared by Gulewitsch (92) in 1899. Two years earlier Schulze and Winterstein (93) had established the structure of arginine by alkaline hydrolysis to ornithine and urea. Confirmation of its structure was provided by Sörensen (94), who synthesized arginine from benzoylornithine in 1910. Arginine can be converted to citrulline by alkaline hydrolysis (95). The cleavage of arginine to ornithine during acid hydrolysis of proteins explains the occasional presence of ornithine in protein hydrolysates.

In addition to its occurrence in a large number of proteins (including all protamines), arginine, in the free state, is widely distributed in mammalian tissues and small amounts (a few milligrams per 24 hours) are commonly found in human urine. Invertebrate muscle contains phosphoarginine, a molecule which assumes a metabolic role analogous to phosphocreatine in higher organisms (96, 97). Naturally occurring compounds, in addition to urea-cycle intermediates, that bear a structural relationship to arginine include homoarginine, canavanine, and octopine.

Methods applicable to amino acids in general have been used for determination of arginine. These include isotope dilution technics (98, 99) and separation by paper (100) or ion-exchange chromatography (101) followed by identification with ninhydrin. Ion-exchange chromatography has been successfully adapted for measuring arginine and other amino acids in body fluids (56, 68). Several methods that are specific for arginine are available. Arginine has been precipitated as the diflavinate from protein hydrolysates and then purified by recrystallization as the monoflavinate (102). Arginine and other guanidine compounds can be measured with the Sakaguchi color reaction, in which arginine in alkaline solution gives a red color when treated with 1-naphthol and sodium hypochlorite or hypobromite (103, 104). The Sakaguchi method is sensitive but has the disadvantage that other guanidine compounds interfere. The arginase method involves quantitation of the urea produced when arginine is cleaved by added enzyme. This method is reasonably specific, and a variety of sensitive procedures is available for quantitating urea (*vide supra*).

REFERENCES

1. Rouelle, F. M. (1773). *J. Med.*, November.
2. Fourcroy and Vauquelin, M. (1798) *Ann. Chim.* 80.
3. Proust, M. (1820). *Ann. Chim. Phys.* **14**, 257.
4. Woehler, F. (1828). *Ann. Chim. Phys.* **37**, 330.
5. Dumas, J. B. (1830). *Ann. Chim. Phys.* **44**, 129, 273.

6. Regnault, V. (1838). *Ann. Chim. Phys.* **69**, 180.
7. Hofmann, A. W. (1850). *Quat. J.C.S.* **2**, 300.
8. Basarov, A. (1868). *J.C.S.* **2**, 6, 194.
9. Werner, E. A. (1913). *J.C.S.* **103**, 1010.
10. Werner, E. A. (1923). *The Chemistry of Urea*, Longmans, New York.
11. Hagel, P., Gerding, J. J. T., Fieggen, W., and Bloemendal, H. (1971) *Biochim. Biophys. Acta* **243**, 366.
12. Pelikan, V., Kalab, M., and Tichy, J. (1964). *Clin. Chim. Acta* **9**, 141.
13. Hunter, A., and Downs, C. E. (1944). *J. Biol. Chem.* **155**, 173.
14. Van Slyke, D. D., and Archibald, R. M. (1946). *J. Biol. Chem.* **165**, 293.
15. Reynold, J., Follette, J. H., and Valentine, W. N. (1957). *J. Lab. Clin. Med.* **50**, 78.
16. Archibald, R. M. (1944). *J. Biol. Chem.* **156**, 121.
17. Fearon, W. R. (1939). *Biochem. J.* **33**, 902.
18. Beale, R. N., and Croft, D. (1961). *J. Clin. Path.* **14**, 418.
19. Veniamin, M. P., and Vakirtzi-Lemonias, C. (1970). *Clin. Chem.* **16**, 3.
20. Lugosi, R., Thibert, R. J., Holland, W. J., and Lam, L. K. (1972). *Clin. Biochem.* **5**, 171.
21. Ormsby, A. A. (1942). *J. Biol. Chem.* **146**, 595.
22. Siest, G., and Vigneron, C. (1968). *Clin. Chim. Acta* **20**, 373.
23. Kitamura, M., and Iuchi, I. (1959). *Clin. Chim. Acta* **4**; 701
24. Hunninghake, D., and Grisolia, S. (1966). *Anal. Biochem.* **16**, 200.
25. Bohuon, C., Delarne, J. C., and Comoy, E. (1967). *Clin. Chim. Acta* **18**, 417.
26. Guthohrlein, G., and Knappe, J. (1968). *Anal. Biochem.* **26**, 188.
27. Wheatley, V. R. (1948). *Biochem. J.* **43**, 420.
28. Geyer, J. W., and Dabich, D. (1971). *Anal. Biochem.* **39**, 412.
29. Momose, T., Ohkura, Y., and Tomita, J. (1965). *Clin. Chem.* **11**, 113.
30. Prescott, L. M., and Jones, M. E. (1969). *Anal. Biochem.* **32**, 408.
31. Archibald, R. M. (1945). *J. Biol. Chem.* **157**, 507.
32. Holden, H. F. (1959). *Aust. J. Exptl. Biol. Med. Sci.* **37**, 177.
33. Moore, R. B., and Kaufmann, N. J. (1970). *Anal. Biochem.* **33**, 263.
34. Gray, D. O., and Weitzman, P. D. J., (1969). In R. M. C. Dawson et al., Eds., *Data for Biochemical Research*, Oxford University Press, London, p. 1.
35. Jones, M. E., Spector, L., and Lipmann, F. (1955). *J. Am. Chem. Soc.* **77**, 819.
36. Metzenberg, R. L., Marshall, M., and Cohen, P. P. (1960). in *Biochem. Prep.* **7**, Wiley, New York, p. 23.
37. Jones, M. E., and Lipmann, F. (1960). *Proc. Natl. Acad. Sci. US* **46**, 1194.
38. Halmann, M., Lapidot, A., and Samuel, D. (1962). *J. Chem. Soc.* 1944.
39. Allen, C. M., and Jones, M. E. (1964). *Biochemistry* **3**, 1238.
40. Raijman, L. (1974). *Biochem. J.* **138**, 225.
41. Herzfeld, A., Hager, S., and Jones, M. E. (1964). *Arch. Biochem. Biophys.* **107**, 544.
42. Spector, L., Jones, M. E., and Lipmann, F. (1957). in S. P. Colowick and N. O. Kaplan, Eds., *Methods Enzymol.* **3**, Academic, New York, p. 653.
43. Grisolia, S., and Cohen, P. P. (1952). *J. Biol. Chem.* **198**, 561.
44. Grisolia, S., and Cohen, P. P. (1953). *J. Biol. Chem.* **204**, 753.

45. Hall, L. M., Metzenberg, R. L., and Cohen, P. P. (1958). *J. Biol. Chem.* **230**, 1013.

46. Shigesada, K., and Tatibana, M. (1971). *J. Biol. Chem.* **246**, 5588.

47. Nicolet, B. (1930). *J. Am. Chem. Soc.* **52**, 1192.

48. Battersby, A. R., and Craig, L. C. (1952). *J. Am. Chem. Soc.* **74**, 4019–4023.

49. Paladini, A., and Craig, L. C. (1954). *J. Am. Chem. Soc.* **76**, 688.

50. Consden, R., Gordon, A. H., Martin, A. J. P., and Synge, R. L. M. (1947). *Biochem. J.* **41**, 596.

51. Schwyzer, R., and Sieber, T. (1957). *Helv. Chim. Acta* **40**, 624.

52. Knoche, H. W., and Shively, J. M. (1972). *J. Biol. Chem.* **247**, 170.

53. Wilkinson, S. G. (1972). *Biochim. Biophys. Acta* **270**, 1.

54. James, W. O. (1949). *New Phytologist* **48**, 172.

55. Tallin, H. H., Moore, S., and Stein, W. H. (1954). *J. Biol. Chem.* **211**, 927.

56. Hamilton, P. B., (1970). In H. A. Sober, Ed., *Handbook of Biochemistry*, Chemical Rubber Co., Cleveland, Ohio, p. B-88.

57. Jaffe, M. (1877). *Ber.* **10**, 1925.

58. Manske, R. H. F. (1937). *Can. J. Res.* **B15**, 84.

59. Fowden, L. (1958). *Nature* **182**, 406.

60. Tigier, H., Kennedy, J., and Grisolia, S. (1965). *Biochim. Biophys. Acta* **110**, 423.

61. Koga, Y., and Odake, S. (1914). *J. Tokyo Chem. Soc.* **35**, 519.

62. Wada, M. (1930). *Biochem. Z.* **224**, 420.

63. Ackermann, D. (1931). *Z. Physiol. Chem.* **203**, 66.

64. Krebs, H. A., and Henseleit, K. (1932). *Z. Physiol. Chem.* **210**, 33.

65. Steinert, P. M., Harding, H. W. J., and Rogers, G. E. (1969). *Biochim. Biophys. Acta* **175**, 1.

66. Harding, H. W. J., and Rogers, G. E. (1971). *Biochemistry* **10**, 624.

67. Stein, W. H., and Moore, S. (1954). *J. Biol. Chem.* **211**, 915.

68. Armstrong, M. D., and Stave, U. (1973). *Metabolism* **22**, 561.

69. Westall, R. G., (1962). In J. T. Holden, Ed., *Amino Acid Pools*, Elsevier, New York, p. 204.

70. Knivett, V. A., (1953). In *Biochemical Preparations*, **3**, Wiley, New York. p. 104.

71. Hamilton, P. B., and Anderson, R. A., (1953). In *Biochemical Preparations*, **3**, Wiley, New York, p. 100.

72. Grisolia, S., (1970). In *Enciclopedia della Chimica*, **3**, p. 479.

73. Cohen, P. P., (1957). In S. P. Colowick and N. O. Kaplan, Eds., *Methods in Enzymology*, **3**, Academic, New York, p. 651.

74. Borsook, H., and Dubnoff, J. W. (1941). *J. Biol. Chem.* **141**, 717.

75. Ratner, S., and Petrack, B. (1951). *J. Biol. Chem.* **191**, 693.

76. Ratner, S., Petrack, B., and Rochovansky, O. (1953). *J. Biol. Chem.* **204**, 95.

77. Westall, R. G. (1960). *Biochem. J.* **77**, 135.

78. Ratner, S., and Kunkemueller, M. (1966). *Biochemistry* **5**, 1821.

79. Petrack, B., and Ratner, S. (1958). *J. Biol. Chem.* **233**, 1494.

80. Kowalsky, A., and Ratner, S. (1969). *Biochemistry* **8**, 899.

81. Ratner, S., (1973). In A. Meister, Ed., *Advances in Enzymology*, Wiley, New York, 1–90.

82. Havir, E. A., Tamir, H., Ratner, S., and Warner, R. C. (1965). *J. Biol. Chem.* **240**, 3079.

83. Ratner, S., Morell, H., and Carvalho, E. (1960). *Arch. Biochem. Biophys.* **91**, 280.

84. Cusworth, D. C., and Westall, R. G. (1961). *Nature* **192**, 555.
85. Shih, V. E., Efron, M. L., and Mechanic, G. L. (1967). *Anal. Biochem.* **20**, 299.
86. Davison, D. C., and Elliott, W. H. (1952). *Nature* **169**, 313.
87. Walker, J. B. (1953). *J. Biol. Chem.* **204**, 139.
88. Campbell, J. W. (1965). *Nature* **208**, 1299.
89. Campbell, J. W., and Speeg, K. V., Jr. (1968). *Comp. Biochem. Physiol.* **25**, 3.
90. Schulze, E., and Steiger, E. (1886). *Ber.* **19**, 1177.
91. Hedin, S. G. (1895). *Z. Physiol. Chem.* **20**, 186.
92. Gulewitsch, W. (1899). *Z. Physiol. Chem.* **27**, 178.
93. Schulze, E., and Winterstein, E. (1897). *Ber.* **30**, 2879.
94. Sorensen, S. P. L. (1910). *Ber.* **43**, 643.
95. Fox, S. W. (1938). *J. Biol. Chem.* **123**, 687.
96. Meyerhof, O., and Lohmann, K. (1928). *Biochem. Z.* **196**, 49.
97. Ennor, A. H., Morrison, J. F., and Rosenberg, H. (1956). *Biochem. J.* **62**, 358.
98. Foster, G. L. (1945). *J. Biol. Chem.* **159**, 431.
99. Shemin, D. (1945). *J. Biol. Chem.* **159**, 439.
100. Greenstein, J. P., and Winitz, M. (1961). *Chemistry of the Amino Acids*, Wiley, New York.
101. Spackman, D. H., Moore, S., and Stein, W. H. (1958). *Anal. Chem.* **30**, 1190.
102. Vickery, H. B. (1940). *J. Biol. Chem.* **132**, 325.
103. Sakaguchi, S. (1925). *J. Biochem. (Tokyo)* **5**, 25.
104. MacPherson, H. T. (1942). *Biochem. J.* **36**, 59.

Discussion

The morning discussion was started by Dr. Krebs. However, since some of his comments appear in his paper, to avoid duplication this part of the discussion, at his request, is omitted. Nevertheless, since in the succeeding discussions people often referred to his remarks, it seems appropriate, in order not to confuse the reader, to indicate briefly that Sir Hans addressed his remarks mostly to some evolutionary comparisons.—Editor

Dr. Cohen: The impact on biological and medical sciences of Wöhler's synthesis of urea by heating ammonium cyanate was so great that for over 100 years no serious consideration was given to any other possible mechanism of urea biosynthesis. In addition to searching for cyanate in blood and other tissues, the French and German investigators of the late 1800s reported studies which were directed to the demonstration of cyanate formation from protein when the latter was subjected to high temperatures. It is of interest that one of the experiments reported in the classical paper of Krebs and Henseleit in 1932 was that of testing the ability of liver slices to convert ammonium cyanate to urea. It proved to be toxic, and in effect this experiment served to put an end to the idea that ammonium cyanate might be a metabolic intermediate in urea biosynthesis. On the other hand the significance of the new concept developed by Krebs and Henseleit of the cyclic nature of the urea biosynthetic pathway and the roles of ornithine, citrulline, and arginine was such that a more scientifically advanced group of biochemists and physiologists could fruitfully exploit the new concept so effectively that within a period of about 20 years (excluding the war years) the details of the pathway of urea biosynthesis would be described in terms of individual enzymatic steps.

Dr. Illiano: Dr. Cedrangolo and I have a question for Dr. Cohen. Sometime ago (1965) we showed in our laboratory an L-aspartate deaminating activity, which has been called aspartase, in the liver of two elasmobranchs, *Schyllum conicula* and *Scyllum stellare*. We would like to ask whether you know of other studies carried out comparing the general deaminating activity of L-amino acids respectively in ammoniotelic and ureotelic animals and whether you know if there is a difference when ammoniotelic fishes switch to ureotelic metabolism. A difference would be important, we think, in clarifying the physiological role of ammonia in urea biosynthesis.

Dr. Cohen: I must confess I am not sure which L-amino acid oxidase is being referred to. I assume that is the system that oxidizes L-amino acids by a combination of transamination and glutamic dehydrogenase?

Dr. Illiano: We mean L-amino acid oxidase, oxidation of the amino acid to the imino acid and then hydrolysis to the keto acid and to ammonia.

Dr. Cohen: Is this a specific enzyme that oxidizes any L-amino acid to ammonia and an α-keto acid?

Dr. Illiano: It should not be specific, but common for all L-amino acids. Editor: Dr. Illiano is probably referring to the enzyme first described by Blanchard, M., Green, D. E., Nocito, V., and Ratner, S. (*J. Biol. Chem.*, **155**, 421, 1944). This enzyme has been obtained in crystalline form from rat kidney mitochondria (M. Nakano and T. S. Danowski, *J. Biol. Chem.*, **241**, 2075, 1966), but it has a greater activity with L-α-hydroxy acids than with L-amino acids.

Dr. Cohen: To answer your direct question, we have not investigated this, but I am not aware that a physiologically important L-amino acid oxidase with a high enough activity to convert any amino acid to ammonia and a keto acid actually exists other than in snake venom. I do not have that information.

Dr. Raijman: I wanted to make two points, first of all that carbamyl phosphate has been determined in neurospora, as I recall, by Dr. Rowland Davies. His recoveries are not very good, but many of us who have tried to determine carbamyl phosphate in tissues are familiar with that. Nevertheless, it has been done and was done before the determination in liver. Second point I wanted to make has something to do with, in a distant way, evolution and concerns the early need for a system for the detoxication of ammonia. Campbell did some rather beautiful thinking which has been confirmed to the effect that uricotelic species that do not have an intramitochondrial carbamyl phosphate synthetase must have something else to take care of that ammonia. So he reasoned that glutamine synthetase, which is cytoplasmic in the ureotelic animals, might be mitochondrial in the uricotelic animals. So he looked for it in the pigeon liver and he found it, and I think that is quite elegant.

Dr. Cohen: I would like to offer the comment that there is less need at this time for examining the levels of the enzymes of urea biosynthesis in additional species, even if they are exotic, than there is for looking into the factors involved in evolutionary genetics and how these operate in determining the presence, function, and regulation of urea biosynthesis.

Enzymes

5

The Regulation of Glutamate Dehydrogenase

Carl Frieden
Department of Biological Chemistry
Division of Biology and Biomedical Sciences
Washington University School of Medicine
St. Louis, Missouri

In 1932, Krebs and Henseleit demonstrated that liver tissue slices could synthesize urea when suspended in a saline media containing ammonium salts (1). Several years later it was shown that glutamate oxidation, first observed in 1920 (2, 3) required the pyridine nucleotide coenzymes and that, in fact, for animal tissues, the enzyme glutamate dehydrogenase could use either NAD or NADP (4) to produce α-ketoglutarate and ammonia. Since the oxidative deamination of glutamate yields ammonia and since the enzyme is present in the liver mitochondria at high concentration, it was logical to assume that this enzyme provided ammonia for urea biosynthesis from glutamate, which in turn arose from the transamination of other amino acids. The now traditional view is that glutamate dehydrogenase provides the ammonium necessary for the synthesis of carbamyl phosphate, which in turn is used to provide one of the nitrogen atoms of urea. Other possible sources of ammonia have been discussed by Cohen and Brown (5), by Krebs et al. (6), and by McGiven and Chappell (7). Although it is still perhaps not clear what the sole source of the ammonia for urea biosynthesis is or whether in fact there is a single major source, there is much to argue for the central role of glutamate dehydrogenase.

Glutamate dehydrogenase is essentially a ubiquitous enzyme occurring in all animal tissues and other organisms. Within any given animal, no major

Portions of the work described were supported by USPHS research grant AM-13332.

isozymes of glutamate dehydrogenase which are significantly different in kinetic properties appear to exist. Even enzymes from tissues of different animal species have many of the same characteristics both qualitatively and quantitatively as that from liver (8). Since these tissues carry out rather different metabolic roles, one presumes that the role of glutamate dehydrogenase varies in different tissues. For this to occur, it is necessary that the enzyme be under very tight regulation so that its function may be controlled by the particular environment. Indeed this seems to be the case for glutamate dehydrogenase, and it is the complex regulatory effects which I would like to describe in detail along with a description of some of the characteristics of the liver enzyme.

PHYSICAL PROPERTIES OF THE ENZYME FROM BOVINE LIVER

Glutamate dehydrogenase was crystallized from bovine liver by Olson and Anfinsen in 1951 (9). In their studies of the physical properties of the enzyme, they noticed an unusual characteristic in that the sedimentation coefficient decreased as the concentration of the protein decreased. They interpreted this result to mean either that the protein was dissociating or markedly unfolding at low protein concentration. This observation was then taken up in our own study of the enzyme a few years later and it was conclusively shown, in 1959, that the enzyme undergoes a rapid polymerization-depolymerization reaction (10). Furthermore it was shown that the equilibrium of this polymerization was influenced by the coenzyme of the reaction as well as by ADP or ATP when examined in the presence of coenzyme (11). It was later shown that other purine nucleotides can influence the sedimentation behavior of the enzyme-coenzyme complex (12). Furthermore all these nucleotides influence the kinetic behavior of the enzyme. I shall have more to say of the kinetic effects of these metabolites later.

It is now known that active enzyme is comprised of six identical polypeptide chains, each of molecular weight 56,000. The 500 residue amino acid sequence has been established by Smith and his collaborators (13). This six-mer unit of 336,000 molecular weight polymerizes to higher molecular weight forms giving rise to the type of sedimentation behavior originally noted by Olson and Anfinsen (9). This polymerization has been extensively investigated in a number of different laboratories, and although there is still some disagreement as to the details of polymerization (14–19), it is most probably of the type which produces long rods of indefinite length. In the absence of purine nucleotides, all molecular weight forms of the enzyme appear to have the same specific activity (20, 21). It has been clearly shown that GTP in the presence of

NADH or NADPH binds preferentially to active enzyme of the lower molecular weight forms (i.e., 336,000) (21) and consequently causes depolymerization of the enzyme.

RAT LIVER GDH

The enzyme from rat liver was first isolated and crystallized by King and Frieden (22). It was observed then, and studied in greater detail later (23), that compared with the bovine enzyme there is much less tendency to polymerize beyond the stage of the active enzyme containing (probably) six subunits. However, there was early evidence that there is considerable similarity between the sequence of the rat and bovine liver enzymes (22). We have now found that in fact there is probably over a 95% homology in these two enzymes, and many of the replacements are of a conservative type (C. J. Coffee and C. Frieden, unpublished data). Because of the close homology between the rat and bovine enzyme, it seems almost certain that the three-dimensional structures of these proteins will be similar. Thus one might predict that only a small change in the primary sequence leads to the difference in polymerizing ability. It is known, for example, that rat liver enzyme can apparently form hybrid polymers with the bovine enzyme (23). It may be of interest that the *N*-terminal group of the rat liver enzyme is a threonine but it is alanine in the bovine enzyme and is CyS-Glu-Ala in the chicken liver enzyme (24). These enzymes all polymerize to different extents, with the chicken liver enzyme having a degree of polymerization intermediate between the rat and bovine liver enzyme (25). Although a lot of sequence data are currently available, it will probably not be possible to determine which residues are involved in the polymerization process until X-ray crystallographic data are obtained. This latter has been hampered by the lack of ability to produce the proper crystals, although several laboratories have tried to do so. On the basis of sequence homology to those dehydrogenases for which the structure is known, some predictions have been made with respect to the nature of the coenzyme binding site (26, 27).

GENERAL KINETIC CHARACTERISTICS

It was demonstrated over 35 years ago that animal glutamate dehydrogenase can use either NADP or NAD as coenzyme (4). More recently, it was observed that, in addition to glutamate, the enzyme may also deaminate a number of monocarboxylic amino acids (28). However, the pH dependence of the reaction using the monocarboxylic acids is markedly shifted toward alkaline pH values, and although at high pH's the velocity of the reaction can be an appreciable fraction of that using glutamate as substrate, it is unlikely that there can be much physiological importance attached to these other substrates. Not only is their activity quite low at physiological pH values, but the Michaelis constants are much poorer than the normal substrates.

One unusual substrate is trinitrobenzene sulfonate (29). This compound may be reduced by either NADH or NADPH to yield the corresponding oxidized coenzyme and trinitrobenzene. The reaction proceeds with the same stereospecificity involved in the oxidation of reduced coenzyme using normal substrates, and therefore a similar mechanism is suggested. Of particular interest is the fact that this reaction may also occur nonenzymatically, again apparently by a similar mechanism (30), and therefore the system serves as an excellent model reaction for dehydrogenases.

There have been exhaustive kinetic studies of the bovine liver enzyme, and I do not wish to go into any detail with respect to these studies. However, the mechanism of the reaction probably involves random addition of substrates although a preferred pathway of reaction. The kinetic studies are complicated by the formation of "abortive" complexes of several types. Both enzyme-NADPH-glutamate and enzyme-NADP-α-ketoglutarate complexes have been postulated (31, 32). In the normal reaction, it is probable that the rate-limiting step in the overall reaction is the rate at which coenzyme dissociates from one of these abortive complexes. It is of interest that NADH, at high levels, serves to inhibit the coenzyme oxidation but NADPH shows little of this tendency. This inhibition, which is a consequence of NADH binding to a nonactive site, as discussed later, shows an interesting characteristic in that it is a time-dependent phenomenon (33). This is shown in Fig. 1, which represents the complete oxidation 485 μM NADH and shows three clear phases. The phases represent, respectively, the uninhibited rate of the reaction, the slower inhibited rate, and the partial return to the uninhibited rate as the NADH concentration decreases and finally is depleted. From these and similar data, one may calculate the rate at which NADH binding to the nonactive site induces inhibited enzyme. Under the conditions shown the half time for this conformational change is about 1 second (33). I believe that such time-dependent phenomena (that is, hysteretic) have real physiological significance, as has been discussed in detail elsewhere (34).

Measuring the initial velocity as a function of NAD or NADP yields curves which are characteristic of NAD activation (that is, negative cooperativity). In particular Engel and Dalziel (35) observed, at pH 7, that double reciprocal plots at high glutamate levels showed three or four linear regions with different slopes as the coenzyme concentration was increased. They interpreted these data in terms of negative homotropic interactions between enzyme subunits. Evidence for negative cooperativity of NAD and NADP binding

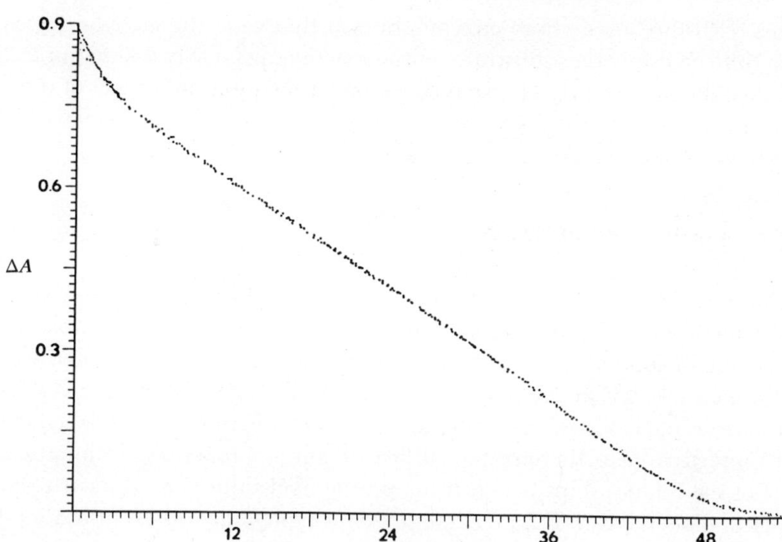

Figure 1. Data for the oxidation of 485 μM NADH catalyzed by glutamate dehydrogenase. The data represent the output of a stopped flow experiment performed at 365 nm using a cell of 0.5 cm path length. Experiment performed at 10° in 0.1 M Tris-acetate, pH 7.4, buffer containing 1 mM phosphate and 0.1 mM EDTA. The enzyme concentration was 0.05 mg/ml at α-keto-glutarate and ammonium chloride concentrations of 5 mM and 100 mM respectively. The data were taken from ref. 33, which also described the mechanism to explain the data.

was also observed in the presence of glutarate (36). However, the exact correlation between binding and kinetic data still remains somewhat equivocal, and it would appear that the presence of abortive complexes, as well as negative cooperativity, may have something to do with the nature of the kinetic results observed. It could be noted in passing that concepts such as negative cooperativity or half-site reactivity have still not been satisfactorily analyzed for many dehydrogenase systems, including glutamate dehydrogenase.

It would be useful, since many metabolic studies are performed using rat tissue, to know some of the kinetic characteristics of the rat liver enzyme. However, it is not commercially available, and although it is not difficult to prepare (22), extensive kinetic studies have not been carried out with it. Although it could be argued that one should carefully examine the kinetics of rat liver glutamate dehydrogenase, it might be useful to note here that, under assay conditions, this enzyme is quite unstable, and the velocity falls off sharply in a short period of time (seconds). This appears to be due to coenzyme (NADH or NADPH) induced instability (22). Some instability occurs with the bovine liver enzyme, but, in that case, the enzyme appears to be protected by other substrates of the reaction, primarily α-ketoglutarate, and thus the problem is not nearly so serious. I mention this aspect of the rat liver enzyme because it could lead to erroneous estimates of its activity in liver homogenates or sonicates.

REGULATORY PROPERTIES

Bovine liver glutamate dehydrogenase activity is affected by an almost bewildering number of compounds. Those metabolites which are probably most important in influencing the activity of the enzyme *in vivo*, by binding to sites which are not active sites, are ADP, ATP, GDP, GTP, and NADH. The purine nucleotides, in particular, have been found to bind specifically and quite tightly to the enzyme surface. Figure 2 shows diagrammatically some of the relationships between the purine nucleotides and the coenzymes at pH 8. The diagram is based roughly on a type of ligand exclusion model proposed by Cross and Fisher (37). In their model, they assume that there are several subsites used to explain the observed kinetic properties. They further assume considerable overlap in specificity of these subsites and that in some cases binding of one ligand would help to sterically hold another ligand but in other cases binding of one ligand sterically blocks the binding of another ligand. Thus in Fig. 2, for example, the GTP molecule would bind to different portions of the enzyme surface than does the ADP molecule, but because the two molecules would interfere with one another, they cannot bind simultaneously to the surface of the enzyme. Similarly ADP competes with the second NADH binding site, thus preventing excess NADH inhibition. The principal advantage of this concept is that only a small area of the protein surface is needed to accommodate several ligands. However,

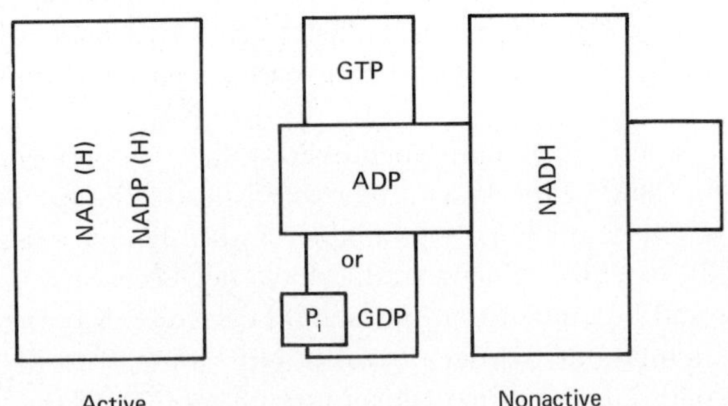

Figure 2. Diagrammatic representation of some relationships between nucleotides. (See text for description.)

the proposal is probably oversimplified, since it is based only on direct steric effects and should be extended to include ligand-induced conformational changes.

There is, of course, a common active site for all coenzymes, and Fig. 2 can be used in a rough way to describe the types of kinetic observations that have been made at pH 8. Thus GTP and GDP do not interfere with NADH binding to either the active or nonactive site. In fact they enhance such binding, and excess NADH inhibition is stronger in the presence of the guanine nucleotides than in its absence. On the other hand ADP blocks NADH binding to the second site and thus prevents inhibition. Further, ADP, GTP, GDP (and ATP which is not represented) all compete for each other with respect to their kinetic effects. However, the sites to which these nucleotides bind are not strictly the same, since, for example, phosphate competes with GTP or GDP but not with ADP.

GTP and GDP enhance binding of the coenzyme and serve also to inhibit the enzymatic reaction. It is probable that the inhibition is a consequence of decreasing the rate at which the coenzyme product dissociates from some enzyme form (or decreasing the rate of an isomerization step which precedes the dissociation). ADP, on the other hand, weakens coenzyme binding to the active site as well as blocking NADH binding to the nonactive site. Aside from relieving NADH inhibition, ADP activates the reaction—probably by increasing the dissociation rate of the coenzyme product from some enzyme form (or increasing the rate of an isomerization step).

Given, however, that purine nucleotides do compete with each other, regardless of the details of that competition, it becomes important to examine the effects of those nucleotides first alone and then together on enzymatic activity. One overall generalization is that the exact nature of their effects depends on which coenzyme is used. This is shown by Table 1 (32). Some typical kinetic data using these nucleotides with NADH and NADPH are

Table 1 Nature of the Purine Nucleotide Effects at pH 8 in 0.01 *M* Tris-Acetate Buffer at 25°

	Purine nucleotide					
Coenzyme	GTP	GDP	ATP	ADP	ITP	IDP
NADH	I[b]	I	I[c]	A	I	I
NADPH	I	I	±	A	I	I
NAD[a]	I	I	A	A	I	I
NADP	I	I	A	A	I	I

[a] Concentrations less than 5×10^{-4} *M*.
[b] I = inhibitor, A = activator, ± = no effect.
[c] Activation at very low NADH concentrations.

given in Table 2. Examination of these tables illustrates the complexity of the system as well as the behavior using different coenzymes. For example, ADP activation is markedly dependent on which coenzyme is used, the concentrations of coenzyme, and the pH (latter data not shown).

Table 2 Properties of Various Purine Nucleotides Affecting Bovine Liver Glutamate Dehydrogenase

Kinetic experiments performed at 25° and pH 8 in 0.01 *M* tris-acetate buffer, 10^{-5} *M* EDTA at α-ketoglutarate and ammonium chloride levels of 5×10^{-3} and 5×10^{-2} *M* respectively. Direct binding experiments in 0.1 *M* tris-acetate, pH 7.2, 1 m*M* phosphate, 10^{-4} *M* EDTA, 5°. v'/v is the ratio of velocity in the presence of nucleotide to that in the absence of nucleotide.

	GDP	GTP	ADP	ATP
NADH 100 μM	$K_i = 2\ \mu M$ $v'/v = 0.05$	$K_i = 0.1\ \mu M$ $v'/v \leqslant 0.02$	$K_a = 30\ \mu M$ $v'/v = 2$	$K_i = 25\ \mu M$ $v'/v = 0.6$
NADPH 100 μM	$K_i = 5\ \mu M$ $v'/v = 0.05$	$K_i = 0.5\ \mu M$ $v'/v = 0.05$	$K_a = 15\ \mu M$ $v'/v = 4$	$K_a = 40\ \mu M$ $v'/v = 1.0$
No coenzyme (calculated from kinetics)	$K_i = 5\ \mu M$	$K_i = 10\text{–}20\ \mu M$	$K_a = 2\text{–}5\ \mu M$	$K_a = 20\text{–}30\ \mu M$
Direct binding (no coenzyme present)		18 μM	2 μM	

However, since ADP both activates and also relieves NADH inhibition, the extent of activation using NADH as coenzyme depends on whether high levels of NADH are present. In contrast, this does not occur with NADPH, since this coenzyme does not appreciably inhibit the reaction at high levels. Furthermore, the excess NADH inhibition is potentiated in the presence of GTP or GDP. Should one believe that these effects are simple, it should be noted that ATP shows quite complex behavior, as shown by Table 3. ATP can serve either as activator or inhibitor, depending on the pH, nature of the coenzyme, and the coenzyme concentration. Thus it can activate the reaction using 100 μM NADH at pH 7 but inhibit it at pH 8. At the latter pH and low NADH levels, ATP has little or no effect but is a good inhibitor at high NADH levels. With NADPH, ATP hardly affects the velocity at pH 8 but activates the reaction at pH 7. At either pH 7 or 8, ATP activates both NAD and NADP reduction.

The above few observations are intended to show that the effects of purine nucleotides can be complex even when examined individually on the separate

Table 3 ATP Effects

Experiments performed in 0.01 *M* tris-acetate buffers, 25° at indicated pH and coenzyme concentration in the presence of 5×10^{-3} *M* α-ketoglutarate and 5×10^{-2} *M* NH_4Cl. v_0 is the velocity in the absence and v' is the velocity in the presence of the concentration of ATP indicated.

	v'/v_0					
	20 μM		100 μM		320 μM	
	NADH	NADPH	NADH	NADPH	NADH	NADPH
pH 7						
20 μM ATP	1	1.2	0.75	1.5	0.54	2.2
320 μM ATP	1.1	1.2	1.4	1.5	0.55	2.0
pH 7.5						
20 μM ATP	1	1.1	0.65	1.3	0.48	1.3
320 μM ATP	1.1	1	0.84	1.7	0.36	2.0
pH 8.0						
20 μM ATP	1	1	0.58	1	0.5	1.1
320 μM ATP	1	0.8	0.42	1.1	0.24	1.3

Velocity ratios of 1 are ±10%.

coenzymes. Now imagine the case within the mitochondria where there is a good possibility that purine nucleotides are in high enough concentration to affect activity profoundly. Here, these nucleotides are competing for each other, as are also the coenzymes for the enzymatic reaction for the active site of the enzyme. ATP might not markedly influence NADPH oxidation alone, but since it does bind to the enzyme, it would serve to activate any GTP-inhibited enzyme. On the other hand, ADP activation will be relieved by ATP, and thus the effect of ATP, even though it has little effect per se could serve either to activate or inhibit the reaction in the presence of other nucleotides. Such arguments can be carried out almost ad infinitum.

Clearly, however, since we do not know the concentration of nucleotides available to the intramitochondrial glutamate dehydrogenase, we cannot predict what will happen. However, these effects form the basis for my assumption that the enzyme is tightly regulated by purine nucleotides. As described above, the fact that the nucleotide effects differ depending on which coenzyme is being used leads me further to conclude that the metabolic role of glutamate dehydrogenase, that is, the direction of the reaction and which coenzyme is used, is also controlled by the levels of purine nucleotides. I shall return to this point after a discussion of the role of the enzyme as a function of complexity or need of the organism.

GLUTAMATE DEHYDROGENASE IN LOWER ORGANISMS

Since a key issue concerning glutamate dehydrogenase has been its role in metabolic processes, one way to approach this question is to examine some of the properties of the enzymes from lower organisms. We have surveyed elsewhere (39) some information concerning nonanimal and animal glutamate dehydrogenases. The nonanimal enzymes form a remarkable class of proteins which, although carrying out the same overall reaction, differ markedly from organism to organism. However, I will illustrate that information here with one example and some simplifications. The nonanimal enzymes are, for all practical purposes, specific either for NAD or NADP. Thus rather than having a single enzyme which can use both coenzymes, as is true in animal tissues, the organism possesses either one coenzyme specific enzyme or two different enzymes which differ in their coenzyme specificity. This latter is true in *Neurospora crassa*, for example, where sequence data with respect to both these enzymes have been obtained (40, 41). The data are of interest because there are portions of the sequences which show considerable homology between the NADP-dependent *Neurospora* enzyme and the bovine liver enzyme, particularly around a lysine residue which is strongly implicated in the reactive site of the enzyme (40). This particular lysine residue and its surrounding sequence do not appear in the NAD-dependent enzyme (41). The sequence of the NADP-dependent *Neurospora* enzyme also has regions of the sequence which are different from the bovine liver enzyme. Wootton et al. (40) have shown that the bovine enzyme contains about 48 more residues than the *Neurospora* enzyme, 13 more residues at the N-terminal end, and 36 more at the C-terminal end. There is a lack of homology in the last 100 residues near the C-terminal end. This difference may have something to do with the fact that the NADP-dependent *Neurospora* enzyme is not subject to the same type of regulation as the bovine liver enzyme with respect to purine nucleotides. Thus neither the *Neurospora* enzymes nor for that matter any of the nonanimal enzymes are affected to any significant effect by purine nucleotides binding to a site distinct from the active site. Instead of regulation of enzyme activity by purine nucleotides, in *Neurospora* and yeast, the relative concentrations of the NAD or NADP-dependent enzymes appear important for metabolic control. For example, in *Neurospora*, it has been shown that ammonia represses the NAD enzyme and induces the NADP-dependent enzyme but glutamate has the opposite effect (42). Similar results are observed in yeast (43, 44). These and similar arguments, which have been made in detail elsewhere (39), could lead to the conclusion that lower organisms use the NAD specific enzyme as a mechanism for generating α-ketoglutarate and ammonia but the NADP specific enzyme is used for generating glutamate.

THE METABOLIC ROLE OF GLUTAMATE DEHYDROGENASE

It is a very long extrapolation from this type of data to the metabolic role of glutamate dehydrogenase in animal tissues. It may, however, not be unreasonable to conclude that purine nucleotides serve to control the rate of use of one coenzyme with respect to another and that this in turn serves to control the particular metabolic function of the enzyme. There have been many excellent studies of the metabolic fate of substrates of the glutamate dehydrogenase system. These have ranged from experiments involving isolated mitochondria, liver homogenates, or liver slices to whole-animal experiments. In spite of these studies and the wealth of information provided by both *in vivo* and *in vitro* studies, it still is not possible to define the role of this enzyme in animal tissues.

Krebs et al. (5), in recent experiments using perfused liver, do show that alanine nitrogen appears mainly in urea and ammonia. They discuss the regulation of glutamate dehydrogenase and suggest that, because it is such a high concentration, the various metabolites, ammonia, glutamate, α-ketoglutarate, and aspartate are near equilibrium as defined by redox levels and by transaminase. Examination of Table 1 shows that many purine nucleotides may inhibit glutamate dehydrogenase. One could postulate that in fact this enzyme is greatly inhibited in the liver (as reflected by the low rates observed relative to transamination) and that in fact it may not be in equilibrium with respect to both the NAD/NADH and NADP/NADPH couples.

In the enormous literature related to the regulation of glutamate metabolism, there are only occasional references to the fact that glutamate dehydrogenase is directly affected by purine nucleotides. Indeed there is always considerable discussion of the energy requirements as a distinct event from the regulation of this enzyme. However, the control of glutamate dehydrogenase may occur as the consequence of a particular energy state and not necessarily by the mediation of, for example, ATP hydrolysis. It is unfortunate that the complexity of the nucleotide effects makes it difficult to decide on a single metabolic role of this enzyme. It is possible, on the other hand, that there is not such a single role. Thus the enzyme may be active primarily with only one coenzyme, it may be closely linked via nucleotide effects, with the flux of substrates through the Krebs cycle, it may serve to regulate the levels of α-ketoglutarate, glutamate, ammonia, or coenzyme, or it may perform several of these functions simultaneously. These remain unanswered questions. This dilemma arises because of the central importance of those metabolites which influence the catalytic activity, the lack of knowledge of the concentrations of purine nucleotides in the immediate environment of the enzyme in mitochondria, and the fact that many metabolic studies are almost certainly complicated by the fact that purine nucleotide levels are altered from those which exist *in vivo*. Furthermore, it is probably true that the enzyme function would not be the same in such diverse tissues as liver, muscle, brain, and kidney or that it performs only a single function in any given tissue. Since different tissues do not appear to possess different isozymes, one is tempted to speculate that the complexity of the kinetic parameters is in fact related to the metabolic role of the enzyme and to the possibility that the enzyme performs many different roles, depending on the particular circumstance. Certainly one of those roles is the production of ammonia for the biosynthesis of urea.

REFERENCES

1. Krebs, H. A., and Henscheit, K. (1932). *Z. Physiol. Chem.* **210**, 33.
2. Thunburg, T. (1920). *Skand. Arch. Physiol.* **40**, 1.
3. Thunburg, T. (1921). *Biochem. Z.* **206**, 109.
4. Von Euler, H., Adler, E., Gunther, G., and Das, N. B. (1938). *Z. Physiol. Chem.* **254**, 61.
5. Cohen, P. P., and Brown, Jr., G. W., (1960). In M. Florkin and H. S. Mason, Eds., *Comparative Biochemistry*, **2**, Academic, 361–377.
6. Krebs, H. A., Hems, R., and Lund, P. (1973). *Adv. Enz. Regul.* **11**, 361–377.
7. McGiven, J. D., and Chappell, J. B. (1975). *FEBS Lett.* **52**, 1–7.
8. Frieden, C. (1965). *J. Biol. Chem.* **240**, 2028–2035.
9. Olson, J. A., and Anfinsen, C. B. (1951). *Fed. Proc.* **10**, 230; (1952), *J. Biol. Chem.* **197**, 67.
10. Frieden, C. (1959). *J. Biol. Chem.* **234**, 809–814.
11. Frieden, C. (1959). *J. Biol. Chem.* **234**, 815–820.
12. Frieden, C. (1963). *J. Biol. Chem.* **238**, 3286–3299.
13. Moon, K., Piszkiewica, D., and Smith, E. L. (1972). *Proc. Natl. Acad. Sci. US* **69**, 1380–1383.
14. Reisler, S., Pouyet, J., and Eisenberg, H. (1970). *Biochem.* **9**, 3095–3102.
15. Sund, H., Pilz, I., and Herbst, M. (1969). *Eur. J. Biochem.* **7**, 517–525.
16. Reisler, E., and Eisenberg, H. (1971). *Biochem.* **10**, 2659–2663.
17. Malencik, D. A., and Anderson, S. R. (1972). *Biochem.* **11**, 3022–3027.
18. Chun, P. W., Williams, S. J., Cope, W. T., Tang, L. H., and Adams, E. T. (1972). *Biopolymers* **11**, 197–214.
19. Thusius, D., Dessen, P., and Jallon, J. M. (1974). *J. Mol. Biol.* **92**, 413–432.
20. Fisher, H. F., Cross, D. G., and McGregor, L. L. (1962). *Nature* **196**, 895.
21. Frieden, C., and Colman, R. F. (1967). *J. Biol. Chem.* **242**, 1705–1715.
22. King, K. S., and Frieden, C. (1970). *J. Biol. Chem.* **245**, 4391–4396.
23. Ifflaender, V., Markau, K., and Sund, H. (1975). *Eur. J. Biochem.* **52**, 211–220.
24. Moon, K., Piszkiewicz, D., and Smith, E. C. (1973). *J. Biol. Chem.* **248**, 3093–3103.
25. Frieden, C. (1962). *Biochim. Biophys. Acta* **62**, 421–423.
26. Rossman, M. G., Mores, D., and Olsen, K. W. (1974). *Nature* **250**, 194–199.
27. Wootton, J. C. (1974). *Nature* **252**, 542–546.

28. Fisher, H. F. (1969). In G. W. Schwert and A. D. Winer, Eds., *The Mechanism of Action of Dehydrogenases*, The University Press, Kentucky, 221–254.

29. Bates, D. J., Goldin, B. R., and Frieden, C. (1970). *Biochem. Biophys. Res. Comm.* **39**, 502–507.

30. Kurz, L. C., and Frieden, C. (1975). *J. Am. Chem. Soc.* **97**, 677–679.

31. Cross, D. G., McGregor, L. L., and Fisher, H. F. (1972). *Biochim. Biophys. Acta* **289**, 28–36.

32. diFranco, A., and Iwatsubo, M. (1972). *Eur. J. Biochem.* **30**, 517–532.

33. Bates, D. J., and Frieden, C. (1973). *J. Biol. Chem.* **248**, 7885–7890.

34. Frieden, C. (1970). *J. Biol. Chem.* **245**, 5788–5799.

35. Engel, P. C., and Dalziel, K. (1969). *Biochem. J.* **115**, 621–631.

36. Dalziel, K., and Egan, R. R. (1972). *Biochem. J.* **126**, 975–984.

37. Cross, D. G., and Fisher, H. F. (1970). *J. Biol. Chem.* **245**, 2612–2621.

38. Frieden, C. (1969). In H. Kalckar, Ed., *Role of Nucleotides for the Function and Conformation of Enzymes*, Munksgaard, Copenhagen, 194–217.

39. Goldin, B. R., and Frieden, C. (1971). Current Topics in Cellular Regulation **4**, 77–117.

40. Wootton, J. C., et al. (1974). *Proc. Natl. Acad. Sci. USA* **71**, 4361–4365.

41. Veronese, F. M., Degoni, Y., Nye, J. F., and Smith, E. L. (1974). *J. Biol. Chem.* **249**, 7936–7941.

42. Sanwal, B. D., and Lata, M. (1962). *Arch. Biochem. Biophys.* **97**, 582–588.

43. Holzer, H. (1966). *Biochem. J.* **98**, 37P.

44. Hollenberg, C. P., Riks, W. F., and Borst, P. (1970). *Biochim. Biophys. Acta* **201**, 13–19.

6

Glutamate Dehydrogenase from Rat Brain

P. González
E. Ventura
and T. Caldes
Departmento de Bioquimica
Facultad de Farmacia
Madrid, Spain

The importance of glutamate dehydrogenase for the assimilation of ammonium ion in plants as well as its role in animal tissues has long been emphasized. This mitochondrial enzyme has a widespread distribution (1–7). The enzyme is regulated by many metabolites and in many cases can react with either diphospho- or triphosphopyridine nucleotides (3). The enzyme was crystallized from bovine liver by Strecker and by Olson and Anfinsen (8, 9). Because of its ready accessibility, the bovine liver enzyme is most often studied. Although the literature on glutamate dehydrogenase is very large, thus far few studies have been made with the brain dehydrogenase; a partial purification of the enzyme has been described by Grisolia et al. (10).

Changes in glutamate dehydrogenase activity during the development of rat cerebral cortex were reported by Kuhlman and Lowry (11). The enzyme increased in brain during the period in which the contribution of cell bodies to the total volume was decreasing; they concluded that glutamate dehydrogenase is generally present in low concentrations in nerve cell bodies.

We present here the purification and some properties of the brain glutamate dehydrogenase.

We wish to thank Dr. Grisolia for his help and advice in this work.

LOCATION AND ENZYME PURIFICATION

White albino rats were anesthetized, killed, and the brains were quickly removed and washed with cold 9‰ NaCl.

The brains were homogenized with 9 vol of 0.32 *M* sucrose (in a mechanical Potter-Elvehjem Tri-R Stir-R, model K-43, at 600 rpm) for 1 minute. The homogenate was then centrifuged at 900 × g for 10 minutes. This supernatant was centrifuged at 11,000 × g for 20 minutes. The precipitate was washed twice with 0.32 *M* sucrose and then homogenized at high speed (6000 rpm) for 1 minute with 5 vol of 0.2 *M* potassium phosphate buffer, pH 7.4, containing 0.1% of Triton-X-100 and then centrifuged at 55,000 × g for 15 minutes. The supernatant is the crude fraction.

PROPERTIES OF THE PURIFIED ENZYME

When studying the cellular distribution of the enzyme, we found the bulk of activity in the nuclear-heavy mitochondrial fraction. As illustrated in Table 1, brain homogenate showed almost no glutamate dehydrogenase activity, indicating enzyme inactivation or the presence of an inhibitor.

Table 1 Localization of Glutamate Dehydrogenase in Rat Brain

Fraction	Units
Homogenate	Traces
11,000 × g supernatant	0
55,000 × g supernatant	300

The brain was homogenized as indicated in the text. 50 μl of each fraction was taken to measure activity with α-ketoglutarate as substrate.

To each 100 ml of the crude fraction 5 g of Na_2SO_4 was added, and after solution it was heated in a water bath at 60°C for 5 minutes. It was then rapidly chilled and centrifuged and the precipitate discarded. To each 100 ml of supernatant fluid (fraction I) 10 g of $(NH_4)_2SO_4$ was added, and the mixture kept with stirring at room temperature until in solution. After cooling for 15 minutes it was centrifuged at 10,000 × g for 15 minutes, and the precipitate discarded. Each 100 ml of supernatant fluid (fraction II) was mixed with 20 g of $(NH_4)_2SO_4$. The mixture was kept in the cold overnight and then centrifuged at 10,000 × g for 15 minutes. The supernatant was discarded, and the precipitate was taken in about 3 vol of 0.2 *M* potassium phosphate buffer, pH 7.4 (fraction III). This fraction was used in all cases. For comparative purposes, crystalline glutamate dehydrogenase from bovine liver in $(NH_4)_2SO_4$ was also used. The suspension was centrifuged and the precipitate dissolved in 0.2 *M* potassium phosphate buffer, pH 7.4.

ENZYME ASSAY

Routinely the following components were mixed in a final volume of 1.3 ml: 43 μmole of potassium phosphate buffer, pH 7.4; 0.55 μmole of NAD or NADP; 16 μmole of potassium glutamate and 10 μl of enzyme. All measurements were carried out in an Unicam SP-500 spectrophotometer at 340 nm and 37°C.

When oxidation of NADH or NADPH was measured, 200 μmole of potassium phosphate buffer, pH 7.4; 5.2 μmole of EDTA-Na; 210 μmole of ammonium acetate; 0.3 μmole of NADH or NADPH; 20 μmole of α-ketoglutarate and 10 μl of enzyme.

A unit is defined as the amount of enzyme that causes a change in absorbancy of 0.001 per minute under the standard conditions of assay. Specific activity is the number of units per mg of protein. Proteins were measured by the method of Lowry et al. (12).

Table 2 illustrates the purification procedure. We obtained about 30 times purification with respect to the crude fraction.

Table 2 Purification of Glutamate Dehydrogenase from Rat Brain[a]

Fraction	Volume (ml)	Mg protein/ml	S.A.	Yield
Crude	19.0	5.1	82	100
Fraction I	12.6	1.1	610	100
Fraction II	2.7	2.4	1222	100
Fraction III	2.6	1.7	1818	96

[a] Activity was measured with α-ketoglutarate as substrate.

Studies on coenzyme specificity are shown in Table 3. Glutamate dehydrogenase from different sources shows differing specificity for coenzymes. The enzyme from bovine liver appears to use either NADP or NAD (13), although the velocity with the NADP nucleotides is somewhat less than with NAD nucleotides (14). Glutamate dehydrogenase from rat brain shows very low activity with NADP, one-fifth of that with NAD, in contrast to the liver enzyme.

Table 3 Coenzyme Specificity

	Units/mg protein	
Coenzyme	Rat brain	Bovine liver
NAD	14 (1.8)	193 (5.7)
NADP	3 (0.4)	106 (3.2)
NADH	770 (100)	3344 (100)
NADPH	243 (32)	2213 (66)

The relative activity is also expressed as percent in the table.

No careful kinetic studies as a function of pH have been performed. Such a study is difficult not only because of the number of substrates involved but also because ionizable groups complicate the interpretation of the pH variation. Strecker (15) observed, with liver enzyme, an optimum pH of 8.3 to 8.5 for oxidation of glutamate, using either Tris, phosphate, or pyrophosphate buffers. Most workers find the optimum pH between 8 and 8.5 for the bovine enzyme as well as the enzyme from other sources (7, 16, 17). The rat brain enzyme shows a pH optimum between 7.4 to 7.8 with glutamate as substrate, NAD as coenzyme, and potassium phosphate buffer.

Values listed in Table 4 are apparent Michaelis constants.

Table 4 Kinetic Parameters for Rat Brain Glutamate Dehydrogenase at pH 7.4 and 37°C in 0.2 *M* Potassium Phosphate Buffer

Substrate	Coenzyme	Apparent K_m for substrate
α-ketoglutarate	NADH	8.6×10^{-4} *M*
α-ketoglutarate	NADPH	5.9×10^{-4} *M*
Ammonium acetate	NADH	10.4×10^{-2} *M*
Glutamate	NAD	2.9×10^{-3} *M*

The nucleotides seem to have higher activation affect on the enzyme with α-ketoglutarate than with glutamate as substrate (Table 5). According to Frieden (18) ADP, with bovine liver glutamate dehydrogenase, increases the initial velocity of NADH oxidation because it displaces the second inhibitory molecule of NADH, thereby preventing the dissociation of the enzyme by NADH to its inactive form. As a result, stimulation by ADP becomes greater at higher concentrations up to the point in which the NADH concentration is so high that it displaces the ADP. In direct contrast to ADP, ATP, when bound to the nonactive site with NADH as coenzyme, enhances enzyme dissociation.

Table 5 shows that AMP, ADP, and ATP activate the glutamate dehydrogenase from rat brain enzyme when it was measured with glutamate as substrate and using NAD as coenzyme. The higher activator was ADP, although AMP and ATP activate slightly. It has been postulated that ADP competes with NADH or NADPH for the nonactive coenzyme binding sites, leaving the active sites still accessible for the reduced coenzyme (19,20). Moreover, ADP abolishes the negative cooperativity between these remaining NADH or NADPH binding sites and causes a fourfold increase of the dissociation constants of both NADH and NADPH.

Table 5 Effect of AMP, ADP, and ATP on Glutamate Dehydrogenase from Rat Brain

	Units	
Additions	With glutamate	With α-ketoglutarate
None	34 (100)	17 (100)
1 μmole AMP[a]	42 (123)	— —
10 μmole AMP	39 (114)	80 (470)
1 μmole ADP[a]	64 (188)	— —
10 μmole ADP	59 (173)	100 (588)
1 μmole ATP[a]	43 (126)	— —
10 μmole ATP	44 (129)	57 (335)

Experimental conditions: 0.1 ml of purified glutamate dehydrogenase was incubated at 37°C for 30 minutes with 1 or 10 μmole of indicated nucleotides in a total volume of 0.2 ml. The activity was checked with glutamate and with α-ketoglutarate as substrate. NAD or NADH was used.

[a] Measured after 1 hour in the cold.

The activator effect of these reagents was similar with enzyme from liver as with the brain enzyme.

A number of metal chelating reagents were observed to inhibit the rat brain enzyme. It has been postulated that liver glutamate dehydrogenase is a zinc enzyme. With rat brain glutamate dehydrogenase, 42 μM zinc inhibited entirely. Mg^{2+} up to 830 μM had no effect. Higher concentrations produced inhibition.

A large number of sulfhydryl reagents, including Ag^{2+} and Hg^{2+} (14, 21, 22), have been tested with the liver enzyme. None, with the sole exception of Ag^{2+}, is a truly good inhibitor.

Carbamylphosphate and cyanate inactivate the rat brain glutamate dehydrogenase. Inorganic phosphate was a slight activator, and azide has no effect (Table 6). The inactivation by cyanate was higher than with carbamylphosphate, and the azide protected against both. This fact, as pointed out by Carreras et al. (28), indicates that the mechanism of carbamylphosphate carbamylation would follow decomposition into cyanate. Grisolia (23) reported that glutamate dehydrogenase from liver was inactivated by carbamylphosphate with the formation of approximately one residue of homocitrulline per subunit, suggesting that such an inactivation might be of regulatory significance (24). Moreover, although the inactivation proceeds via cyanate, carbamyl phosphate is the physiological substance made by the liver and as such is the natural chemotropic reagent (25). Studies of Veronese with crystalline bovine liver glutamate dehydrogenase (26) confirmed this.

Table 6 Effect of Carbamylphosphate, Cyanate, and Inorganic Phosphate on Rat Brain Glutamate Dehydrogenase

Additions	Units
None	24 (0)
C ~ P	8 (66)
KCNO	4 (85)
P_i	30 (0)
Azide	24 (0)
C ~ P + azide	11 (53)
KCNO + azide	7 (70)

0.2 ml enzyme was incubated with 20 μmole of the indicated reagents, in 0.4 ml, 30 minutes at 37°C. The inactivation is expressed in the table also as percent (numbers between parentheses).

Although NAD and glutamate protected against inactivation by cyanate and carbamylphosphate on rat brain enzyme (Table 7), further studies are necessary in order to clarify the mechanism.

In confirmation of the work of Grisolia et al. (27), urea inactivates the rat brain glutamate dehydrogenase; when used at relatively low concentrations, NADPH was an inactivator and greatly increased the inactivation produced by urea (Table 8).

At the beginning of this paper we mentioned the possible presence of an inhibitor in the supernatant fraction of rat brain homogenates. Table 9 presents experimental evidence for this. As illustrated, the inhibitor is not a protein but a hitherto unidentified acetone soluble substance.

Table 7 Protection of NAD and Glutamate against Cyanate and Carbamylphosphate Inactivation

Additions	Units	
	Rat brain	Bovine liver
None	21 (0)	54 (0)
C ~ P	13 (39)	12 (79)
KCNO	15 (28)	0 (100)
NAD	20 (5)	37 (32)
C ~ P + NAD	19 (10)	29 (46)
KCNO + NAD	17 (19)	22 (59)
Glutamate	18 (14)	— —
C ~ P + glutamate	18 (14)	— —
KCNO + glutamate	18 (14)	— —

0.1 ml of enzyme was incubated with 5 μmole of KCNO or 10 μmole of C ~ P, 2.5 μmole of NAD or 5 μmole of glutamate, or the indicated amounts of NAD or glutamate in 0.25 ml for 30 minutes at 37°C. The inactivation is expressed in the table also as percent (numbers between parenthesis).

Table 8 Effect of Urea, NADPH, and Urea plus NADPH on Rat Brain Glutamate Dehydrogenase When NAD or NADH Were Used as Coenzymes

Additions	Percent inhibition	
	With NAD	With NADH
None	0	0
50 μmole urea	0	0
100 μmole urea	4	0
200 μmole urea	22	0
400 μmole urea	30	8
0.05 μmole NADPH	60	56
50 μmole urea + NADPH	80	62
100 μmole urea + NADPH	80	66
200 μmole urea + NADPH	94	77
400 μmole urea + NADPH	100	100

0.1 ml enzyme was incubated with the indicated reagents in 0.35 ml for 30 minutes at 37°C. 0.05 μmole NADPH was used in all cases. The enzymatic activity was measured in 100 μl of the incubation mixtures.

Table 9 Effect of Tissue Fractions on the Glutamate Dehydrogenase of Rat Brain

Conditions	Units	
	Rat brain	Bovine liver
Control	20.4	21.2
900 × g supernatant	11.0	—
+ Protein fraction	23.0	—
+ 10 μl soluble fraction	19.2	—
+ 25 μl soluble fraction	3.8	15.4
+ 50 μl soluble fraction	0.0	9.2

10 ml supernatant fraction, obtained as indicated in the text, was treated with 10 volumes of acetone at −20°C; then it was quickly filtered to yield a "protein fraction" (acetone-insoluble material) and one acetone-soluble fraction. The acetone-water fraction was evaporated by vacuum distillation at room temperature. This residue was taken into 1.5 ml. This is the soluble fraction. The "protein fraction" was taken into 1.5 ml of 0.2 *M* potassium phosphate buffer, pH 7.4. The enzymatic activity was measured as follows: 43 μmole of potassium phosphate buffer, pH 7.4; 0.55 μmole of NAD; 16 μmole of potassium glutamate; 25 μl of fraction III and 300 μl distilled water or 300 μl of 900 × g supernatant fraction, 300 μl of protein fraction or the indicated μl of soluble fraction to total 1.3 ml.

REFERENCES

1. Cohen, P. P., and Sallach, H. J. (1961). In D. M. Greenberg, Ed., *Metabolic Pathways*, 2nd. ed., **2**, Academic, New York, 1–78.

2. Schmidt, E., Schmidt, F., Horn, H. D., and Gerlach, U. (1963). In H. U. Bergmeyer, Ed., *Methods of Enzymatic Analysis*, Academic, New York, p. 668.

3. Frieden, C. (1965). *J. Biol. Chem.* **240**. 2028.

4. Copenher, J. H., Jr., McShan, W. H., and Meyer, R. K. (1950). *J. Biol. Chem.* **180**, 73.

5. Frieden, C., in *The Enzymes*, P. D. Boyer, H. A. Lardy, and K. Myrback, Eds., 2nd ed., **7**, Academic, New York, p. 18.

6. Dewan, J. G. (1938). *Biochem. J.* **32**, 1378.

7. Gaull, G., Hagerman, D. D., and Vilee, C. A. (1960). *Biochim. Biophys. Acta* **40**, 552.

8. Olson, J. A., and Anfinsen, C. B. (1951). *Fed. Proc.* **10**, 230.

9. Olson, J. A., and Anfinsen, C. B. (1952). *J. Biol. Chem.* **197**, 67.

10. Grisolia, S., Quijada, C. L., and Fernandez, M. (1964). *Biochim. Biophys. Acta* **81**, 61.

11. Kuhlman, R. E., and Lowry, O. H. (1956). *Neurochem.* **1**, 173.

12. Lowry, O. H., Rosenrough, N. J., Farr, A. L., and Randall, R. J. (1951). *J. Biochem.* **193**, 265.

13. Von Euler, H., Adler, E., Gunther, G., and Das, N. B. (1938). *Z. Physiol. Chem.* **254**, 61.
14. Olson, J. A., and Anfinsen, C. B. (1953). *J. Biol. Chem.* **202**, 841.
15. Strecker, H. J. (1953). *Anal. Biochem. Biophys.* **46**, 128.
16. Damodaran, M., and Nair, K. R. (1938). *Biochem. J.* **32**, 1064.
17. Snoke, J. E. (1956). *J. Biol. Chem.* **223**, 271.
18. Frieden, C. (1959). *J. Biol. Chem.* **234**, 815.
19. Koberstein, R., and Sund, H. (1973). *Eur. J. Biochem.* **36**, 545.
20. Pantaloni, D., and Dessen, P. (1969). *Eur. J. Biochem.* **11**, 510.
21. Hellerman, L., Schellenberg, K. A., and Reiss, O. K. (1958). *J. Biol. Chem.* **233**, 1468.
22. Pfleiderer, G., Jeckel, D., and Wieland, T. (1956). *Biochem. Z.* **328**, 187.
23. Grisolia, S. (1968). *Biochem. Biophys. Res. Comm.* **32**, 56.
24. Chabas, A., and Grisolia, S. (1972). *FEBS Letters* **21**, 25.
25. Grisolia, S., and Hood, W. (1972). In E. Kun and S. Grisolia, Eds., *Biochemical Regulatory Mechanisms in Eukaryotic Cells*, Wiley-Interscience, New York, p. 138.
26. Veronese, F. M., Piszkiewicz, D., and Smith, E. L. (1972). *J. Biol. Chem.* **247**, 754.
27. Grisolia, S. (1964). *Physiological Rev.* **44**, 657.
28. Carreras-Barnes, J., Diederich, D. A., and Grisolia, S. (1972). *Eur. J. Biochem.* **27**, 103.

7

The Krebs Cycle Depletion Theory of Hepatic Coma

Samuel P. Bessman
and Nandita Pal
Department of Pharmacology
University of Southern California
School of Medicine
Los Angeles, California

Hepatic coma is a mental syndrome arising when severe damage occurs to the liver or when the liver has been removed.

Pavlov and others found in 1893 (1) that when the blood draining the intestinal tract is shunted around the liver and the animal is fed a diet containing normal or large amounts of nitrogen, hepatic coma ensues. This has also been found to occur in humans. In the presence of liver disease the intake of ammonium salts or food substances able to produce ammonium has led experimentally and clinically to the syndrome of hepatic coma.

Recknagel and Potter, in 1951 (2), demonstrated, *in vitro*, that excess ammonium salts caused increased formation of ketone acids by rat liver slices. They postulated the ketogenic effect of ammonium salts to be through the reductive amination of α-ketoglutaric acid to produce a depletion of this Krebs cycle intermediate and, subsequently, all the succeeding substances. This would lead to a diminished turnover of the Krebs cycle, with a consequent lack of capability to handle the normal production of acetyl CoA from fats. The excess acetyl CoA would condense to form ketoacids.

In 1952 this principle was applied to the observations of Davidson's (3) group that a syndrome of incipient hepatic coma could be evoked by feeding patients with liver cirrhosis urea, glycine, protein, or ammonium chloride.

UPTAKE OF AMMONIA BY BRAIN

It was postulated by Bessman and Bessman, in 1955 (4), that ammonia, elevated in content in the arterial blood, would enter the brain at an excessive rate, causing reductive amination of α-ketoglutarate and resulting in lowered turnover in the Krebs cycle, lowered oxygen uptake, and decreased energy generation by brain. Evidence was presented which showed the blood leaving the brain contained less ammonia than the blood entering the brain. The uptake of ammonia by human brain is directly related to the arterial level, with an equilibrium point at about 0.05 m*M*. This agrees reasonably well with the equilibrium conditions of the glutamate dehydrogenase reaction. Other studies in human beings showed that the oxygen uptake of the brain was diminished about 50% during hepatic coma. Experiments showed a 40% decrease in α-ketoglutarate in the brains of rats given a series of small ammonia injections over several hours.

Further clinical investigation did not show a clear correlation between blood level of ammonia and symptomatology. This relation was not postulated, however, in the original hypothesis, for the concentration of ammonia would only affect the rate of depletion of the Krebs cycle. It was the degree of depletion or attenuation of the Krebs cycle that should parallel the degree of attenuation of mental function. The hypothesis was that a toxic agent like ammonia did not exert its effect directly, by its presence, but by causing the depletion of some central pathway or intermediate.

CO_2 FIXATION

Important studies by Berl et al., in 1962 (5), addressed the question of whether ammonia administration intracerebrally to rats caused changes in the fixation of CO_2 into tricarboxylic acid cycle intermediates. They showed that the increase in CO_2 fixation did occur, which would be expected if there were a drain on the cycle. They interpreted their results as indicating that, even though there might be a depletion of the cycle, the increased rate of CO_2 fixation should replenish any deficit. Their data were used by Garfinkel, in 1966 (6), in a computer analysis of a model of the cycle activities in brain. This model was based on two pools of α-ketoglutarate, one large, presumably extramitochondrial pool and a small, presumably intramitochondrial pool. The data of Berl et al., when tested in the model, showed a depletion of about 50% in the intramitochondrial pool. This corresponds well with the clinical data, which show a 50% drop in cerebral oxygen consumption.

Analysis of the effect of ammonia on α-ketoglutarate, glutamate, ATP, and creatine phosphate in brain have been attempted by several laboratories, with conflicting results. All agree that glutamine increases markedly in concentration after administration of ammonia. Clinically this is reflected in a large increase in spinal fluid glutamine. This finding is used for the diagnosis of hepatic coma.

ATP LEVELS

Schenker and Mendelson, in 1964 (7), using whole brain from rats frozen in liquid nitrogen found no significant change in whole brain ATP content up to 100 minutes after administration of 5 mmoles/kg NH_4 acetate IP. With this dose rats exhibited some symptoms of toxicity by 7 minutes. Subsequently Schenker et al., in 1967 (8), using a dose of NH_4 citrate 50% larger and assaying cortex and "basilar" areas, found about 50% fall in ATP in the basilar tissue with no change in the cortical tissue. They also found a 75% fall in basilar phosphocreatine with no significant change in the cortical phosphocreatine.

Hindfelt and Siesjö, in 1971 (9), found marked changes in creatine phosphate content of the cortex, cerebellum, and brain stem of rats treated with ammonium acetate. They found no change in ATP. There were marked increases in lactate and pyruvate, with no significant alteration in α-ketoglutarate or glutamate. The rise in pyruvate in the presence of an increase in lactate reveals a severe failure in oxidation which correlates well with the reported marked drop in oxygen uptake by brain in hepatic coma. The values reported for α-ketoglutarate are more than an order of magnitude greater than the values for the intramitochondrial pool reported by Srere, in 1972 (10). These values are consistent with the large extramitochondrial pool shown by Garfinkel, in 1966 (6), to be unchanged in ammonia intoxication at the same time that the small pool is diminished by half. These data therefore do not provide information relevant to the question of intramitochondrial depletion.

Hawkins et al., in 1973 (11), using a technique of blowing the brain tissue rapidly out of the cranium onto a freezing plate, reported a slight fall in creatine phosphate and no change in ATP, α-ketoglutarate, or glutamate. Their high α-ketoglutarate values also provide no information about the intramitochondrial pool, but they also report a more than 20% drop in oxalacetate. Since the basal concentration of oxalacetate is near the intramitochondrial level, it is likely that this smaller pool more closely reflects the intramitochondrial situation than the observed data for α-ketoglutarate. One puzzling observation reported in this paper is that the oxygen uptake of brain in animals poisoned with ammonia rises by 35%. This is not consistent with clinical ammonia intoxication (Weschsler, Crum, and Roth, 1954) (12) and may suggest that acute experiments are not relevant to the problem of hepatic coma, which is a rather chronic, slowly developing syndrome characteristically associated with a marked decrease in cerebral oxygen uptake.

ORGANIC PHOSPHATE INTERMEDIATES

We should like to report some preliminary observations on ammonia toxicity in mice, using a new method for automatically separating and quantitating phosphorylated intermediates (Bessman, 1974) (13). To stop metabolic activity in the brain, the heads of the animals were subjected to a focused beam of microwaves which delivered 1.5 kW of energy, according to the procedure of Stavinoha et al., 1970 (14). Three seconds of this irradiation suffice to stop all enzyme activity in the brain. The following experiment shows that there is no significant interconversion of phosphorous compounds in mouse brain after irradiation for 3 seconds.

A 35 g mouse was irradiated as above and decapitated. The head was bisected in a vertical sagittal plane, and the brain was removed immediately from the left half of the cranium. The brain tissue was homogenized in 10% trichloroacetic acid. This entire procedure took about 30 seconds. The right half of the head was left at room temperature for 30 minutes; then its brain tissue was also homogenized as above. The neutralized trichloroacetic acid filtrates were analyzed for phosphorylated intermediates by placing aliquots corresponding to 40 mg wet weight of tissue on the columns of the automatic phosphate analyzer (Bessman et al., 1974) (15). Figure 1 shows the records obtained on the two

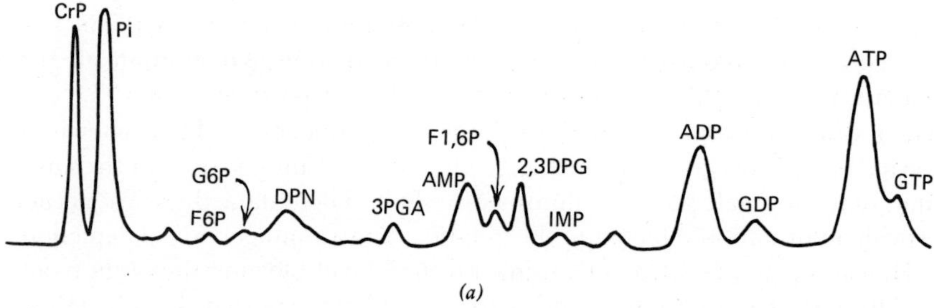

(a)

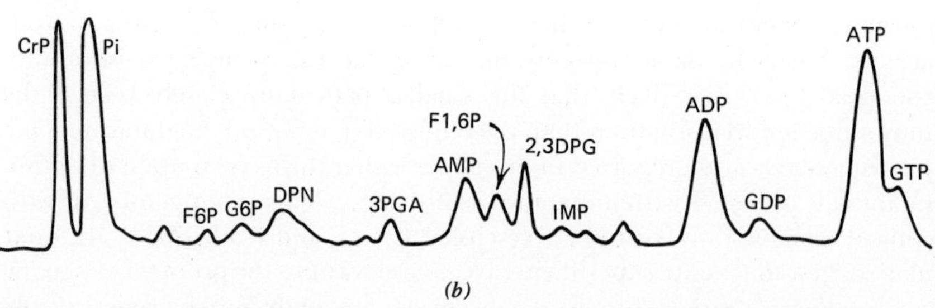

(b)

Figure 1. Brain organic phosphate analysis after 3 seconds of microwave irradiation: (a) "Left half" immediately after exposure; (b) "Right half" 30 minutes after exposure

filtrates. The upper chart is the record of the control half of the brain made into a filtrate immediately after microwave treatment of the animal, and the lower chart shows the data on the half of the brain left to stand at room temperature. The difference between the two records is no greater than is seen between the two halves of a brain prepared at the same time. It is clear that all enzyme activity involving phosphate compounds detectable by this method has been halted by the 3 second microwave treatment.

The microwave treatment does not destroy the architecture of the brain, nor is there a detectable change in its consistency. This permits us to separate the cortex, brain stem, and cerebellum with ease. It is also possible to dissect out smaller portions of the brain, but this was not done in the experiments to be reported.

In the experiments we wish to report 35 g mice were injected intraperitoneally with 0.8 mmole/100 g of ammonium or sodium acetate in isotonic solution. Twelve minutes after injection the mice were subjected to microwave irradiation to the head for 3 seconds. Trichloroacetic acid filtrates of brain were prepared and neutralized, and aliquots equivalent to 40 mg of wet weight were analyzed for glycolytic and nucleotide intermediates. Table 1 shows the results obtained for cerebrum, brain stem, and cerebellum. Anatomically what is designated as brain stem was the tissue left after removal of cerebrum and cerebellum. It includes the hypothalamus, basal gauglia, medulla, and pons.

Table 1 Organic Phosphate (mmole/200 g protein)[a]

	Cortex		Cerebellum		Midbrain and brain stem	
	Control	Rx	Control	Rx	Control	Rx
CrP	2.99	2.16	5.25	2.43*	3.96	2.59
F6P	.17	.15	.34	.20	.28	.29
G6P	.37	.35	.83	.37	.48	.38
NAD	1.55	1.30	2.49	1.25**	1.78	1.67
3PGA	.41	.32	.68	.34*	.51	.40
AMP	1.53	1.00	2.00	1.00***	1.57	1.27
F1, 6P	.75	.58	1.28	.57**	1.02	.70
2, 3DPG	.58	.48	.99	.62*	.43	.29
ADP	3.80	2.31	4.99	2.51***	3.88	2.83
ATP	6.80	5.32	10.31	4.25***	8.30	6.50
GDP	1.08	.65	1.31	.81	1.17	.86
GTP	1.74	1.39	2.71	.80***	1.81	1.75

* $p < .05$.
** $p < .01$.
*** $p < .001$.
[a] Each value represents three runs, each on two combined brains.

There are statistically significant changes in many of the intermediates in the cerebellum. Although changes in the same direction are seen in all areas of the brain, the variability among determinations prevents statistical significance in the small number of animals. There is a marked fall in both phosphocreatine and ATP. Of particular interest are the changes in 2,3-diphosphoglyceric acid, which we interpret as changes in the volume of blood in the brain. This may be a reflection of the blood flow through the tissue, for there is very little of this intermediate in any other tissue than red blood cells. In all areas this value goes down. Together with the evidence of the slow development of clinical hepatic coma this leads us to believe that the acute toxicity of ammonium salts may not be a good model for the syndrome of ammonium poisoning which manifests itself as hepatic coma.

ACUTE VERSUS CHRONIC TOXICITY

To approach the question of acute versus chronic toxicity the effect of a single dose of 0.8 mmole/100 g was compared to the effect of six doses of 0.6 mmole/100 g given intraperitoneally at 60 minute intervals. In this experiment the mice were irradiated 12 minutes after the last dose. Whole brains were analyzed. There were three controls, four single-dose-treated animals and two given six doses. The results for this experiment, Table 2, show values consistent with Table 1 when the single-dose-treated animals are compared. The "long-

Table 2 Whole Brain Organic Phosphate (mmole/200 g protein)

	Control	1 dose	6 doses
CrP	4.91	3.45*	3.65*
F6P	.11	.15	.21
G6P	.39	.36	.61***
NAD	1.39	1.45	2.39**
3PGA	.40	.40	.65**
AMP	1.48	1.53	2.36*
F1, 6P	1.00	.86	1.14
2, 3DPG	.85	.59	.94
IMP	.30	.30	.58**
ADP	4.59	3.55	4.35
ATP	9.54	7.16**	7.52
GDP	.74	.72	1.60***
GTP	2.22	1.79	1.33*

* $p < .05$.
** $p < .01$.
*** $p < .001$.

term" treated animals show results which, in many cases, are in the opposite direction. In the first place, the 2,3-diphosphoglycerate has returned to normal. The glycolytic intermediates have increased, and AMP, inosinic acid, and GDP have increased significantly. The increase in GDP together with the decrease in GTP is of special interest in view of the postulated deficiency of α-ketoglutarate in chronic ammonium intoxication.

Perhaps extension of these experiments will provide more insight into the question of whether the syndrome of hepatic coma is indeed due to the depletion of the mitochondrial Krebs cycle, proposed more than 20 years ago.

REFERENCES

1. Hahn, M., Massen, O., Nencki, M., and Pavlov, J. (1893). *Arch. Exper. Path. Pharmakol.* **32**, 161.
2. Recknagel, R. O., and Potter, V. R. (1951). *J. Biol. Chem.* **191**, 263.
3. Gabuzda, G. J., Phillips, G. B., and Davidson, C. S. (1952). *N. Engl. J. Med.* **246**, 124.
4. Bessman, S. P., and Bessman, A. N. (1955). *J. Clin. Invest.* **34**, 622.
5. Berl, S., Takagaki, G., Clarke, D. D., and Waelsch, H. (1962). *J. Biol. Chem.* **237**, 2570.
6. Garfinkel, D. (1966). *J. Biol. Chem.* **241**, 3918.
7. Schenker, S., and Mendelson, J. H. (1964). *Am. J. Physiol.* **206**, 1173.
8. Schenker, S., McCandless, D. W., Brophy, E., and Lewis, M. S. (1967). *J. Clin. Invest.* **46**, 838.
9. Hindfelt, B., and Siesjö, B. K. (1971). *Skand. J. Clin. Lab. Invest.* **28**, 365.
10. Srere, P. A. (1972). In *Energy Metabolism and the Regulation of Metabolic Processes in Mitochondria*, Academic Press, 79–82.
11. Hawkins, R. A., Miller, A. L., Nielson, R. C., and Veech, R. L. (1973). *Biochem. J.* **134**, 1001.
12. Wechsler, R. L., Crum, W., and Roth, J. L. A. (1954). *Clin. Res. Proc.* **2**, 74.
13. Bessman, S. P. (1974). *Anal. Biochem.* **59**, 524.
14. Stavinoha, W. B., Weintraub, S. T., and Modak, A. T. (1973). *J. Neurochem.* **20**, 361.
15. Bessman, S. P., Geiger, P. J., Lu, T. C., and McCabe, E. R. B. (1974). *Anal. Biochem.* **59**, 533.

8

Some Considerations Regarding the Feasibility of Bessman's Hypothesis for Ammonia Toxicity

Vicente Rubio
and S. Grisolia
Department of Biochemistry
University of Kansas Medical Center
Kansas City, Kansas

Bessman and Bessman (1) suggested that an increase of ammonia in the brain will concomitantly increase the rate of glutamate dehydrogenase (GDH) in the direction of glutamate synthesis. In the normal state, the concentration of ammonia in brain is much below the K_M for glutamate dehydrogenase ($K_M = 5.7 \times 10^{-2}\ M$) (2). Higher concentrations of ammonia would divert significant amounts of α-ketoglutarate from the citric acid cycle to synthesize glutamate, decreasing the production of ATP.

There are some experimental data supporting the hypothesis, in particular the markedly lower concentrations of ammonia found in blood from the jugular vein than in that from the carotid artery in patients with ammonia intoxication (1). However, the hypothesis should be consistent with the activity of glutamate dehydrogenase in brain and other related parameters.

It seemed of interest to assess some available biochemical data in relation to Bessman's hypothesis. The following calculations must be regarded only as approximated; more accurate calculations must wait for further knowledge on the biochemistry of brain under normal and under pathological conditions.

This work was supported in part by Grant AM-01855 from the National Institutes of Health.

GDH ACTIVITY IN RAT BRAIN HOMOGENATES

As shown previously, beef brain contains appreciable quantities of GDH (3). Rat brain was extracted with CTAB as described by Fahien et al. (4). We found 24.8 u (3) GDH per gram. The assay was carried out at 340 nm and 30° C in 3.0 ml, containing 130 μmole of potassium phosphate buffer (pH 7.6), 90 μmole of α-ketoglutarate, 171 μmole of NH_4Cl, and 0.5 μmole of NADH. Thus, under these conditions, 72 μmole of glutamate can be formed per gram of brain per hour, or $\cong$ 200 μmole when corrected for temperature (37°) and V_{max} (the concentration of ammonia used in the assay is $\cong$ the K_M).

NORMAL RATE FOR GDH IN BRAIN

The concentrations of components for GDH activity in brain (2) are α-ketoglutarate $\cong$ 1.36 mM ($\cong 10 \times K_M$), NADH $= 132\ \mu M$ ($\cong 7.3 \times K_M$), and $NH_4^+ = 340\ \mu M$ ($\cong 0.006\ K_M$); thus the rate-limiting factor is the $[NH_4^+]$. The calculated rate is

$$V = \frac{V_{max}[S]}{K_M + [S]} = \frac{200 \times 3.4 \times 10^{-4}}{0.057 + 3.4 \times 10^{-4}} = 1.19\ \mu\text{mole glutamate/g/hour}$$

CALCULATED GDH RATE UNDER PATHOLOGICAL CONDITIONS

Warren and Schenker (5) have shown that up to $6.6 \times 10^{-3}\ M\ NH_4^+$ may be present in brain under experimentally induced ammonia intoxication. Therefore

$$V = \frac{200 \times 6.6 \times 10^{-3}}{0.057 + 6.6 \times 10^{-3}} = 20.75\ \mu\text{mole glutamate/g/hour}$$

α-KETOGLUTARATE CONSUMPTION

Assuming that rat brain produces ~800 μmole ATP/g/hour, based on the Q_{O_2} value of 10 ml O_2/g dry weight/hour for rat brain (2), and an R.Q. of 1, approximately 42 μmole of α-ketoglutarate will be used per gram of brain per hour during normal operation of the citric acid cycle (2 moles of α-ketoglutarate $\cong$ 1 mole of glucose = 38 moles of ATP). Since the K_M for α-ketoglutarate for the α-ketoglutarate dehydrogenase is $8.5 \times 10^{-6}\ M$, while the K_M for glutamate dehydrogenase is $1.2 \times 10^{-4}\ M$ (2), very high ammonia concentration will be required to impair extensively α-ketoglutarate use via the tricarboxylic acid cycle. Nevertheless, it is apparent that the kinetic data are congruent and within the order of magnitude of Bessman's hypothesis.

REFERENCES

1. Bessman, S. P., and Bessman, A. N. (1955). *J. Clin. Invest.* **34**, 622.
2. *Biochemists' Handbook* (1961). Cyril Long, Ed., Van Nostrand, Princeton, N.J.
3. Grisolia, S., Quijada, C. L., and Fernandez, M. (1964). *Biochim. Biophys. Acta* **81**, 61.
4. Fahien, L. A., Strmecki, M., and Smith, S. (1969). *Arch. Biochem. Biophys.* **130**, 449.
5. Warren, K. S., and Schenker, S. (1964). *J. Lab. Clin. Med.* **64**, 442.

9

Acetylglutamate Synthetase

Masamiti Tatibana
Katsuya Shigesada*
and Masataka Mori
Department of Biochemistry
Chiba University School of Medicine
Inohana, Chiba, Japan

Research on acetylglutamate started almost simultaneously with the enzymological studies on the urea cycle. As early as 1948, Cohen, Hayano, and Grisolia (1, 2) found that glutamate or its derivatives were necessary to promote the enzymatic fixation of ammonia and CO_2 into ornithine to form citrulline. After the discovery and identification of carbamyl phosphate as the intermediate in the reaction of citrulline formation in 1955 by Jones, et al. (3), increasing attention was focused on the role of *N*-acylglutamate in the reaction catalyzed by the carbamyl phosphate synthetase from mammalian liver. Although several acyl derivatives of a five-carbon dicarboxylic acid were active *in vitro* (4, 5), *N*-acetylglutamate has been generally considered to be the principal natural cofactor, since it was found in mammalian liver (6, 7) and has the highest affinity for the enzyme (4).

At early stages of the studies on the role of acetylglutamate the compound was considered to function as a coenzyme to the carbamyl phosphate synthetase I. But all the subsequent studies along this line gave negative evidence (5, 8–11). It became eventually clear that acetylglutamate does serve as an activator, exactly an allosteric activator for the enzyme (12–19). The first indications for this possibility were given by Caravaca and Grisolia (12) in 1959

* Present address, Institute for Virus Research, Kyoto University, Kyoto, Japan.

and by Marshall et al. (13) in 1961. Recent developments in kinetical analysis of the effects of acetylglutamate as well as in studies on the molecular events involved in the activation process will be discussed in the subsequent sessions on carbamyl phosphate synthetase I by Jones, Elliott, and Marshall.

On the other hand, the physiological significance of the activation of carbamyl phosphate synthetase by acetylglutamate remained obscure until recently. In 1971, we presented evidence for the notion that acetylglutamate is a messenger of metabolic signal and does coordinate the rate of synthesis of carbamyl phosphate and amino acid metabolism in the liver (7). As extension of the work, we detected an enzyme activity responsible for the synthesis of acetylglutamate in the extracts of liver mitochondria and elucidated some properties of the enzyme (20). Although the presence of the acetylating activity in mammalian liver was reported by Grisolia and Marshall as early as in 1955 (21), no further studies were reported. This report will review the results of previous studies on tissue distribution of acetylglutamate and of its biosynthetic activity and then describe purification and properties of acetylglutamate synthetase obtained from rat liver.

Table 1 Acetylglutamate Levels in Various Tissues

From Shigesada and Tatibana (7). Sample tissues obtained from mice and rats maintained on a normal laboratory chow were pooled and processed for analysis of acetylglutamate content. Preparation of subcellular fractions of rat liver and assay of succinic acid dehydrogenase were performed as described (7).

Animal	Tissue or subcellular fraction	Acetylglutamate content	Succinic dehydrogenase activity
		(nmole/g tissue)[a]	(u/g tissue)[a]
Rat (2)[b]	Liver	28	
	Small intestine	10	
	Kidney	0	
	Spleen	0	
	Heart	0	
Rat (1)[b]	Liver		
	Whole homogenate	25	280
	Fraction		
	Nuclear	6	55
	Mitochondrial	14	166
	Cytoplasmic	2	0
Mouse (4)[b]	Liver	30	

[a] All values represent amounts per gram of original tissue, wet weight.
[b] Figures in parentheses indicate the number of animals used.

TISSUE DISTRIBUTION AND BIOSYNTHETIC ACTIVITY OF ACETYLGLUTAMATE

A rat given a normal laboratory chow ad libitum, liver was found to contain about 30 nmole of the compound per gram of wet tissue in the morning. A lesser amount was detected in the small intestine, but none was found in kidney, spleen, and heart (Table 1). In the rat liver, acetylglutamate is present exclusively in mitochondria; its distribution is coincidental with that of succinic dehydrogenase, a marker enzyme of mitochondria. For the assay of the compound, tissues were extracted with trichloracetic acid, and the extract was chromatographed on Dowex 50, Dowex 1, and on paper. It was then determined based on its activator function of carbamyl phosphate synthetase I. Correction for the yield was made based on the recovery of ^{14}C-labeled acetylglutamate added in a minute amount at the step of tissue homogenization.

The site of the synthesis of this compound was examined using isolated tissues, which were cut to slices or sectioned and incubated with [^{14}C] glutamate. [^{14}C]Acetylglutamate formed was purified, and its radioactivity was determined. As far as studied, liver and small intestine were shown to have the acetylglutamate synthesizing activity (Table 2). The intestinal contents, obtained from the same length of the tissue, showed a relatively low activity, indicating that the activity of sections of washed intestine was not due to

Table 2 Activity of Various Tissues in Vitro to Synthesize Acetylglutamate from [^{14}C] glutamate

From Shigesada and Tatibana (7). Slices or sections (300 mg, wet weight) of each tissue were incubated with [^{14}C]glutamate (3.0×10^6 cpm, 1.0 μmole) for 30 minutes at 37°. Acetyl[^{14}C]-glutamate formed was purified and radioactivity was determined as described (7).

Animal	Tissue	Radioactivity in acetyl-glutamate (cpm)
Mouse	Liver	573
	Small intestine[a]	212
	Contents of intestinal tract[b]	167
	Kidney	21
	Spleen	15
Rat	Liver	722

[a] Sections of small intestine used were 40 mm in total length. They were carefully washed with 0.9% NaCl before use.

[b] Total contents of small intestine (40 mm long) were suspended in 1 ml of 0.9% NaCl and used.

contamination by intestinal flora. Thus acetylglutamate can be synthesized only in those few specific organs, liver and intestine, where it is distributed. It was further shown that in the liver the mitochondria were the site of the synthesis, in experiments with the isolated mitochondria incubated either with $[^{14}C]$glutamate or $[^{14}C]$ acetate as tracer (7). The presence of both glutamate and acetate were necessary for the synthesis. Pyruvate could replace acetate as a substrate.

Intramitochondrial localization of the synthetic activity was determined by further disruption and fractionation of the mitochondria. The synthetase was localized in the matrix together with carbamyl phosphate synthetase I and ornithine transcarbamylase (Table 3). The data for the latter enzymes were well in accord with the results by Gamble and Lehninger (25). It is also notable that the localization of the synthetase is distinct from that of arginase, which, if present in mitochondria, degrades arginine, the important activator for the synthetase as shown later.

Table 3 Enzyme Distribution in Submitochondrial Fractions

Mitochondria were isolated from livers of Donryu strain rats by the procedure of Hogeboom (22). Subfractionation of mitochondria was performed according to Schnaitman and Greenawalt (23) and Chan et al. (24).

Enzymes	Outer membrane + intermembrane space	Inner membrane + matrix	Inner membrane fraction	Matrix fraction
	(% recovery)		(% recovery)	
Acetylglutamate synthetase[a]	7	93	16	84
Carbamyl phosphate synthetase I	20	80	11	89
Ornithine transcarbamylase	19	81		
Succinic dehydrogenase	4	96	100	0
Monoamine oxidase	100	0		
Arginase[b]	70	30		
Arginase[c]	70	30	86	14

[a] Assayed in the presence of 1 m*M* arginine.

[b] About 10% of total hepatic arginase activity was recovered in the mitochondrial fraction.

[c] Data for the mitochondria which were further washed three times with 0.125 *M* sucrose–0.07 *M* KCl and contained about 1% of total hepatic arginase activity.

ENZYME ASSAY

The enzyme was incubated at 37° for 30 minutes with 1.0 mM [^{14}C]glutamate (about 5 μCi/μmole) and 0.5 mM acetyl-CoA in 50 mM tris-Cl buffer at pH 8.2, containing 1.0 mM EDTA. The enzyme is not saturated with the substrates under these conditions. The product, [^{14}C]acetylglutamate, was purified to be free from ^{14}C-labeled metabolites of glutamate, prior to counting, by means of successive chromatographies on Dowex 50 and Dowex 1 and further on paper (7). The procedure was essential with crude enzyme. [^{14}C]Acetyl-CoA could be used as the labeled substrate for the assay of partially purified enzyme; the reaction was stopped by the addition of 2 mg of activated charcoal suspended in 50 μl of 1 N HCl containing 50% (v/v) ethanol, and the whole mixture was centrifuged. [^{14}C]Acetylglutamate in the supernatant was purified by paper chromatography. One unit of enzyme is defined as the activity which produces 1 nmole of AGA in 30 minutes under the standard conditions.

ENZYME PURIFICATION

The enzyme is unstable, particularly at low protein concentrations, so that care was taken in purification to minimize the time of exposure of the enzyme to such conditions. In Table 4 is shown a protocol of purification of the synthetase starting from 67 g of rat liver. The mitochondria were prepared and treated with sonic disruption at 20 kHz. The extract was subjected to two

Table 4 Purification of Acetylglutamate Synthetase from Rat Liver Mitochondria

The mitochondria were prepared from 67 g of rat liver. The detailed procedures are described elsewhere (K. Shigesada and M. Tatibana).

Step	Protein (mg)	Activity (u)	Specific activity (u/mg)
1. Sonic extracts of mitochondria 100,000 xg supernatant	1140	1070	0.94
2. $(NH_4)_2SO_4$ Fraction (0–40%)	310	2280	7.4
3. Fraction (0–35%)	150	1940	13.0
4. Hydroxyapatite	28	1410	50.4
5. DEAE-Cellulose	4.9	870	196
6. Sephadex G 100	2.0	670	335

successive factionations with ammonium sulfate, chromatography on hydroxyapatite and on DEAE-cellulose, and then gel filtration on Sephadex G-100. The procedure gave about 300-fold purification, although the activity in the initial crude extract appeared to be underestimated. The preparation obtained was concentrated by ultrafiltration and stored at −70°. The enzyme kept about 50% of the activity after 6 months under the storage conditions. In solution the loss in activity was 10 to 20% or more for 24 hours. The final preparation had a specific activity of 335 u/mg; that is, it catalyzed the formation of 11 nmole of acetylglutamate per minute per milligram of protein. From this value it appears that several hundred fold further purification is necessary, at least, to obtain a homogenous preparation.

SUBSTRATES AND INHIBITORS

The reaction is dependent on both glutamate and acetyl-CoA and required no addition of metals. The enzyme has a high substrate specificity for L-glutamate and acetyl-CoA, as shown in Table 5.

Table 5 Substrate Specificity of Acetylglutamate Synthetase

From Shigesada and Tatibana (20). The purified rat liver enzyme (55 μg of protein) was incubated either with [^{14}C] glutamate (1.0 mM, 1.0 × 10^6 cpm) and an acyl-CoA derivative, 0.5 mM (Experiment A) or with [^{14}C] acetyl-CoA (1.0 mM, 2.0 × 10^5 cpm) and an amino acid, 4.0 mM, structurally related to L-glutamate (Experiment B). The radioactive acylamino acid formed was isolated by treatment with a Dowex 50 column followed by paper chromatography. The area of the resulting chromatogram corresponding to the respective acylamino acid was cut out and counted. The results presented here were corrected for blank values which were obtained in the incubation without added acyl-CoA or amino acid in Experiment A or Experiment B, respectively.

Experiment	Labeled substrate	Addition		[^{14}C] Acylamino acid formed (cpm)
A	[^{14}C] Glutamate	Acetyl-CoA,	0.5 mM	3310
		Propionyl-CoA,	0.5 mM	220
		Butyryl-CoA,	0.5 mM	40
		Benzoyl-CoA,	0.5 mM	32
B	[^{14}C] Acetyl-CoA	L-Glutamate,	4 mM	2110
		D-Glutamate,	4 mM	57
		L-Glutamine,	4 mM	0
		L-Aspartate,	4 mM	0
		Glycine,	4 mM	250

As acyl donor, only acetyl CoA was active, and only a low activity was shown with 0.5 m*M* propionyl CoA. Acetylcarnithine, acetylphosphate, or N^2-acetylornithine at 0.5 m*M* did not serve as acyl donor. As the acceptor, L-glutamate was active, and a low activity was shown with 1 m*M* glycine. 2 m*M* DL-Aminoadipic acid also showed an activity similar to that of glycine. It is not clear whether the minute activities observed with propionyl-CoA, glycine, or aminoadipic acid were catalyzed by the same enzyme. It is notable that the enzyme was not active with aspartate and thus distinguishable from acetylaspartate synthetase known to occur in the brain (26).

The enzyme reaction does not require addition of metals. Rather free Mg^{2+} inhibited the reaction, about 75% at a concentration of 5 m*M*, and the inhibition was reversed by EDTA. The enzyme is highly susceptible to SH blocking agents, such as *p*-hydroxymercuribenzoate, dithiobisnitrobenzoate, and heavy metal ions, including Ag^{+}, Hg^{2+}, Cu^{2+}, Pb^{2+}, and Zn^{2+}. These agents caused 70 to 90% inhibition at 0.1 m*M*. The enzyme activity was inhibited at a high ionic strength. Potassium chloride at 100 and 200 m*M* exerted 30 and 50% inhibition, respectively.

The enzyme is subject to inhibition by the reaction product, acetylglutamate. Under the standard assay conditions 0.2 m*M* acetylglutamate effected a 50% inhibition. Kinetic analyses showed that the effect of acetylglutamate was competitive with glutamate with a K_i value of 0.12 m*M*. On the other hand, acetylglutamate did not affect binding of acetyl CoA. Structural analogs of acetylglutamate, such as *N*-acetyl-α-amino adipate, propionylglutamate, and acetylglutamine, were also highly inhibitory. Free CoA was only a very poor inhibitor.

ACTIVATION BY ARGININE

The activity of the synthetase was increased in the presence of arginine. The reaction proceeded linearly with time under the conditions. When acetyl-CoA or glutamate was omitted from the tubes containing arginine, there was no synthesis of acetylglutamate, suggesting that arginine itself is not involved in the reaction. The effect of arginine is highly specific, as shown in Table 6. D-Arginine and other amino acids, usual components of protein or other intermediates of the urea cycle such as ornithine, citrulline, and argininosuccinate, were not effective. Guanidine compounds so far tested, canavanine, agmatine, creatine, L-arginine methyl or ethyl ester, or L-homoarginine, were also without effect. Only L-argininic acid was slightly effective. The K_a value for L-arginine is about 10 μM, although it appears to vary with assay conditions.

Table 6 Effect of Arginine and Related Compounds on Acetylglutamate Synthetase Activity

Assay was performed under the standard conditions with the enzyme preparation at the step of DEAE-cellulose. The details are published elsewhere.

Addition (1 mM)	Activity (%)
None	100
L-Arginine	610
D-Arginine	89
L-Homoarginine	85
L-Canavanine	95
N-Acetylarginine	78
L-Arginine methyl ester	103
L-Arginine hydroxamate	95
Agmatine	89
L-Argininic acid	130
L-Ornithine	88
L-N-Acetylornithine	106
L-Lysine	86
L-Citrulline	83

KINETIC PROPERTIES

Varying concentrations of L-glutamate on reaction velocity were examined at different levels of acetyl-CoA. All the plots fit hyperbolic kinetics, and the same apparent K_m value, 3.0 mM, was obtained (Fig. 1). From the kinetics in the presence of 10 μM arginine with 0.5 mM acetyl CoA, it was shown that arginine increases the V_{max} without affecting the K_m for acetyl-CoA. It is

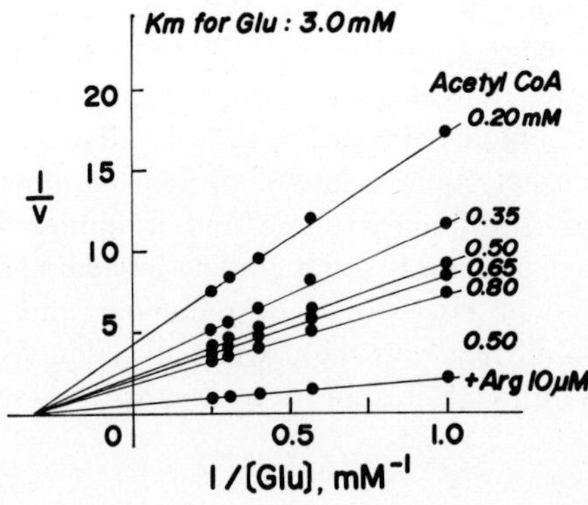

Figure 1. Effect of glutamate concentration on acetylglutamate synthetase activity. The enzyme preparation (2 μg of protein) at the step of DEAE-cellulose was used. Enzyme assay was performed by the standard procedure except for variations indicated. The details are published elsewhere (K. Shigesada and M. Tatibana).

also notable that the K_m for glutamate is relatively high. In a similar manner, the apparent K_m value for acetyl-CoA was determined with different levels of glutamate. A similar kinetics as with glutamate was obtained, and the K_m was 0.7 mM. Arginine did not affect the K_m for acetyl CoA. The observations indicate that the effects of the two substrates and arginine are independent from each other and are compatible with a rapid equilibrium random sequential mechanism (27).

The enzyme reaction has a broad pH dependency, with an optimum at pH 8. In the presence of arginine, an allosteric activator for the enzyme, the optimum slightly shifted toward a higher pH range of about 8.5, and the peak of the velocity–pH curve became much steeper than in its absence.

The enzyme activity showed nonlinear dependence, concave upward, with concentration. The nonlinearity became more pronounced at a higher pH and in the presence of arginine. Although enzyme inactivation did occur at lower protein concentrations in the assay mixture, the extent was relatively small, and the reaction proceeded linearly with time. Thus it was possible that concentration-dependent dissociation and association of oligomeric proteins took place with concomitant change in catalytic activity. Concerning this possibility, the effects of varying arginine concentrations on the enzyme activity were examined at different protein concentrations (Fig. 2). With 4 μg protein in 0.1 ml, the K_a value for arginine was found to be about 10 μM. It is notable that all the curves are slightly sigmoidal at low concentrations of arginine. The extent of stimulation by arginine varies, depending on enzyme protein concentration; the extent of stimulation is larger at higher enzyme concentration. These results suggest concentration-dependent and arginine-dependent changes in enzyme conformation. Thus a preliminary study was

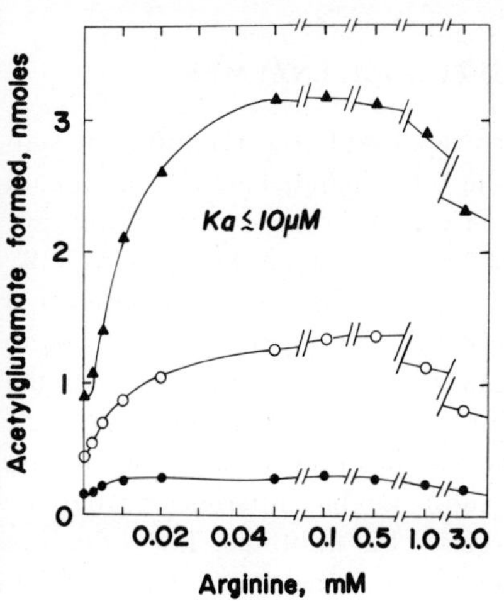

Figure 2. Effect of arginine concentration on acetylglutamate synthetase activity at different protein concentrations. Standard assay procedure was used with the enzyme at the DEAE-cellulose step. Amounts of the enzyme protein in the 0.1 ml system were 0.8 μg (●—●), 2 μg (○—○), or 4 μg (▲—▲). The details are published elsewhere (K. Shigesada and M. Tatibana).

made on the behavior of the enzyme on a column of Sephadex G-200 in the presence and absence of arginine (Fig. 3). The upper part of the figure shows the elution pattern of rat liver acetylglutamate synthetase in the absence of arginine. The enzyme apparently behaves as a single peak. When the central portion of the enzyme forming a peak was recovered and then rechromatographed on the same column in the presence of 5 m*M* arginine, as shown in the lower part of the figure, the enzyme was eluted faster than in the absence of arginine. The molecular weight was estimated to be 100,000 in the absence of arginine and 200,000 in the presence of arginine, though the peak has a small shoulder.

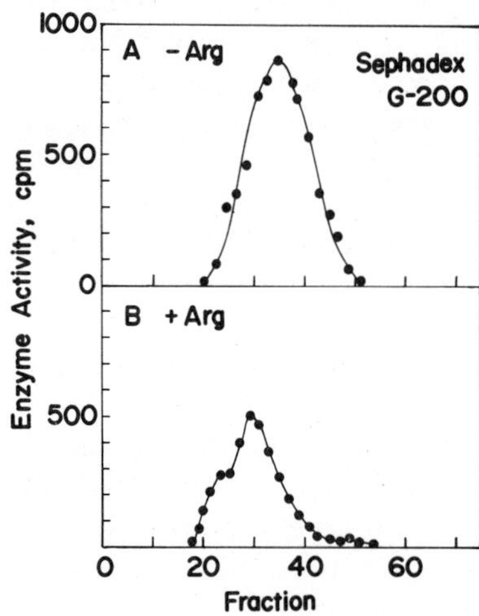

Figure 3. Arginine-dependent association of acetylglutamate synthetase. (*A*) The enzyme at the DEAE-cellulose step, 1 mg of protein, was subjected to gel filtration at a flow rate of 10 ml per hour on a Sephadex G-200 column (1.5 × 40 cm) in potassium phosphate buffer, pH 7.5. (*B*) Two enzyme fractions in the central part of the peak in *A* were resubjected to gel filtration on the same column in the presence of 5 m*M* arginine.

COMPARISON WITH BACTERIAL AND FUNGAL ENZYMES

It appears that no other studies have been reported for acetylglutamate synthetase from mammalian sources. On the other hand, bacterial or fungal synthetase has a relatively long history of research because of the important role of acetylglutamate as the precursor in arginine synthesis, as initially found by Vogel in 1953 (28). The enzyme catalyzing its synthesis was detected in *Escherichia coli* (29) and later found to be subject to repression and also to inhibition (30) by arginine. Recently Haas et al reported studies on the acetylglutamate synthetases from *Pseudomonas aeruginosa* (31) and *E. coli* (32). The enzymes have properties similar to those of the rat liver enzyme, including the strict substrate specificities. But the most notable difference is that the bacterial enzymes are not activated by arginine but inhibited. The difference in regulatory response between the bacterial and mammalian enzymes is apparently related to their distinct metabolic roles. Molecular basis for this difference offers another interesting problem regarding acetylglutamate synthetase.

REFERENCES

1. Cohen, P. P., and Hayano, M. (1948). *J. Biol. Chem.* **166**, 687–699.
2. Cohen, P. P., and Grisolia, S. (1948). *J. Biol. Chem.* **174**, 389–390.
3. Jones, M. E., Spector, L., and Lipmann, F. (1955). *J. Am. Chem. Soc.* **77**, 819–820.
4. Grisolia, S., and Cohen, P. P. (1953). *J. Biol. Chem.* **204**, 753–757.
5. Schooler, J. M., Fahien, L. A., and Cohen, P. P. (1963). *J. Biol. Chem.* **238**, 1909.
6. Hall, L. M., Metzenberg, R. L., and Cohen, P. P. (1951). *J. Biol. Chem.* **230**, 2229.
7. Shigesada, K., and Tatibana, M. (1971). *J. Biol. Chem.* **246**, 5585–5595.
8. Marshall, M., Metzenberg, R. L., and Cohen, P. P. (1958). *J. Biol. Chem.* **233**, 102–105.
9. Metzenberg, R. L., Marshall, M., Cohen, P. P., and Miller, W. G. (1959). *J. Biol. Chem.* **234**, 1534–1537.
10. Jones, M. E., and Spector, L. (1960). *J. Biol. Chem.* **235**, 2897–2901.
11. Allen, C. M., Jr., and Jones, M. E. (1966). *Arch. Biochem. Biophys.* **114**, 115–122.
12. Caravaca, J., and Grisolia, S. (1959). *Biochem. Biophys. Res. Commun.* **1**, 94–97.
13. Marshall, M., Metzenberg, R. L., and Cohen, P. P. (1961). *J. Biol. Chem.* **236**, 2229–2232.
14. Raijman, L., and Grisolia, S. (1961). *Biochem. Biophys. Res. Commun.* **4**, 262–265.
15. Fahien, L. A., and Cohen, P. P. (1964). *J. Biol. Chem.* **239**, 1925–1934.
16. Fahien, L. A., Schooler, J. M., Gehred, G. A., and Cohen, P. P. (1964). *J. Biol. Chem.* **239**, 1935–1941.
17. Grisolia, S., and Raijman, L. (1964). *Advan. Chem., Ser.* **44**, 128–149.
18. Novoa, W. B., Tiger, H. A., and Grisolia, S. (1964). *Biochim. Biophys. Acta.* **113**, 84–94.
19. Guthöhrlein, G., and Knappe, J. (1968). *Eur. J. Biochem.* **7**, 119–127.
20. Shigesada, K., and Tatibana, M. (1971). *Biochem. Biophys. Res. Commun.* **44**, 1117–1124.
21. Grisolia, S., and Marshall, R. O. (1955). In W. D. McElroy and B. Glass, Eds., *A Symposium on Amino Acid Metabolism*, Johns Hopkins, Baltimore, pp. 258–276.
22. Hogeboom, G. H. (1955). *Methods Enzymol.* **1**, 16–19.
23. Schnaitman, C., and Greenawalt, J. M. (1968). *J. Cell. Biol.* **38**, 158–175.
24. Chan, T. L., Greenawalt, J. W., and Pedersen, P. L. (1970). *J. Cell Biol.* **45**, 291–305.
25. Gamble, J. G., and Lehninger, A. L. (1973). *J. Biol. Chem.* **248**, 610–618.
26. Goldstein, F. B. (1959). *J. Biol. Chem.* **234**, 2702–2706.
27. Cleland, W. W. (1963). *Biochim. Biophys. Acta* **67**, 104–137.
28. Vogel, H. J. (1953). *Proc. Natl. Acad. Sci. USA.* **39**, 578–583.
29. Maas, W. K., Novelli, G. D., and Lipmann, F. (1953). *Proc. Natl. Acad. Sci. U.S.A.* **39**, 1004–1007.
30. Vyas, S., and Maas, W. K. (1963). *Arch. Biochem. Biophys.* **100**, 542–546.
31. Haas, D., Kurer, V., and Leisinger, T. (1972). *Eur. J. Biochem.* **31**, 290–295.
32. Leisinger, T., and Haas, D. (1975). *J. Biol. Chem.* **250**, 1690–1693.

10

Partial Reactions of Carbamyl-P Synthetase: A Review and an Inquiry into the Role of Carbamate

Mary Ellen Jones
Department of Biochemistry
University of Southern California
School of Medicine
Los Angeles, California

The frog and mammalian liver ligase located in the mitochondria that synthesizes carbamyl phosphate (carbamyl-P) has the *in vitro* stoichiometry (1)

$$2ATP^{4-} + HOCO_2^- + NH_4^+ \xrightarrow{AGA,\ Mg^{2+}} 2ADP^{3-} + NH_2CO_2PO_3^{2-} + HOPO_3^{2-} + H^+ \quad (1)$$

Acetylglutamate (AGA) is not a substrate but an activator that produces the most active conformation of the enzyme (2, 3), which is a dimer of two weight-homogeneous subunits (2, 4–6). The enzyme can also catalyze two partial reactions (7, 8)*:

$$ATP^{4-} + H_2O \xrightarrow{Mn^{2+}\ (\text{or } Mg^{2+}),AGA,HCO_3^-} ADP^{3-} + HOPO_3^{2-} + H^+ \quad (2)$$

Research reported in this publication was supported by grants from the National Institutes of Health (now numbered HD-06538) and the National Science Foundation (now numbered GB-31537).

* Data supporting partial reaction 3 catalyzed by a beef liver acetone powder were reported orally for the first time by M. E. Jones, L. Spector, and F. Lipmann in a symposium talk presented at the Third International Congress of Biochemistry in Brussels, Belgium, in August 1955. However, these authors failed to revise the paper they had submitted, prior to the meeting, to include this material. The published paper which appeared in the Proceedings of the Third

$$ADP^{3-} + NH_2CO_2PO_3^{2-} \xrightarrow{Mn^{2+}\ (\text{or } Mg^{2+}),\text{AGA}} ATP^{4-} + (NH_2CO_2^{1-}) \underset{}{\overset{\pm H_3{}^{+}O}{\rightleftharpoons}} NH_4{}^{+} + HCO_3{}^{-} \quad (3)$$

Reaction 2 is not reversible unless $NH_4{}^{+}$ is present (8), in which case ^{32}P-orthophosphate can be rather slowly incorporated into ATP, even though the majority of the ATP is being used to produce carbamyl phosphate via the parent reaction 1. As will be shown below, reaction 3 is not reversible either, but produces ammonium ion in quantities equivalent to ATP, and presumably also produces an equivalent amount of bicarbonate.

Experiments with ^{18}O-labeled bicarbonate have shown that the role of the 2 moles ATP required for the parent reaction are distinct (9). One mole of ATP is used to dehydrate bicarbonate so that ^{18}O is transferred from $HC^{18}O_3{}^{-}$ without dilution from the surrounding water to the orthophosphate produced from this mole of ATP. This lack of dilution from water suggested that the transfer of the oxygen atom from bicarbonate might be direct, perhaps occurring through the intermediate formation of an enzyme-bound carbonyl phosphate. Tests for such an intermediate with vertebrate synthetases have not been positive (10), although an abstract has appeared reporting that an enzyme-P intermediate is formed prior to the addition of bicarbonate to the enzyme (11). The second mole of ATP donates a phosphoryl group to an intermediate equivalent to carbamate to form carbamyl-P, such that ^{18}O from bicarbonate is the source of the bridge oxygen between the carbon and phosphorus atoms of the carbamyl-P produced. These experiments plus partial reactions 2 and 3 lead to the following formulation:

$$E + ATP^{4-} + HCO_3{}^{-} \rightarrow E\cdot CO_3PO_3^{3-} + ADP^{3-} + H^{+} \qquad (a)$$

$$E\cdot CO_3PO_3^{3-} + NH_4{}^{+} \rightarrow E\cdot CO_2NH_2^{1-} + HOPO_3^{2-} + H^{+} \qquad (b)$$

$$E\cdot CO_2NH_2^{1-} + ATP^{4-} \rightarrow E + NH_2CO_2PO_3^{2-} + ADP^{3-} \qquad (c)$$

Mg^{2+} and AGA^{2-} increase the rate or are essential for all three steps, but they have been omitted from the reaction schemes for simplification.

For the moment, and for the purpose of a new hypothesis to be presented, I would like to consider this formulation possible, even though kinetic results with rat liver synthetase do not as yet support such a sequence (8, 12). If one considers the partial reactions 2 and 3, partial reaction 2 would presumably represent the decomposition of carbonyl phosphate formed by the hypothetical reaction *a*. Reaction 2 presumably occurs because reaction *b* cannot fix $HCO_3{}^{-}$ into a semistable enzyme-bound carbamate in the absence of am-

---

International Congress of Biochemistry, published by Academic Press, New York, in 1956, under the editorship of C. Lieberg, therefore did not report this experiment, and it was never published elsewhere.

monia. The requirement of ammonia for the ^{32}P:ATP exchange (8) would be predicted by hypothetical reactions *a* and *b* above. Partial reaction 3 would presumably be the reverse of reaction *c* above. However, reaction 3 could equally well be the reverse of reaction *a* if carbamyl-P can act as an analog of carbonyl-phosphate.

Several model reactions in which substrate analogs are used are also catalyzed by this enzyme. Acetate (Ac) and acetyl-P (Ac ~ P) participate in the following reactions (13–15):

$$Ac \sim P^{2-} + ADP^{3-} \xrightarrow{\text{AGA,Mn}^{2+}\text{ (or Mg}^{2+})} ATP^{4-} + Ac^{-} \qquad (i)$$

$$ATP^{4-} + Ac^{-} \text{ (or } HCOO^{-}) \xrightarrow{\text{AGA, Mn}^{2+}\text{ (or Mg}^{2+})} Ac \sim P^{2-} + ADP^{3-} \qquad (ii)$$

Presumably reactions *i* and *ii* are the forward and reverse steps of the same reaction. I have drawn them separately, because reversibility has not been tested under identical conditions. In particular, very high concentrations of acetate ion are required for reaction *ii*; the K_m for acetate is estimated to be 0.4 *M*. The optimum pH for reaction *i* has not been determined; the reaction has usually been measured near pH 7.4. The pH optimum for reaction *ii* is 6.5, but the reaction occurs at pH 7.4. At pH 7.5 the relative rates of reaction *i* to reaction *ii* = 37 (i.e., 0.37 μmole ATP/minute/mg frog liver carbamyl phosphate synthetase/0.01 μmole acethydroxamate/minute/mg frog liver synthetase). Reaction *ii* was not measured directly but was observed when the acetyl phosphate formed was trapped by hydroxylamine added to the reaction mixture to form acethydroxamate, the measured product. In the absence of hydroxylamine to "pull" reaction *ii*, the formation of acetyl phosphate may even be considerably slower than the rate reported using hydroxylamine. The two model reactions can represent either or both of the theoretical reactions *a* or *c*, depending on whether acetyl-P is a carbamyl-P or a carbonyl-P analog and on whether the enzyme recognizes acetate as a bicarbonate or carbamate analog.

Other substrate analogs that have been used are NH_2NH_2 and NH_2OH (10, 16, 17). Hydrazine can substitute for ammonia, in which case the product is *N*-aminocarbamyl-P (10); the other products are those of the parent reaction 1. The role of NH_2OH is unclear. It is a competitive inhibitor (10) versus NH_4^+ in the parent reaction 1. One would conclude therefore that it binds to the ammonium ion site. It does not react with carbamyl-P to form hydroxyurea (10). The only reaction observed is the release of ADP and orthophosphate. It has been suggested therefore that NH_2OH can replace NH_3 to form an enzyme-bound carbamate analog, HONH—COO^-, which is more unstable than the theoretical enzyme-carbamate of reaction *b* or *c*. Another possibility is that such an analog is not bound as tightly to the enzyme as carbamate, and when hydroxy-carbamate is liberated to the solution, it is rapidly hydrolyzed. So far, the model reactions using NH_2NH_2 or NH_2OH

have not seemed to offer an opportunity to probe further into the parent reaction.

A number of years ago, Suzanne McKinley and I carried out a series of experiments (some of which are reported in Appendix I of this paper). These experiments were designed to inquire into the characteristics of partial reaction 2. Two general questions were asked. First, we asked if carbamate could serve as a direct substrate by combining hypothetical reaction *c* with another reaction *d*, namely, the direct binding of carbamate to the enzyme:

$$E + NH_2CO_2^- \xrightarrow{AGA} E{\cdot}NH_2CO_2^- \qquad (d)$$

$$E{\cdot}NH_2CO_2^- + ATP^{4-} \xrightarrow{AGA,Mg^{2+}} E + ADP^{3-} + NH_2CO_2PO_3^{2-} \qquad (c)$$

$$\text{Sum: } NH_2CO_2^- + ATP^{4-} \xrightarrow{E,AGA,Mg^{2+}} ADP^{3-} + NH_2CO_2PO_3^{2-} \qquad (e)$$

There is a significant difference between reaction *e* and the parent reaction 1, namely, the stoichiometry between ATP and carbamyl-P. For reaction *e* the ratio of ATP used to carbamyl-P formed is 1, but for reaction 1 it is 2. We ran innumerable experiments using ammonium carbamate as substrate, which we shall not document here but would gladly make available to anyone who is interested. In all experiments a parallel vessel was run with ammonium carbonate as substrate. Variables included concentrations of ammonium carbamate or carbonate (between 12 and 50 m*M*), very short incubation intervals (1 minute), large amounts of enzyme, and inclusion of carbonic anhydrase inhibitors. In addition, analysis of ammonium carbamate was done to ensure that on addition more than 90% of the solid material added was carbamate. In all trials the amounts of carbamyl-P formed were of the same order of magnitude, whether ammonium carbonate or carbamate was used as the N- and C-source for carbamyl-P. The rate of carbamyl-P formation when carbamate was substrate was somewhat less than with carbonate; significantly, it was never greater than the rate with carbonate. Equilibrium mixtures of carbamate and carbonate produced rates of carbamyl-P formation intermediate between that produced by carbonate and that produced by carbamate. These experiments were run at pH 8.5 or 9.0.

The significant fact was not the above results but the fact that the stoichiometry was always 2 moles of ATP utilized per mole of carbamyl-P formed, whether the substrate added was carbamate or carbonate. Since a solution of ammonium carbamate always has ammonium carbonate the ATP/carbamyl-P stoichiometry clearly suggests that only ammonium carbonate (or bicarbonate) served as substrate and that reaction *d*, the binding of carbamate to the enzyme, either does not occur or is so slow that reaction *e* cannot compete with the parent reaction 1.

To complete our studies, we investigated reaction 3 in more detail. We did not try to ascertain if the second product was carbamate. This could not be done with the carbamate methods available then but could probably now be done with the improved carbamate method devised by Michael Caplow (18). We did, however, ask if the amount of ammonia produced, either directly or after decomposition of carbamate, was equivalent to the ATP formed. We also asked two additional questions by carrying out the incubations in the presence of $^{15}NH_4$ ion. First, we asked if the reaction was readily reversible. To answer this question fairly, we had to have a control with ATP, HCO_3^-, and NH_4^+ equal to the amounts of these compounds formed from carbamyl-P and ADP. Second, we asked if an exchange between $^{15}NH_4^+$ and carbamyl-P could occur in the absence of ADP. In other words, could the following types of reversible reaction (*f* or *g* plus *h* = *i*) occur?

$$\text{Enz-XH} + NH_2CO_2PO_3^{2-} + H^+ \rightleftharpoons \text{Enz-X-}CO_2PO_3^{2-} + NH_4^+ \qquad (f)$$

$$(\text{Enz-X-}CO_2PO_3^{2-} + ADP^{3-} + H_2O \rightarrow \text{Enz-XH} + HOCO_2^{1-} + ATP^{4-})$$

$$\text{Sum: } (NH_2CO_2PO_3^{2-} + ADP^{3-} + H_3O^+ \rightarrow NH_4^+ + HCO_3^{1-} + ATP^{4-})$$

or

$$\text{Enz-XH} + NH_2CO_2PO_3^{2-} \rightleftharpoons \text{Enz-X-}PO_3^{2-} + NH_2CO_2^{1-} + H^+ \qquad (g)$$

$$NH_2CO_2^{1-} + H_3O^+ \rightleftharpoons NH_4^+ + HOCO_2^{1-} \qquad (h)$$

$$\text{Sum: Enz-XH} + NH_2CO_2PO_3^{2-} + H_3O^+ \rightleftharpoons \text{Enz-X-}PO_3^{2-} + NH_4^+ + HOCO_2^{1-} + H^+ \qquad (i)$$

Enz represents carbamyl-P synthetase; —XH is a group at the active center that can form a phospho- or carbonyl phosphate derivative.

The answer to the second question was a resounding no. Absolutely no $^{15}NH_4^+$ was incorporated into carbamyl-P in the absence of ADP. Since the $^{15}NH_4$ ion added had an atom percent excess of 90, since equal amounts of NH_4^+ and carbamyl-P were present at zero time, and since we could readily detect as little as 0.05 atom percent excess (over the normal value of 0.4 atom of ^{15}N for N_2), that is, 1 part in 900 in the product, we can rather firmly state that carbamyl phosphate does not directly react with the synthetase in the absence of ADP, other than to bind to the synthetase at a substrate site.

The results for the first question are conditional. There is some incorporation of ^{15}N from $^{15}NH_4^+$ into carbamyl-P. At a maximum the rate of the incorporation is 2×10^{-4} μmole/minute compared with a rate of formation of ATP and NH_4^+ from carbamyl-P and ADP of 1×10^{-1} μmole/minute. If there is a reversal of reaction 3, it can at a maximum occur at a rate one two-hundredth that of ATP formation. If, however, our controls in which we incubated the amounts of ATP, $^{14}NH_4^+$, and HCO_3^- with the usual amount of $^{15}NH_4^+$ are subtracted from the observed value, then the ratio of the rate of ATP formation from carbamyl-P and ADP to the rate of carbamyl-P formation from $^{15}NH_4^+$ by a reversal of reaction 3 = 600/1. (See Appendix I for a summary of these experiments.) I am inclined to believe that the formation of ATP from carbamyl-P and ADP is not reversible for one of two reasons. If the product is enzyme-bound carbamate, it dissociates from the enzyme as long as the ammonium and bicarbonate ion concentrations are low, and when it is dissociated from the enzyme, it is hydrolyzed more rapidly than it can react with ATP to reform carbamyl-P. If the product is $NH_4^+HCO_3^-$ rather than carbamate, that is, the reaction is

$$NH_2CO_2PO_3^{2-} + ADP^{3-} + H_3O^+ \xrightarrow{Mg^{2+},AGA} ATP^{4-} + NH_4^+ + HOCO_2^-$$

then the only return reaction is, of course, the parent reaction 1.

It seems to the author that the partial reactions catalyzed by carbamyl-P synthetase may not necessarily be the "apparent" reaction and that one cannot assign the partial reactions so readily as has been previously done.

Where do I think we stand? The solid facts seem to be the stoichiometry of the parent reaction and the fact that the two ATP's that are required for the parent reaction have distinct functions. Not so clear is whether there are two ATP binding sites. This is a missing point of information which should be easy to answer with the many ATP analogs that can usually be bound in place of ATP but which cannot be cleaved to ADP and orthophosphate. Knowledge of the number of ATP sites and of the substrates required for ATP binding is an extremely important piece of information to validate all the kinetic schemes (8, 12) which would have to be modified if the enzyme had only one ATP binding site per catalytic site. The AGA-requiring carbamyl-P synthetases appear to have two weight-homogeneous subunits which must be combined to have an active enzyme (2, 4–6). If the subunits are chemically identical and if there are two distinct ATP binding sites per subunit, kinetic studies must be carried out under conditions where one is sure that both sites are indeed functioning; that is, one must check to see if both orthophosphate and carbamyl phosphate formation have been affected in parallel. Since most kinetic experiments have measured ADP produced and since this is a product, whether both or only one ATP site(s) are participating in ADP production, I think the conclusions one draws (in contrast to the results one obtains) can be fallacious. Finally, there has been a tendency to overlook the effect of reagents on the conformational state of the enzyme. Results of Annamarie Herzfeld and myself, which we hope to publish shortly, indicate that the substrates such as ATP and HCO_3^- may affect this parameter; Guthöhrlein and Knappe (2) have already noted the ATP effect.

I do not wish to cover the conclusions of kinetic analysis, which will be more ably handled by Dr. Elliott. However, I would like to conclude by directing your attention to new work to be published by Carol K. Sauers, William P. Jencks, and Susan Groh (19). They have carried out structure reactivity studies on the equilibria for formation of alkyl monocarbonates and on the rate of decomposition of these carbonates in dilute alkali which allow them to predict the equilibrium for the hypothetical carbonyl phosphate and H_2O versus bicarbonate and orthophosphate:

$$K_{eq} = \frac{[CO_3PO_3^{3-}][H_2O]}{[HCO_3^-][HOPO_3^{2-}]} = 0.13$$

They can also calculate the nonenzymatic rate of decomposition, $k_d \geq 10$ second^{-1}. This rate is compatible with partial reaction 2, the bicarbonate-dependent ATPase which has $k \geq 0.3$ second^{-1}, being due to the enzymatic formation of carbonyl phosphate which decomposes chemically to bicarbonate and phosphate dianion in the absence of NH_4^+. They suggest, however, that the existence of carbonyl phosphate on the enzyme may be short and that the important enzyme intermediate for this synthetase, as well as many biotin enzymes, may be an enzyme-bound CO_2 which has a low entropy, because it is enzyme-bound and has a high Gibbs free energy as a consequence of its localization in the correct position to react with the amine acceptor (ammonia for carbamyl-P synthetase and biotin for the biotin enzyme). The steps they propose are

$$\underbrace{\text{ATP}\cdot{}^-\text{O—}\overset{\overset{\text{O}}{\|}}{\text{C}}\text{—OH}\cdot\text{HNR}_2}_{\text{I}} \rightleftharpoons \underbrace{\text{ADP}\cdot{}^=\text{O}_3\text{PO—}\overset{\overset{\text{O}}{\|}}{\text{C}}\text{—O}^-\cdot\text{HNR}_2}_{\text{II (carbonyl-P)}} \rightleftharpoons$$

$$\underbrace{\text{ADP}\cdot{}^=\text{O}_3\text{PO}^-\cdot\overset{\overset{\text{O}}{\|}}{\underset{\underset{\text{O}}{\|}}{\text{C}}}\cdot\text{NHR}_2}_{\text{III ("CO}_2\text{")}} \;\rightleftharpoons\; \underbrace{\text{ADP}\cdot{}^=\text{O}_3\text{PO}^-\cdot{}^-\text{O—}\overset{\overset{\text{O}}{\|}}{\text{C}}\text{NR}_2}_{\text{IV (carbamate)}}$$

$$\text{III} \rightleftharpoons \underbrace{\text{ADP}\cdot{}^=\text{O}_3\text{PO}^- \quad \text{HNR}_2}_{\text{V}} + \text{O=C=O}$$

Intermediate III is the Sauers-Jencks-Groh intermediate. In the carbamyl-P synthetase reaction, intermediate IV would be the intermediate that allows the P^{32}-ATP exchange (8) to occur and which normally reacts with the second mole of ATP required in parent reaction 1 to form carbamyl-P. Intermediate V, without NH_3 or, on occasion, with NH_3, is the intermediate that accounts for the bicarbonate-dependent ATPase (and perhaps the hydroxylamine-dependent ATPase). This formulation suggested by Sauers, Jencks, and Groh allows one to consider new experiments; they have suggested that under experimental conditions where the rate of the bicarbonate-carbonate exchange is slow, a high concentration of $^{14}CO_2$ might serve as a direct substrate for ^{14}C-carbamyl-P synthesis with the exclusion of unlabeled $HOCO_2$. I can envision that such a reaction might use only 1 mole of ATP. This is a very intriguing suggestion, particularly when one considers that this synthetase is located in the mitochondrial matrix where CO_2 levels should be high (unless the matrix has an extremely active carbonic anhydrase or unless the other matrix conditions promote rapid chemical hydration of the CO_2 being produced by the Krebs citrate cycle).

SUMMARY

New data presented show that one of the partial reactions catalyzed by frog liver mitochondrial carbamyl phosphate synthetase, namely, carbamyl-P + ADP → ATP + NH_4^+ (+HCO_3^-?) is not a reversible reaction under the usual conditions used. In addition, carbamate (added in place of ammonium bicarbonate as substrate) is not apparently used as a substrate. If it were, carbamyl phosphate synthesis should occur with the use of a single mole of ATP per mole of carbamyl phosphate formed. The synthesis of carbamyl phosphate that does occur when 90% pure carbamate is provided as the N and C source requires 2 moles of ATP per mole of carbamyl-P produced. A comparison is made between the parent reaction, the partial reactions, and the model reactions catalyzed by this enzyme, and certain conclusions and suggestions for future work are presented. Among these is a suggestion that the requirement for 2 moles of ATP and the use of bicarbonate as a substrate may be a phenomenon of the *in vitro* assay but *in vivo*, within the mitochondrial matrix, CO_2 and a single mole of ATP may suffice.

Appendix I Experiments Concerned with N Exchange between $^{15}NH_4^+$ and $^{14}NH_2CO_2PO_3^{2-}$ Using Frog Liver Carbamyl-Phosphate Synthetase I

Suzanne McKinley
and Mary Ellen Jones

Preliminary experiments gave a pH optimum of 6.5 for the formation of ATP from carbamyl-P and ADP as catalyzed by the frog liver mitochondrial carbamyl-P synthetase in the presence of acetylglutamate (AGA) with Mg^{2+} or Mn^{2+} ion. This pH was used for all but one exchange experiment which was carried out at pH 7.5. In addition, it could be shown that an equivalent of ammonia was formed for every mole of ATP (see Table 2). Whether this mole of ammonia is formed directly by the enzyme (reaction *A* below) or by subsequent chemical hydrolysis of 1 mole of carbamate or an equivalent enzyme intermediate (reaction *B* below), we do not know:

$$NH_2CO_2PO_3^{2-} + ADP^{3-} + H_3O^+ \xrightarrow{AGA, Mg^{2+}} ATP^{4-} + NH_4^+ + HCO_3^- \qquad (A)$$

$$NH_2CO_2PO_3^{2-} + ADP^{3-} \underset{}{\overset{AGA, Mg^{2+}}{\rightleftharpoons}} ATP^{4-} + HN_2CO_2^- \quad \overset{\pm H_3O^+}{\rightleftharpoons} \quad NH_4^+ + HOCO_2^- \qquad (B)$$

We wished to see if this reaction was reversible by using $^{15}NH_4^+$ and $^{14}NH_2CO_2PO_3^{2-}$ as substrates in the presence of enzyme, acetylglutamate, and Mg^{2+}. We measured the loss of ^{15}N from $^{15}NH_4^+$, due to the formation of $^{14}NH_4^+$ from carbamyl-P, as well as a possible exchange due to incorporation of ^{15}N into carbamyl-P by reversal of reaction 1 or 2. We subsequently measured, using the same aliquot, the ^{15}N in the remaining carbamyl-P, as described in the following section. Experiments were carried out in the presence of ADP and in the absence of ADP. An exchange reaction in the absence of added ADP would suggest that a covalent enzyme phosphate (reaction *g* of the main body of this paper) or a covalent enzyme carbonyl phosphate (reaction *f* of the main body of this paper) is formed when carbamyl-P synthetase and carbamyl-P are incubated with AGA and Mg^{2+}. When ADP was absent, (1) no ATP was formed, (2) the NH_4^+ present did not increase due to an enzyme reaction, (3) no ^{15}N-nitrogen from added $^{15}NH_4^+$ was incorporated into carbamyl-P, and (4) the atom % excess of ^{15}N in NH_4^+ was exactly the value for the $^{15}NH_4^+$ added. No reaction had occurred in the 10 minute incubation at 37° except chemical hydrolysis of about 10% of the added carbamyl-P to cyanate and phosphate (20) and probably the binding of a small fraction of the carbamyl-P to the enzyme.

When both ADP and carbamyl-P were added, ATP, NH_4^+, and probably HCO_3^- are formed (Table 2). Since these represent all the substrates needed to form carbamyl-P in the parent reaction, that is, $2ATP + NH_4^+ + HCO_3^- \rightarrow 2ADP + NH_2CO_2PO_3^{2-} + HOPO_3^{2-}$, we used a control vessel which contained no carbamyl-P but had exactly the amounts of ATP, $^{14}NH_4^+$, and HCO_3^- formed in the experimental vessel (A, Table 2) plus the amount of ADP remaining in the experimental vessel at the end of incubation. In addition, the regular amounts of $^{15}NH_4^+$, AGA, $MgCl_2$, and enzyme of the experimental vessel were also present in the control vessel. This vessel therefore measures the maximum amount of ^{15}N incorporated into carbamyl-P by the parent reaction above. The difference between the control and experimental vessels represents the minimum exchange that has occurred in the half reaction. The data and methods are detailed in the following section. The main body of this paper discusses the significance of the results.

Search for An ^{15}N-Exchange Between $^{15}NH_4^+$ and $^{14}NH_2CO_2PO_3^{2-}$ Catalyzed by Carbamyl-Phosphate Synthetase I

Purified frog liver mitochondrial carbamyl-P synthetase I (10) with a specific activity of 1.33 μmole of carbamyl-P formed/minute/mg protein, when assayed by the procedure of Marshall et al. (1), was used for all experiments.

A prereaction mixture (18.4 ml total volume) was prepared so that it contained 55 m*M* imidazole buffer, pH 6.5; 5.5 m*M* $MgCl_2$; 3.3 m*M* L-acetylglutamate; and 3.3 m*M* $^{15}NH_4Cl$ (the $^{15}NH_4Cl$ was a gift from Dr. Julius Marmur). For experimental vessels, column A of Table 2, 5.5 m*M* ADP was present; the control vessels of column B contained $^{14}NH_4Cl$, $KHCO_3$, ATP, and ADP in the amounts shown.

Table 1 Preparation of the Control Vessels

Time vessel was incubated (minutes)	Substrates added to control vessels			
	$^{14}NH_4Cl$ (m*M*)	$KHCO_3$ (m*M*)	ATP (m*M*)	ADP (m*M*)
7.5	1.0	1.0	1.0	4.5
15	1.67	1.67	1.67	3.9
20	2.2	2.2	2.2	3.3

These mixtures, minus enzyme and carbamyl-P, were incubated in flasks at 37° for 7 minutes (i.e., incubation started at 0 time minus 8 minutes). After the 7 minutes, 0.6 ml of solution was added which contained 60 μmole of carbamyl-P for the A vessels; an equivalent volume of H_2O was added to the B vessels. After mixing and at time = −0.5 minute, a 2 ml aliquot was removed in order to determine 0 time levels of ATP, carbamyl-P, and NH_4^+. One milliliter of enzyme (18.5 mg of protein) was added at 0 time, and the incubation continued for the desired interval. At the end of the incubation the flask was rapidly immersed in an ice water bath and swirled vigorously until the temperature was 0° (cooling occurred in less than 1 minute). Aliquots were again removed for an estimation of the amounts of ATP, carbamyl-P, and NH_4^+.

Of the remaining reaction mixture, 15 ml was used for distillation of the $^{15}NH_4^+$ and the N of carbamyl-P. For the B vessels, carbamyl-P had to be added just prior to distillation in amounts to make them equivalent to their respective A vessels. Specially designed, custom-made, large lucite Conway vessels were used (the design can be provided to anyone wishing it by Dr. Jones). The center well of the vessel contained 1 ml of 0.36 N H_2SO_4 and 5 ml of freshly boiled and chilled water. The outer well was empty until the 15 ml sample was added to the bottom side of a slightly tilted vessel. Then 0.4 ml of 10 N KOH was added on the upper side of the vessel, a cover which had a film of K_2CO_3 was placed on top, and the contents of the outer well (sample and KOH) were mixed. The Conway vessels were placed in a covered desiccator with some water in it to maintain a moist environment, and the desiccator was placed in a 37° incubator. The distillation of the $^{15}NH_3$ continued for 5 hours at 37°. The contents of the center well were pipetted into the Y tube to be used for the conversion of NH_4 to N_2 for analysis in the mass spectrometer (21). The center well was rinsed with two 0.5 ml water washes which were added to and mixed with the original solution. The volume was reduced by lyophilization to less than 1 ml. To ensure that all the $^{15}NH_4Cl$ was completely removed from the sample in the outer well before the CNO^- (formed from carbamyl-P when the sample was exposed to the KOH of the outer well) could be converted to NH_4^+ and then distilled, two NH_4^+ washes were carried out as the initial distillation. The amount of $^{14}NH_4Cl$ added to the outer chamber was 440 μmole for the first wash and 44 μmole for the second. Fresh H_2SO_4 was placed in the center well for each wash.

For conversion of CNO^- (from carbamyl-P) to NH_4^+, the outer well contents were quantitatively pipetted into a beaker using two 2 ml washes. Using a magnetic stirrer, the pH was lowered to 1, with 12 N perchloric acid: $KClO_4$ precipitates. The mixture was allowed to stand at room temperature for 30 minutes. It was filtered by suction with small washes; the filtrate was measured. A measured aliquot of the filtrate was placed in the lower portion of the outer well of a tilted Conway vessel which contained the usual amount of H_2SO_4 in the center well. The distillation of NH_4^+ was initiated by the addition of 0.7 ml of 10 N KOH to the raised portion of the outer well. Distillation was as described above. The ammonia (from CHO^-) that was trapped in the center well was transferred to the Y tube to be used for mass spectrometry. All samples for mass spectrometry were reduced to less than 1 ml volume by lyophilization. The NH_4^+ samples were converted to N_2 gas (21) which was analyzed on a Consolidated 21–103C mass spectrometer which the Department of Chemistry at Harvard University kindly made avilable to us. The N^{28}- and N^{29}-mass peaks were measured, as was the A^{40} -mass peak, which allowed us to check for air contamination.

NH_4^+ concentrations were determined by the distillation method of Seligson and Seligson (22); ATP was determined by the firefly or luciferin assay (23); carbamyl-P was determined by differential phosphate analysis (24), using base hydrolysis to convert carbamyl-P to orthophosphate and cyanate.

Experiments shown were run at pH 6.5. Similar experiments at pH 7.5 for 10 minutes showed more ^{15}N incorporation in the control vessel, B, but the difference between the control and experimental vessels was smaller, indicating more exchange for the parent reaction and less exchange for the half reaction.

Table 2 Results of the Experiments

Analysis	Vessel: Experimental (A)	Vessel: Control (B)	(A − B)
ΔCAP (μmole/ml)			
After 7.5 min	−0.85	+0.03	
15.0 min	−1.45	+0.02	
20.0 min	−1.81	+0.02	
ΔATP (μmole/ml)			
After 7.5 min	+0.82	+0.04	
15.0 min	+1.32	+0.06	
20.0 min	+1.98	+0.01	
ΔNH_4^+ (μmole/ml)			
After 7.5 min	+0.90	—	
15.0 min	+1.48	+0.04	
20.0 min	+1.75	+0.04	
Atom % excess ^{15}N in carbamyl-P			
After 7.5 min	0.034	0.014	0.02
15.0 min	0.15	0.10	0.05
20.0 min	0.19	0.094	0.096
Atom % excess ^{15}N in NH_4^+			
After 7.5 min	71.9	72.2	
15.0 min	62.7	62.1	
20.0 min	57.0	57.4	

Appendix II History of the Discovery of Carbamyl-P: A Boston Collaboration

In the spring of 1954, I was one of a lively group of about 10 research associates working with Fritz Lipmann in the Biochemical Research Laboratories of the Massachusetts General Hospital. For the preceding 3 years I had worked on the "acetate-activating" enzyme of yeast (the first ligase shown to cleave ATP into AMP and inorganic pyrophosphate) with Simon Black, Ruth Flynn Souza, and Dr. Lipmann (25). Dr. Lipmann encouraged me to take on a new degree of independence, and I had read with interest the experimental data on citrulline synthesis with rat liver. This was an extremely challenging reaction. Grisolia and Cohen (26) had shown that two enzyme reactions were required for citrulline synthesis from ATP, ammonium bicarbonate, and ornithine: an ATP-using step which produced an unstable carbamyl intermediate, called compound X, and a transfer step which placed a carbamyl group from compound X on ornithine to yield citrulline. In addition, these authors had shown that the enzyme required an *N*-acylglutamate as cofactor (26). The ATP enzyme interested me, because it appeared to have too many substrates (ATP, NH_4^+, HCO_3^-, and acetylglutamate), and the product, compound X, presumably had 1 mole each of NH_2-, CO_2-, and PO_3-residues plus acetylglutamate. I could not devise a suitable carbamyl donor from so many reactants. In addition, although the reaction presumably used only 1 mole of ATP (26) to form compound X, or citrulline, most experiments indicated that more phosphate than one equivalent of phosphate was formed per mole of citrulline formed (27). Having been involved with the discovery of the first pyrophosphate ligase, it seemed possible that pyrophosphate rather than orthophosphate might be a product of ATP. I therefore started the studies that would lead to the discovery and synthesis of carbamyl phosphate using the rat liver acetone powder described by Grisolia and Cohen (27).

After having become familiar with this enzyme preparation and the rather "tricky" assay for citrulline (28), I began to study the stoichiometry between citrulline and the inorganic phosphate formation. Fortunately, at this point, Dr. Lipmann attended the excellent symposium on amino acid metabolism at Johns Hopkins University held in June 1954. Here papers by Oginsky, Korzenovsky, and Slade (29–31) on the arginine dihydrolase system suggested to Lipmann that the animal enzyme was unduly complex. He recommended the bacterial enzyme system which did not require acetylglutamate, for he thought it might be a more amenable system in which to isolate an intermediate. The arginine dihydrolase system is now known to consist of three enzymes. Crude extracts were the source of these three enzymes. The first enzyme cleaves arginine to citrulline and ammonia; the second and third enzymes catalyze the phosphorolysis of citrulline in the presence of ADP. The phosphorolysis of citrulline occurred at pH 6 and had the stoichiometry

$$H^+ + HOPO_3^{2-} + ADP^{3-} + \text{citrulline} \rightarrow \text{ornithine} + NH_4^+ + CO_2 + ATP^{4-}$$

Dr. Lipmann encouraged me to be sure to consider ornithine-δ-phosphate as a potential intermediate and provided me with a reference on phosphoamidate chemistry. Hoping that the bacterial intermediate would function in the rat liver system, I was inclined toward carbamyl-P as an intermediate, since it should be more related to compound X.

As detailed elsewhere (32), the first essential step in our search to find a labile phosphate intermediate was to find conditions to reverse the phosphorolysis, that is, to form citrulline from ATP, HCO_3^-, NH_4^+, and ornithine. If carbamyl-P was the intermediate, then conditions for carbamate formation from ammonium bicarbonate would be ideal for the reversal of the phosphorolysis. Studies of the literature on carbamates were mainly concerned with hemoglobin; however, such references led to the work of Faurholt in the early 1920s on ammonium carbamate (33) and the fact that alkaline pH's favored carbamate formation from ammonium carbonate solutions. Also, Faurholt had shown that the carbamate ion was more stable than carbamic acid (which decomposes extremely rapidly). By working at pH 8.5 and using phosphoenol-pyruvate, ADP, and pyruvate kinase to provide ATP, we observed a good yield of an unstable phosphate intermediate when only ammonium bicarbonate was added (34). This intermediate yielded citrulline if ornithine was also added. The intermediate should be carbamyl-P, since ATP, NH_4^+, and HCO_3^- were all essential for its formation. Studies on the stability of the labile phosphate showed that it was more acid-labile than ATP, that it was more heat-labile than ATP at neutral pH, and that it was decomposed by dilute alkali (pH 11) in a few minutes at room temperature. These characteristics distinguished it from all known phosphate compounds important in metabolism that were then known (see 35).

Isolation of the enzyme product seemed like a long task, and we therefore decided to seek help for its synthesis. I first approached my brother, Elmer E. Jones, an organic chemist, who was working with Sigmund Thannhauser. He was hesitant to try the synthesis of carbamyl-P and wished to try to synthesize phenylcarbamyl-P from phenyl isocyanate and orthophosphate using some sulfuric acid as a catalyst. Our attempts, made one Saturday morning, were to no avail, probably due to the use of the sulfuric acid as catalyst. However, since carbamyl-P was more stable than acetyl-P, Dr. Lipmann and I were convinced that a successful synthesis could be found.

Leonard Spector, an excellent organic chemist working in the Huntington Memorial Laboratory at the Massachusetts General Hospital, was approached. The personal and physical relationship of the Biochemical Research and the Huntington Laboratory was close. The Huntington Laboratory occupied the floor below the Biochemical Research Laboratory. We shared many seminars, lent each other chemicals, and enjoyed shared social events. Dr. Spector was informed of the stability characteristics of the intermediate. Fortunately, during the few days he was considering alternatives, he attended a performance by the Boston Symphony which included "Prelude to the Afternoon of a Faun," by Debussy. Dr. Spector does not enjoy this particular music, so he spent this interval considering carbamyl-P synthesis. He decided that an incubation of equimolar amounts of a concentrated solution of potassium cyanate and potassium acid phosphate might be ideal. Indeed, the very first incubation of such a mixture at 37° yielded, in only 30 minutes, carbamyl phosphate as 50% of the phosphate added. This compound was identical with the enzyme intermediate, as indicated by its hydrolysis characteristics and its ability to form citrulline with ornithine and the bacterial extract. Only 2 months had passed since the first production of the bacterial extract and the synthesis of carbamyl-P. These 2 months were a very exciting period. All experiments had been carefully planned, but we had the good fortune to have selected a proper solution and to have the right talent.

As it was near Christmas, Dr. and Mrs. Lipmann had a party for the Biochemical Research Laboratory staff, both husbands and wives, for Dr. Spector and other friends from the Huntington Laboratory, as well as for a few additional Boston scientists. I was presented with a corsage which had attached to the pin by ribbons a small model *car*, then a musical staff or *bar* (made by our glassblower, Elliott Lane), and a small bag of meal (or *myl*). It was truly an occasion and a year to remember.

REFERENCES

1. Marshall, M., Metzenberg, R. L., and Cohen, P. P. (1958). *J. Biol. Chem.* **233**, 102–105.
2. Guthöhrlein, G., and Knappe, J. (1968). *Eur. J. Biochem.* **7**, 119–127.
3. Chabas, A., Grisolia, S., and Silverstein, R. (1972). *Eur. J. Biochem.* **29**, 333–342.
4. Virden, R. (1972). *Biochem. J.* **127**, 503–508.
5. Strahler, J. R. (1975). *Federation Proceed.* **34**, 604.
6. Raijman, L., and Jones, M. E. (1976). *Arch. Biochem. Biophys.* In press.
7. Metzenberg, R. L., Marshall, M., Cohen, P. P., and Miller, W. G. (1959). *J. Biol. Chem.* **234**, 1534–1537.
8. Guthöhrlein, G., and Knappe, J. (1969). *Eur. J. Biochem.* **8**, 207–214.
9. Jones, M. E., and Spector, L. (1960). *J. Biol. Chem.* **235**, 2897–2901.
10. McKinley, S., Anderson, C., and Jones, M. E. (1967). *J. Biol. Chem.* **242**, 3381–3390.
11. Rubio, V., Feijoo, B., and Grisolia, S. (1975). *Federation Proc.* **34**, 680.
12. Elliott, K. R. F., and Tipton, K. F. (1974). *Biochem. J.* **141**, 789–805; 806–816; 817–824.
13. Novoa, W. B., and Grisolia, S. (1962). *J. Biol. Chem.* **237**, PC 2710–2711.
14. Grisolia, S., and Raijman, L. (1964). *Adv. Chem. Ser.* **44**, 128–149.
15. Raijman, L., and Grisolia, S. (1964). *J. Biol. Chem.* **239**, 1272–1276.
16. Caravaca, J., and Grisolia, S. (1960). *J. Biol. Chem.* **235**, 684–693.
17. Marshall, M., Metzenberg, R. L., and Cohen, P. P. (1961). *J. Biol. Chem.* **236**, 2229–2237.
18. Caplow, M. (1968). *J. Amer. Chem. Soc.* **90**, 6795–6803.
19. Sauers, C. K., Jencks, W. P., and Susan Groh (1975). *J. Amer. Chem. Soc.*, **97**, 5546–5553.
20. Allen, C. M., and Jones, M. E. (1964). *Biochemistry* **3**, 1238–1247.
21. San Pietro, A. (1957). *Methods Enzymol.* **4**, 473–488.
22. Seligson, D., and Seligson, H. (1951). *J. Lab. Clin. Med.* **38**, 324–330.

23. Strehler, B. L., and McElroy, W. D. (1957). *Methods Enzymol.* **3**, 871–873.

24. Spector, L., Jones, M. E., and Lipmann, F. (1957). *Methods Enzymol.* **3**, 643–655.

25. Lipmann, F., Jones, M. E., Black, S., and Flynn, R. M. (1952). *J. Am. Chem. Soc.* **74**, 2384.

26. Grisolia, S., and Cohen, P. P. (1952). *J. Biol. Chem.* **198**, 561–571.

27. Grisolia, S., and Cohen, P. P. (1951). *J. Biol. Chem.* **191**, 189–202.

28. Archibald, R. M. (1944). *J. Biol. Chem.* **156**, 121–142.

29. Oginsky, E. L. (1955). In W. D. McElroy and B. Glass, Eds., *Amino Acid Metabolism*, Johns Hopkins, Baltimore, pp. 300–308.

30. Korzenovsky, M. (1955). In W. D. McElroy and B. Glass, Eds., *Amino Acid Metabolism*, Johns Hopkins, Baltimore, pp. 309–320.

31. Slade, H. D. (1955). In W. D. McElroy and B. Glass, Eds., *Amino Acid Metabolism*, Johns Hopkins, Baltimore, pp. 321–334.

32. Raijman, L., and Jones, M. E. (1973). In P. D. Boyer, Ed., *The Enzymes*, 3rd ed., **9**, Part B, Academic, New York, pp. 97–119.

33. Faurholt, C. (1921). *Z. anorg. allgem. Chem.* **120**, 85–102.

34. Jones, M. E., Spector, L., and Lipmann, F. (1955). *J. Amer. Chem. Soc.* **77**, 819–820.

35. Cardini, C. E., and Leloir, L. F. (1957). *Methods Enzymol.* **3**, 835–850.

11

Kinetic Studies on Mammalian Liver Carbamyl Phosphate Synthetase

Keith R. F. Elliott
Biological Laboratory
University of Kent
Canterbury, Kent, England

Since the discovery that carbamoyl phosphate is an intermediate in the biosynthesis of citrulline from ornithine, carbon dioxide, and ammonia (1–3) a considerable amount of research has been directed toward attempts to elucidate the mechanism of action of the enzyme carbamoyl phosphate synthetase. The enzyme prepared from frog liver mitochondria (4) or mammalian liver mitochondria (3) has been shown to catalyze the synthesis of carbamoyl phosphate with the following stoichiometry:

$$2ATP + HCO_3^- + NH_4^+ = 2ADP + H_2N\,COO\,PO_3^{2-} + H_3PO_4$$

For many years it has been known that a derivative of glutamate is involved in the synthesis of citrulline by mitochondria (5), and it was finally demonstrated that *N*-acetyl-L-glutamate is required for carbamoyl phosphate synthetase activity (6). Mg^{2+} and K^+ have also been shown to be necessary for enzymic activity (7).

Little was published on mammalian liver carbamoyl phosphate synthetase between 1957 and 1968, mainly due to difficulties involved in purification of the enzyme because of its extreme instability (3). During this period carbamoyl phosphate synthetase was purified to near homogeneity from frog liver (4), and attention became focused on this enzyme, which was found to be more stable than that from mammalian liver. The first published results of kinetic investigations of purified carbamoyl phosphate synthetase were obtained by

Fahien and Cohen (8), using this preparation. Unlike all previous investigations in which the carbamoyl phosphate synthetase activity was assayed by measuring the amount of citrulline formed in the presence of ornithine and ornithine transcarbamylase (4), these used a continuous spectrophotometric assay that coupled ADP formation to NADH oxidation with pyruvate kinase and lactate dehydrogenase (8). This investigation was incomplete in that HCO_3^- and Mg^{2+} were kept at fixed concentrations throughout. The essential features of the results were (*a*) that the K_m for *N*-acetyl-L-glutamate (1 m*M*) had the same numerical value as the dissociation constant of the enzyme–*N*-acetyl-L-glutamate complex (9) and was not affected by variation in the concentrations of NH_4^+ or ATP Mg and (*b*) that reciprocal plots curved upward when ATP Mg was the varied substrate. This suggested that the reaction velocity might depend on the square of the ATP Mg concentration, since two molecules of ATP Mg are used in the reaction, and it was possible to demonstrate two kinetically active ATP Mg-binding sites (8).

The breakthrough in the study of mammalian liver carbamoyl phosphate synthetase came when Guthöhrlein and Knappe (10) produced a stable and essentially homogeneous preparation from rat liver by making use of previous observations that β-mercaptoethanol (11), KCN (11), and glycerol (11, 12) all exerted a stabilizing effect on the enzyme. A limited kinetic study was performed using this preparation (13) in an attempt to clarify the mechanism of HCO_3^- activation (14). Having shown that the enzyme contained no biotin, a study of the kinetic effects of varying NH_4^+ and HCO_3^- concentrations was reported, which showed that the K_m values for each substrate were independent of the concentration of the other substrate.

Most authors seem to have considered a study involving variation of all substrates and activators to be too complex and so have kept the concentration of at least one substrate and one activator constant throughout their experiments. It was my belief that this approach in fact complicated the situation, as the systematic variation of all substrate and activator concentrations would allow greater distinction between possible mechanisms. The aim of this presentation is to summarize recent results of investigations of carbamoyl phosphate synthetase from bovine liver (15–17) and to compare them with earlier results obtained with the enzyme from rat and frog liver.

PREPARATION OF BOVINE LIVER CARBAMOYL PHOSPHATE SYNTHETASE

Carbamoyl phosphate synthetase was purified from bovine liver (Table 1) by a method similar to that used for rat liver (10) but with modifications made in order to enable more convenient handling of the larger volumes involved (15). This preparation was essentially homogeneous, containing less than 1% kinetically contaminating enzymes (16).

Table 1 Purification of Carbamoyl Phosphate Synthetase from Bovine Liver[a]

	Activity (u)	Sp. act (u/mg)	Yield (%)	Purification
Homogenate	600	0.006		
CTAB extract of mitochondria	500	0.075	84	125
50–55% $(NH_4)_2SO_4$ ppt desalted	300	0.15	50	250
DEAE-cellulose column	150	1.14	25	1900
G200 column	111	1.5	18.5	2500

[a] 1 u = amount of enzyme producing 1 μmole ADP/minute at 30°C with 5 m*M* ATP Mg, 10 m*M* NH_4^+, 100 m*M* HCO_3^-, 1 m*M* Mg^{2+}, and 10 m*M* *N*-acetyl-L-glutamate, pH 7.6.

KINETIC STUDIES

Initial Rate Studies

Throughout the investigations carbamoyl phosphate synthetase activity was assayed using the coupled spectrophotometric assay (18), although the colorimetric assay for citrulline (4) was used when exogenous ADP was added to the system.

Tables 2 and 3 summarize the results of reciprocal plots obtained at high concentrations of nonvaried substrates (16) well above the previously determined K_m value for the rat liver enzyme (18). Most combinations of variable and fixed variable substrates gave patterns of reciprocal plots that intersected

Table 2 Summary of Reciprocal Plots[a]

Fixed variable substrate	Variable substrate				
	AG	Mg^{2+}	ATP Mg	HCO_3^-	NH_4^+
AG	—	Int.	Int.	Int.	Par.
Mg^{2+}	Int.	—	Int. on $1/v$ axis	Int.	Int.
ATP Mg	Int.	Int.	—	Int.	Par.
HCO_3^-	Int.	Int.	Int.	—	Int.
NH_4^+	Par.	Int.	Par.	Int.	—

[a] Int. = patterns of reciprocal plots intersect on or near the 1/[S] axis.
Par. = patterns of reciprocal plots appear parallel. Concentrations of nonvaried substrates kept at 5 m*M* ATP Mg, 10 m*M* NH_4^+, 100 m*M* HCO_3^-, 1 m*M* Mg^{2+}, and 10 m*M* *N*-acetyl-L-glutamate.

Table 3 Apparent K_m Values[a]

Fixed Variable Substrate	K_m^{AG}	K_m^{Mg2+}	$K_m^{ATP\ Mg}$	$K_m^{HCO_3}$	$K_m^{NH_4^+}$
AG	—	0.18 mM	0.80 mM	4.5 mM	0.25 mM
Mg^{2+}	80 μM	—	0.71 mM	3.4 mM	0.25 mM
ATP Mg	90 μM	0	—	10 mM	0.76 mM
HCO_3^-	50 μM	0.19 mM	0.50 mM	—	0.40 mM
NH_4^+	125 μM	0.19 mM	0.52 mM	4.0 mM	—

[a] Concentrations of nonvaried substrates kept at 5 mM ATP Mg, 10 mM NH_4^+, 100 mM HCO_3^-, 1 mM Mg^{2+}, and 10 mM N-acetyl-L-glutamate.

on or near the 1/[S] axis, but there were important exceptions. When ATP Mg was the variable substrate and free Mg^{2+} the fixed variable substrate, the lines intersected on the $1/v$ axis. This pattern is indicative of a mechanism in which ATP Mg binds to the enzyme after free Mg^{2+}, which binds to the enzyme in equilibrium (19). The appearance of parallel lines when the NH_4^+ concentration was varied at a number of fixed ATP Mg or N-acetyl-L-glutamate concentrations is at first sight anomalous. As true parallel lines on reciprocal plots are indicative of an essentially irreversible step between binding of the two substrates to the enzyme, it seemed anomalous that no parallel line patterns were observed when HCO_3^- was the varied substrate.

There seemed to be two possible explanations for the parallel lines. The first possibility was that the irreversible step was in fact due to a "saturating" concentration of HCO_3^- (19). When the concentration of HCO_3^- was decreased to near its K_m value, there was no effect on the type of pattern observed when NH_4^+ and ATP Mg were the variable and fixed variable substrates. However, decreasing the HCO_3^- concentration resulted in an intersecting pattern of reciprocal plots when NH_4^+ and N-acetyl-L-glutamate were the variable and fixed variable substrates. This indicates that HCO_3^- binds to the enzyme between N-acetyl-L-glutamate and NH_4^+.

The second possible explanation was that the lines were not truly parallel but were in fact weakly converging. By performing experiments with low concentrations of ATP Mg and NH_4^+ as the variable and fixed variable substrates it was possible to obtain an apparently intersecting pattern of reciprocal plots, indicating no essentially irreversible step between the binding of these two substrates to the enzyme. Reciprocal plots against both ATP Mg and NH_4^+ were nonlinear at low concentrations. Both these were not unexpected, as an upward curvature of reciprocal plots against ATP Mg had been noted with frog liver carbamoyl phosphate synthetase (8), and the enzyme is also able to act as an HCO_3^- dependent ATPase in the absence of NH_4^+ (8), thus tending to cause a downward curvature of reciprocal plots against NH_4^+.

The curvature of reciprocal plots against ATP Mg instead of being a complication was in fact a useful tool in the elucidation of the kinetic mechanism. Nonlinear plots will only occur if there is no irreversible step between binding of the two ATP Mg molecules to the enzyme (20). The secondary plot of the intercepts on the $1/v$ axis, when ATP Mg was the fixed variable and HCO_3^- the variable substrate, against 1/[ATP Mg] was linear. This is equivalent to a reciprocal plot against ATP Mg at an infinite, and thus "saturating" (19), HCO_3^- concentration. As only this secondary plot, and not those at infinite concentrations of other substrates, was linear, it would appear that only HCO_3^- binds compulsorily to the enzyme between the two ATP Mg molecules.

Product Inhibition Studies

The results of product inhibition studies are summarized in Table 4. The patterns of inhibition by inorganic phosphate and carbamoyl phosphate indicate that carbamoyl phosphate is the first product and inorganic phosphate the last product to leave the enzyme and that ATP Mg is the first substrate to bind to the enzyme (20). The nonlinearity of inhibition by ADP Mg was not unexpected, as two ADP Mg molecules are released by the enzyme.

Table 4 Product Inhibition Patterns

Product	Substrate		
	ATP Mg	HCO_3^-	NH_4^+
ADP Mg	Mixed Nonlinear K_i = 1.0 mM	Mixed Nonlinear K_i = 0.6 mM	Mixed Nonlinear K_i = 1.3 mM
Carbamoyl phosphate	Mixed K_i = 10 mM	Mixed K_i = 13 mM	Mixed K_i = 19 mM
Inorganic phosphate	Competitive K_i = 25 mM	Mixed K_i = 60 mM	Mixed K_i = 75 mM

The product inhibition patterns were obtained at 10 mM N-acetyl-L-glutamate, 1 mM Mg^{2+}, and approximately K_m values of the nonvariable substrates, i.e., 0.5 mM ATP Mg, 1 mM NH_4^+, and 10 mM HCO_3^-.

As with substrate binding, nonlinear inhibition indicates that there is no irreversible step between loss of the two product molecules, which is consistent with both ADP Mg molecules being released by the enzyme between carbamoyl phosphate and inorganic phosphate. The fact that inhibition by ADP Mg was mixed respect to all substrates was at first sight anomalous, as it would be expected to be uncompetitive (20). However, both AMP and adenosine were shown to be competitive inhibitors with respect to ATP Mg and mixed inhibitors with respect to NH_4^+ and HCO_3^-. It would seem likely that ADP Mg would also show the same competitive effect with respect to ATP Mg, which would mean that the overall pattern of inhibition would be mixed.

Summary of Kinetic Results

The conclusions drawn from the kinetic study of bovine liver carbamoyl phosphate synthetase may be summarized as follows:

a. There is no essentially irreversible step (including product release) before all substrates are bound to the enzyme.

b. Only HCO_3^- binds compulsorily to the enzyme between the two ATP Mg molecules.

c. HCO_3^- binds to the enzyme between *N*-acetyl-L-glutamate and NH_4^+.

d. Mg^{2+} must bind to the enzyme in equilibrium before ATP Mg.

e. The order of product release is first carbamoyl phosphate, followed by the two ADP Mg molecules, and last the inorganic phosphate.

Unfortunately there are no published results of substrate binding studies for mammalian liver carbamoyl phosphate synthetase, although there are some for the frog liver enzyme (7). Frog liver carbamoyl phosphate synthetase is capable of binding independently one molecule of ATP Mg or *N*-acetyl-L-glutamate in the absence of all other substrates (7). If the bovine liver enzyme is assumed to show the same effects, then the following mechanism will account for all the results presented above, including those for the rat (13) and frog (8) liver enzymes:

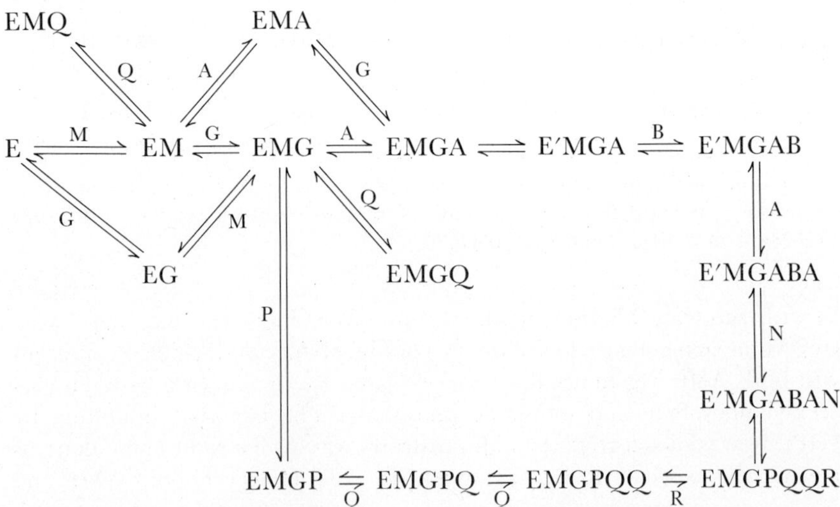

where A = ATP Mg, B = HCO_3^-, G = *N*-acetyl-L-glutamate, M = Mg^{2+}, N = NH_4^+, P = inorganic phosphate, Q = ADP Mg, and R = carbamoyl phosphate.

The initial binding of *N*-acetyl-L-glutamate, Mg^{2+}, and one molecule of ATP Mg is assumed to be in equilibrium, so that the method of Cha (21) in combination with that of Vol'kenshtein and Gol'dshtein (22) was used to derive the following complex rate equation for the mechanism:

$$v = \frac{V}{\begin{aligned} 1 &+ \frac{K_m^{A_1} + K_m^{A_2}}{A} + \frac{K_m^B}{B} + \frac{K_m^G}{G} + \frac{K_m^N}{N} + \frac{K_m^{A_1}K_{m'}^B + K_s^B(K_m^{A_2} + K_{m'}^{A_2})}{AB} + \frac{K_s^G K_m^{A_1}}{AG} + \frac{K_s^M K_m^{A_1}}{AM} + \frac{K_s^{A_2}K_m^N}{AN} \\ &+ \frac{K_s^G K_{m'}^B}{GB} + \frac{K_s^G K_s^M K_m^{A_1}}{AMG} + \frac{K_s^{A_1}K_s^G K_{m'}^B + K_s^G K_s^B K_{m'}^{A_2}}{AGB} + \frac{K_s^{A_1}K_s^M K_{m'}^B}{AMB} + \frac{K_s^{A_2}K_s^B(K_m^N + K_{m'}^N)}{ABN} \\ &+ \frac{K_s^{A_1}K_s^M K_s^G K_{m'}^B}{AMGB} + \frac{K_s^{A_2}K_s^G K_s^B K_{m'}^N}{AGBN} + \frac{K_s^{A_1}K_s^B K_{m'}^{A_2}}{A^2B} + \frac{K_s^{A_1}K_s^G K_s^B K_{m'}^{A_2}}{A^2GB} + \frac{K_s^{A_1}K_s^M K_s^B K_{m'}^{A_2}}{A^2MB} \\ &+ \frac{K_s^{A_1}K_s^{A_2}K_s^B K_{m'}^N}{A^2BN} + \frac{K_s^{A_1}K_s^M K_s^G K_s^B K_{m'}^{A_2}}{A^2MGB} + \frac{K_s^{A_1}K_s^{A_2}K_s^M K_s^B K_{m'}^N}{A^2MBN} \\ &+ \frac{K_s^{A_1}K_s^{A_2}K_s^G K_s^B K_{m'}^N}{A^2GBN} + \frac{K_s^{A_1}K_s^{A_2}K_s^M K_s^G K_s^B K_{m'}^N}{A^2MGBN} \end{aligned}}$$

Mechanisms have been proposed for the kinetic mechanisms of carbamoyl phosphate synthetase from frog (8) and rat liver (13) but on the basis of less extensive studies. The studies on rat liver carbamoyl phosphate synthetase (13) were superficial and only showed that there was no essentially irreversible step between binding of HCO_3^- and NH_4^+ to the enzyme. This only made possible a distinction between different groups of mechanisms rather than allowing the formulation of an overall mechanism.

The studies on frog liver carbamoyl phosphate synthetase (8) were far more complete, although HCO_3^- and Mg^{2+} concentrations were kept constant throughout the experiments. In general the results with the bovine liver enzyme (16, 17) are very similar to those with the frog liver enzyme (8). Both enzymes have the ability to use hydrazine and hydroxylamine as substrates or inhibitors, depending on the assay technique employed (23, 24), and are inhibited by monovalent cations (17, 25). There are, however, differences in detail, for example, numerical value of K_{ms} or degree of curvature of reciprocal plots.

Fahien and Cohen (8) postulated two kinetic mechanisms for frog liver carbamoyl phosphate synthetase, both involving random-order equilibrium binding of *N*-acetyl-L-glutamate and one molecule of ATP Mg. In the first mechanism this is followed by steady-state binding of NH_4^+ with release of inorganic phosphate before the binding of the second ATP Mg. In the light of more recent theoretical considerations (19–21) it would seem probable that this mechanism (without complex side reactions) would not account fully for the results obtained, as it will produce linear reciprocal plots against

ATP Mg. The second proposed mechanism involves random-order steady-state binding of NH_4^+ and the second ATP Mg before any products are released and will therefore account for the nonlinear reciprocal plots.

Obviously it is very dangerous to try to superimpose results from one animal on those from another, especially when the animals are as diverse as the frog, rat, and bull. However, from the results obtained with carbamoyl phosphate synthetase from these animals, it would seem quite possible that all three enzymes obey essentially the same kinetic mechanism, possibly intermediate between those postulated. Species variation in detail may then account for the minor differences observed.

STUDIES AT PHYSIOLOGICAL CONCENTRATIONS

All the kinetic results discussed above are of obvious academic interest but were in general obtained at very nonphysiological substrate and activator concentrations. In an attempt to obtain results which may have more relevance to the situation within the liver, a number of experiments were performed at substrate concentrations approximating those *in vivo*. The concentrations used were 100 μM *N*-acetyl-L-glutamate, 0.7 mM NH_4^+, 2.5 mM ATP, 10 mM HCO_3^-, 0.4 mM Mg^{2+}, 150 mM K^+, 1 mM ADP, and 10 mM inorganic phosphate (23). (With the exception of *N*-acetyl-L-glutamate these are whole-cell concentrations and as such must be viewed with a great deal of caution, as carbamoyl phosphate synthetase is a mitochondrial enzyme.) Under these conditions the apparent K_m values were 0.5 mM ATP Mg, 0.25 mM NH_4^+, 5 mM HCO_3^-, 0.6 mM Mg^{2+}, and 100 μM *N*-acetyl-L-glutamate. All these concentrations are similar to those *in vivo*, suggesting that carbamoyl phosphate synthetase has a great capacity to react to changes in substrate concentrations.

A final point of interest emerging from these studies is the calculation of the *in vivo* activity of carbamoyl phosphate synthetase. By examination of the K_m values in Table 3, the initial rate equation, and the approximate *in vivo* substrate concentrations, it would appear that the activity of the enzyme *in vivo* may only be about 5% of that measured under ideal conditions *in vitro*. Thus, although argininosuccinate synthetase may have the lowest activity *in vitro* (26), complex kinetic considerations may make that of carbamoyl phosphate synthetase considerably lower *in vivo*.

SUMMARY

Results of kinetic investigations of bovine liver carbamoyl phosphate synthetase are reported which indicated (*a*) the order of substrate binding to the enzyme is ATP Mg, HCO_3^-, ATP Mg, and last NH_4^+; (*b*) the order of product release from the enzyme is carbamoyl phosphate, ADP Mg, ADP Mg, and last inorganic phosphate. These results and the proposed kinetic mechanism are discussed in relation to those obtained with the enzyme from other sources. The possible *in vivo* significance of the kinetic results is discussed.

ACKNOWLEDGMENT

I wish to thank Dr. K. F. Tipton (Department of Biochemistry, University of Cambridge, England) for his support and encouragement throughout and Dr. W. Ferdinand (Department of Biochemistry, University of Sheffield, England) for his helpful comments on the formulation of the kinetic equation.

REFERENCES

1. Jones, M. E., Spector, L., and Lipmann, F. (1955). *J. Amer. Chem. Soc.* **77**, 819–820.
2. Marshall, R. O., Hall, L. M., and Cohen, P. P. (1955). *Biochim. Biophys. Acta* **17**, 279–281.
3. Metzenberg, R. L., Hall, L. M., Marshall, M., and Cohen, P. P. (1957). *J. Biol. Chem.* **229**, 1019–1025.
4. Marshall, M., Metzenberg, R. L., and Cohen, P. P. (1958). *J. Biol. Chem.* **233**, 102–105.
5. Cohen, P. P., and Grisolia, S. (1948). *J. Biol. Chem.* **174**, 389–390.
6. Hall, L. M., Metzenberg, R. L., and Cohen, P. P. (1958). *J. Biol. Chem.* **230**, 1013–1021.
7. Marshall, M., Metzenberg, R. L., and Cohen, P. P. (1961). *J. Biol. Chem.* **236**, 2229–2237.
8. Fahien, L. A., and Cohen, P. P. (1964). *J. Biol. Chem.* **239**, 1925–1934.
9. Fahien, L. A., Schooler, J. M., Gehred, G. A., and Cohen, P. P. (1964). *J. Biol. Chem.* **239**, 1935–1941.
10. Guthöhrlein, G., and Knappe, J. (1968). *Eur. J. Biochem.* **7**, 119–127.
11. Novoa, W. B., and Grisolia, S. (1964). *Biochim. Biophys. Acta* **85**, 274–282.
12. Skinner, D. M., and Jones, M. E. (1961). *Fed. Proc.* **20**, 240.
13. Guthöhrlein, G., and Knappe, J. (1969). *Eur. J. Biochem.* **8**, 207–214.
14. Jones, M. E. (1965). *Ann. Rev. Biochem.* **34**, 381–418.
15. Elliott, K. R. F., and Tipton, K. F. (1973). *FEBS Letters* **37**, 79–81.
16. Elliott, K. R. F., and Tipton, K. F. (1974). *Biochem. J.* **141**, 807–816.
17. Elliott, K. R. F., and Tipton, K. F. (1974). *Biochem. J.* **141**, 817–824.
18. Kerson, L. A., and Appel, S. H. (1968). *J. Biol. Chem.* **243**, 4279–4285.
19. Cleland, W. W. (1970). In Boyer, P. D., Ed *The Enzymes*, **2**, Academic, New York. pp. 1–65.
20. Elliott, K. R. F., and Tipton, K. F. (1974). *Biochem. J.* **141**, 789–805.
21. Cha, S. (1968). *J. Biol. Chem.* **243**, 820–825.
22. Vol'kenshtein, M. V., and Gol'dshtein, B. N. (1966). *Biokhimiya (Eng.)*, **31**, 473–478 (Eng).
23. Elliott, K. R. F. (1973). Ph.D. Thesis, University of Cambridge.
24. McKinley, S., Anderson, C. D., and Jones, M. E. (1967). *J. Biol. Chem.* **242**, 3381–3390.
25. Kennedy, J., and Grisolia, S. (1965). *Biochim. Biophys. Acta* **96**, 102–113.
26. Schimke, R. T. (1962). *J. Biol. Chem.* **237**, 459–468.

12

Carbamyl Phosphate Synthetase I from Frog Liver

Margaret Marshall
Department of Physiological Chemistry
University of Wisconsin
Madison, Wisconsin

The early work of Grisolia and Cohen (1) on citrulline biosynthesis was done with extracts of rat liver mitochondria from which they separated ornithine transcarbamylase free of carbamyl phosphate synthetase but not the reverse. The presence of adenylate kinase and ATPase in their preparations obscured the stoichiometry of the reaction. Metzenberg et al. (2) purified the enzyme from dog and bovine liver sufficiently to establish that 2 moles of ATP is required per mole of citrulline synthesized, but further purification of carbamyl phosphate synthetase from mammalian livers was unsuccessful with the techniques then available because of its extreme lability. On the other hand, the enzyme from frog liver proved to be stable enough to be chromatographed on cellulose ion exchangers. The disconcerting feature of the attempt to purify it was that no further increase in specific activity beyond threefold over that of the mitochondrial extract seemed possible (3). Moving-boundary electrophoresis and ultracentrifugation, the primary criteria of homogeneity in use at the time, suggested that the preparation was 90% pure (4), which implied that approximately 10% of the mitochondrial protein is carbamyl phosphate synthetase. More sensitive criteria, in particular gel electrophoresis in sodium dodecyl sulfate (5), have since confirmed the estimate of the purity; and Melnick et al. (6), applying the same technique directly to the matrix fraction of rat liver mitochondria, where carbamyl phosphate synthetase is known to

The studies reported in this paper from the author's laboratory were supported in part by Grant CA-03571 from the National Institutes of Health.

be localized (7), report that 30% of the protein has a molecular weight of 130,000, approximately that of the subunit of rat carbamyl phosphate synthetase and much larger than that of most proteins. The concentration of carbamyl phosphate synthetase in the matrix space must be, therefore, ~ 5% in both frog and rat liver mitochondria.

With the observation that glycerol stabilizes rat carbamyl phosphate synthetase (8), Guthöhrlein and Knappe (9) were able to isolate it and to establish that its instability is due to dissociation induced by low temperature. This also occurs with the frog enzyme to a more limited extent (5, 10, 11). The essential similarity of the two enzymes is documented in Table 1.

Table 1 Comparison of Carbamyl Phosphate Synthetase I from Frog and Rat

Property	Frog	Rat	Ref.
(1) Subunits	2	2	(5, 12)
(2) $\bar{M}_w$ subunit	156,000	160,000	(5, 12)
(3) Turnover number at 37°	7×10^2 minute^{-1}	5.5×10^2 minute^{-1}	(5, 9)
(4) Cofactor requirements	Acetylglutamate, K^+, Mg^{2+} in excess of ATP	Same	(4, 13–15)
(5) Partial reactions catalyzed:			
(*a*) Bicarbonate-dependent ATPase	+	+	(16, 14)
(*b*) CP + ADP → ATP	+	+	(16, 17)
(6) Inhibited by monovalent antibody to frog enzyme	+	+	(18)

PHYSICAL PROPERTIES

The remaining impurities in the preparation of frog carbamyl phosphate synthetase were removed by chromatography on CM-Sephadex (5). This did not change the specific activity but did remove the lower-molecular-weight impurities observed on electrophoresis in gels containing sodium dodecyl sulfate. The molecular weight estimated by this technique, 145,000, is essentially the same as the value determined by sedimentation equilibrium in 6 *M* Gnd-HCl (Table 2). Performic acid oxidation or reduction in urea followed by alkylation did not cause further dissociation. In the absence of dissociating agents and at 20°, $\bar{M}_w$, determined by sedimentation equilibrium, is 314,000, in agreement with an earlier value, determined by the Archibald "approach to equilibrium" method at 4° (4). At 20° there is no evidence of dissociation; however, more recent measurements at 4°, done at <0.1 the initial protein concentration at which the earlier value was obtained, give $\bar{M}_w$ 260,000 in the absence of acetylglutamate and 188,000 in its presence (Table 2). These values

Table 2 Molecular Weight of Frog Carbamyl Phosphate Synthetase I (5, 19)

Medium[a]	$\bar{M}_w$[b]
(1) 6 *M* Gnd-HCl	156,000
(2) 0.1 *M* KCl, pH 7.5; 20°	314,000
(3) 0.1 *M* KCl, pH 7.5; 10 m*M* acetylglutamate; 20°	300,000
(4) 0.1 *M* KCl, pH 7.5; 4°	260,000
(5) 0.1 *M* KCl, pH 7.5; 10 m*M* acetylglutamate; 4°	187,000

[a] 1 m*M* dithiothreitol present in all solutions.
[b] Determined by sedimentation equilibrium.

were determined in the presence of dithiothreitol, in the absence of which dissociation was accompanied by aggregation (5, 19)*.

An equilibrium between monomer and dimer is confirmed by electron micrographs of the enzyme. The enzyme, incubated at 0° in 0.1 *M* sucrose, 10 m*M* Tris-HCl (pH 7.4), 1 m*M* dithiothreitol, and 5 m*M* Mg-acetate, then stained with ammonium molybdate, shows a mixture of two sizes of particles, (75 Å × 85 Å) and (75 Å × 200 Å) (Fig. 1a). Under these conditions $\bar{M}_w$ is 248,000. In the presence of 10 m*M* acetylglutamate all the particles are spherical (diam. 75 Å) (Fig. 1b), and $\bar{M}_w$ is 162,000 (5, 19).

The mammalian enzyme is partially dissociated in the absence of acetylglutamate even at 20°, in keeping with its greater instability. Failure to take this into account probably accounts for the discrepancies in the literature with respect to the size of the subunits of rat (9, 12) and bovine (21) carbamyl phosphate synthetase. Virden (12) obtained a value for the molecular weight of the rat enzyme at 5° in 10 m*M* acetylglutamate of 170,000, which he estimated set a lower limit for the molecular weight of the monomer of 150,000, when corrected for the small amount of dimer present under these conditions. A value of 158,000 was obtained for the bovine enzyme in 6 *M* Gnd-HCl (22).

CHEMICAL PROPERTIES

Serine was the only NH_2-terminal residue identified by reaction with dansyl chloride. Quantitation by the cyanate method was difficult because of the extensive destruction of serine and because the large size of the polypeptide chain resulted in a high value for serine in the noncarbamylated control. No COOH-terminal residue was detected on digestion with carboxypeptidase A or B, with or without sodium dodecyl sulfate or urea, nor was any identified

* The loss of activity induced by acetylglutamate in the cold, reported by Raijman and Grisolia (20), results from the aggregation and not from the dissociation per se, which is a reversible process.

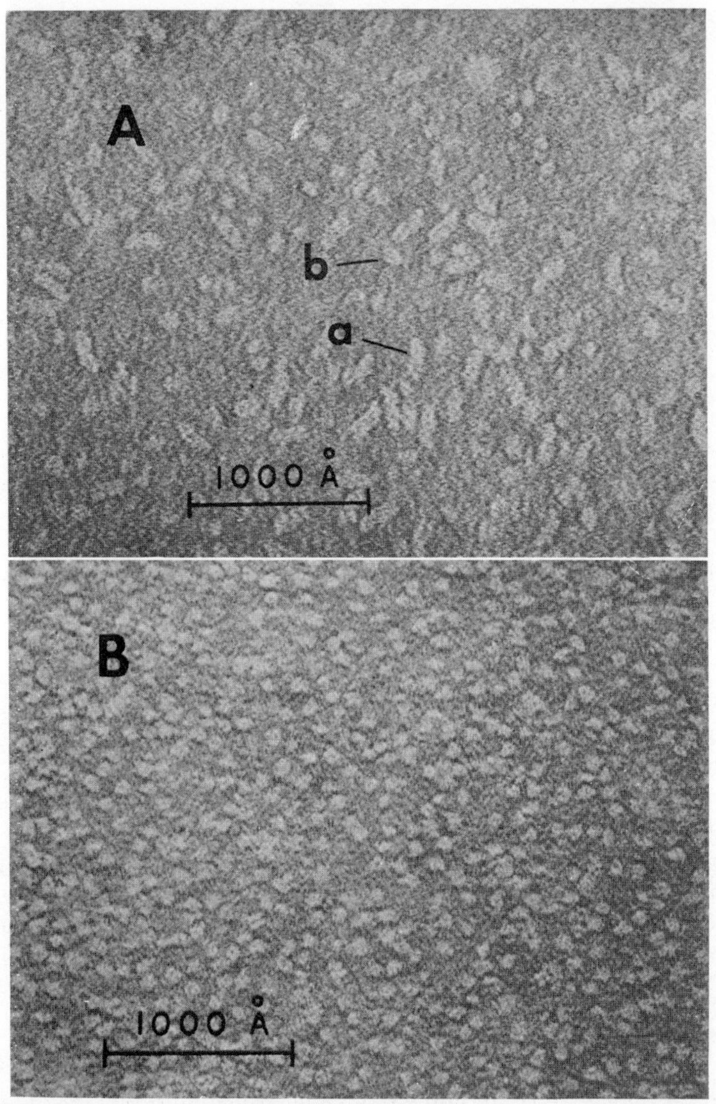

Figure 1. Electron micrographs of frog carbamyl phosphate synthetase. *A*: (*a*) 75 Å × 200 Å particle: (*b*) 75 Å × 85 Å particle. *B*: incubated with acetylglutamate (10 m*M*) at 0°.

by hydrazinolysis (5). The amino acid composition is recorded in Table 3 (5). Frog, bovine (23), and rat (9) carbamyl phosphate synthetase I and, contrary to an earlier report (24), carbamyl phosphate synthetase II from *E. coli* (23) do not contain biotin.

Table 3 Amino Acid Composition of Frog Carbamyl Phosphate Synthetase I (5)

Amino acid	Residues per 156,000 g
Lysine	96
Histidine	27
Arginine	58
Aspartic acid	129
Threonine	78
Serine	96
Glutamic acid	142
Proline	70
Glycine	111
Alanine	113
Half-cysteine	20
Valine	114
Methionine	38
Isoleucine	94
Leucine	117
Tyrosine	39
Phenylalanine	56
Tryptophan	18

ROLE OF ACETYLGLUTAMATE

Since the original observation of Grisolia and Cohen (25) that carbamyl-L-glutamate was required for the synthesis of citrulline by extracts of rat liver mitochondria, many of its analogs have been tested. The only variation permitted is in the substituent at the α position. Formyl-L-glutamate and acetyl-L-glutamate (13) are active and also acetoxyglutarate (26). Acetylglutamate has the lowest K_m and appears to be the natural cofactor (27).

Much effort has been expended in the search for a covalent intermediate of acetylglutamate. Grisolia et al. (28), with the use of deuterium-labeled carbamylglutamate, established that there is no exchange of the stable hydrogens during the reaction. Since acetoxyglutarate is active (26) and tritium from TOH is not incorporated into the methyl group of the acetyl moiety (29), substitution on the α substituent is unlikely. Reichard (30) found that ^{18}O from carboxy-labeled acetylglutamate is not transferred to P_i or carbamyl phosphate; Metzenberg et al. (31) established that ^{18}O is not incorporated into acetylglutamate from water or the terminal phosphate of ATP; and Jones and Spector (32) demonstrated that there is no loss of ^{18}O from the acetyl group. These results do not, however, exclude anhydrides with carbonic or phosphoric acid as long as acetylglutamate retains its original oxygen atoms. Grassl and Bach (33) tested a mixture of acetylglutamylphosphates, probably contaminated with unreacted acetylglutamate; it was less active than the same amount of acetylglutamate. Also Metzenberg et al. (3) reported lack of exchange between ATP and ADP in the presence of acetylglutamate, which might be expected to occur if acetylglutamylphosphate were an intermediate but, as Caravaca and Grisolia point out, might not be detectable if the equilibrium were extremely unfavorable. The current view, based on these negative results, is that the only role of acetylglutamate is as an allosteric effector.

When frog carbamyl phosphate synthetase is incubated with acetylglutamate alone, it is converted to an active form that decays with a half-life of approximately 2.5 minutes on dilution into the assay medium minus acetylglutamate (4, 5). Grisolia and Towne (34) first reported this phenomenon with rat carbamyl phosphate synthetase, but only if the preincubation at 38° was done in the presence of both acetylglutamate and ATP plus Mg^{2+}, a result that suggested that an intermediate might be formed. Subsequently Raijman and Grisolia (20) compared the enzyme from the two sources and found that ATP appeared to be essential only for the rat enzyme. More recently Guthöhrlein and Knappe (9) have measured the rate of activation of rat carbamyl phosphate synthetase by acetylglutamate in an optical assay at 10° where the half-life for activation is 1.25 minutes. They found that if the rat enzyme is preincubated at $\geqslant$ 25° with acetylglutamate alone, there is no lag phase in the assay. The conflicting results may be due to the presence in the latter experiment of mercaptoethanol, in the absence of which ATP plus Mg^{2+} may serve to prevent the inactivation induced by acetylglutamate (17). At 10° preincubation with acetylglutamate converted the rat enzyme to a species with a lower sedimentation constant, presumably the monomer, which was only slowly reactivated in the assay medium. ATP plus Mg^{2+} did prevent this inactivation and, therefore, probably the dissociation, whereas the combination had no effect on the dissociation of the frog enzyme induced by acetylglutamate in the cold (5). Guthöhrlein and Knappe (9) conclude that acetylglutamate shifts the equilibrium between active and inactive dimer toward the active conformation, which dissociates more readily. This implies that at least a small fraction of the enzyme may exist in the active conformation even in the absence of acetylgultamate, and, according to these authors, rat carbamyl phosphate synthetase does have 2% of the activity in the absence of acetylglutamate that it has when saturated with the cofactor. This is not detectable (<0.5%) with the frog enzyme (35), perhaps because the equilibrium lies even further in the direction of the inactive dimer, which would account for its greater stability.

Quantitative correlation between the binding of acetylglutamate and the change in the conformation requires direct measurement of the binding. The high molecular weight of carbamyl phosphate synthetase, a relatively high dissociation constant, and competition by other anions make this experimentally difficult (4). The increased rate of inactivation by heat that occurs in the presence of acetylglutamate, although evidence of a conformational response, is not necessarily proportional to the fraction of enzyme in the binary complex (36), as was assumed (37).

Both rat and frog carbamyl phosphate synthetase are slowly and irreversibly inactivated in the absence of a mercaptan, and this is accelerated in the presence of acetylglutamate (see above). The large number of cysteine residues per polypeptide chain (ca. 20) makes it difficult to determine the particular residues that are essential. Novoa et al. (38), working with the frog enzyme, have reported a distinction between those that react with hydroxymercuribenzoic acid and are protected by acetylglutamate (4) and those that react with 5,5′-dithiobis-2-nitrobenzoic acid and react more rapidly in the presence of acetylglutamate.

ADDITIONAL COFACTORS

Besides acetylglutamate carbamyl phosphate synthetase requires K^+ and Mg^{2+} in excess of Mg-ATP (4). NH_4^+ can substitute for K^+ as well as serving as a substrate, so that in the earlier experiments done in the absence of K^+ the requirement for NH_4^+ seemed unphysiologically high. The Michaelis constants for the substrates and activators determined in the presence of 50 m*M* K^+ and 0.1 *M* Na^+ (4) are listed in Table 4. Mg^{2+} binds to the free enzyme;

Table 4 Michaelis Constants for Substrates and Activators of Frog Carbamyl Phosphate Synthetase

Substrate[a] or activator	K_m at 37° (m*M*)
(1) NH_4^+	2
(2) ATP	0.5
(3) Acetylglutamate	0.1
(4) Mg^{2+}	ca. 0.6

[a] $K_{m,\,HCO_3^-}$ was not determined under these conditions. Caravaca and Grisolia (17) reported a value of 9 m*M* under somewhat different conditions.

this step precedes the binding of ATP (4). Mn^{2+} will substitute for Mg^{2+} and appears to be bound much more tightly (4). Synthesis will also proceed in the presence of Co^{2+}, but the ratio of P_i released to carbamyl phosphate synthesized is > 1 (16).

REACTIONS CATALYZED

The requirement for two ATP for the synthesis of carbamyl phosphate suggested that an intermediate might be formed. The bicarbonate-dependent hydrolysis of ATP catalyzed by the frog enzyme in the absence of NH_4^+ was, therefore, interpreted as evidence for the formation of "active CO_2" as a first step (16). Subsequently the intermediate was formulated as enzyme-bound carboxyphosphate, since bicarbonate[†] was found to contribute one oxygen atom to phosphate (32, 39) (equation 1):

$$E + HCO_3^- + ATP \xrightarrow{AGA,\ Mg^{2+},\ K^+} E\text{-}CO_3PO_3^{2-} + ADP \quad (1)$$

$$E\text{-}CO_3PO_3^{2-} + NH_4^+ \longrightarrow E\text{-}NH_2CO_2^- + P_i \quad (2)$$

$$E\text{-}NH_2CO_2^- + ATP \underset{AGA,Mg^{2+}}{\rightleftharpoons} NH_2CO_2PO_3^{2-} + ADP \quad (3)$$

In the case of carbamyl phosphate synthetase II from *E. coli*, which also catalyzes a bicarbonate-dependent ATPase but does not require acetylglutamate (40), pulse-labeling with $H^{14}CO_3^-$ gave direct evidence of such an enzyme-bound intermediate (41). However, McKinley et al. (42) mentioned that a similar experiment with the frog enzyme was negative. The results of the most recent kinetic studies of the overall reaction, to be discussed by others, have been interpreted as evidence that not only is there no irreversible step between the addition of HCO_3^- and NH_4^+ to the enzyme (14) but that the first product released is carbamyl phosphate; that is, the reaction is concerted with no intermediate (43, 44). These results are not necessarily in conflict. The rate of the ATPase reaction is slow compared with that of the overall reaction, so that in the presence of ammonia, ADP may not dissociate before the addition of ammonia.

Carbamyl phosphate synthetase also catalyzes the synthesis of ATP from carbamyl phosphate and ADP. Since only 1 mole of ATP is formed per mole of carbamyl phosphate used (16) and $^{32}P_i$[‡] is not incorporated (16), it was proposed that this reaction is the reverse of equation 3 (16, 39). However, as formulated, it does not fit into Elliott and Tipton's scheme for ordered product

† HCO_3^-, as opposed to CO_2, is the true substrate (32).

‡ Incorporation of $^{32}P_i$ into ATP has been observed with rat carbamyl phosphate synthetase, but only in the presence of acetylglutamate, NH_4^+, and HCO_3^-, that is, when carbamyl phosphate is being synthesized (14).

release (44). Grisolia and Raijman (45) suggest that, in the apparently analogous but reversible reaction between acetyl or formyl phosphate and ADP to form ATP (46, 47), acetate and formate may be replacing bicarbonate in equation 1.

All these partial reactions except the synthesis of formyl and acetyl phosphate require acetylglutamate, although it is not certain that the requirement is absolute. The synthesis of formyl and acetyl phosphate by rat carbamyl phosphate synthetase is only slightly stimulated by acetylglutamate (46).

Both hydroxylamine and hydrazine are competitive inhibitors of ammonia (42) and might be expected to replace ammonia as substrates. *N*-Aminocarbamyl phosphate is formed from hydrazine (42); on the other hand, hydroxylamine only stimulates the bicarbonate-dependent ATPase (48, 42). Possibly *N*-hydroxycarbamyl phosphate decomposes too rapidly to accumulate.

REFERENCES

1. Grisolia, S., and Cohen, P. P. (1952). *J. Biol. Chem.* **198**, 561–571.
2. Metzenberg, R. L., Hall, L. M., Marshall, M., and Cohen, P. P. (1957). *J. Biol. Chem.* **229**, 1019–1025.
3. Marshall, M., Metzenberg, R. L., and Cohen, P. P. (1958). *J. Biol. Chem.* **233**, 102–105.
4. Marshall, M., Metzenberg, R. L., and Cohen, P. P. (1961). *J. Biol. Chem.* **236**, 2229–2237.
5. Strahler, J. R. (1974). Ph.D. Thesis, University of Wisconsin, Madison, Wisconsin.
6. Melnick, R. L., Tinberg, H. M., Maguire, J., and Packer, L. (1973). *Biochim. Biophys. Acta* **311**, 230–241.
7. Gamble, J. G., and Lehninger, A. L. (1973). *J. Biol. Chem.* **248**, 610–618.
8. Novoa, W. B., and Grisolia, S. (1964). *Biochim. Biophys. Acta* **85**, 274–282.
9. Guthöhrlein, G., and Knappe, J. (1968). *Eur. J. Biochem.* **7**, 119–127.
10. Raijman, L., and Grisolia, S. (1961). *Biochem. Biophys. Res. Commun.* **4**, 262–265.
11. Skinner, D. M., and Jones, M. E. (1961). *Fed. Proc.* **20**, 240.
12. Virden, R. (1972). *Biochem. J.* **127**, 503–508.
13. Grisolia, S., and Cohen, P. P. (1953). *J. Biol. Chem.* **204**, 753–757.
14. Guthöhrlein, G., and Knappe, J. (1969). *Eur. J. Biochem.* **8**, 207–214.
15. Kerson, L. A., and Appel, S. H. (1968). *J. Biol. Chem.* **248**, 4279–4285.
16. Metzenberg, R. L., Marshall, M., and Cohen, P. P. (1958). *J. Biol. Chem.* **233**, 1560–1564.
17. Caravaca, J., and Grisolia, S. (1960). *J. Biol. Chem.* **235**, 684–693.
18. Marshall, M., and Cohen, P. P. (1961). *J. Biol. Chem.* **236**, 718–724.
19. Strahler, J. R. (1975). *Fed. Proc.* **34**, 604.
20. Raijman, L., and Grisolia, S. (1961). *Biochem. Biophys. Res. Commun.* **4**, 262–265.
21. Elliott, K. R. F., and Tipton, K. F. (1973). *Fed. Eur. Biochem. Soc. Lett.* **37**, 79–81.
22. Huston, R. B., and Cohen, P. P., unpublished experiment.
23. Huston, R. B., and Cohen, P. P. (1969). *Biochemistry* **8**, 2658–2661.
24. Wellner, V. P., Santos, J., and Meister, A. (1968). *Biochemistry* **7**, 2848–2851.

25. Grisolia, S., and Cohen, P. P. (1950). *J. Biol. Chem.* **191**, 189–202.

26. Schooler, J. M., Fahien, L. A., and Cohen, P. P. (1963). *J. Biol. Chem.* **238**, PC 1909.

27. Hall, L. M., Metzenberg, R. L., and Cohen, P. P. (1958). *J. Biol. Chem.* **230**, 1013–1021.

28. Grisolia, S., Burris, R. H., and Cohen, P. P. (1954). *J. Biol. Chem.* **210**, 761–764.

29. Allen, C. M., and Jones, M. E. (1966). *Arch. Biochem. Biophys.* **114**, 115–122.

30. Reichard, P. (1959). *Fourth Internat. Congr. Biochem.*, Vienna, 1958, **13**, Pergamon, New York, p. 119.

31. Metzenberg, R. L., Marshall, M., Cohen, P. P., and Miller, W. G. (1959). *J. Biol. Chem.* **234**, 1534–1537.

32. Jones, M. E., and Spector, L. (1960). *J. Biol. Chem.* **235**, 2897–2901.

33. Grassl, M., and Bach, S. J. (1960). *Biochim. Biophys. Acta* **42**, 154–157.

34. Grisolia, S., and Towne, J. C. (1957). *Biochim. Biophys. Acta* **25**, 224–225.

35. Strahler, J. R., and Cohen, P. P., unpublished observation.

36. Citri, N. (1973). *Adv. Enzymol.* **37**, 397–648.

37. Fahien, L. A., Schooler, J. M., Gehred, G. A., and Cohen, P. P. (1964). *J. Biol. Chem.* **239**, 1935–1941.

38. Novoa, W. B., Tiger, H. A., and Grisolia, S. (1966). *Biochim. Biophys. Acta* **113**, 84–94.

39. Jones, M. E. (1965). *Ann. Rev. Biochem.* **34**, 381–411.

40. Anderson, P. M., and Meister, A. (1966). *Biochemistry* **5**, 3157–3163.

41. Anderson, P. M., and Meister, A. (1965). *Biochemistry* **4**, 2803–2809.

42. McKinley, S., Anderson, C. D., and Jones, M. E. (1967). *J. Biol. Chem.* **242**, 3381–3390.

43. Elliott, K. R. F., and Tipton, K. F. (1974). *Biochem. J.* **141**, 807–816.

44. Elliott, K. R. F., and Tipton, K. F. (1974). *Biochem. J.* **141**, 817–824.

45. Grisolia, S., and Raijman, L. (1964). *Adv. Chem. Ser.* **44**, 128–149.

46. Novoa, W. B., and Grisolia, S. (1962). *J. Biol. Chem.* **237**, PC 2711.

47. Raijman, L., and Grisolia, S. (1964). *J. Biol. Chem.* **239**, 1272–1276.

48. Grisolia, S., and Marshall, R. O. (1955). W. D. McElroy and H. B. Glass, Eds., *Symposium on Amino Acid Metabolism*, Johns Hopkins, Baltimore, pp. 258–276.

Discussion

Dr. Jones: I wanted to say earlier and, being somewhat flustered with the thought of having to give my own talk later, failed to, but Dr. Krebs mentioned this morning that some observations are made before their time, and the role of acetylglutamate with this enzyme was certainly an observation before its time. It may have been one of the first allosteric regulators to be discovered, and it came at a time when no one could conceive of an activator not participating in a turnover fashion in an enzyme reaction, so that its now accepted role as an allosteric effector it could not be appreciated at the time when it was discovered. I'd like to open the discussion to any questions that anyone would have, if you still have energy to ask them.

Dr. Illiano: I have a question about the interesting scheme you showed. You spoke about the right position that two molecules of ATP, one molecule of bicarbonate, and one molecule of ammonia must have on the active sites of the carbamyl phosphate synthetase in the mitochondria. Do you have any experimental evidence that ammonia reaches the right position on the enzyme, not as free ammonia, but delivered, by a carrier, or in a kind of activated state?

Dr. Jones: Are you asking me the difference between ammonia and ammonium?

Dr. Illiano: No. I was asking whether ammonia can be delivered on the carbamyl phosphate synthetase by a particular carrier, or does it get into position on the enzyme already as free ammonia?

Dr. Jones: No, it doesn't seem to require a carrier; it just needs to be produced somewhere presumably by glutamic dehydrogenase in the mitochondrial matrix. In fact, one of the aspects of the enzyme is that it does bind ammonium ion, which is the compound for which the K_m values have been determined very well—I think just about as well as any enzyme I know in the literature. So there seems to be no need for a carrier. Now whether it's ammonium that binds, which is what is presumed, or ammonia is not known. If the enzyme has a group that removes a proton, it could be ammonium that binds, and the proton would then be removed by the enzyme; but it would make more sense chemically for it to be ammonia when it reacts with whatever it reacts with.

Dr. Cohen: I wish to ask Dr. Tatibana whether hydroxyglutarate behaves in a competitive fashion with glutamate for acetyl CoA. In other words, will the system that synthesizes *N*-acetylglutamate also make 2-acetoxyglutarate?

Dr. Tatibana: I will try acetoxyglutarate as an inhibitor for the synthetase.

Dr. Cohen: You did try it?

Dr. Tatibana: No, but I would like to try it.

Dr. Krebs: I have a simple question, Mary Ellen. You mentioned on one of the slides that bicarbonate rather than CO_2 was the reagent *in vitro*. I wonder why you stated *in vitro*, because *in vivo* we always have both CO_2 and bicarbonate present at the physiological pH, and I think it would even be difficult *in vitro* to have only one of them present. What was your reason for putting on this slide *in vitro*, which implied that perhaps there could be a difference between *in vivo* and *in vitro*?

Dr. Jones: I was suggesting that possibly the reaction *in vivo* can use CO_2 directly. I feel a little uncertain about how fast bicarbonate and CO_2 do interchange *in vivo* in different spots other than in the blood, where we have carbonic anhydrase and relatively well known amounts there to catalyze the reaction, and we even know the various types of carbonic anhydrase (there are at least two) and their maximal velocities and other data. It seems possible to me that the level of CO_2 would be high in the matrix, that it might not go so rapidly to bicarbonate as it does in the blood, and also one has to consider how fast it permeates the membranes, since it is a very lipid-soluble compound. So, I think there is a slight question as to how much CO_2 or bicarbonate you might have in the matrix unless you know you have a carbonic anhydrase there, and I don't think anybody does.

Dr. Chappell: Yes, there is—I don't know how much—carbonic anhydrase in the mitochondrial matrix.

Dr. Jones: How is it determined?

Dr. Chappell: This is done deviously, I'm afraid, but by adding CO_2 outside the mitochondria, CO_2 goes across the membrane, and then the CO_2 is very rapidly converted into an anion inside which presumably is bicarbonate. This is done by looking at potassium uptake in the presence of valinomycin. It's all right from a physical-chemical point of view, so that there is a carbonic anhydrase there and it appears to be active, but I can't say how much.

Dr. Ratner: I wanted to ask whether you have some idea as to how much arginine is present inside the mitochondria.

Dr. Tatibana: I couldn't assay the content of arginine. I'd like to ask someone in the British school, since they are working on the concentration of the intermediate in liver cells. It is difficult to isolate the mitochondria without changing the concentrations of metabolites in them.

Dr. Ratner: It might possibly be higher in mitochondria than in cytosol. I just wondered if you had tested for it.

Dr. Tatibana: I had a discussion on this problem with Dr. Chappell yesterday. I tried to see whether arginine can get into the isolated mitochondria, but very little was incorporated into them. The synthetase has a very low K_a value for arginine, so that I think a low concentration of arginine would be enough to activate the synthetase.

Dr. Chappell: Is it possible that the arginine is on a neighboring protein normally? That is, that you are looking with an amino acid *in vitro*, but in the mitochondrial matrix a neighboring protein is presenting an arginine group for purposes of activating that enzyme.

Dr. Tatibana: No, I didn't examine this possibility, but I have done some experiments for the activation effect of arginine. I incubated isolated mitochondria with arginine, and the synthesis of acetylglutamate was increased remarkably.

Dr. Grisolia: I was very much interested in the remarks of Dr. Frieden regarding the stability of glutamate dehydrogenase. I was not sure whether in all cases the data he presented referred to rat liver. Anyway, the comment I was trying to make is that obviously he, as well as many others, including work from my laboratory (Grisolia, S., *Phys. Rev.* **44**, 657, 1964) has shown that enzyme stability depends very much on the conditions, ligands, etc., and I was wondering whether you referred to stability in the presence of NADH or to stability in general.

Dr. Frieden: The enzyme itself is fairly stable. It is when you try to assay the enzyme in the presence of coenzymes that it is very unstable. I could make the same remarks to Dr. Gonzalez about the brain enzyme. I would ask if inactivation of that enzyme would be noticed during the assay, since we had a lot of trouble with the liver enzyme, and there are probably no differences between the brain and liver enzymes. I wondered whether you saw that problem with the brain enzyme as well. You didn't?

Dr. Grisolia: Dr. Gonzalez says that the enzyme is quite stable during the measurements she has made. I should like to add that Dr. Salinas in our laboratory compared the half-life of the enzyme with the half-life of stability *in vivo* and *in vitro* under several conditions. She found that the enzyme was usually stable in rat liver, both in homogenates and in the whole tissue (M. Salinas, R. Wallace, and S. Grisolia, *Eur. J. Biochem.*, **44**, 375, 1974) but much less so than *in vivo*. On further examination, she showed that the cytosol, when added to mitochondria, inactivates the enzyme. These findings, in addition to stressing the effect of environmental factors on stability, clearly show for the first time that a protein, the bulk of which is located intramitochondrially and which is synthesized in the cytosol, is likely also degraded there.

Dr. Cohen: I wanted to comment that convincing other scientists that *N*-acetylglutamate was an activator rather than a coenzyme was probably less difficult than convincing some that 2 moles of ATP was required for the synthesis of carbamyl phosphate. Distinguished investigators in the area of ATP-requiring systems suggested that our enzyme preparations were probably contaminated with other ATP using or hydrolyzing enzymes to account for the stoichiometry. However, the confidence that I had in the scientific competence of my colleagues Dr. Metzenberg and Dr. Marshall, who were doing the experiments, was more than enough to convince me that these criticisms were invalid. This comment bears on the point that Dr. Krebs made earlier this morning.

Dr. Grisolia: In view of Dr. Frieden's remarks, I would like to make another comment. I was wondering, in the case of Dr. Gonzalez, if the enzyme is stable only in the purified preparation. This may account for the low activity of the supernatants. In other words she may have sufficient guanosine triphosphate or a similar reagent which may modify the stability in the supernatant during the isolation procedure. What do you think about that possibility, Dr. Frieden?

Dr. Frieden: You mean whether there might be some GTP around?

Dr. Grisolia: Yes. Sufficiently so as to inactivate the enzyme.

Dr. Frieden: It all depends on how much dilution there is before you assay it.

Dr. Grisolia: I just mention this as a word of caution, because it may very well be that NADH or other cofactors are responsible for the stability of the rat liver and other enzymes. I think one should be aware and on the lookout for something like that in view of what is already known of the properties of the enzyme from other sources (S. Grisolia, *Phys. Rev.*, **44**, 657, 1964; S. Grisolia and W. Hood in *Biochemical Regulatory Mechanisms in Eukaryotic Cells*, E. Kun and S. Grisolia, Eds., Wiley, New York, 1972).

Dr. Krebs: If I understood Dr. Frieden correctly, he questioned whether glutamate dehydrogenase is actually at equilibrium, because it is inhibited by allosteric effectors. Am I correct in assuming that you base this criticism on purely theoretical considerations? In reply I would state that glutamate dehydrogenase is, in fact, in the liver under many conditions at equilibrium. This statement is based on the direct observation that the NAD/NADH ratio, calculated from the β-hydroxybutyrate dehydrogenase system and calculated from the glutamate dehydrogenase system, is identical within the limits of error. It is, of course, possible by making all sorts of totally unwarranted assumptions to explain this coincidence as an accident. But such assumptions are much more speculative than the straightforward assumption that if two different dehydrogenases which operate in the same compartments—the mitochondrial matrix—give the same NAD/NADH ratio, as calculated from the equilibrium constant, they are in equilibrium with the same NAD pool. So the point I wish to make is that, whatever the possibilities are of inhibitions of glutamate dehydrogenase by allosteric effectors, there is enough activity left *in vivo* under most conditions to keep the reactants of this dehydrogenase at near-equilibrium.

Dr. Frieden: I would hate to argue with you, Dr. Krebs. I know the paper to which you are referring, and I wondered whether it is also known whether the NADP/NADPH couple is also in equilibrium, since the enzyme presumably uses both coenzymes. I have recently seen some data at the Federation meeting by Dr. Williamson which indicated that that couple was not in equilibrium.

Dr. Krebs: I am most interested to see John Williamson's evidence. As far as I am aware, there are no ways and means of measuring the NADP/NADPH ratio—I refer of course to the free and not to the bound nucleotides, because what matters in the present context are the concentrations of the free purine nucleotides. We definitely know that in the intact mitochondria glutamate dehydrogenase reacts with both NAD and NADP, and this strongly supports the hypothesis that the NAD and NADP couples are at equilibrium. Often much confusion is caused by not paying sufficient attention to the difference between the concentrations of total and free pyridine nucleotides. We know that many dehydrogenases bind NAD, NADP, and NADPH, and therefore the total pyridine nucleotide concentrations do not necessarily bear any relation to the concentrations of the free compounds. To clarify the point under discussion, it is necessary to have independent data from two different dehydrogenase systems. We have these for the NAD-linked dehydrogenases but unfortunately not for the NADP ones.

Dr. Cohen: I would like to add to this discussion, not with specific reference to the nucleotides, but to the possibility of other ligands influencing GDH activity. We reported a few years ago that there were striking kinetic and other differences in the GDH preparations from liver of the tadpole and that from the adult frog. Because of the relatively low level of GDH activity in premetamorphic tadpole liver, a different method of isolation and crystallization was used than that for the frog. The two enzyme preparations showed differences in substrate specificity and molecular size. When Dr. Karen King came to my laboratory as a postdoctoral fellow, she applied some of the techniques that she developed in Dr. Frieden's laboratory for crystallizing rat liver GDH. When exactly the same conditions were used for isolation and crystallization of GDH from liver of tadpole and frog, the enzymes proved to be indistinguishable (see K. S. King and P. P. Cohen, *Comp. Biochem. Physiol.* **B51**, 113–117 1975).

Dr. Grisolia: If there are no more comments, I would like to congratulate Dr. Jones for the unusual and intriguing mechanism she has proposed. Even if it is not right, I think it should be, because it is very elegant.

Dr. Tatibana: When the urea cycle is functioning maximally, my calculation showed nearly 50% of the carbon dioxide produced in the mitochondria may be converted to urea. It means that there must be a high activity of anhydrase in the mitochondria. So, I would like to know whether the carbonic anhydrase in mitochondria is high enough to convert CO_2 to bicarbonate quickly.

Dr. Chappell: I am sorry not to be able to give you numbers, and I must now give you the experimental basis on which I made my statement. The situation is this. We are looking at the swelling of mitochondria, and before that can happen, in say a solution of potassium bicarbonate, it will be necessary for both the potassium and the bicarbonate to cross this membrane. That should meet neutrality and osmotic consideration requirements. Now the situation is that mitochondria do not swell in potassium bicarbonate, from which we conclude that bicarbonate will not penetrate. If we present CO_2 in the presence of potassium ions, CO_2 goes into the mitochondrion, and we can use CO_2 here. Now the half-time for hydration of CO_2 is about 18 seconds, as I recall, something of this sort. In fact the mitochondria instantaneously swell to their final value, which means that there must be in here sufficient carbonic anhydrase to be as much as is necessary. Now, you see, these are not numbers; it is merely saying that it is relatively very rapid. Now, Dr. Holten, at Veterinary College in London, has been studying the properties of carbonic anhydrase in the matrix. But, again, I can't give you any numbers, because I don't have them. I believe I am right in saying that CO_2 is the usual end product in decarboxylation reactions. Is that right? The thing which enzymes produce. And the mitochondrial membrane is certainly very permeable to CO_2. It is a small molecule. And any CO_2 that is produced in here would immediately escape into the much larger volume outside of the cell. To account for your results that CO_2 is going to be used here preferentially, there would have to be a carbonic anhydrase here to keep it in, so that subsequently it can be used as CO_2 or whatever. There is a way that we can answer your problem, by doing an experiment. If mitochondria make citrulline from ornithine, so that if mitochondria are given ornithine and bicarbonate or CO_2 and ammonia, then the synthesis of carbamyl phosphate and the synthesis of citrulline will require ATP. We can determine from the stimulation of respiration how much ATP that is equivalent to, and since there is a factor of 2 difference in your mechanism than in the usual one, it should be a perfectly soluble problem. Does the synthesis of citrulline in mitochondria, with say succinate providing the energy, require one ATP or two?

Dr. Jones: I was stimulated by this meeting to try to devise something novel, and I also wanted to bring to your attention Dr. Jencks' new mechanism, which, by the way, I should also point out holds for the biotin carboxylases and is not limited to the carbamyl phosphate synthetase reaction.

I was unaware of any data on a carbonic anhydrase in the mitochondrial matrix. Certainly if I had known it was there, I would not have made the suggestion. But thinking that that was an open area, it then became possible to make this suggestion. I do think your information on the transport of potassium sounds to me as very reasonable support for a carbonic anhydrase within the matrix of the mitochondria. I also think the fact that you point out that its presence would hold CO_2 and bicarbonate within the matrix of the mitochondria for a longer time is a very potent idea for having it within the matrix. So I really don't wish to fight against the idea of having a carbonic anhydrase within the matrix. I was unaware that one existed, and I am glad that this discussion brought out your data for it.

Dr. Krebs: I would like to comment on the question whether carbonic anhydrase is really necessary. There is in the liver, as in every other mammalian tissue, a large pool of both CO_2 and bicarbonate, the concentration of CO_2 being about 1 mM and that of bicarbonate about 15 to 20 mM. The presence of carbonic anhydrase would see to it that the equilibrium between bicarbonate and CO_2 is established more rapidly than in the absence of the enzyme, but I think this is of no importance for urea synthesis, because the pools of CO_2 and bicarbonate are large enough. It has also to be borne in mind that the rates at which CO_2 and bicarbonate have to interact are of an entirely different order of magnitude in the lungs (where carbonic anhydrase is indispensable) and in the liver. Compared with the gas exchange in the lungs the rates of CO_2—involving reactions in the liver are relatively slow. I am also not fully convinced that mitochondria contain carbonic anhydrase, because there are numerous substances, among them several inorganic ions, which have effects similar to carbonic anhydrase. So my question is whether carbonic anhydrase is really essential in connection with urea synthesis.

Dr. Jones: I don't know whether it's essential or not. I think it is hard to answer unless you know the rates at which the two systems making carbamyl phosphate proceed, that is, what are the rates of its formation from bicarbonate and both CO_2 (if this reaction occurs). The relative rates of these reactions would determine if there is any need at any given time for the use on one reaction versus the other, and since nobody has as yet observed the reaction from CO_2 to carbamyl phosphate, I don't see how you can really tackle that. I do think the suggestion of Dr. Chappell that it could be determined how many ATP's were being used is a very nice idea and would love to see it done.

Dr. Grisolia: Many years ago, if I remember correctly, and probably Dr. Cohen remembers too, Phil Siekevitz and Van Potter (*J. Biol. Chem.*, **201**, 1, 1953) did experiments along these lines. They showed with liver mitochondrial preparations increased respiration and decreased ATP yield when citrulline synthesis occurred and that the rate of citrulline synthesis was markedly inhibited by 2,4-dinitrophenol or hexokinase/glucose. However, the importance of their discovery was unappreciated, due to the lack of understanding of mitochondrial control mechanisms at the time. Thus, there is at least a first experimental model for this type of approach.

Dr. Frieden: I want to make one comment that is pertinent to this point, perhaps the glutamic dehydrogenase as well, that you have to be careful that you don't confuse equilibrium with rate constants and with the kinetics of the process. Although I do not want to argue about metabolic processes, because I am not involved in that directly, it seems to me that there are situations where you poise the situation such that it will be at equilibrium, whereas *in vivo* those equilibrium conditions will not occur, and what you will find are rate effects rather than equilibrium effects.

Dr. Krebs: Or did you want to finish really?

Dr. Jones: I had hoped this would be provocative, and I see I have succeeded.

Dr. Krebs: We are fully aware that in general thermodynamic considerations (and equilibria are a matter of thermodynamics) have nothing directly to do with rates but when rates are sufficiently rapid, we can expect that equilibria are maintained. Only in this respect are rates relevant to the maintenance of equilibria in a dynamic system such as a living cell represents. Now, Mary Ellen, I am really not sure whether you said that you had any data on the rates of carbamyl phosphate synthesis *in vivo*.

Dr. Jones: No.

Dr. Krebs: If we know the maximum rates of urea synthesis in the liver (and these can be obtained from data on freshly prepared isolated rat liver cells) and if we know the rate of urea synthesis in the whole organism, we can compare these rates with the rate of CO_2 production from cell respiration (which can also be accurately measured). Our figures are very similar to those mentioned by Professor Tatibana. The maximum rate of urea synthesis is found when the tissue is saturated with ammonium chloride, ornithine, and a precursor of the carbon skeleton of the aspartate, for example, lactate; under such conditions we obtain rates of up to 5 μmoles of urea formed per gram tissue per minute. Under the same conditions the oxygen consumption is up to twice this value. This leads to the conclusion that up to 50% of the CO_2 produced from respiration could be used for urea synthesis.

There are also many data on the rate at which the noncatalyzed interconversion of bicarbonate and CO_2 takes place. When I worked with Roughton many years ago on carbonic anhydrase I was very much impressed how difficult it is to demonstrate any action of this enzyme at body temperature. We devised a class experiment in which we added bicarbonate to phosphate buffer at a low temperature in a Warburg manometer. Under these conditions the effect of carbonic anhydrase was striking. In the absence of bicarbonate it took about 30 minutes for CO_2 to equilibrate with bicarbonate, but in the presence of excess enzyme the reaction was finished in 1 or 2 minutes. But if we tried to do the experiment at 37°C, then it is very difficult to show any effects. Thus the noncatalyzed reaction is very fast at body temperature, though not fast enough to liberate CO_2 in the lung, where the conversion of bicarbonate to CO_2 has to occur within seconds.

Dr. Grisolia: In regard to the discussion of CO_2 production and maximal rates of urea production discussed by Krebs and Tatibana, we have shown recently (S. Grisolia and J. Mendelson, *Biochem. Biophys. Res. Commun.*, **58**, 968, 1974) that the CO_2 produced in mitochondria may be limiting insofar as urea synthesis is concerned; this will be shown Saturday by Ms. Mendelson. Since her presentation will be rather extensive, I should like to refrain from discussion on this issue at the present time, except for this brief comment. However, I would like to point out that our calculations are possibly simpler, more physiological, and probably more trustworthy than those mentioned by others in this discussion, for they are based on measurements made with man.

Dr. Tatibana: I did some experiments on whole animals to see how much urea is synthesized maximally. I used mice, totally homogenized, with their excreta and then assayed urea on the extract. With 60% casein diet, for 2 to 3 hours after the start of feeding, they produced urea at a rate of 200 to 300 μmole/hour. This is the same value that Sir Hans mentioned. This means that CPS is maximally functioning *in vivo* at the same rate as you assayed *in vitro*. There may be some critical reasons for the difference between the *in vivo* and *in vitro* activities of the CPS and some other urea cycle enzymes.

Dr. Jones: I think Elliott also mentioned today that the rate you get with the purified enzyme at substrate concentrations that supposedly exist *in vivo* was lower. Did I read you right on that one?

Dr. Elliott: No, what I actually said was, taking into account kinetic considerations from the complex nature of the rate equation and assuming that acetylglutamate concentration seems to be similar to the K_m and possibly also the ammonia concentration would appear to be similar to the K_m, it is possible that the *in vivo* activity is considerably lower than that measured *in vitro*. It may even be as low as 5%, but that is an extreme. I would suspect more like 20%.

Dr. Cohen: Dr. Kennan, working in my laboratory, calculated the daily urea biosynthesis potential of the human adult based on our assays of urea biosynthesis enzymes in liver preparations from human adults (see A. L. Kennan and P. P. Cohen, *Proc. Soc. Exp. Biol. Med.*, **106**, 170–173, 1961). His calculations indicated that, on an average protein intake, only one-fifth of potential urea synthesis ability is used for the daily synthesis of urea.

It should be mentioned that the concentration of mitochondrial carbamyl phosphate synthetase is higher than that of any other protein making up the mitochondrial matrix, reaching a level as high as 20% in the frog. Enhancement of urea biosynthesis can be demonstrated in acute experiments by administration of a lethal dose of an ammonium salt and measuring the protective effect of a mixture of carbamyl glutamate plus arginine. It has been demonstrated that rats can be protected 100% against a lethal dose of ammonium acetate (see S. Kim, W. K. Paik, and P. P. Cohen, *Proc. Nat. Acad. Sci. USA*, **69**, 3530–3533, 1972), if given a suitable treatment with a mixture of carbamyl glutamate plus arginine. A similar effect can be observed with sheep (P. P. Cohen, unpublished studies). The rationale for the use of carbamyl glutamate in place of the naturally occurring *N*-acetylglutamate activator of mitochondrial carbamyl phosphate synthetase is that *N*-acetylglutamate is rapidly hydrolyzed by tissue acylaminoacid acylases but carbamyl glutamate is not hydrolyzed at all. Arginine serves as a ready source of ornithine as a result of liver arginase activity. These experiments clearly indicate that the amount of carbamyl phosphate synthetase activity available for conversion of ammonia to carbamyl phosphate is dependent on the extent of its activation by *N*-acetylglutamate, or an analog such as *N*-carbamylglutamate, and that the total amount of carbamyl phosphate synthetase present in mitochondria exceeds by a considerable factor the amount of enzyme normally activated by the *N*-acetylglutamate synthesized in the mitochondria by the system reported by Shigesada and Tatibana (*J. Biol. Chem.*, **246**, 5588–5595, 1971).

Dr. Grisolia: I disagree somewhat in the assumed large excess of enzyme activity for urea synthesis in relation to the maximum capacity for nitrogen use by man. There are pioneer experiments done in man by Cohn et al. (*Amer. J. Med. Sci.*, **231**, 394, 1956) showing clearly that within 6 to 10 hours after ingestion of ~200 to 250 g of protein, for example, ~3 times what is normally considered the average daily protein intake, the blood urea level can triple and does not return to the basal level for ~24 hours. It can be calculated from that data and from the data of Schimke and Ratner that one might have possibly as much as two to three times excess for the enzymatic activity needed to take care of the quantities of protein normally consumed in the American diet. A main goal of modern biochemistry is to correlate measurements *in vitro* with physiological parameters *in vivo*. We are beginning to recognize that concentration of enzymes and substrates and the general characteristics of the catalytic environment, although difficult to assess, must be known and taken into consideration. Nevertheless, many of the measurements done *in vitro* have been and can often only be done, even with the relative sophistication of present technology, at unphysiological conditions. For example, most enzyme measurements are made of necessity, convenience, or routine under more or less "best" conditions, for example, very high substrate concentration, initial, or nearly so, velocity, and low enzyme, but in the cell the opposite conditions, that is, high enzyme and low substrate, are prevalent. A possible approach to this problem is to force conditions *in vivo* to the maximum capability of the organism for a certain metabolic reaction and then compare values for this reaction with *in vitro* measurements for the same reaction made at or near maximum rate conditions. It seemed to us that an area eminently suitable for such an approach is the urea synthesis system. Therefore, we have attempted to study this by injection of large amounts of protein to animals (S. Grisolia, R. Wallace, and J. Mendelson, *Physiol. Chem. Phys.*, **7**, 219, 1975). By this approach, we have shown that the parameters as measured *in vivo* and *in vitro* agree reasonably well. It is quite interesting, then, that the enzyme levels are very close to the maximum level of nitrogen that can be handled by the whole animal and that the rather unphysiological and possibly dangerous use of high protein diets with comparatively high urea blood levels seems not to have been noted. It should not go unnoticed also that urea is in equilibrium with CNO^- and that the latter chemotropic reagent has a number of deleterious effects on tissues, including modification of brain proteins. High protein diets have been favored by many, particularly in countries where the standard of living is high. Indeed, some of the diets for weight reduction favored by some so-called diet experts use protein almost exclusively, and these may be not only expensive and unphysiological but dangerous.

Dr. Soberon: We ran several experiments a few years ago, to explore the capacity of the urea cycle under conditions of reduced enzyme activities. We submitted animals to different treatments in order to lower the enzyme levels of the urea cycle. We succeeded in doing so by dietary means, by partial hepatectomy, and by the administration of carbon tetrachloride. In all cases, except by chronic poisoning with CCl_4, the ammonium concentration in blood was not altered in spite of the great diminution of enzyme levels. The decreased capacity of the animal to cope with ammonium loads revealed the insufficiency of the cycle to deal with extra work. Accordingly, one has to be very careful to extrapolate results obtained by the study of enzyme activity *in vitro* to *in vivo* particularly when multienzyme systems are involved; the systems have other factors of complexity. One thing that was very striking in our work is the following: by anoxia we lowered the concentration of ATP in liver something to less than 10%, which immediately caused a sharp increase in ammonia concentration in blood. So ATP is very critical in limiting the capacity of the liver to convert ammonia to urea.

Dr. Grisolia: The catalytic effect, as well as the activating (allosteric) effect, of acetylglutamate acid (S. Grisolia and P. P. Cohen *J. Biol. Chem.*, **204**, 753, 1953; S. Grisolia and J. C. Towne, *Biochim. Biophys. Acta*, **25**, 224, 1957) was not the only first for acetylglutamate. The acetylglutamate-induced decreased instability of CPS (J. Caravaca and S. Grisolia, *Biochem. Biophys. Res. Commun.*, **1**, 94, 1959) led to the discovery of the generally occurring substrate-induced decrease of enzyme stability by substrates and thus to the elastoplastic theory (S. Grisolia, *Phys. Rev.*, **44**, 657, 1964; R. Silverstein and S. Grisolia, *Physiol. Chem. Physics*, **4**, 37, 1972). Studies with this reagent led, in addition, to the discovery of possibly (the other being Racker's ATPase) the first example of cold-labile enzymes (L. Raijman and S. Grisolia, *Biochem. Biophys. Res. Commun.*, **4**, 262, 1961).

Dr. Jones: I think we are all tired and I believe this is a good time to end.

Dr. Grisolia (*Added in proof*): We conducted a number of experiments to detect an intermediate possibly bound to carbamyl phosphate synthetase (Chabas et al., *Eur. J. Biochem.*, **29**, 333, 1972; Raijman, Lizarralde, and Grisolia, unpublished). We thought that our inability to detect such an intermediate was due also to the almost impossibility of obtaining ammonia-free reaction mixtures. We have finally been able to demonstrate (Rubio and Grisolia, unpublished) the formation of a CO_2 intermediate initially using pulse-label experiments similar to those described by Anderson and Meister (*Biochemistry*, **4**, 2803, 1965) with the *E. coli* enzyme and then by a refined technique based on volume dilution. For the successful demonstration of the intermediate *it is essential to get rid of trace amounts of ammonia.* We solved this by the very simple technique of conversion into carbamyl phosphate by a short incubation with nonradioactive bicarbonate, before addition of $NaH^{14}CO_3$.

Guthöhrlein and Knappe (*Eur. J. Biochem.*, **8**, 207, 1969) indicate that the elusive "active CO_2" may have a half-life of ~0.6 to 2 seconds, and that pulse-label experiments successful with the *E. coli* enzyme may not be so with the animal enzyme due to the extremely unfavorable kinetic situation. Their calculations are in agreement with the recent work of Sauers et al. (*J. Am. Chem. Soc.*, **97**, 19, 1975) referred to by Dr. Jones.

The properties of the putative carbonyl phosphate (CO_2 precursor of carbamyl phosphate) are as follow: It is formed upon reaction of enzyme with acetylglutamate, Mg^{++}ATP, and bicarbonate until a maximal steady state concentration is reached, the magnitude of which depends on the amount of enzyme. In the absence of ammonia it is rapidly decomposed; the half-life in aqueous solutions at 25°C is ~0.7 seconds, and at 0°C ~5 seconds. Experiments with dilution in acetone have shown an increase in the precursor half-life, suggesting hydrolysis as main mechanism of decomposition. The rate of decomposition remains constant between pH 5.5 to 10.5. Above pH 10.5 base catalysis occurs, and below pH 5.5 the reaction is acid catalyzed. The precursor reacts rapidly with ammonia to produce carbamyl phosphate. Kinetic experiments indicate that the reaction requires NH_3, and therefore the K_m for $[NH_3 + NH_4^+]$ varies with pH. K_m values range from 1 mM at pH 7.4 to 0.24 mM at pH 8, in very good agreement with those already reported for the overall reaction (0.71 mM, Caravaca and Grisolia, 1960; ~0.63 mM, Mackinley et al., 1967; 0.66 mM, Fahien et al., 1964), but lower than given by Marshall (2 mM) at the Meeting. From our data such K_m would occur at pH of 7.1 to 7.2.

The reaction with ammonia also takes place in acetone. While under these conditions the enzyme does not show activity and it seems that the reaction of the precursor with ammonia to produce carbamyl phosphate does not require an active enzyme. This suggestion is further supported by the finding that the reaction proceeds to a large extent (50 to 75%) in 1.3 M KOH. It must be remembered that at such extremely high pH values the decomposition of the precursor is very fast as a consequence of the base-catalyzed reaction but slower than its reaction with ammonia. If the reaction with ammonia had been enzymatic the enzyme might have been fixed by acetone on possibly the most reactive conformation for reactivity. This could be tested by the addition of low levels of ammonia, glutamine, and/or the use of enzyme inhibitors.

We have recalculated the ATPase activity versus the activity for carbamyl phosphate synthesis of the *E. coli*, rat liver, and frog enzyme and found it to be ~8 to 10% in all cases. This indicates the similarity if not the identity of the "carbonyl phosphate" presumably formed with these three enzymes, and therefore presents new and interesting problems.

The scheme of Sauers, et al. (1975, op. cit.) seems incompatible, as given, with the pulse experiments of Anderson and Meister, and particularly with ours insofar as the presence of ammonia or glutamine bound to the enzyme; that is, carbonyl phosphate synthesis occurs in the absence of these acceptors. Moreover, we have demonstrated experimentally by the use of high concentrations of acetone and ^{32}P-ATP that all the ATP is used before reaction with ammonia. Thus the bypass of ATP utilization for CO_2 activation by high concentration of CO_2, postulated by Sauers and co-workers and further elaborated by Jones, is unlikely *in vitro*. Our findings prove the postulated existence of a carbonyl phosphate.

Another unsuspected role for N-acetylglutamate has just been uncovered (Rivas, Reglero, and Grisolia, unpublished). N-acetylglutamate stimulates markedly the proteolysis *at neutral* pH of rat liver mitochondria. There is no need for SH reagents. Although much remains to be clarified, it is interesting to point out the possibility that control or activation of proteases *in vivo* at neutral pH may be a general phenomenon and that the initiator or controller(s) for protein degradation may often be acyl amino acids and/or related compounds. If this is the case, the role of formyl methionine in protein synthesis initiation may be more than coincidental.

Finally, the requirement for acetylglutamate with frog liver carbamyl phosphate synthetase has more recently been shown by Forman and Grisolia

(*Physiol. Chem. Phys.* **6**, 213, 1974) to be absolute. On the other hand, Forman et al. (*Biochem. J.*, **143**, 63, 1974) have shown that, at very high concentrations, acetyl aspartate can activate carbamyl phosphate synthetase of both frog and rat liver.

13

Carbamylphosphate Phosphatase

Giampietro Ramponi
Istituto di Chimica Biologica
Universita di Firenze
Florence, Italy

Carbamylphosphate phosphatase catalyzes the hydrolysis of carbamylphosphate according to the following reaction:

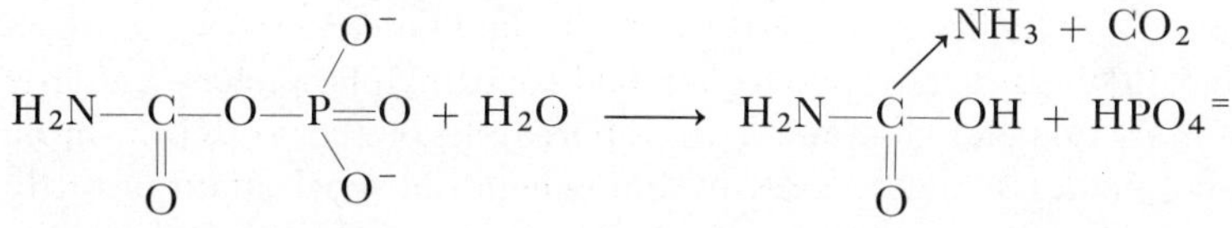

$$H_2N{-}\underset{\underset{O}{\|}}{C}{-}O{-}P(O^-)_2{=}O + H_2O \longrightarrow H_2N{-}\underset{\underset{O}{\|}}{C}{-}OH + HPO_4^{=}, \quad H_2N{-}COOH \rightarrow NH_3 + CO_2$$

This enzyme is, therefore, a phosphatase which acts on the carboxyl-phosphate bond.

The history of carbamylphosphate phosphatase starts with the history of its substrate, which was known as compound X until the structure of carbamyl phosphate was defined. In fact, Grisolia made the observation, in 1949, that skeletal and cardiac muscle homogenates, when added to mitochondrial preparations of rat liver, inhibited the reaction ornithine → citrulline. The advances in understanding the mechanism of the reaction ornithine → citrulline permitted Grisolia and his colleagues (1) to reinvestigate, in 1954, the action of tissue extracts on compound X and to demonstrate that preparations from a variety of tissues enzymatically decomposed this substrate and thus to conclude that this enzyme was a phosphatase.

Later, Jones et al. (2) confirmed the carbamylphosphate nature of compound X, and Grisolia (3) showed that carbamylphosphate was inactivated enzymically as compound X by a splitting enzyme. This last result may be considered the first certain evidence of a carbamylphosphate phosphatase activity. Successively, several other carbamylphosphate phosphatase activities were found in many tissues. Krebs in 1958 (4) described a carbamyl phosphate phosphatase activity in *E. coli* and in rat tissue homogenates. However, from the first preliminary studies on the purification of this enzyme, in brain tissue (5) by Grisolia and in muscle by Ramponi (6), it became evident that acetyl phosphatase, described in 1946 by Lipmann (7) and subsequently given the name acylphosphatase, and carbamylphosphate phosphatase were the same enzyme, having group specificity as follows:

$$\mathrm{R{-}\underset{\underset{\displaystyle O}{\|}}{C}{-}O{-}\overset{\displaystyle O^-}{\overset{/}{P}}\!\!=\!O \underset{\displaystyle O^-}{\overset{}{}}} + H_2O \longrightarrow \mathrm{R{-}\underset{\underset{\displaystyle O}{\|}}{C}{-}OH} + HPO_4^{=}$$

It is well known that acylphosphatase, purified from many sources and called *heat-stable* because of its characteristic property of retaining the bulk of its activity at acid pH and at a temperature near 80°C, is less active on carbamylphosphate than on other acyl phosphates, and it is thus unlikely that it may account for all the carbamylphosphatase activity found in various tissues.

The presence of two kinds of acylphosphatase was demonstrated by Lipmann (7), Shapiro and Wertheimer (8), and Guerritore (9); it appeared also that the heat-denatured enzyme had carbamylphosphatase activity. In 1961 a study from our laboratory (10) reported the presence of a heat-unstable acyl phosphatase activity in the postmitochondrial supernatant of rat liver, with a ratio C—P phosphatase/acetyl phosphatase of about 0.4, that is, much higher than the ratios generally observed with the enzymes purified from other sources. Recently Grisolia and Hood (11) showed that kidney homogenates from various species have an acyl phosphatase activity with higher ratios between acetyl phosphatase and carbamylphosphatase (from ~3/1 to 1/1) and that centrifugation at 15,000 g for 30 minutes results in a greater ratio in the supernatant than in the pellet, indicating the presence of a double activity, a soluble and a pellet fraction with high specificity for C—P.

Furthermore, the hydrolysis of C—P does not appear to be accomplished by enzymes having only acyl phosphatase activity. In fact, Yoshida et al. (12) obtained from guinea pig microsomes a K^+-dependent phosphatase which split *p*-nitrophenylphosphate, carbamylphosphate, and acetylphosphate with an optimum pH around 7.8, but Herzfeld and Knox (13) found in rat tissue preparations a phosphate ester hydrolase that destroys C—P. This activity was similar to that shown by alkaline phosphatase, both having the same molar activities, pH curves, and particulate localization.

All these findings suggest that the hydrolysis of C—P may be carried out by several enzymes; whether or not an enzyme specific only for C—P exists in some tissues remains to be elucidated.

However, considering that the acyl phosphatases which up to now have been purified, with the exception of erythrocyte acyl phosphatase, show a considerable hydrolytic activity on C—P, this paper will be essentially concerned with the structural and functional properties of these enzymes.

C—P phosphatase has been purified from different sources, such as horse muscle (6) and liver (14), bovine brain (5), pig heart muscle (15). The purification of horse muscle carbamylphosphate phosphatase consisted of an acid extraction with 0.11 *N* HCl, a precipitation of foreign proteins by adjusting the pH of the extract to 4.8, a $ZnCl_2$ precipitation, and, after the enzyme was dissolved by the removal of Zn ions, two subsequent steps on chromatographic columns of CM Sephadex C-25. The homogeneity of the final product was established by starch gel and polyacrylamide gel electrophoresis, gel filtration on Sephadex G-75, and ultracentrifugation. The specific activity of the enzymatic solution thus obtained indicated a 4225-fold purification with a yield of 5.8%, with respect to the acid extract used in the beginning.

The preparation of the horse liver enzyme was based on extraction with 1.5 *N* acetic acid, isoelectric precipitation of foreign proteins at pH 4.9, and three chromatographic steps on CM-Sephadex C-25 columns, the third chromatography being performed with a specific elution using 0.05 *M* phosphate buffer (phosphate is a well-known inhibitor of acylphosphatase activity and was therefore chosen for attempting a specific elution of the enzyme). The final preparation was purified 16,000 times with respect to the acid extract and the yield was 6.7%. The purification of carbamylphosphate phosphatase from bovine brain, as reported by Diederich and Grisolia, consisted of homogenization in sodium acetate pH 4, precipitation with acetone, and chromatography on Bio-Rex 70 column. The enzyme was purified about 25,000 times with a yield of 5% and was pure by acrylamide disc electrophoresis. Similar techniques were employed to purify C—P phosphatase from pig heart muscle. Four species of this enzyme were separated by ion-exchange chromatography; the major component of these fractions was extensively purified and obtained as a homogeneous product when subjected to acrylamide gel electrophoresis.

The enzymatic proteins obtained as pure products have been studied in their physicochemical properties (Table 1) and analyzed for their amino acid composition and the nature of their end groups. All carbamyl phosphatases

Table 1 Some Molecular Properties of Carbamylphosphate Phosphatase from Various Sources

Source	mol. wt.	pI	pH Optimum
Bovine brain	12,000[a] 13,800[b]	—	7.4–7.6
Horse muscle	9,400	11.4	5.3
Horse liver	8,300	—	5.5
Pork heart	11,095	7.25	5.4

[a] Determined on Sephadex G-75.
[b] Determined on Bio-Gel.

exhibit a very low molecular weight, ranging from a minimum of 8,300, determined for the horse liver enzyme by gel filtration on Sephadex G-75, to a maximum of 13,800, found for the bovine brain enzyme using Bio-gel column. Greater differences were found in the isoelectric points; however, the data available up until now indicate that C—P phosphatases are generally basic proteins, as is also indicated by their amino acid composition. The optimal pH on C—P as substrate for horse muscle and liver enzymes are 5.3 and 5.5 respectively, but the optimal pH for bovine brain enzyme is 7.4 to 7.6, and the optimum pH range for the hydrolysis of the heart enzyme is 5.4 to 5.6.

Table 2 presents a comparison of the amino acid composition of C—P phosphatases obtained from various sources. Horse muscle and liver enzymes show a very close amino acid composition, but the brain enzyme appears to be slightly different.

Horse muscle and liver C—P phosphatases and the bovine brain enzyme have been the object of a preliminary approach to the study of the primary structure of these proteins. As regards the horse muscle enzyme, it has been shown that the native protein is a mixed disulfide with glutathione. The inability to detect the NH_2-terminal amino acid even after the removal of glutathione by carboxy-methylation suggested the hypothesis that this residue was blocked. In order to verify this possibility, after tryptic digestion of the carboxymethylated protein the NH_2 terminal peptide, recognized by the alkaline hydrolysis method, was isolated, analyzed for its amino acid composition, and sequenced by an enzymatic method using carboxypeptidase B and A. The kinetic of amino acid release indicated the sequence X-NH-Ser-Thr-Ala-Arg. Further studies allowed the identification of the agent blocking the NH_2-terminal amino acid as an acetyl group (17). The inability to detect any NH_2-terminal amino acid derivative for the horse liver and bovine brain

Table 2 Comparison of Amino Acid Composition of Carbamylphosphate Phosphatase from Various Sources

Amino Acid	Horse Muscle	Horse Liver	Bovine Brain
Tryptophan	1[a]	—	2[b]
Lysine	8	8	7
Histidine	1	1	2
Arginine	5	5	3
Cystine (half)	2	1	—
Aspartic acid	6	7	7
Threonine	5	5	5
Serine	10	13	4
Glutamic acid	9	10	11
Proline	3	3	3
Glycine	7	9	7
Alanine	3	4	4
Valine	8	8	7
Methionine	2	1	1
Isoleucine	2	3	3
Leucine	3	3	5
Tyrosine	3	2	2
Phenylalanine	3	3	4

[a] Determined by enzymic hydrolysis.
[b] Determined spectrophotometrically.

enzymes suggests that in these cases also the NH_2-terminal residue is probably substituted. In horse muscle C—P phosphatase experiments with hydrazynolysis, selective tritiation, digestion with carboxypeptidase A, and the analysis of tryptic peptides indicate that the COOH-terminal sequence is Arg-Tyr-COOH. Tyrosine is also the COOH-terminal amino acid in horse liver C—P phosphatase, but lysine is the COOH-terminal residue of the bovine brain enzyme. The data concerning the end groups of C—P phosphatase purified from various sources are summarized in Table 3.

Table 3 End Group Analysis of Carbamylphosphate Phosphatase

End Group	Horse Muscle	Horse Liver	Bovine Brain
C terminal	Glycine, tyrosine	Tyrosine	Lysine
N terminal	Glutamic acid, acetyl-serine	Not known	Not known

As mentioned above, C—P phosphatase hydrolyzes, in addition to C—P, other acyl phosphates, such as acetyl phosphate, 1, 3-diphosphoglycerate, benzoyl phosphate, and *p*-nitrobenzoyl phosphate, but shows very low hydrolytic activity on esteric, phosphoamidic, and pyrophosphoric bonds.

In Table 4 are reported the relative rates of enzymic hydrolysis and the K_m values with respect to different substrates. The hydrolytic activity toward synthetic compounds such as benzoyl and *p*-nitrobenzoyl phosphate has been used to develop easy and sensitive assay methods (18).

Table 4 Activity of Carbamylphosphate Phosphatase from Different Sources toward a Number of Compounds[a]

Substrate	Horse Muscle	Horse Liver	Bovine Brain	Pork Heart
Carbamyl phosphate	9×10^{-2}	1×10^{-1}	1×10^{-1}	9×10^{-2}
Acetyl phosphate	1	1	1	1
Benzoyl phosphate	6	7.5	—	—
3-Phosphoglyceryl phosphate	1.2	7×10^{-1}	11×10^{-2}	9×10^{-2}
Phosphocreatine	0	—	—	—
Adenosine-5′-triphosphate	3.7×10^{-4}	—	—	—
Pyrophosphate	1.3×10^{-4}	—	—	—
p-Nitrophenylphosphate	0	—	—	—
Phosphoenol pyruvate	0	—	—	—
K_m (*M*) carbamyl phosphate	5.2×10^{-2}	3.2×10^{-3}	6.4×10^{-3}	—
K_i (*M*) inorganic phosphate	1×10^{-3}	1.6×10^{-3}	—	—

[a] The rate of hydrolysis with acetyl phosphate is taken as unit.

Some metabolic compounds have been shown to inhibit C—P phosphatase activity in less pure preparations and also in the pure enzyme. The inhibitory effect exerted by many phosphorylated compounds indicates the presence of a site in the enzyme for the phosphate group. Horse muscle enzyme is moderately inhibited by simple phosphate esters such as methyl phosphate and benzyl phosphate with a competitive mechanism. The strong inhibition produced by inorganic phosphate could arise through hydrogen bonding of the free –OH group of phosphate to a base. Adenosine triphosphate and fructose 1, 6-diphosphate show noncompetitive inhibition. Horse liver C—P phosphatase is inhibited competitively by inorganic phosphate and noncompetitively by orotic acid (19). As regards the bovine brain enzyme, it is not inhibited by $HgCl_2$ nor by iodoacetate at a concentration of 40 and 4 m*M* respectively. Under certain conditions 8 *M* urea produces an inhibition of enzyme activity of 100%. An enhancement of the enzyme activity of this enzyme by 15 m*M* KCl and 5 m*M* $MgCl_2$ has been demonstrated.

In a recent study we showed that horse muscle enzyme is inhibited by pyridoxal-5′-phosphate (20). We were led to study the effect of this compound by the relative abundance of lysine residues found in the enzyme; in fact, pyridoxal phosphate has been extensively used to study the role of lysyl ε-amino groups in the action of many enzymes, because it can easily react with these groups to form Schiff bases. The result of our study indicated that the inhibition by pyridoxal phosphate is purely competitive and pH-dependent, reaching a maximum at pH 7.6. At this pH we found that free lysine, although it has little effect on normal enzyme activity, when it is added to pyridoxal phosphate–inhibited enzyme it can produce a time-dependent reversal of the inhibition. This indicates that lysine competes with the enzyme for pyridoxal phosphate and represents an indirect evidence that the action of this compound is related to its binding with lysine residues of the enzyme. Furthermore, the formation of an enzyme–pyridoxal phosphate complex was confirmed by spectral data (which demonstrated absorption maxima in the 410 to 430 nm region attributable to the presence of a protonated Schiff base) and by the identification of ε-pyridoxyl lysine residues in the modified enzyme. All these data suggest that there is at least one lysine residue at the active site of the enzyme which is susceptible to the interaction with pyridoxal phosphate. Since no inhibition was observed when pyridoxal was incubated with the enzyme, the phosphate group of pyridoxal phosphate appears to play an essential role in the inhibition process, probably directing the molecule to the active site whereupon a Schiff base is formed between lysine residues and the inhibitor.

In regard to the hydrolysis mechanism of C—P (21), Allen and Jones demonstrated that the decomposition of C—P in solution proceeds by different paths, depending on the pH; these pathways have been referred to as acid-base–catalyzed hydrolysis. In acid medium the reaction is the following: $NH_2CO_2PO_3H_2 + H_3O^+ \rightarrow NH_4^+ + CO_2 + H_3PO_4$; whereas, in a more alkaline medium, the hydrolysis of C—P may be represented as $NH_2CO_2PO_3^= + OH^- \rightarrow NCO^- + HPO_4^= + H_2O$. As is evident, the first type of hydrolysis leads to the simultaneous production of ammonia, carbon dioxide, and inorganic phosphate, but the second type of hydrolysis yields cyanate and phosphate. In the latter case, since the half-life of cyanate is seven to eight times longer than that of C—P, there is an initial accumulation of cyanate, and, as this compound decomposes, ammonia is produced. Thus, at low pH, the rates of ammonia and phosphate production are equal, but at higher pH values the release of ammonia is lower with respect to the release of inorganic phosphate. For example, at pH 6, the release of ammonia is about 15% of the rate of phosphate formation. In work performed in 1971 with Diederich and Grisolia (22), we attempted to determine whether the enzymatic hydrolysis of C—P involved the production of cyanate, having in mind that in the pH range of C—P phosphatase activity there could be the intermediate formation of this compound. In order to ascertain this possibility we incubated C—P at pH 5.0, 7.4, and 8.1 with enzyme preparations from bovine brain, bovine liver, and horse muscle. We also investigated the production of cyanate by the semidine modification of the diacetyl monoxime method, after conversion to hydroxyurea and the formation of ammonia with a large excess of glutamate dehydrogenase. The results of this study indicated that all the enzymatic preparations, at all the pH values assayed, hydrolyze C—P without the intermediate formation of cyanate. It remains to be clarified whether there is intermediate production of carbamate or a direct splitting of C—P into ammonia and carbon dioxide.

As regards the physiological function of C—P phosphatase, it is obvious that this enzyme, by splitting C—P and other acylphosphates, can play a regulatory role in all the metabolic pathways in which these kinds of compounds are involved. Keeping in mind that this enzyme is prevalently located in the cytosol, glycolysis and pyrimidine biosynthesis appear to be the pathways more susceptible to this control.

With regard to pyrimidine biosynthesis, we demonstrated in our laboratory that orotic acid inhibits the hydrolysis of C—P by the liver enzyme with a noncompetitive mechanism and with a K_i of 0.9×10^{-3} *M*.

In addition, some *in vivo* experiments (23) indicated that a diet enriched by orotic acid produced an increase in liver aspartate transcarbamylase and dihydroorotase, in spite of the high levels of free pyrimidine nucleotides. All these data suggest that orotic acid may be involved in a complicated feedback mechanism in which this compound stimulates the enzymes related to its synthesis and inhibits C—P phosphatase, which by splitting C—P, could act as a rate-limiting factor in this metabolic pathway. Both these actions could contribute to the same effect: the maintenance of a high rate of orotic acid biosynthesis. However, in addition to playing a regulatory role in specific metabolic pathways, C—P phosphatase could have a more general function, by controlling the intracellular levels of an important chemotropic effector, such as C—P. As reported by Grisolia and Hood (24) in their excellent and comprehensive review, carbamylation is one or the main chemotropic modifications which may affect several kinds of proteins, including enzymes (glutamic dehydrogenase, pepsin, subtilisin, pyruvate carboxylase), hormones (insulin, glucagon, oxytocin), and other proteins (casein, hemoglobin, eye lens α-crystallin and histones). Since carbamylation can result in a modification of the functional properties of proteins (generally with an inhibitory effect), C—P phosphatase may perhaps exert a physiological function in preventing the accumulation of deleterious levels of carbamyl-phosphate, especially in sensitive areas, that is, in those tissues which lack most, if not all, the enzymes for urea synthesis.

As regards more specifically the histones, the modification of these proteins by group substitution reaction appears to be one of the major new aspects of nuclear protein metabolism. Most of these reactions, such as acetylation, methylation, and phosphorylation are enzymatically catalyzed and may have biological significance, due to the capacity of histones to inhibit ribonucleic acid biosynthesis.

A few years ago, Grisolia and myself (25) demonstrated that 1, 3-diphosphoglycerate and carbamylphosphate can readily react with histones, especially with F_1 histones, to give 3-phosphoglyceryl and carbamyl histones, and we suggested that these nonenzymatic modifications might be of great biological interest, owing to the large metabolic occurrence of such compounds.

In a subsequent study we demonstrated homocitrulline formation following nonenzymatic carbamylation of histones. The carbamyl groups were attached solely to the ε-amino nitrogen of lysine residues (26).

In order to ascertain the biological significance of this modification, we studied the effect of carbamylated lysine-rich and arginine-rich histones on *in vitro* mammalian RNA biosynthesis in comparison with the corresponding native histones (27).

For this research, DNA-dependent RNA polymerase was purified from rat liver; native and carbamylated histones were preincubated with denaturated DNA before the addition of nucleotide triphosphate.

The results obtained indicated that the carbamylation of arginine-rich histones can lower their effectiveness as inhibitors of RNA biosynthesis (Fig. 1). A similar effect, though less marked, was observed with F_1 histone.

In other work (28) carried out with the collaboration of the Institute of Anatomy of the University of Florence, we studied the effect of native and carbamylated histones on the development of embryonic chicken lung cultured *in vitro*. From the results, it appeared that native histones, when added to the culture medium, changed the normal histological development of embryonic tissues. The modifications were essentially represented by a lesser degree of bronchial branching and by a decrease of epithelial and mesenchimal mytoses. On the other hand, when carbamylated histones were added, the cultures showed histological features more similar to the controls (Fig. 2).

Finally, it should be pointed out that cyanate carbamylates extensively amino acid *t*RNA from rat liver and from yeast with preferential carbamylation of arginine (29).

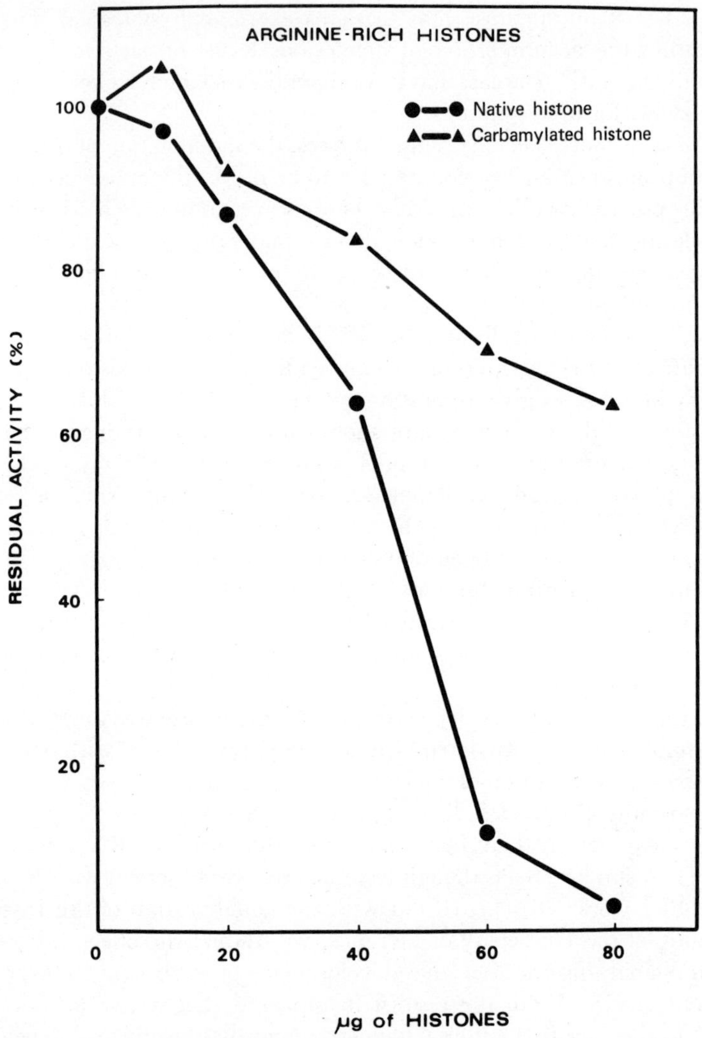

Figure 1. Preincubation of histories with DNA

In conclusion, on the basis of the data issuing from our research and from the studies of other authors, we believe that the carbamylation of histones may represent an important mechanism of modulation in their effect on gene expression and protein biosynthesis and that carbamyl-phosphate phosphatase through control of the concentration of this subtrate could prevent carbamylation with all the consequent implications.

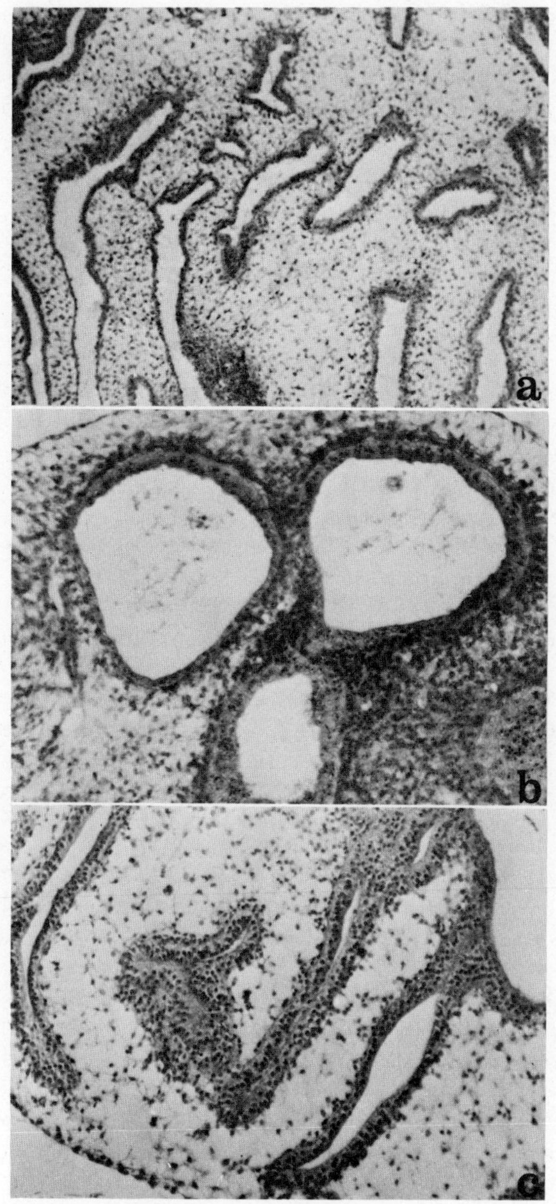

Figure 2. Embryonic chicken lung cultured *in vitro*. (*A*) Standard culture medium. (*B*) Standard culture medium plus lysine-rich histone. (*C*) Standard culture medium plus carbamylated lysine-rich histone. (From Balboni, Tedde Piras, and Ramponi.)

REFERENCES

1. Grisolia, S., and Marshall, R. O. (1954). *Biochim. Biophys. Acta* **14**, 446.
2. Jones, M. E., Spector, L., and Lipmann, F. (1955). *J. Amer. Chem. Soc.* **77**, 819.
3. Grisolia, S., Wallach, D. P., and Grady, H. J. (1955). *Biochim. Biophys. Acta* **17**, 150.
4. Krebs, H. A., Jensen, P. K., and Eggleston, L. V. (1958). *Biochem. J.* **70**, 397.
5. Diederich, D. A., and Grisolia, S. (1969). *J. Biol. Chem.* **244**, 2412.
6. Ramponi, G., Guerritore, A., Treves, C., Nassi, P., and Baccari V. (1969). *Arch. Biochem. Biophys.* **130**, 362.
7. Lipmann, F. (1946). *Adv. Enzymol.* **6**, 231.
8. Shapiro, S., and Wertheimer, E. (1945). *Nature* **156**, 690.
9. Guerritore, A., Zanobini, A., and Ramponi, G. (1959). *Boll. Soc. It. Biol. Sper.* **35**, 2163.
10. Melani, F., Ramponi, G., and Guerritore, A. (1961). *Boll. Soc. It. Biol. Sper.*, **37** 1268.
11. Grisolia, S., and Hood, W., (1974). *FEBS Letters*, **42** 246.
12. Yoshida, H., Izumi, F., and Nagai, K. (1966). *Biochim. Biophys. Acta* **120**, 183.
13. Herzfeld, A., and Knox, W. E. (1972). *Cancer Research*, **32**, 1837.
14. Ramponi, G., Nassi, P., Cappugi, G., Treves, C., and Manao, G. (1972). *Biochim. Biophys. Acta*, **284** 485.
15. Diederich, D., and Grisolia, S. (1971). *Biochim. Biophys. Acta*, **227**, 192.
16. Ramponi, G., Cappugi, G., Treves, C., and Nassi, P. (1971). *Biochemistry* **10**, 2082.
17. Cappugi, G., Chellini, P. C., Nassi, P., and Ramponi, G. (1976). *Int. J. Peptide Protein Res.*, in press.
18. Ramponi, G., Treves, C., and Guerritore, A., (1966). *Experientia*, **22** 705.
19. Nassi, P., Cappugi, G., Niccoli, A., and Ramponi, G. (1973). *Physiol. Chem. Physics*, **5**, 109.
20. Ramponi, G., Manao, G., Camici, G., and White, G. F. (1975). *Biochim. Biophys. Acta*, **391**, 486.
21. Allen, Jr., C. M., Richelson, E., and Jones, M. E. (1966). In N. O. Kaplan and E. P. Kennedy, Eds., *Current Aspects of Biochemical Energetics*, p. 401.
22. Diederich, D., Ramponi, G., and Grisolia, S. (1971). *FEBS Letters*, **15**, 30.
23. Bresnich, E., Mayfield, E. D., and Mosse, H. (1968). *Mol. Pharmacol.*, **4**, 173.
24. Grisolia, S., and Hood, W. (1972). In E. Kun and S. Grisolia, Eds., *Biochemical Regulatory Mechanisms in Eukaryotic Cells*, p. 138.
25. Ramponi, G., and Grisolia, S. (1970). *Biochem. Biophys. Res. Commun.*, **38**, 1056.
26. Ramponi, G., Leaver, J. L., and Grisolia, S. (1971). *FEBS Letters*, **16**, 311.
27. Ramponi, G., Treves, C., Manao, G., Camici, G., and Cappugi, G. (1973). *9th International Congress of Biochemistry*, Stockholm, 1–7 July, 3i22.
28. Balboni, G. C., Tedde Piras, A., and Ramponi, G. (1975). *Bull. Ass. Anat. 59ᵉ Congr. Bordeaux.*
29. Rudzinska, M., Wallace, R., Escarmis, C., and Grisolia, S. (1972). *Physiol. Chem. Phys.* **4**, 527.

14

Ornithine Transcarbamylase from Bovine Liver

Margaret Marshall
Department of Physiological Chemistry
University of Wisconsin
Madison, Wisconsin

Grisolia and Cohen (1) first separated ornithine transcarbamylase from carbamyl phosphate synthetase by heat inactivation* of the synthetase in an extract of the particulate fraction of rat liver and demonstrated that the transcarbamylase catalyzes the transfer of the carbamyl group from an intermediate, subsequently identified as carbamyl phosphate (2), to the δ -amino group of ornithine. Reichard (3) isolated essentially pure ornithine transcarbamylase from rat liver; however, there are only a few tenths of a milligram of enzyme per gram liver, so that rat liver is not a convenient source from which to isolate sufficient enzyme to study its properties in detail. A procedure for the isolation of a highly active preparation of the enzyme from bovine liver was reported by Burnett and Cohen (4); this has been used frequently as a source of the enzyme for the assay of carbamyl phosphate. The starting point for both of these preparations was an acetone powder of the particulate fraction of liver. Marshall et al. (5) found that both the transcarbamylase and synthetase could be extracted from the particulate fraction in higher yield and with less extraneous protein by the detergent, cetyltrimethylammonium bromide, and this procedure was then used by Brown and Cohen (6) to prepare an

The studies reported in this paper from the author's laboratory were supported in part by Grant CA-03571 from the National Institutes of Health.

* The relative stability to heat (60 to 65°) is a property common to all ornithine transcarbamylases that have been tested, including those from bacteria and plants.

extract from whole liver for the quantitative assay of all the urea-cycle enzymes. Pure ornithine transcarbamylase has been isolated from a detergent extract of the particulate fraction of bovine liver (7). It is this enzyme that has been studied in the greatest detail of any from a higher animal and, therefore, will be discussed here.

QUATERNARY STRUCTURE

Isoelectric focusing and polyacrylamide gel electrophoresis revealed that the preparation of ornithine transcarbamylase from bovine liver is composed of a number of enzymatically active species, possibly arising from hydrolysis of amide groups, since they are present in progressively decreasing amounts with increase in the net negative charge and have isoelectric points consistent with this interpretation (pI of major component 6.8) (7). These have been separated by chromatography on Whatman DE-52 (Fig. 1) and shown to all have the same specific activity. The unfractionated enzyme is homogeneous with respect to molecular weight (108,000 by sedimentation equilibrium in nondenaturing media and 36,000 in 6 *M* Gnd-HCl (7)) and only a single component, molecular weight 37,000, was detected on gel electrophoresis in

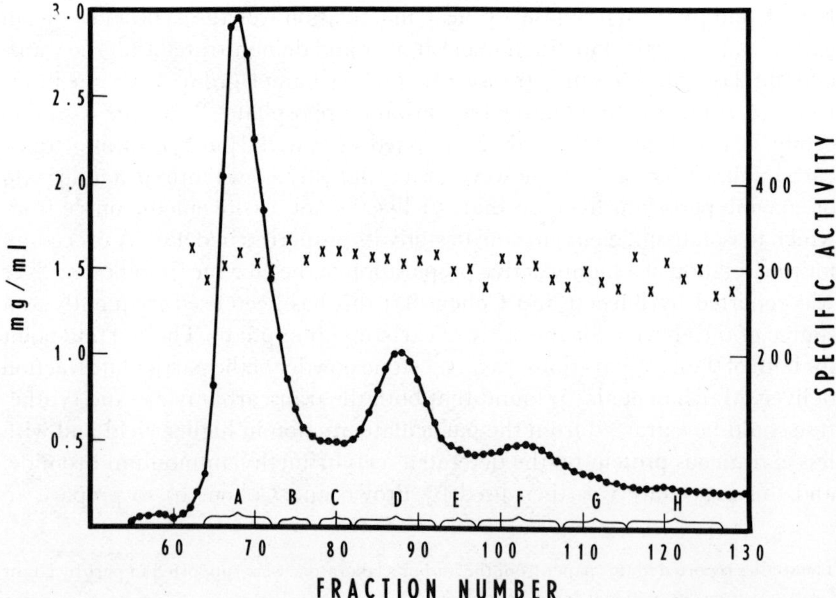

Figure 1. Chromatogram of mono-S-carboxamidomethyl bovine ornithine transcarbamylase on Whatman DE-52 at pH 8.3 (8).

sodium dodecyl sulfate of the separated fractions. Furthermore, asx and lys-phe were recovered quantitatively from the N and C terminus, respectively. Therefore, the inhomogeneity is not due to proteolysis. The major component, which has the lowest mobility, was homogeneous on polyacrylamide gel electrophoresis in the presence of urea; the next, in order of increasing negative charge, contained a second, faster migrating component in addition to the one with the mobility of the major fraction; and the remaining fractions contained, in varying proportions, one or two additional components equally spaced and, therefore, presumably differing by equal increments of charge, a result consistent with the hydrolysis of one or two amide groups from individual subunits (8).

The conclusion that bovine ornithine transcarbamylase is a trimer (7) was disconcerting, because at the time there was only one other well-documented example of a trimer; since then ornithine transcarbamylase from *E. coli* (9) and yeast (10) have been shown to be trimers with subunit molecular weights ranging from 35,000 to 36,000. Rat ornithine transcarbamylase has the same $s_{20,w} = 5.5$ S as the bovine enzyme. Its quaternary structure has not been reported.

BINDING OF LIGANDS

Binding was measured at 25° by gel filtration through a column of Sephadex equilibrated with the desired concentration of ligand, a procedure that could be carried out rapidly enough to ensure minimal decomposition of carbamyl phosphate ($<3\%$). The enzyme forms tight binary complexes with carbamyl phosphate but not with ornithine ($K_{diss} > 5$ mM) or its competitive inhibitor norvaline. The binding of carbamyl phosphate is a completely reversible process not accompanied by hydrolysis of carbamyl phosphate or carbamylation of the enzyme, a result consistent with Reichard's observation that rat ornithine transcarbamylase does not catalyze an exchange between carbamyl phosphate and ^{14}C-citrulline (3).

In the presence of norvaline carbamyl phosphate is bound more tightly than in the binary complex, so that it is possible to measure the stoichiometry directly. The result, 1 mole of carbamyl phosphate bound per 38,000 g, is consistent with one site per subunit. Both in the presence and absence of norvaline the titration curves for the binding of carbamyl phosphate have more than one inflection point and require a minimum of three dissociation constants to fit the data (Fig. 2). On the other hand, the data for the binding of norvaline in the presence of carbamyl phosphate can be fitted by a single dissociation constant, 20 μM at pH 7.9. Ornithine, but not norvaline, is bound to the enzyme-phosphate complex (11).

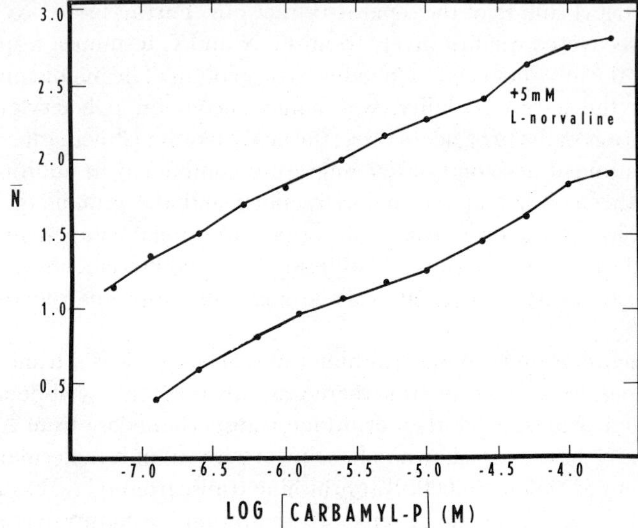

Figure 2. Titration of bovine ornithine transcarbamylase with [^{14}C]carbamyl phosphate at pH 7.9, μ = 0.11. Lower curve calculated for K_1 = 0.18 μM, K_2 = 25 μM, and K_3 ~ 3.0 mM; upper curve calculated for K_1 ~ 2 nM, K_2 = 0.25 μM, and K_3 = 30 μM (11).

KINETICS

The kinetics of bovine ornithine transcarbamylase have been studied in detail only at 25° (11). V_m and K_{orn}, determined at a saturating concentration of carbamyl phosphate as a function of pH, are listed in Table 1. The variation of V_m conforms to a titration curve for a group with pK 6.65. K_{orn^0}, calculated for the species of ornithine with zero net charge, remains constant up to pH 8.0 and then increases with further increase in pH.* Essentially the same result was obtained by Snodgrass (13) with human ornithine transcarbamylase studied at 37°. If the reactive species of ornithine is that with zero net charge, it is probably the form with the δ-amino group unionized (10% of the total isoelectric form), since only analogs with a charged α-amino group are effective competitive inhibitors (11).

The ordered binding of the substrates suggests that the reaction proceeds by an ordered, sequential mechanism, which gives linear, intersecting double reciprocal plots of activity against the concentration of substrates, provided

* The apparent excess of ornithine transcarbamylase activity in livers of ureotelic animals (12) relative to that of carbamyl phosphate synthetase (~ 10 fold), which is found when each enzyme is assayed under optimal conditions, is deceiving, since the transcarbamylase must function both below its pH optimum and at concentrations of its substrates probably less than their K_m's.

Table 1 Effect of pH on V_m and K_{orn} for Bovine Ornithine Transcarbamylase[a]

pH	K_{orn} (mM)	K_{orn^0} (μM)	$V_{f,\,exptl}$ (s^{-1})	$V_{f,\,calc}$ (s^{-1})
6.14	10	22	41	42
6.75	2.85	25	101	105
7.10	1.4	27	146	138
7.88	0.24	26	180	177
8.47	0.14	45	187	187
8.90	0.14	78	187	187

[a] Rates, expressed as turnover per subunit, were measured at 10 mM carbamyl phosphate in 0.1 M tetramethylammonium methylsulfonate. $V_{f,calc}$ are the rates relative to that at pH 8.90, calculated on the assumption that the active species is the unprotonated form of a group with pK 6.65 (11).

the active sites are identical and function independently. However, the necessity for three dissociation constants to describe the binding of carbamyl phosphate means that this restriction does not apply to bovine ornithine transcarbamylase. And, in fact, when the concentration of carbamyl phosphate was varied at a concentration of ornithine close to saturating at a particular pH, reciprocal plots of activity against the concentration of carbamyl phosphate were concave downward at every pH (Fig. 3). This was also true of the reciprocal plots of rates versus the concentration of carbamyl phosphate at several fixed, nonsaturating concentrations of ornithine *. There is direct evidence, to be presented below, that each site is potentially active and has approximately the same turnover number at saturating concentrations of substrates.

Phosphate inhibits competitively with respect to carbamyl phosphate both through the binary complex and the ternary complex, E-P_i-orn, which predominates at high concentrations of ornithine. The combination of the inhibition by ornithine and an equilibrium that lies overwhelmingly in the direction of citrulline synthesis† makes it difficult to study the reverse reaction. However, the arsenolysis of citrulline, first observed by Krebs et al. (14) and subsequently found to be catalyzed by ornithine transcarbamylase (3), has been used clinically as an assay for the enzyme (15).‡

* These results were obtained with the unfractionated enzyme, but identical results have since been obtained with an electrophoretically homogeneous preparation.

† K_{Eq} = (cit°) (HPO_4^{2-}) (H^+)/(orn^+) (CP^{2-}) is 7mM, so that at pH 7.9 $K_{app\ Eq} = 6 \times 10^5$ (11).

‡ First evidence for the arsenolysis of citrulline with rat liver ornithine transcarbamylase was presented by S. Grisolia and R. O. Marshall (*Amino Acid Metabolism*, 258, 1955).—Editor

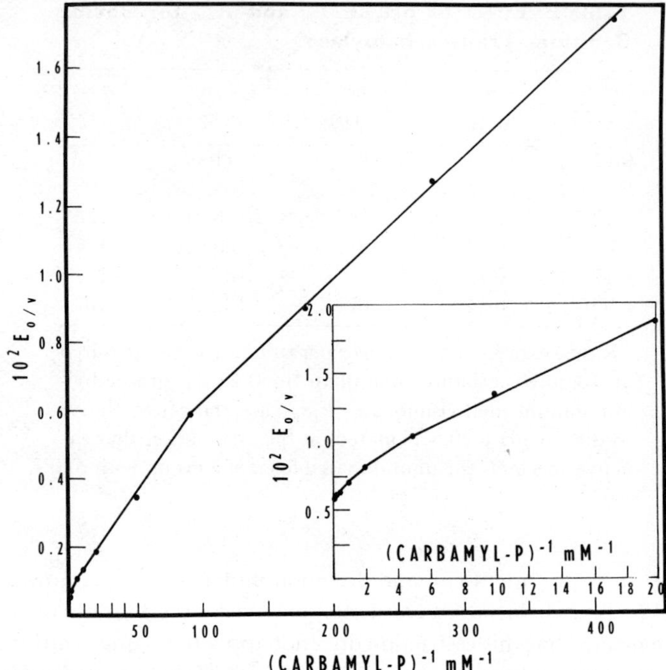

Figure 3. Double reciprocal plot of the velocity against the concentration of carbamyl phosphate for bovine ornithine transcarbamylase. Concentration of ornithine, 5 m*M*; buffer, 0.1 *M* tetramethylammonium methylsulfonate, 20 m*M* triethanolamine methylsulfonate, pH 7.9 (11).

SPECIFICITY

At pH 8.0 no analog of ornithine tested, including L-2, 4-diaminobutyric acid, L-lysine, and D-ornithine, is carbamylated, but at pH > 9.0 homocitrulline is synthesized slowly from lysine. L-Norvaline is by far the most effective competitive inhibitor of L-ornithine. One less methyl group (L-α-aminobutyric acid) or an OH group on the δ-carbon interfere with binding (11). Acetyl phosphate, which will replace carbamyl phosphate as a substrate (16), yielding δ-acetylornithine, is not an effective inhibitor with respect to carbamyl phosphate at high concentrations of ornithine, possibly because it does not provide the tight site for ornithine that phosphate does.

SULFHYDRYL GROUPS

Bovine ornithine transcarbamylase contains three classes of sulfhydryl groups, the stoichiometry of which agrees with one of each per subunit (17). The residues in the first category, termed "nonessential," react readily with all the reagents tested without affecting the activity and appear to be easily oxidized to form the dimer, so that in the absence of a thiol a second component, $s_{20,w} = 8.0$, is seen on ultracentrifugation. They can be alkylated with iodoacetamide under mild enough conditions to confine the reaction to these residues, and most of the experiments on chemical modification to be discussed here have been done with this derivative. The residues in the second category, termed "essential," react more slowly and selectively than those in the non-essential group, in general with loss of activity that is prevented by the presence of carbamyl phosphate. They can be reversibly protected by oxidation with cystamine. Substitution of the thioethylamine group did not affect the binding of carbamyl phosphate but did prevent that of norvaline; whereas, substitution of 2-nitro-5-thiobenzoate by reaction with Ellman's reagent both increased the dissociation constants for carbamyl phosphate and prevented the binding of norvaline. The *S*-thioethanol derivative was active, although the concentrations of the substrates required for saturation were greatly increased, and, therefore, these sulfhydryl groups cannot be required for catalysis. The residues in the third category, termed "buried," react only in the presence of denaturing agents.

The thiocyano-derivative of the monoalkylated enzyme was cleaved in 6 *M* Gnd-HCl at pH 9.0 (8) to form the three primary peptides expected if the subunits are identical plus the two overlap peptides that were present because the derivatization was not complete, particularly at the buried sulfhydryl groups. The peptides were separated both by gel electrophoresis in sodium dodecylsulfate and by gel filtration through Sephadex G-100 in 6 *M* Gnd-HCl. The sum of the amino acid compositions of the primary peptides is in good agreement with that of the uncleaved enzyme. Their alignment in the sequence, shown in the diagram, is based on the following results: (*a*) Only peptides A and B + C were present after alkylation of the "essential" sulfhydryl group (17); therefore, it is located between B and C. (*b*) All of the peptides except A were labeled in the derivative prepared with [cyano-^{14}C] 2-nitro-5-thiocyanobenzoic acid; therefore, A is at the N-terminus.* (*c*) Peptide C contains the C-terminal lys-phe.

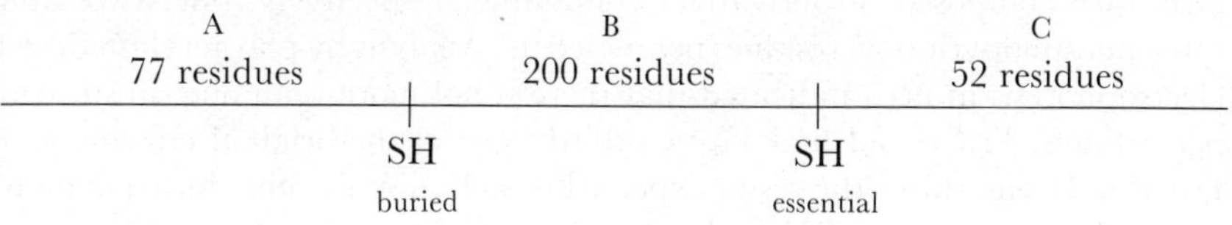

* Cleavage leaves the label attached to the NH_2-terminal residue of the peptide (18).

REACTION WITH PYRIDOXAL PHOSPHATE

Substitution of a single phosphopyridoxyl residue per subunit on a lysyl residue in peptide A caused nearly complete loss of activity ($>98\%$), and the presence of a saturating concentration of carbamyl phosphate prevented the reaction from occurring (8).* Figure 4 shows a chromatogram of the derivative

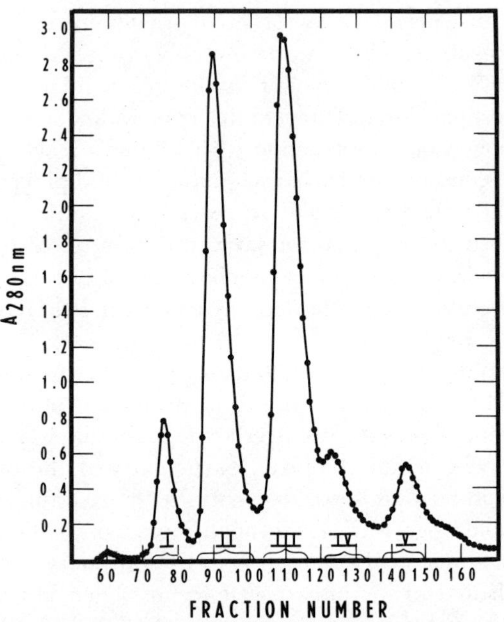

Figure 4. Chromatogram of the phosphopyridoxyl derivative of bovine ornithine transcarbamylase (Fraction A) on Whatman DE-52 at pH 8.3 (8).

prepared from an electrophoretically homogeneous species of ornithine transcarbamylase by reduction of the Schiff's base with $NaBH_4$. The first three peaks are composed of derivatives containing, respectively, one, two, and three phosphopyridoxyl residues per molecule. Analysis by polyacrylamide gel electrophoresis in urea indicated that there is not more than one substituent per subunit. Fraction I had V_m two-thirds that of the original enzyme and Fraction II one-third, the result expected if each subunit contributed equally to the activity. Fraction III had only 4% of the original activity at low con-

* This is not due to reaction of pyridoxal phosphate with the cyanate present in the carbamyl phosphate, which does occur but much too slowly to account for the protection.

centrations of ornithine (probably due to contamination with underivatized subunit) but was active at much higher concentrations of ornithine ($K_{orn} \sim$ 66 mM compared with 0.16 mM for the underivatized enzyme and Fractions I and II). It bound carbamyl phosphate so poorly that it was not possible to determine the stoichiometry (assuming a single site, $K = 50\ \mu M$). Fraction II had one site for carbamyl phosphate, $K = 2.3\ \mu M$, presumably at the underivatized subunit. If the binding sites for carbamyl phosphate were intrinsically different because of their location within the quaternary structure, then, provided pyridoxal phosphate cannot discriminate among them (see below), one might expect a population of molecules with structurally dissimilar binding sites in the derivative even though any one molecule has only a single site. On the other hand, if the complexity of the binding is due to anticooperativity then the population of molecules with only a single site remaining should be homogeneous, as appears to be the case.

The kinetics and equilibrium of inactivation were measured after reduction of the Schiff's base with $NaBH_4$.* Since the inactivation is due to reaction at a single, apparently unique lysyl residue per subunit and each site contributes approximately equally to the activity, the loss of activity should be a measure of the extent of Schiff's base formation at that lysyl residue. This assumption was supported by the constant values found for K_{eq} of the Schiff's base at several concentrations of pyridoxal phosphate. The rates of inactivation, which were constant for more than three half-lives, showed saturation kinetics with respect to the concentration of pyridoxal phosphate (Fig. 5). This suggests that the reaction proceeds according to equation 1:

$$\mathrm{E + P} \overset{K_1}{\rightleftharpoons} \mathrm{E{\cdot}P} \underset{k_r}{\overset{k_f}{\rightleftharpoons}} \mathrm{EP} \qquad (1)$$

where E·P is probably a noncovalent complex (19). K_1 ($10^3\ M^{-1}$) is approximately constant between pH 6.0 and 8.0, as is k_r (0.01 minute^{-1}), either measured directly or calculated from $K_{eq} = K_1 k_f / k_r$; whereas, K_{eq} ($10^5\ M^{-1}$ at pH 7.0) and k_f increase with increase in pH. The results suggest at least two reasons for the specificity of the reaction: the noncovalent binding, which probably occurs primarily through the phosphate, since both P_i and carbamyl phosphate are competitive inhibitors of the reaction; and the slow rate of hydrolysis of the imine.† It is also possible that the pK of the lysyl residue is unusually low,‡ although there is no evidence for this in the reaction of the

* The rate of the reverse reaction is negligible compared with the rate of reduction.

† At pH 7.4 this is 2% of the rate of hydrolysis of the Schiff's base of α-aminobutyrate with pyridoxal (20) and 20% of that of the Schiff's base of lys-97 in glutamate dehydrogenase with pyridoxal phosphate (19).

‡ A value of 8.4 gives an approximate fit to the variation of K_{eq} with pH if values for the pK's of the pyridinium-N and phosphate groups in the imine, which are also unknown, are assumed to be the same as those calculated for the Schiff's base of glutamate dehydrogenase (19).

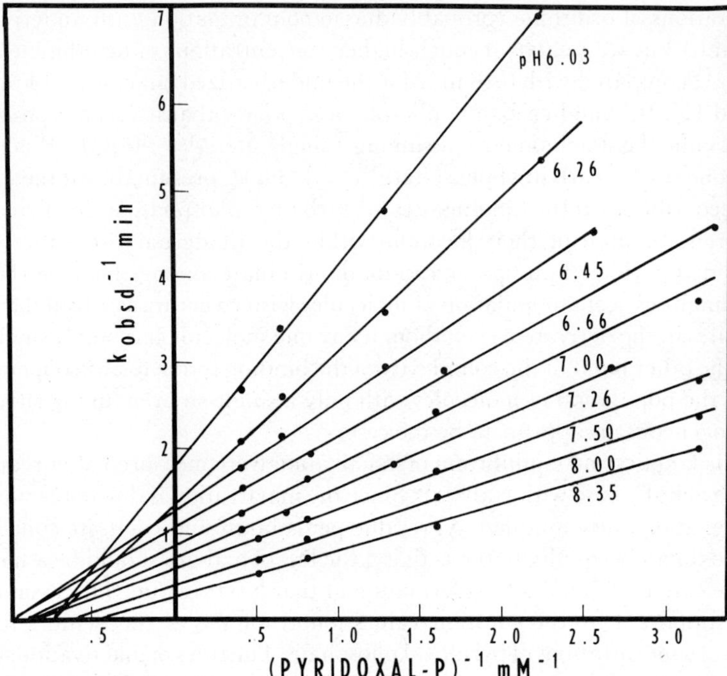

Figure 5. Double reciprocal plot of the rate constant for the inactivation of bovine ornithine transcarbamylase by pyridoxal phosphate against the concentration of pyridoxal phosphate at the indicated pH values (8).

enzyme with other amino group reagents. Hydroxymercuribenzoate, which reacts with the essential sulfhydryl group (11) prevents the inactivation by pyridoxal phosphate, even though the lysyl and sulfhydryl groups are at opposite ends of the polypeptide chain.

COMPARATIVE ASPECTS

Ornithine transcarbamylase from *Streptococcus faecalis* is in considerable detail similar to the bovine enzyme except for those features dependent on the quaternary structure (11, 17). In particular, it has a single essential sulfhydryl group at the C-terminal end of the polypeptide chain with similar reactivity and is inactivated by reaction of a single lysyl residue at the N-terminal end with pyridoxal phosphate (8). The catalytic subunit of aspartate transcarbamylase from *E. coli* also possesses a single essential sulfhydryl group (21) and a single lysyl residue per polypeptide chain that reacts with pyridoxal phosphate with loss of activity (22).

REFERENCES

1. Grisolia, S., and Cohen, P. P. (1952). *J. Biol. Chem.* **198**, 561–571.
2. Jones M. E., Spector, L., and Lipmann, F. (1955). *J. Am. Chem. Soc.* **77**, 819–820.
3. Reichard, P. (1957). *Acta Chem. Scand.* **11**, 523–536.
4. Burnett, G. H., and Cohen, P. P. (1957). *J. Biol. Chem.* **229**, 337–344.
5. Marshall, M., Metzenberg, R. L., and Cohen, P. P. (1958). *J. Biol. Chem.* **233**, 102–105.
6. Brown, G. W., and Cohen, P. P. (1959). *J. Biol. Chem.* **234**, 1769–1774.
7. Marshall, M., and Cohen, P. P. (1972). *J. Biol. Chem.* **247**, 1641–1653.
8. Marshall, M., and Cohen, P. P., unpublished results.
9. Legrain, C., Halleux, P., Stalon, V., and Glansdorff, N. (1972). *Eur. J. Biochem.* **27**, 93–102.
10. Simon, J. P., Penninckx, M., and Wiame, J. K., unpublished; quoted in (9).
11. Marshall, M., and Cohen. P. P. (1972). *J. Biol. Chem.* **247**, 1654–1668.
12. Brown, G. W., and Cohen, P. P. (1960). *Biochem. J.* **75**, 82–91.
13. Snodgrass, P. J. (1968). *Biochemistry* **7**, 3047–3051.
14. Krebs, H. A., Eggleston, L. V., and Knivett, V. (1955). *Biochem. J.* **59**, 185–193.
15. Reichard, H., and Reichard, P. (1958). *J. Lab. Clin. Med.* **52**, 709–717.
16. Grisolia, S., and Harmon, P. (1962). *Biochim. Biophys. Acta* **59**, 482–483.
17. Marshall, M., and Cohen, P. P. (1972). *J. Biol. Chem.* **247**, 1669–1682.
18. Catsimpoolas, N., and Wood, J. L. (1966). *J. Biol. Chem.* **241**, 1790–1796.
19. Piszkiewicz, D., and Smith, E. L. (1971). *Biochemistry* **10**, 4544–4552.
20. Olivo, F., Rossi, C. S., and Silipeandi, N. (1963). In E. E. Snell, P. M. Fasella, A. Braunstein, and A. Rossi Fanelli, Eds., *Chemical and Biological Aspects of Pyridoxal Catalysis*, Pergamon, New York, pp. 91–101.
21. Vanaman, T. C., and Stark, G. R. (1970). *J. Biol. Chem.* **245**, 3565–3573.
22. Greewell, P., Jewett, S. L., and Stark, G. R. (1973). *J. Biol. Chem.* **248**, 5994–6001.

15

Enzymes of Arginine and Urea Synthesis

Sarah Ratner
The Public Health Research Institute
of the City of New York, Inc.
New York, New York.

HISTORICAL INTRODUCTION

The addition of NH_3 and CO_2 to ornithine to form citrulline, and of NH_3 to citrulline to form arginine, and the hydrolysis of arginine to ornithine and urea constituted the sequence originally proposed in 1932 by Krebs and Henseleit as successive steps in the formation of urea by a process of cyclical turnover in the mammalian liver. In respiring liver slices, NH_3 was by far the most active nitrogen source in the presence of lactate and small amounts of ornithine or citrulline or arginine (1, 2). The role of ornithine and citrulline as intermediates in the cycle was corroborated further when Gornal and Hunter showed that ornithine and citrulline were equally effective catalysts as theory required and that, under certain conditions, citrulline could accumulate (3).

In 1941 studies by Borsook and Dubnoff on respiring kidney slices revealed that either aspartic or glutamic acid, but not NH_3, were effective nitrogen donors for the conversion of citrulline to arginine. These investigators proposed an oxidative mechanism of nitrogen transfer involving glutamic acid (4). The nitrogen donor specificity in liver was reappraised in 1946 when Cohen and Hayano succeeded in obtaining arginine synthesis in respiring liver homogenates. On finding that glutamate was the preferred nitrogen donor, they

Research reported from the author's laboratory and preparation of this article were supported by Public Health Service Grant AM-03428 from the National Institute of Arthritis and Metabolic Diseases.

considered aspartate to be active only insofar as it could be converted to glutamate and therefore adopted the oxidative "transimination" mechanism of Borsook and Dubnoff to explain the behavior of liver homogenates. Their reports raised questions as to why the transfer of nitrogen from glutamate to citrulline should require a dehydrogenation step and as to how amino nitrogen from the general amino acid pool would enter the ornithine cycle at step 2.

It became evident that the conversion of citrulline to arginine was more complex with respect to both nitrogen donor and mechanism of nitrogen transfer than could be inferred from the nitrogen requirements for urea formation in liver slices. We hoped that further insight into the nature of the mechanism of nitrogen transfer might be gained through the study of the isolated system.

Early Evidence for Role of Aspartate and ATP

With a soluble liver preparation, free of pyridine nucleotides, it was possible to show that arginine was formed anaerobically from citrulline and aspartic acid in the presence of Mg^{++}, ATP, and an ATP-generating system. Under these conditions (Table 1) glutamic acid was active only when oxaloacetic acid was present simultaneously (6). Thus one-half of the urea nitrogen originates in aspartic acid, and the formation of the new C—N bond required the stoichiometric participation of phosphate bond energy in the form of ATP.

Table 1 Anaerobic Synthesis of Arginine from Citrulline in Extracts of Bovine Liver

Micromoles Substrate Added					Micromoles Found	
DL-Aspartate	L-Glutamate	Oxaloacetate	L-Citrulline	L-PGA[a]	Urea[b]	L-Malic acid
40			20	30	9.3	9.5
	20		20	30	0.2	0.2
	20	30	20	30	8.6	9.0
		30	20	30	0.1	0.0
	20	30		30	0.0	0.2
40				30	0.0	0.2
40			20		0.9	

[a] When L-phosphoglyceric acid was replaced by 8 μmole of ATP, 3.2 μmole of urea were found. Each tube also contained 2.2 mM $MgSO_4$, 0.33 mM ATP, 33 mM KPO_4, pH 7.5, muscle extract, and liver extract. From Ratner (6).

[b] Arginine was estimated as urea.

The urea-forming activity found anaerobically in soluble liver extracts was of the same order as had been found in respiring liver slices with NH_3 as the sole nitrogen source or in respiring homogenates with glutamate as the donor and therefore represented a major step in the pathway of urea formation (7, 8).

Along with arginine, fumaric acid (found as malic acid, due to the presence of fumarase) was formed as the second product of the reaction with aspartate. This observation and the fact that the rate curves showed an induction period, suggesting accumulation of an intermediate, implied that two enzymatic steps were involved in the conversion of citrulline to arginine (7, 8). After separating the participating enzymes, it became possible to show that one enzyme catalyzes the ATP-using condensation of citrulline with aspartate to form an intermediary condensation product and a second enzyme catalyzes the cleavage of the intermediary condensation product, argininosuccinic acid, to arginine and fumarate (6–8). The latter was originally estimated as malate.

Elucidation of this stepwise mechanism of nitrogen transfer involved also the demonstration that the energy used to form the new C—N bond in arginine is derived from phosphate bond energy, specifically in the form of ATP, and that the nitrogen is specifically derived from aspartate (reactions 1 and 2).

Arginine Synthesis and Interrelationships between the Citric Acid and Ornithine Cycles

The new findings were at variance with a number of published observations to the effect that (*a*) glutamate was about four times as effective as aspartate in respiring liver homogenates (5, 9); (*b*) both glutamate and aspartate were poor precursors in liver slices (1, 5); (*c*) the general nature of nitrogen transfer from the amino acid pool to form aspartate was not immediately apparent. The results of experiments with respiring liver homogenates obtained in other laboratories at this time have been reviewed and also the complexities introduced by the limited permeability of aspartate and glutamate to liver cells as compared to freely permeable NH_3 and glutamine.

Ratner and Pappas (7) undertook studies on arginine synthesis in respiring liver homogenates, in the presence and absence of malonate, and the presence and absence of fumarate and the keto acids of the citric acid cycle, in order to reconcile the various observations gained under aerobic conditions with the reaction requirements shown in reactions 1 and 2. Essentially this work showed that when the conversion of citrulline to arginine is examined in respiring homogenates, the requirements for ATP and aspartate are masked. When glutamate was used as nitrogen donor, it functioned in a triple capacity as an oxidizable substrate for the generation of ATP, as a precursor of oxaloacetate, and as nitrogen donor in the formation of aspartate by transamination.

Taking cognizance of the reactions that occur in respiring homogenates due to the operations of the citric acid cycle as described by Krebs and to phosphorylation coupled to respiration, it was proposed in 1949 that the ornithine cycle must be linked to the citric acid through dependence on

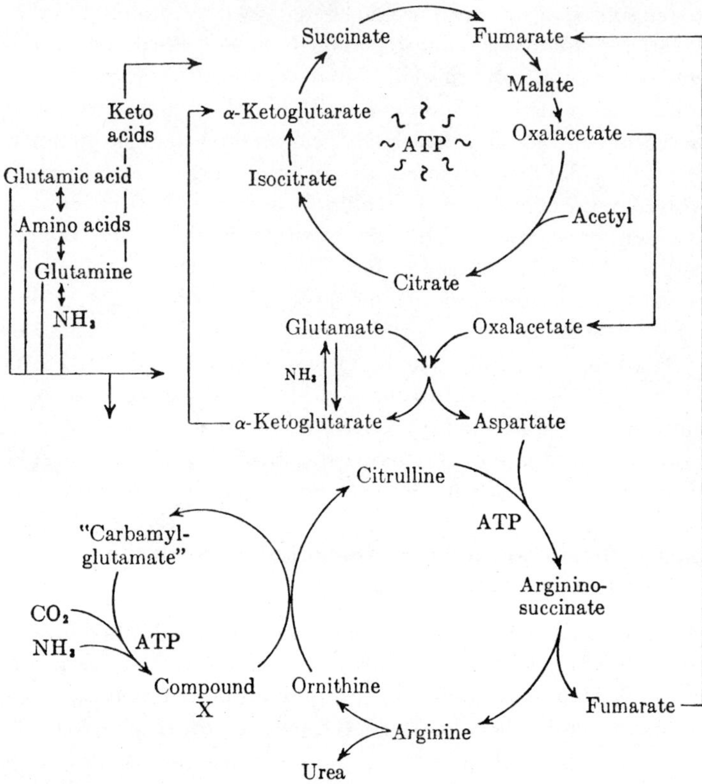

Figure 1. The relationship of the ornithine cycle to the citric acid cycle and to transamination. Pathway of amino nitrogen transfer to form citrulline, arginine, and urea. Ratner et al. (7, 92).

glutamic dehydrogenase, transamination, and coupled phosphorylation. These metabolic interrelations are shown in Fig. 1 as originally formulated. The scheme explained the experimental observations discussed above and described the relationship of the pathways of nitrogen transfer in relation to the oxidative processes of the cell under physiological conditions (7, 10). Further elaboration of the reaction sequence of the ornithine cycle pathway, elucidation of mitochondrial functions, and full appreciation of the role of glutamine in nitrogen transfer have not basically altered the interlocking metabolic relationships as they are shown here.

Distribution and Metabolic Function

The conversion of citrulline to arginine through the formation and cleavage of argininosuccinate is obligatory for all species in the biosynthesis of the arginine used for tissue proteins, for creatine formation via transamidination, and for creatine phosphate and other muscle phosphagens. In the livers of ureotelic species, the presence of arginase diverts this pathway to urea formation through the cyclical turnovers of the ornithine cycle. The two enzymes occur together in the soluble, cytoplasmic fraction of liver tissue (7, 11, 12). Values for their activity in liver tissue are given elsewhere in this volume.

The two enzymes occur also in a number of extrahepatic tissues (Table 2), such as kidney (13) pancreas and muscle (14) and brain (15), not capable of forming citrulline from ornithine (15, 16). Presumably these tissues can derive citrulline via the blood circulation. The presence of both enzymes in the kidneys of the chick (17) carries the phylogenetic implication that in uricotelic species their presence is vestigial in nature. More recently enzyme activities have been determined in cell lines grown in tissue culture (Table 2).

General Features of Nitrogen Transfer from Aspartic Acid in Formation of C—N Bonds

Formation of the Guanidino Group In the formation of the guanidino group of argininosuccinate this pair of reactions brings about a transfer of amino nitrogen without the participation of free NH_3. Two enzymatic steps are needed to complete the transfer of nitrogen; condensation of a carbon atom of the acceptor substrate with the aspartate nitrogen gives rise to an intermediate, as shown in (*a*), which retains the carbon chain of aspartate. In the second step chain detachment, as shown in (*b*), occurs so as to eliminate fumaric acid (7, 8, 10, 13):

$$\mathrm{(H)HN{\diagdown}\!\!\overset{}{C}{-}OH\,(HN{-}R) + H_2N{-}\overset{\diagdown}{C}H{-}\overset{\diagup}{C}H{-} \;\underset{}{\overset{ATP}{\rightleftharpoons}}\; HN{=}C{-}NH{-}\overset{\diagdown}{C}H{-}\overset{\diagup}{C}H{-}\,(HN{-}R)} \qquad (a)$$

$$\mathrm{HN{=}C{-}NH{-}\overset{\diagdown}{C}H{-}\overset{\diagup}{C}H\,(HN{-}R) \;\rightleftharpoons\; HN{=}C{-}NH_2\,(HN{-}R) + \overset{\diagdown}{C}H{=}CH{\diagdown}} \qquad (b)$$

No change occurs in the oxidation-reduction level. The condensation is endergonic and must be coupled to the hydrolysis of ATP, whereas the detachment of fumaric acid involves a much smaller energy change.

Formation of Amidine and Amide Groups Within a few years it was shown that the conversion of inosinic acid to adenylic acid proceeds by an analogous pair of reactions (22–25). Soon thereafter Buchanan and collaborators showed that nitrogen atom 1 of the purine ring is derived from aspartate by analogous

Table 2 Activity of Enzymes of Arginine Synthesis in Extrahepatic Tissues and Cell Lines

	Activity					
Tissue	Carbamyl Phosphate Synthetase	Ornithine *trans*-Carbamylase	Argininosuccinate Synthetase	Argininosuccinase	Arginase	Refs.
	(μmole/hour/g wet weight)					
Kidney (rat)	4.43	78.0	12.0	43.2	2240	18
Brain (rat)	4	0	[3.0]	6.9	105	15,16
Erythrocytes (human)				1.3	300	19
Cell lines			(μmole/hour/mg protein)			
HeLA S_3			0.084	0.130	0.5	20
Fibroblasts (mouse)			0.112	0.162	0.007	20
Fibroblasts (human)				0.147		21

steps (26–28). Aspartate functions metabolically as a carrier of nitrogen, the carbon chain being recovered as fumarate.

The nitrogen in positions 3 and 9 of the purine ring, previously thought to come from NH_3, actually come from glutamine (28, 29) and one-half of the pyrimidine ring nitrogens originates in aspartate (30). With clarification of the mechanisms it became increasingly evident that many biosynthetic reaction mechanisms make use of aspartate or glutamine as the immediate nitrogen donor in the transfer of nitrogen.

ARGININOSUCCINATE SYNTHETASE FROM BOVINE LIVER

Introduction

The two enzymes catalyzing the overall conversion of citrulline to arginine were separated by fractionation procedures. At the time that separation of the synthetase was obtained the addition of a protein factor from yeast was necessary for full condensing activity (32). Soon thereafter this factor was identified as inorganic pyrophosphatase (33). The large stimulation stems from the fact that PP_i exerts a strong product inhibition. The phosphatase is very active in almost all tissues. However removal during purification of the synthetase requires that it be provided in the assay mixture.

Assay Procedures

Synthetase activity can be measured colorimetrically from the rate of urea formation in crude preparations containing argininosuccinase and arginase (7, 13), providing that argininosuccinase is added in excess as described by Schimke (34). After arginase and argininosuccinase have been removed the rate at which citrulline is used can be followed colorimetrically (7, 13, 31, 32, 35) without interference from urea which gives full color in the citrulline analysis. Colorimetric citrulline analysis has been applied to crude preparations after decomposing the urea with urease (12). After removal of argininosuccinase, synthetase activity can be measured colorimetrically from the rate of formation or argininosuccinate. The determination involves the colorimetric measurement of arginine obtained after cyclization of argininosuccinate followed by oxidative cleavage (15).

In the course of purification, and for study of kinetic properties, an optical kinetic assay is used in which DPNH oxidation is coupled to the AMP formed in reaction 1. The incubation medium contains the substrates and myokinase, phosphoenol-pyruvate, pyruvate kinase, DPNH, lactic dehydrogenase, and inorganic pyrophosphatase. Since 2 moles of DPNH are oxidized per mole of AMP formed, this is by far the most sensitive of the assays (36–40).

Molecular Properties

Isolation The synthetase has been prepared about 50% pure from acetone-dried powder of hog kidney (36) and bovine liver (37, 38). Recently the arduous acetone drying step has been successfully eliminated, and this has facilitated the preparation of crystalline homogeneous enzyme from bovine liver (39). Although synthetase from both sources has been used for mechanism studies, of the two, the bovine liver enzyme is the more thoroughly described.

Molecular Constitution The crystalline preparation has a specific activity of 230 μmoles/hour/mg at 38° and is homogeneous as judged by sedimentation velocity and gel electrophoresis. The enzyme with a molecular weight of 175,000, determined by Sephadex chromatography, is composed of four subunits equal in size and 45,000 in molecular weight (SDS-gel electrophoresis in the presence of reducing agents). A molecular weight of 90,000 was obtained by SDS-gel electrophoresis in the absence of a reducing agent. Thus the enzyme is a tetramer in which two pairs of monomers are cross-linked by disulfide bonds (39).

Equilibrium and Free-Energy Changes

In the reaction catalyzed by argininosuccinate synthetase, 1 mole of ATP is used in the reversible condensation between citrulline and aspartate (10, 33, 41):

$$\begin{array}{l} NH_2 \\ | \\ C{=}O \\ | \\ NH \\ | \\ (CH_2)_3 \\ | \\ H_3\overset{+}{N}CHCOO^- \end{array} + \begin{array}{l} COO^- \\ | \\ H_3\overset{+}{N}CH \\ | \\ CH_2 \\ | \\ COO^- \end{array} + MgATP^{2-} \rightleftharpoons$$

$$\begin{array}{l} \overset{+}{N}H \\ \| \\ C{-}NH{-} \\ | \\ NH \\ | \\ (CH_2)_3 \\ | \\ H_3\overset{+}{N}CHCOO^- \end{array}\begin{array}{l} COO^- \\ | \\ CH \\ | \\ CH_2 \\ | \\ COO^- \end{array} + AMP^{2-} + MgPP_i^{2-} + 2H^+ \qquad (1)$$

The rate of reaction 1 is strongly dependent on pH: the optimum in the forward direction is 8.7 in the presence of PP_iase and is about the same in its absence; in the reverse direction the optimum lies at pH 6 (33). In order to obtain values for the equilibrium constant and free-energy change at any particular pH, it is desirable to have an expression for K_{eq} that is independent of pH. By assuming that the three amino acids undergo no significant change in charge in the pH range studied and by selecting a complexed species of ATP and PP_i which makes a significant contribution to the total concentration of that reactant at physiological pH, an expression can be formulated for reaction 1 that is independent of pH (36):

$$K_{eq} = \frac{(\text{argininosuccinate})_T(\text{AMP}^{2-})(\text{MgPP}_i^{2-})[\text{H}^+]^2}{(\text{citrulline})_T(\text{aspartate})_T(\text{MgATP}^{2-})} = K'_{eq}[\text{H}^+]^2$$

It then becomes necessary to give consideration to the influence of pH on the apparent equilibrium constant $K_{app'}$ owing to changes in the state of ionization of ATP, AMP, and PP_i and their ionized complexes with magnesium. Then K_{eq} was related to observed concentrations of reactants and to the pH-dependent values for K_{app} through designation of the ionic species at each pH as a fraction of the total concentration of that reactant (36):

$$K_{eq} = K_{app}\frac{\text{AMP}^{2-}\ \text{MgPP}_i^{2-}[\text{H}^+]^2}{\text{MgATP}^{2-}}$$

Values thus obtained for K_{eq} at a particular pH were 0.89, 8.9, and 89 at pH 7.0, 7.5, and 8.0, respectively. The $\Delta F^{\circ\prime}$ values at pH 7.5 and 37° calculated for reaction 1 and for the hydrolysis of MgATP ($\text{MgATP}^{2-} \rightarrow \text{AMP}^{2-} + \text{MgPP}_i^{2-} + 2\text{H}^+$) are −2.1 and −10.3 kcal, respectively (Table 3). The free-energy change associated with the formation of argininosuccinate is therefore +8.2 kcal, about the same (−8.6) as the value calculated for the hydrolysis of arginine to citrulline and NH_3 (cf. also ref. 42).

Table 3 Standard Free-Energy Changes at pH 7.5 and 37°

		$\Delta F^{\circ\prime}$ (kcal)
a.	Reaction 1	−2.1
b.	$\text{MgATP}^{2-} + H_2O = \text{AMP}^{2-} + \text{MgPP}_i^{2-} + 2H^+$	−10.3[a]
c.	Aspartate + citrulline = argininosuccinate + H_2O	+8.2
d.	Argininosuccinate = arginine + fumarate	+2.8
e.	Fumarate + NH_3 = aspartate	−2.4
f.	Arginine + H_2O = citrulline + NH_3	−8.6
g.	$\text{MgPP}_i^{2-} + H_2O = 2HPO_4^{2-} + Mg^{2+}$	−5.1[a]
h.	$\text{MgATP}^{2-} + H_2O = \text{MgADP}^{1-} + HPO_4^{2-} + H^+$	−7.7[b]
i.	$\text{ATP}^{4-} + H_2O = \text{ADP}^{3-} + HPO_4^{2-} + H^+$	−9.3[b]

[a] From Schuegraf et al. (36), calculated on the assumption that, in the absence of Mg, $\Delta F^{\circ\prime}$ is −9.3 kcal (cf. line *i*).

[b] From Benzinger et al. (43).

Because the complexing constant is much higher for $MgPP_i^{2-}$ than for $MgADP^{1-}$, the $\Delta F^{\circ\prime}$ value for hydrolysis of $MgATP^{2-}$ at the α position is -7.7 kcal, somewhat lower than for hydrolysis at the β position (Table 3). The value of $\Delta F'$ for the hydrolysis of $MgPP_i^{2-}$ is -5.1 kcal. Since inorganic pyrophosphatase is highly active in most tissues, it can then be assumed that the synthetase reaction is driven by the energy of hydrolysis of two pyrophosphate bonds. When reaction 1 (-2.1 kcal) is coupled to the hydrolysis of $MgPP_i^{2-}$ to $2P_i$ (-5.1 kcal), the net process is strongly exothermic (-7.2 kcal). Thus a considerable thermodynamic advantage is inherent in hydrolytic cleavage at the β pyrophosphate bond of ATP.

Our equilibrium experiments were limited to a fixed concentration of Mg^{2+}. Alberty (44) has since published contour diagrams conveniently relating ΔG_{obs} to changes in pH and pMg. An informative discussion pertaining to the effects of changes in pH and pMg on the differences in the free-energy of hydrolysis at the α and β pyrophosphate bond of ATP is to be found in the recent review by Stadtman (45).

Steric Orientation at Active Site

Aspartate Analogs as Substrates and Inhibitors The structural and steric requirements for substrate binding to the enzyme have been examined by comparing the affinities of aspartate analogs either as substrates or inhibitors. Analogs were selected with respect to configuration in the alpha and beta positions (38, 46). The D isomer of aspartate is inactive as substrate and has a low affinity for the enzyme as an inhibitor. Methyl substitution in the alpha position prevents catalytic activity but does not interfere with binding to the enzyme or with action of the compound as inhibitor. Substitution in the beta position, by —CH_3 or by —OH, does not interfere with activity if these groupings are in the threo configuration and the erythro hydrogen is unsubstituted (Table 4). Thus threo-β-methyl-L-aspartate and threo-β-hydroxy-DL-aspartate are active as substrates in the condensation reaction with citrulline (38, 46) and compare favorably with aspartate. With the former, the β-methyl analog of argininosuccinate is the product formed; it has been isolated as the barium salt (46).

For activity as substrate it is necessary that the proton in the erythro configuration be free. When the erythro position is substituted, as in erythro-β-hydroxyaspartate, activity as substrate is lost, although the compound binds to the enzyme and is active as inhibitor (cf. the studies by Jenkins (47) and Kagan and Meister (48)). Examination of models suggests that the amino and carboxyl groups are favorably spaced for attachment when in the extended conformation. In this conformation the α-hydrogen points away from the enzyme. The inability of α-methyl aspartate to act as substrate in the condensation may be due to steric hindrance or possibly to a requirement

Table 4 Affinities (K_m and K_i) of Aspartate Analogs for Argininosuccinate Synthetase[a]

Compound	K_m (*M*)	K_i (*M*)
L-Aspartate	3.5×10^{-5}	
α-Methyl-DL-aspartate		1.8×10^{-3}
D-Aspartate		2.0×10^{-2}
Threo-β-methyl-L-aspartate[b]	8.8×10^{-4}	
Threo-β-hydroxyl-DL-aspartate[b]	4.5×10^{-3}	
Erythro-β-hydroxy-DL-aspartate		2.4×10^{-3}
Threo-β-hydroxy-β-methyl-DL-aspartate		4.5×10^{-2}
Erythro-β-hydroxy-β-methyl-DL-aspartate[c]		7.1×10^{-4}

[a] All inhibitions were competitive with L-aspartate, and all inhibitors were inactive as substrates in the condensation reaction. Modified from Rochovansky and Ratner (38, 46).

[b] At saturation, activity was 70% compared with aspartate with the β-methyl analog and 50% with the β-hydroxy analog.

[c] The β-methyl group is threo.

for a free α-hydrogen in the reaction mechanism. The α-methyl analog does, however, promote ATPase activity, as described below. In the extended conformation, a methyl substituent in the erythro position sterically hinders the α-amino group and can thus interfere with activity. This explains why erythro-β-hydroxy-β-methylaspartate is a stronger inhibitor than the threo form (Table 4) and why only the threo forms of mono-β-substituted analogs can act as substrates (46).

Reaction Mechanism

General Considerations In the condensation of citrulline with aspartate, it is the ureido group of citrulline that is activated by ATP. When L-[ureido-^{18}O] citrulline is used as substrate in reaction 1, the ^{18}O is quantitatively transferred to the phosphate group of AMP (37). The activated form of citrulline is therefore considered to be citrulline adenylate, to which the isoureidophosphoryl structure is assigned:

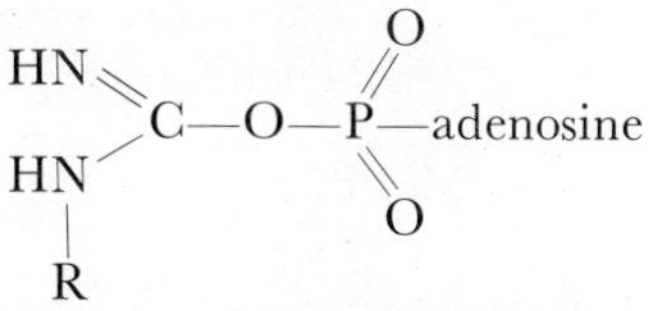

```
HN                O
  \\              //
    C—O—P—adenosine        R = (CH2)3CHNH2COOH
  /          \\
HN            O
 |
 R
```

If this compound is involved as an intermediate, partial reactions are to be expected, such as have been found with acetyl-CoA (49) and amino acid-*t*RNA synthetases (50). Reactions 1*a* and 1*b* should then be detectable either by $^{32}PP_i$-ATP or [^{14}C]AMP-ATP exchange:

$$\text{citrulline} + \text{MgATP}^{2-} \rightleftharpoons \text{citrulline-AMP}^{2-} + \text{MgPP}_i^{2-} + 2\text{H}^+ \quad (1a)$$

$$\text{citrulline-AMP}^{2-} + \text{aspartate} \rightleftharpoons \text{argininosuccinate} + \text{AMP}^{2-} \quad (1b)$$

However, isotope exchange was not found except under conditions in which overall reaction reversal could have occurred (37). Other possibilities as to mechanism were assessed by kinetic analysis (38) according to Cleland (51). Since a family of intersecting reciprocal plots was obtained for each substrate, the results indicated that the three substrates must be bound to the enzyme as a quaternary complex before any product is released.

Product inhibition is appreciable; AMP is a competitive inhibitor with respect to ATP, and argininosuccinate is competitive with respect to citrulline, aspartate, and ATP. Inhibition by inorganic pyrophosphate is unusually high ($K_i = 6.2 \times 10^{-5}$ *M*). It is pertinent to mention, in relation to the kinetic studies, that the assay conditions included an ATP-generating system and allowed zero-order initial rate measurements to be made within the first 2 or 3 minutes, or even sooner. The presence of excess PP_iase was found to be necessary to maintain linearity beyond 1 minute (38).

The kinetic plots expected for an overall reaction involving successive partial reactions, such as 1*a* and 1*b*, where PP_i is released before aspartate binds to the enzyme, should display at least one set of parallel reciprocal plots. Many aminoacyl-*t*RNA synthetases catalyze $^{32}PP_i$-ATP exchange and obey some form of ping-pong kinetics. Argininosuccinate synthetase differs from these with respect to isotope exchange and kinetic behavior, although the reaction catalyzed is formally analogous.

Evidence for Two Partial Reactions The formation of a quaternary complex as indicated by kinetic analysis suggested a concerted mechanism, but the ^{18}O evidence left open the possibility of a stepwise mechanism. Additional evidence for citrulline adenylate participation, obtained with the enzyme at substrate rather than at catalytic levels, supports the mechanism shown in reactions 1*c* and 1*d* (38):

$$\text{E} + \text{citrulline} + \text{MgATP}^{2-} \rightleftharpoons \text{E}\begin{matrix}\diagup \text{AMP-citrulline} \\ \diagdown \text{MgPP}_i\end{matrix} \quad (1c)$$

$$\text{E}\begin{matrix}\diagup \text{AMP-citrulline} \\ \diagdown \text{MgPP}_i\end{matrix} + \text{aspartate} \rightleftharpoons$$

$$\text{E} + \text{argininosuccinate} + \text{MgPP}_i^{2-} + \text{AMP}^{2-} \quad (1d)$$

As proposed, citrulline adenylate is formed in the first partial reaction, tightly bound to the enzyme, and reacts with aspartate in the second reaction. Release of PP_i occurs in the presence of aspartate. Reaction 1*c* is supported by the results of pulse-labeling experiments; when ^{14}C-L-citrulline, ATP, Mg^{2+}, and limiting enzyme were present in the pulse, this led to a marked preferential incorporation of ^{14}C into argininosuccinate (38). The revealing use of pulse labeling was originally introduced by Krishnaswamy et al. (52) in the elucidation of the enzyme-bound reaction sequence catalyzed by glutamine synthetase from sheep brain.

Argininosuccinate synthetase also promotes a citrulline-dependent hydrolysis of ATP, as predicted by reaction 1*c*. The time course is characterized by a burst of activity, 2 or 3 minutes in duration, which comes to a standstill when the amount of ATP hydrolyzed is approximately stoichiometric with enzyme. The standstill implies that neither citrulline adenylate nor PP_i is released from the enzyme. It provides an explanation for the absence of a citrulline-dependent exchange between $^{32}PP_i$ and ATP and suggests that the presence of aspartate is necessary for the release of PP_i, on which detection of the reversibility of reaction 1*c* depends. The affinity of PP_i for the enzyme is evidently much higher in the absence of aspartate than in its presence (in the latter case $K_i = 6.2 \times 10^{-5}\ M$). Aspartate presumably alters the affinity of PP_i through a conformational change. That aspartate affects this site is supported by evidence that the enzyme also catalyzes an intrinsic aspartate-dependent hydrolysis of ATP (Table 5) to AMP and PP_i (38).

Table 5 Analogs of Aspartate and Citrulline as Substrates for AT^{32}P Cleavage and Overall Reaction[a]

Amino Acid Added	AT^{32}P Cleavage (%)	Overall Reaction (%)
L-Aspartate	100	100
α-Methyl-DL-aspartate	48	0
D-Aspartate		0
Threo-β-methyl-L-aspartate	22	70
L-Citrulline	100	100
L-Homocitrulline	20	10

[a] Modified from Rochovansky and Ratner (38).

Because of overall reaction reversal, aspartate cannot be used to test an effect on PP_i binding by $^{32}PP_i$-ATP exchange. The analog, α-methylaspartate, has suitable properties for this purpose, since it promotes ATP cleavage (Table 5) but cannot act as substrate in the condensation with citrulline.

The compound proved to have the anticipated effect of promoting a citrulline-dependent $^{32}PP_i$-ATP exchange. The rate of exchange was more rapid at pH 8 than at pH 7, in accord with the expectation that this rate would be increased by the amount of PP_i formed (Table 6).

Table 6 Effect of α-Methylaspartate on $^{32}PP_i$-ATP Exchange[a]

Amino acids added		pH 7		pH 8	
L-Citrulline	α-Methyl-DL-aspartate	Counts in ATP	$^{32}PP_i$ in ATP	Counts in ATP	$^{32}PP_i$ in ATP
(μmole)		(cpm)	(nmole)	(cpm)	(nmole)
0	0	200	0.20	74	0.07
5.0	0	330	0.32	130	0.13
0	1.25	110	0.11	40	0.04
5.0	1.25	1620	1.6	2900	2.8

[a] Modified from Rochovansky and Ratner (38).

Thus supported by evidence for partial reactions, steps 1*c* and 1*d* are visualized as taking place successively on the enzyme surface without dissociation of the products of the first partial reaction (38). The authors consider that the conformational change necessary for the release of PP_i occurs with those synthetases that have a high affinity for PP_i.

Recent studies of arginyl-*t*RNA synthetase by Papas and Peterkofsky (53) are pertinent to this interpretation. Their kinetic analyses indicate that ATP, arginine, and *t*RNA bind in random order and that all substrates must be bound before product is released. This resembles the kinetic behavior of argininosuccinate synthetase (38). Investigations of isotope exchange rates at equilibrium, as outlined by Boyer (54), confirmed the random order of substrate binding and product release. Exchange rates for $PP_i \leftrightarrow$ ATP, arginine ↔ arginyl-*t*RNA^{arg}, AMP ↔ ATP, and *t*RNA ↔ arginyl-*t*RNA^{arg} varied through a 100-fold range, and therefore the rate-limiting step in the overall reaction involves binding or release of a reactant or product, not interconversions of enzyme-bound quaternary complexes. The relative rates were such as to exclude a compulsory binding order in which *t*RNA binding precedes that of any other substrate, a possibility previously suggested (55).

Mehler and Mitra (56) have observed that arginyl-*t*RNA synthetase does not catalyze an arginine-dependent $^{32}PP_i$-ATP exchange except in the presence of arginine-specific-*t*RNA. Catalysis of the reverse reaction, pyrophosphorolysis of arginyl adenylate, also has this requirement. They suggest that *t*RNA binding is required for conformational reasons in the activation step, that is, for arginyl-adenylate formation. Arginyl-*t*RNA synthetase is strongly inhibited by PP_i (54), and this suggests that *t*RNA may decrease the affinity of the enzyme for PP_i. Similarity with the requirements for isotope exchange shown by argininosuccinate synthetase, and similarity in the kinetic patterns (38, 53), suggest that the isotope-exchange–promoting and pyrophosphorolysis-promoting effects of *t*RNA are due to overall reaction reversal. The first partial reaction 1*c* may be applicable to arginyl-*t*RNA synthetase and other acyl synthetases strongly inhibited by PP_i. Glutamyl-*t*RNA synthetase and glutaminyl-*t*RNA synthetase, which require *t*RNA for isotope exchange (57), may also belong to this group.

Allosteric Properties

Allosteric Regulation Kinetic analyses (39) of enzyme substrate interactions with each of the three substrates give biphasic double reciprocal plots characteristic of negatively cooperative binding. The affinity of ATP increases threefold in the change from high to low concentrations; the increase was observed whether the concentration of the fixed amino acid was high or low. Negative cooperativity also characterizes citrulline and aspartate binding; the affinity of these amino acids increases approximately ten- and twentyfold, respectively, in going from high to low concentrations. Negative cooperativity was found only when the second amino acid was well below saturation and prevailed whether ATP was fixed at high or low concentrations; otherwise the plots were linear. All experiments were carried out in tris buffer at uniform salt concentration.

In general, then, the affinity of the enzyme for either amino acid substrate is not affected by changes in ATP concentration, but a reduction in the concentration of one amino acid increases enzyme affinity for the second amino acid when present in low concentrations. Thus catalytic activity is increased at low substrate concentration by increasing substrate affinity. Regulation of activity in this way greatly extends the lower limits of substrate concentration necessary for significant catalytic activity (39). The changes in substrate affinity are considered to be negatively cooperative in the sense defined by Levitzki and Koshland (58). Although the number of catalytic sites or substrate binding sites is not yet known, the information on the subunit composition given above under "Molecular Properties" suggests that the enzyme has at least two catalytic sites. This reinforces the interpretation that the cooperative substrate effects are allosteric in origin (39).

ARGININOSUCCINASE FROM BOVINE LIVER

Introduction

Argininosuccinase was originally found together with argininosuccinate synthetase (7) in the soluble fraction of rat and bovine liver. On separation from the synthetase it was possible to show that enzyme preparations from liver and kidney of various mammalian species catalyze the reversible cleavage of argininosuccinate to arginine and fumarate (13, 31):

$$
\begin{array}{llcllcl}
\mathrm{NH} & \mathrm{COOH} & & \mathrm{NH} & & & \\
\| & | & & \| & & & \\
\mathrm{C{-}NH{-}CH} & & & \mathrm{C{-}NH_2} & & & \mathrm{COOH} \\
| & | & & | & & & | \\
\mathrm{NH} & \mathrm{CH_2} & \rightleftharpoons & \mathrm{NH} & & + & \mathrm{CH} \\
| & | & & | & & & \| \\
\mathrm{(CH_2)_3} & \mathrm{COOH} & & \mathrm{(CH_2)_3} & & & \mathrm{CH} \\
| & & & | & & & | \\
\mathrm{H_2NCHCOOH} & & & \mathrm{H_2NCHCOOH} & & & \mathrm{COOH}
\end{array}
\qquad (2)
$$

The stoichiometry of reaction 2 and other reaction characteristics were established by Ratner et al. (59), making use of the partially purified enzyme from bovine liver and argininosuccinate prepared by enzymatic synthesis (60). The free-energy change ΔF° for reaction 2 is +2.8 kcal, calculated from $K_{eq} = 11.4 \times 10^{-3}$ (mole/liter) at 38°, pH 7.5, for the expression [arginine] [fumarate]/[argininosuccinate]. For reactions of this order the composition of the equilibrium mixture is highly sensitive to concentration, and a low concentration of argininosuccinate, such as that used during assay, favors the direction of cleavage. Since high concentrations of arginine and fumarate favor the reverse reaction, this has been taken advantage of for preparative purposes (60). The studies discussed in the remaining parts of this section were carried out with highly purified crystalline preparations from bovine liver (61–63) or bovine kidney (64) with a specific activity of 1400 μmole/hour/mg at 38°.

Mechanism of C—N Cleavage by β-Elimination

Stereospecificity of β-Elimination The stereospecificities of cleavage by argininosuccinase, aspartase, and fumarase can be interrelated, since aspartate participates in the synthetase-catalyzed formation of argininosuccinate, and fumarate is a product of cleavage by argininosuccinase. Starting with fumaric acid, labeled stereospecifically with tritium, the stereospecificity of the addition catalyzed by argininosuccinase was shown to be trans by Hoberman et al. (65) and is illustrated in the stereospecific steps of Fig. 2. It is known that aspartase and fumarase catalyze trans additions to fumarate. Addition occurs so that —NH_2 or —OH at C-2 is in the L configuration and the proton at C-3 is in the erythro configuration. Recent reviews on the stereospecificity of these and related lyases have appeared (66–68).

In the sequence of reactions in Fig. 2, the tritium was retained in the step catalyzed by aspartase, by argininosuccinate synthetase, by argininosuccinase,

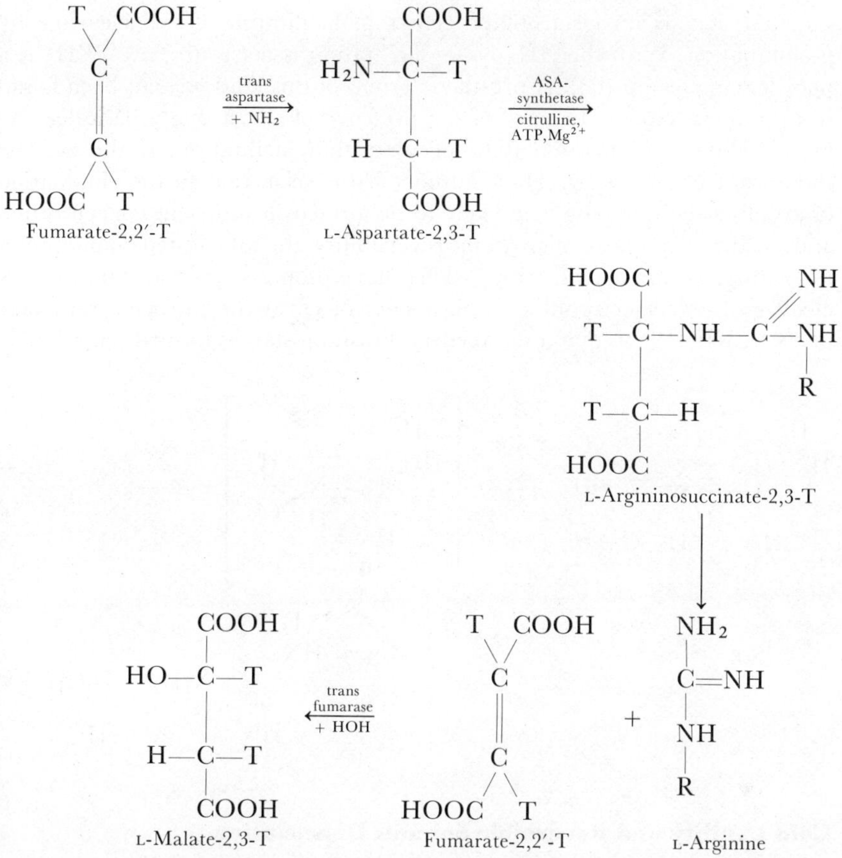

T COOH
C
C
HOOC T
Fumarate-2,2′-T

trans aspartase, $+ NH_2$ →

COOH
H_2N—C—T
H—C—T
COOH
L-Aspartate-2,3-T

ASA-synthetase, citrulline, ATP, Mg^{2+} →

HOOC NH
T—C—NH—C—NH
R
T—C—H
HOOC
L-Argininosuccinate-2,3-T

↓

COOH
HO—C—T
H—C—T
COOH
L-Malate-2,3-T

← trans fumarase, + HOH

T COOH
C
C
HOOC T
Fumarate-2,2′-T

+

NH_2
C=NH
NH
R
L-Arginine

Figure 2. Stereochemical specificity of the elimination reaction catalyzed by argininosuccinase in relation to assigned stereospecificities of elimination catalyzed by aspartase and fumarase. From Hoberman et al. (65).

and by fumarase; the stereospecificity must therefore be the same for the three lyases. A study of the structural requirements of aspartate and its analogs for activity suggests that the extended conformation of the substrate with a free erythro group is preferred by argininosuccinate synthetase. As in the formation of malate or aspartate, the proton added to the succinate moiety of argininosuccinate by trans addition must be in the erythro configuration. If the succinate moiety binds to argininosuccinase in the extended conformation, the erythro proton will then be favorably located for elimination.

Nature of the Leaving Group at C^α Phenylalanine NH_3-lyase (69, 70) and histidine NH_3-lyase (71, 72) contain a dehydroalanine residue at the active site that acts as an electrophilic center in facilitating C—N cleavage by β-elimination. With the NH_3-lyases facilitation is necessary, as $—NH_2$ is a poor leaving group (68). A prosthetic group of this kind has not been found in argininosuccinase. In view of the presence of the strongly basic leaving group, Havir and Hanson (69) speculate that facilitation by the enzyme should not be necessary. These authors (70) reason that in the elimination of arginine the free base is unlikely to be formed in sufficient concentration and, by the principle of microscopic reversibility, the substituted guanidinium ion cannot be the leaving group. They have proposed therefore that C—N cleavage by an electrophilic displacement of C^{α} at the nitrogen must take place (reaction 3) so that a tetrahedral transition state is formed (cf. ref. 67):

H HN C N H H HN R — A H — C H — H C H — B → [H HN C N H HN R δ^+ H $A^{\delta-}$ C^{α}] → H HN C NH H HN R + CH HC (3)

Cold Lability and Reversible Subunit Dissociation

Cold-Induced Dissociation Except in the presence of substrate or phosphate, argininosuccinase (S = 9.3) loses activity on standing at low temperatures, and the loss can be recovered by a brief period of warming (61). Loss of activity is promoted by cations such as Tris and imidazole and by an increase in pH; at pH 8.3, for example, the half-life is about 2.5 hours at 0° in imidazole. Inactivation is accompanied by dissociation into catalytically inactive subunits. Resolution of mixtures of dissociated and undissociated forms by sedimentation velocity at 0° in the dissociating buffers fails to occur because of rapid equilibration between the two species. If, however, the imidazole buffer is replaced at low temperature by phosphate, resolution is achieved, since the presence of phosphate, by diminishing dissociation, prevents equilibration (62). The S value for the cold-dissociated subunit species (A_2) is 5.6. Resolution of the two species by sucrose density gradient sedimentation at 0° in phosphate demonstrates that inactivation and reactivation are consequences of reversible dissociation according to $A_4 \rightleftharpoons 2A_2$. As may be

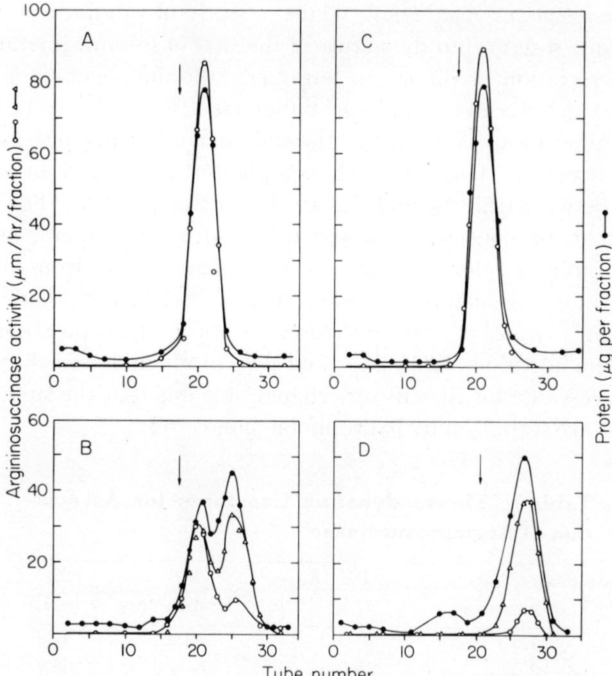

Figure 3. Separation of enzymatically active and inactive forms of argininosuccinase by sucrose gradient sedimentation. Samples A, B, and C were held 0° in imidazole chloride buffer, pH 8.3, for 6 hours, and sample D for 31 hours. (A) Fully active enzyme (held at 0° with 2 mM substrate). (B) Partially inactivated (substrate added just before sedimentation). (C) Thermally reactivated before sedimentation. (D) Extensively inactivated. Protein ●—●. Enzymatic activity before (○—○) and after (△—△) thermal reactivation of each fraction. Numbers on left and right ordinates are the same. Arrows indicate position of catalase in the gradients. Schulze et al. (62).

seen in Fig. 3 activity was very low in the fractions emerging in the 5.6 S peak (panels C and D), but after the same fractions were exposed to a short period of warming, almost all the initial activity was recovered. Although not evident from the figure, it can be shown by very rapid assay that the 5.6 S subunit is entirely inactive before reassociation (62). The protective effect of substrate in preventing dissociation is illustrated in panel A.

Argininosuccinase dissociates in Tris or imidazole buffer at low temperature as a first-order process, independent of protein concentration. The effect of pH is extremely marked, and in the pH range 6.5 to 9.0 the value of the first-order rate constant increases about 12-fold at 0° (62). This increase may be caused by further weakening at hydrophobic sites of interaction due to neutralization of released protons.

Subunit Association and Equilibria A kinetic study of subunit association has provided some insight into the nature of the sites of subunit interaction. The rate of reassociation of dimers to tetramers (thermal reactivation) follows first-order kinetics. In phosphate buffer at 24.5°, $t^{1/2} = 13.3$ minutes. Although subunit association occurs throughout a wide enzyme concentration range, reassociation fails to reach completion at high dilution because equilibria between the A_2 and A_4 species are established. The equilibria were found to be concentration and temperature dependent, and it then became possible to calculate K_{eq} at several temperatures from kinetic data (Table 7). The association constant $K_{eq} = [A_4]/[2A_2]^2$, had the value $3.2 \times 10^7\ M^{-1}$ at 24.5°, and decreased as the temperature was lowered. Thermodynamic values, $\Delta H°$, $\Delta G°$, and $\Delta S°$, are included in the table. The large positive value for the entropy change suggests that the sites of subunit interactions are stabilized by hydrophobic bonds (62).

Table 7 Thermodynamic Constants for Association of Argininosuccinase[a]

Temperature	$10^{-7}K_{eq}$ (M^{-1})	$\Delta G°$ (kcal/mole)	$\Delta S°$ (e.u.)
24.5°	3.2	−10.2	189
19.5°	0.82	−9.3	189
15.0°	0.25	−8.4	189

[a] An Arrhenius plot of log K_{eq} against $1/T$ yielded a straight line; $\Delta H°$, obtained from the slope, had a value of 46 kcal/mole. From Schulze et al. (62).

Primary Structure

The amino acid composition of bovine liver argininosuccinase shows few unusual features; the number of tyrosine and tryptophan residues is somewhat above average, and the number of half-cystine residues somewhat below (Table 8). Also the hydrophobic amino acids, which amount to 41% of the total, lie within the usual range (64). Evidently the hydrophobic character of the enzyme is increased by favorable folding.

Oligomeric Structure

Molecular Weight and Identity of Subunits The molecular weights of the catalytically active enzyme and of the two subunit species obtained by dissociation at pH 10, or in 6 M guanidine HCl, were determined by sedimentation equilibrium (63). These values are summarized with additional data in Table 9. The physical evidence indicates that argininosuccinase is a tetramer

Table 8 Amino Acid Composition of Bovine Kidney and Liver Argininosuccinase[a] (Number of Residues per Mole of 202,000)

Residue	Kidney	Liver	Residue	Kidney	Liver
Lysine	73	76	Half cystine	17	16
Histidine	45	47	Valine	102	98
Arginine	110	116	Methionine	53	51
Aspartic acid	143	138	Isoleucine	60	59
Threonine	96	95	Leucine	209	202
Serine	141	139	Tyrosine	36	36
Glutamic acid	199	191	Phenylalanine	52	51
Proline	62	61	Tryptophan	35	35
Glycine	124	121	Total	1,713	1,683
Alanine	156	151	Total residue mol wt	190,340	187,613

[a] From Bray and Ratner (64).

Table 9 Physical Properties of Bovine Liver Argininosuccinase and Subunits[a]

Species	$s_{20,w}$ (sedimentation velocity) (S)	Molecular Weight (sedimentation equilibrium)
Active enzyme	9.3	202,000
Dimeric subunit		
Cold dissociated	5.6	
pH 10 dissociated	5.6	100,000
Monomeric subunit		
8 *M* urea	1.9	50,400
6 *M* guanidine HCl	1.9	50,400
SDS-electrophoresis		52,000

[a] Compiled from Lusty and Ratner (63).

composed of four subunits of identical molecular weight. Disulfide crosslinking is absent, since all the half-cystine residues can be accounted for by —SH group titration in guanidine HCl and by the fact that the presence of reducing agents has no effect on the molecular weight of the subunit. Thus each subunit of 50,000 molecular weight is a single polypeptide chain.

Studies on the NH_2-terminal residue suggest that this end group may be blocked, perhaps by acylation. The carboxyl-terminal amino acid was found to be glutamine-glutamic acid by carboxypeptidase digestion, and the quantitative end group data corresponded to 4 moles/mole of enzyme. The number of peptide fragments given by cyanogen bromide cleavage and by tryptic cleavage was always one-fourth of the maximum number predicted from the amino acid composition. This detailed examination, indicating that the four polypeptide chains are the same or quite similar in amino acid sequence, appears to exclude the possibility that isozyme forms composed of subunits containing different polypeptide chains can occur (63).

Relation of —SH Groups to Tetrameric Structure The distribution of 16 cysteine residues of argininosuccinase in relation to sites of subunit interaction has been construed from —SH group titrations. All react rapidly with DTNB or *p*-HMB in the presence of denaturing reagents. In the catalytically active enzyme, four —SH groups are available to the sulfhydryl reagent, and the binding of these four groups by the reagent does not interfere with catalytic activity. On cold dissociation to dimeric subunits, four additional —SH groups become available to either reagent. Binding of the four newly exposed —SH groups by *p*-HMB interferes with reassociation. Exposure of previously unavailable —SH groups during cold dissociation suggests that they are located at or near the hydrophobic contact sites between the two dimers. It can also be shown, through DTNB titration in the presence of 0.1% SDS, that the release of —SH groups is stepwise; eight —SH groups (four of which are new) become available within 2 minutes, corresponding to dimer formation; and the next eight —SH groups are released more slowly, as dissociation to monomers proceeds. The last eight to be released may be situated at or near contact sites between pairs of monomers, as shown in Fig. 4.

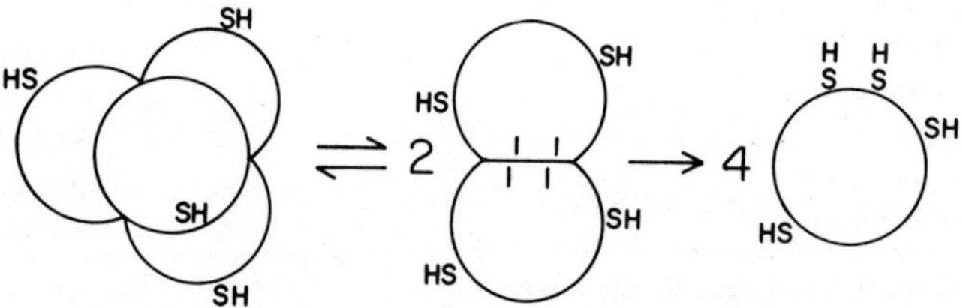

Figure 4. Schematic representation of the argininosuccinase tetramer and postulated sites of subunit interactions. The cysteinyl residues may be located at sites either between the subunits or within the subunits such that dissociation promotes their accessibility. From Lusty and Ratner (63).

Arrangement of Subunits Electron micrographs of negatively stained argininosuccinase support the tetrameric structure (63). In the most common views the enzyme molecules appear as triangles and as slightly distorted squares. These views are most compatible with a geometric configuration in which the subunits are arranged with the centers of mass at the vertices of a tetrahedron. The subunits appear to be arranged very compactly, with limited stain penetration at subunit contact areas. The estimated subunit diameter of 45 Å corresponds reasonably well with a molecular weight of 50,000.

Dihedral symmetry can be deduced from the quaternary structure, since two axes of symmetry are evident in the electron micrograph views. This is in agreement with prediction based on dissociation behavior, the value of which has been pointed out by Klotz et al. (73) and by Haschemeyer (74). In the dissociation of argininosuccinase, two types of subunit contact sites differing in bonding strength are indicated; this is the least number required in D_2 symmetry.

Catalytic Regulation

Cooperative Substrate Binding The four subunits of argininosuccinase have been found to be closely similar or identical in primary structure and molecular weight. Differences were seen among sites of subunit interactions. The preferential dissociation of the tetrameric form of the enzyme to inactive dimers under mild conditions indicated that the weaker bonds are between dimers. To gain further insight into structure-function relationships, substrate binding and catalytic function have been viewed in connection with quaternary structure.

The number of substrate binding sites, as determined by equilibrium dialysis, show negatively cooperative changes. Scatchard plots of data obtained in phosphate buffer indicate that, at low substrate concentrations, argininosuccinate binds at two sites and, at high concentrations, binds at four sites. Similarly arginine binds at two and at four sites at low and at high concentrations (Table 10). The presence of fumarate is not required for arginine binding. In Tris buffer arginine binding exhibits the same cooperative

Table 10 Substrate Binding Sites in Bovine Liver Argininosuccinase[a]

Substrate and Concentration	Phosphate: Number of Sites Bound	Phosphate: K_s (M)	Tris: Number of Sites Bound	Tris: K_s (M)
Arginine				
High	4	1.6×10^{-3}	4	7.3×10^{-4}
Low	2	4.9×10^{-4}	2	7.2×10^{-5}
Argininosuccinate				
High	4	2.9×10^{-4}	4	3.4×10^{-5}
Low	2	1.6×10^{-5}	4	3.4×10^{-5}

[a] Determined by equilibrium dialysis, pH 7.5; phosphate 0.060 M, Tris 0.05 M. Argininosuccinate binding corrected for substrate cleavage. From Rochovansky (75).

behavior as in phosphate, and argininosuccinate binds to four sites in an identical manner (75). The K_s values calculated from the binding data are summarized in Table 10.

Cooperative Kinetic Effects Double reciprocal plots of kinetic data obtained in phosphate buffer exhibit a break in slope also indicative of negatively co-operative behavior, whereas in Tris buffer the plot is linear. For example, in 0.06 *M* potassium phosphate buffer, pH 7.5 ($\Gamma/2 = 0.15$), the K_m for argininosuccinate changes from 1.6×10^{-4} *M* at high substrate concentrations to 4.4×10^{-5} *M* at low concentrations. The increases in affinity are accompanied by velocity increases above expectation had the plot remained linear. The affinity of argininosuccinate is also influenced by ionic strength. Increases in phosphate and ionic strength facilitate cooperative behavior and elevate V_{max} at high substrate concentrations.

The possibility that the negatively cooperative substrate effects stem from changes in dimer-tetramer association has been rigorously excluded by evidence that dissociation of the enzyme in phosphate cannot be detected even at the high dilution of enzyme prevailing during assay. The cooperative effects are allosteric in nature, since the kinetics are modified as substrate becomes bound to additional sites. Negatively cooperative changes at low substrate concentration serve to regulate catalysis by maintaining appreciable activity over wide changes in substrate concentration (75). Also the increased activity response to high substrate concentrations is distinctly advantageous in preventing accumulation of an intermediate in a multienzyme system.

In 0.02 *M* phosphate buffer ($\Gamma/2 = 0.15$) GTP has a marked activating effect brought on by a decrease in K_m from 1.6×10^{-4} *M* to 3.8×10^{-5} *M* at high substrate levels and a small decrease in K_m at low substrate concentrations. Since negative cooperativity was reduced in the presence of GTP, argininosuccinase activity was significantly increased at low concentrations of substrate. The effect is specific for GTP in that stimulation by GDP and adenine nucleotides is appreciably lower. In this way GTP may modulate the rate of arginine formation, possibly in response to the initiation steps of protein synthesis (75).

A considerable degree of conformational flexibility is made evident by the binding data; at least two conformational states are substrate-induced in the presence of phosphate and ionic strength, and possibly a third is induced by GTP (75).

Number of Catalytic Sites Hill plots of the kinetic data obtained in phosphate buffer show a change in slope from 1.0 to 0.45, corresponding to high and low substrate concentrations. Of equal importance are the findings that the K_m values observed at low and high substrate concentrations show changes analogous to those found for K_s values at low and high substrate concentrations. This correspondence indicates that each binding site represents a catalytic site. With a maximum of four binding sites, in agreement with the tetrameric structure, it follows that each monomeric subunit contains one catalytic site (75).

Formation of Paracrystals

The enzyme crystallizes from ammonium sulfate solution in the form of fine needles. As seen in the light microscope (× 1500), the needles have low refractility and poorly defined side and end faces. Studies by Dales et al. (76) show electron micrographs (× 35,000) in which each needle consists of a spindle-shaped bundle of loosely packed hollow tubules. Images of cross sections cut through a pellet (viewed × 360,000) show that these tubules are quite uniform in wall thickness, 65 Å, and in diameter, 210 Å. The wall of the tubule is evidently one molecular layer in thickness, for the 65 Å measurement corresponds approximately to molecular subunit dimension.

At magnification × 160,000 it can be seen that three or four parallel rows run the length of each tubule and a periodicity of 50 Å is clearly resolved. From optical diffraction analysis, elucidated by Moody (77), the repeat units are arranged in the form of a single basic helix having approximately 3.5 repeat units per turn, corresponding to seven longitudinal rows. Where tubules are tightly grouped, the 50 Å periods appear to be in register across almost the entire bundle.

The resemblance in form and dimensions to the intracellular structures, known as microtubules, is particularly striking. Other enzymes have been shown to form tubular structures. Glutamine synthetase tubules of *E. coli* consist of stacked discs (78). Catalase, depending on conditions, can form either a crystal lattice or tubules (79, 80); the latter show helical symmetry. On dissolving argininosuccinase paracrystals in buffer, only one molecular species is obtained, that of the catalytically active enzyme. Numerous attempts to find favorable conditions for crystal formation, a possibility not incompatible with paracrystal formation, have thus far been unsuccessful.

Views taken at early stages suggest that tubular formation begins with aggregation into short, irregular, ribbonlike sheets, one molecular layer in thickness; these coil and elongate and close over to form tubules. At the same time the tubules become grouped in loosely packed bundles which show relatively weak side-to-side interactions (76). This contrasts with the manner of formation and structure of glutamic dehydrogenase tubes (81). However, the protein known as brain tubulin undergoes a temperature-dependent reversible association of subunits to form microtubules (82, 83), many features of which resemble the dimer-tetramer association of argininosuccinase and the assembly from paracrystals.

ARGININOSUCCINASES IN EXTRAHEPATIC ORGANS

Bovine Kidney Argininosuccinase

Among extrahepatic tissues the kidney has the highest capacity to convert citrulline to arginine; the level of activity, relative to the liver, varies among different mammalian species. Arginine has diverse metabolic functions in kidney and liver. Either synthesis or use or both may be independently controlled in the two organs. As the terminal enzyme in arginine biosynthesis, argininosuccinase is a possible site for regulation of arginine formation. Prompted by these considerations, Bray and Ratner (64) have undertaken a comparison of the properties of highly purified argininosuccinase prepared from kidney and liver of the same species.

Comparison of the catalytic, physical, and numerous other properties (Table 11) discloses that the two enzymes are essentially the same. The amino acid compositions are closely similar (Table 8). Nine amino acids are present in essentially the same amount (the differences come within the 2 or 3% error of analysis), the other nine differ by no more than 5 or 6%, and the number of half-cystine residues are the same. The similarity seen in aromatic amino acid composition is reflected also in the identical absorption spectra (64).

Table 11 Comparison of Bovine Liver and Kidney Argininosuccinase[a]

Property	Enzyme Source	
	Liver	Kidney
$s_{20°, w}$	9.3	9.3
Molecular weight	202,000	(202,000)[b]
Molecular weight of dimer	100,000	(100,000)[b]
E_{280} (1%)	13.0	12.5
E_{260} (1%)	7.1	6.8
ε_{280}	2.58×10^5	2.50×10^5
Partial specific volume	0.734	0.734
K_m (66 mM KPO_4)	1.3×10^{-4} M	1.4×10^{-4} M
Antigenic properties		
Precipitin reaction	Cross-reactive	Cross-reactive
Inhibition by antibody	Cross-reactive	Cross-reactive
Amino acid composition	Cf. Table 8	Cf. Table 8
NH_2-Terminal groups	None	None[c]
Peptide fragments (CNBr)	13–14	13–14[c]
Specific activity (μmole/hour/mg)	1300–1400	1300–1400

[a] Modified from ref. 61 to 64.

[b] Assumed.

[c] Determined by C. J. Lusty as in ref. 63. Unpublished.

Judging also from the $s_{20,w}$ value of 9.3 S and similarity in the kinetics of subunit dissociation, the kidney and liver enzymes are thought to be identical in molecular weight and oligomeric structure. As the data in Table 11 indicate, a high degree of similarity in primary structure and in identity among the polypeptide chains is evident.

In further pursuit of structural similarities or differences, since both enzymes behave as effective antigens, examination of the immunological properties shows that the antigenic sites appear to be the same. Complete cross reactivity was found in precipitin formation. The catalytic activity of each enzyme was inhibited appreciably by the antibody. Kinetic measurements showed complete cross reactivity with respect to degree of inhibition and amount of inhibitor (64).

On the basis of these similarities and the identity of the polypeptide chains in the tetramer, it seems justified to consider that the enzyme in the two organs is determined by the same single structural gene. This conclusion does not exclude the possibility that, in the kidney, argininosuccinase may be under the control of a separate regulatory gene. The identity of argininosuccinase from kidney and liver is of importance also in relation to the inborn error of metabolism, argininosuccinic aciduria, in which liver argininosuccinase activity is known to be impaired, for these results suggest that enzymatic activity in the kidney is similarly affected.

Argininosuccinase in Brain

Argininosuccinase occurs in a number of other mammalian tissues in addition to liver and kidney; and in fact, some activity can be found in almost all organs tested (Table 2). Both argininosuccinate synthetase and argininosuccinase are present in brain tissue of mammalian species, including primates (15). The levels of activity found, when considered in terms of the whole organ, indicate that the capacity of the normal brain to convert citrulline to arginine can be of significance in brain metabolism.

Observations that the level of argininosuccinate in the cerebrospinal fluid of individuals afflicted with argininosuccinic aciduria is severalfold higher than that in the plasma (84) are difficult to explain, particularly in view of the poor permeability of this amino acid. If the disorder is associated with a decrease in the activity of argininosuccinase in brain tissue as well as in liver, accumulation of the amino acid in the cerebrospinal fluid can be attributed to the reduced activity of the brain enzyme. The presence of argininosuccinase in normal brain (15, 85) offers presumptive evidence of this explanation. However, the identity of the brain enzyme with liver argininosuccinase has yet to be established. The presence in a single animal species of different proteins having the same catalytic function is not altogether rare. For example, the fructose diphosphatase from rabbit muscle differs in chemical, physical, and immunological properties from the enzyme in kidney and liver (86, 87). Other examples are known, arginase among them.

Argininosuccinase of Neurospora

Introduction There have been few opportunities in animal species to ascertain in a genetic mutation that an abnormal protein is produced as a gene product. *Neurospora* mutants deficient in argininosuccinase have been used to examine mutationally abnormal argininosuccinase. This enzyme was found in *Neurospora crassa* by Fincham and Boylen (88). Partially fractionated extracts of the mycelium afforded an enzyme preparation which catalyzes reaction 2 reversibly under conditions quite similar to those described for the mammalian liver enzyme. Mutants belonging to the *arg-10* group lack this enzyme and accumulate argininosuccinate (88). Genetic analysis by Newmeyer (89) has shown that the group of *arg-10* mutants are alleles which map within a single locus. Two suitably chosen mutants within this locus have recently been successfully combined by interallerlic complementation to produce heterokaryons which contain appreciably reduced levels of argininosuccinase activity, as compared with the wild type (90). According to current theories of genetic complementation, the heterokaryon enzyme represents a hybrid oligomer, formed by reassociations among the polypeptide chains of the two mutants.

Comparison of Wild Type and Mutant Enzymes Cohen and Bishop (90) have purified both the wild type and the heterokaryon enzymes in order to compare their properties. Purification of mycelium extracts from the wild type achieved a 700-fold purification by conventional procedures not unlike those used for the liver enzyme. The purified enzyme was homogeneous, as judged by sedimentation velocity and electrophoretic behavior on starch gel, and had a specific activity of 1000 (μmoles of argininosuccinate formed/hour/mg). The extract from heterokaryon mycelium had only about 2% of the activity of the wild type, and it was possible to purify this enzyme only about 120-fold by following the same procedure as for the wild type. Elution characteristics differed, and the preparation was less stable.

Kinetic properties such as pH optimum and K_m values were roughly the same for both enzymes. For the wild type of enzyme a value for $s_{20,w}$ of 8.7 S was obtained by ultracentrifugation and a molecular weight of 176,000 by sedimentation equilibrium. In the presence of SDS a single component was obtained in the ultracentrifuge with a sedimentation coefficient of 2.3 S, which suggests that the subunits may be identical. When the two enzymes were analyzed by sucrose gradient centrifugation, enzymatic activity appeared within a single peak which was located in both cases at the same position in the gradient. The value calculated for this peak was found to be 8.7 *S*. The electrophoretic mobility was also identical for the two enzymes.

Although similarity was found in some physical and kinetic properties, the failure to obtain a homogeneous preparation of the heterokaryon enzyme by application of the fractional procedures used for the wild type reflects some differences between the two enzymes (90). However, the enzymes from both sources appear to have the same molecular weight. It is consistent with the concept of complementation that the "hybrid" enzyme was found to differ from the wild type of enzyme in properties that reflect a mutation resulting in amino acid substitution.

The application of genetic techniques to establish the sequence of intermediates in a biosynthetic pathway was introduced by Srb and Horowitz in 1944 (91) for the arginine pathway with mutant strains of *Neurospora crassa*. The studies discussed in this section stem from their concepts.

ARGININOSUCCINATE

Properties

Introduction Argininosuccinate was originally prepared enzymatically from citrulline, aspartate, and ATP (reaction 1) and also from arginine and fumarate (reversal of reaction 2) and was isolated as the alcohol-insoluble barium salt (92, 93). It is significant in the proof of structure that the same compound was prepared by either method. The second method lends itself to large-scale preparations. Both procedures have been used for selective ^{14}C-labeling in the aspartate or arginine moiety of argininosuccinate (94). Anhydride I was originally prepared from argininosuccinate by ring closure at pH 3.2 and 25° (93). Anhydride II was originally found by Westall in the course of analyzing, by ion-exchange chromatography, the argininosuccinate excreted in the urine of a child afflicted with argininosuccinic aciduria (95). Both anhydrides have been obtained in crystalline form and have been chemically and physically characterized (93–96). The barium salt of argininosuccinate does not have a defined composition, owing to the strong basicity of the guanidino group; the preparations are therefore amorphous (93).

Neither anhydride is active as substrate for argininosuccinate synthetase or argininosuccinase, but under conditions likely to be encountered during isolation from biological materials argininosuccinate is converted by ring closures to anhydrides I and II. Differences in the properties of the two compounds have prompted investigations of their structure and those of related guanidino anhydrides of biological interest.

Estimation of Argininosuccinate and Anhydrides The instability of argininosuccinate, which otherwise has general amino acid properties, has imposed a requirement for analytical specificity. Several methods depend on enzymatic cleavage by argininosuccinase, followed either by arginine or urea determination (93) or fumarate determination (61). Another involves conversion to anhydride I, followed by oxidative cleavage to arginine (15). The mechanism of cleavage is discussed below in relation to structure.

When present in appreciable amounts, argininosuccinate can be determined in biological fluids and tissue extracts containing other amino acids by modifications of conventional ion-exchange chromatography. Analysis on strongly acidic resins is hampered by overlap with other amino acids and by conversion to anhydrides during passage down the column. Three components may therefore emerge in varying proportions, depending on time, pH, and temperature of elution (84, 94, 95). Modifications have been described by Cusworth and Westall (97) and by Shih et al. (98).

Chromatography on an anionic exchange resin has the advantage that only acidic amino acids are retained and interconversions are avoided during chromatography on the basic resin. In the resolution of a mixture of aspartate, glutamate, argininosuccinate, and the two anhydrides (Fig. 5), the order of

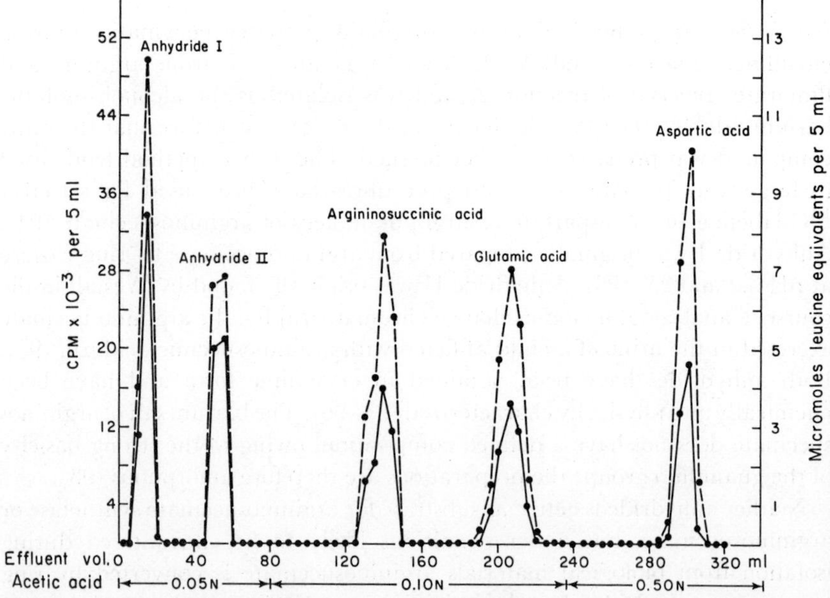

Figure 5. Chromatographic analysis of the indicated mixture of ^{14}C-labeled amino acids on Dowex-l-acetate. Radioactivity ●—●; ninhydrin ●—●. From Ratner and Kunkemueller (94).

elution is determined by the isoionic points, the respective values for anhydride I and II and argininosuccinate being 5.7, 4.2, and 3.2 (93, 94).

Interconversions between Argininosuccinate and Its Two Anhydrides The effects of pH and temperature on interconversion have been studied by Westall, using paper electrophoresis (95), and by Ratner and Kunkemueller, using Dowex-1 (94). Under conditions of pH, temperature, and time, somewhat resembling those prevailing during exposure on acidic resins, a 50% conversion to anhydride I occurs (Fig. 6). The rate increases with increases in acidity and temperature. Anhydride I can be hydrolyzed by raising the pH, but not anhydride II (94, 95). Anhydride II is formed from argininosuccinate more slowly than anhydride I. The effects of pH on the rate of conversion at elevated temperatures are difficult to determine because of degradative side reactions. In spite of the stability of Anhydride II, the conditions that influence the rate of formation or hydrolysis are poorly understood (94).

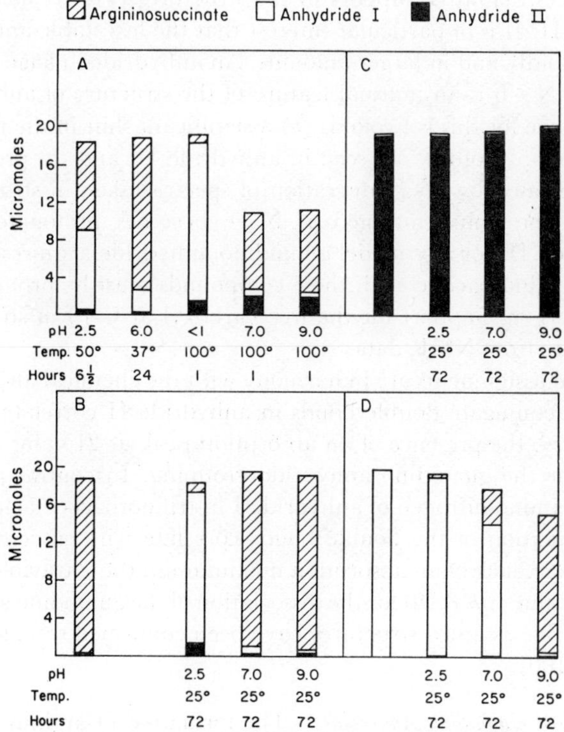

Figure 6. Composition of mixtures formed by exposing argininosuccinate or anhydride I or II to various conditions of temperature and pH. From Ratner and Kunkemueller (94).

Properties and Structures of Anhydrides I and II As has been mentioned, anhydride I is susceptible to mild oxidative cleavage in alkaline solution. According to the mechanism elucidated by means of selective ^{14}C-labeling (94), oxidation takes place at C-8, the α-carbon of the aspartate moiety, to form an unstable "pseudobase" which undergoes hydrolysis at the anhydride linkage. A subsequent rearrangement releases arginine and oxalacetate, and the latter then undergoes β-decarboxylation to form pyruvate and CO_2; the CO_2 liberated originates from C-10. Since anhydride II is entirely resistant to oxidative cleavage, other means were taken to elucidate structure with respect to the nitrogen and carboxyl groups involved in anhydride linkage.

Differences between the anhydrides bearing on structure are (*a*) greater heat lability of anhydride I, (*b*) a higher dissociation constant for the guanidino anhydride grouping of anhydride I, (*c*) an absorption maximum shown by anhydride II in the short ultraviolet region but not shown by anhydride I (93, 94), and (*d*) ease of oxidation of anhydride I.

A study of NMR spectra of these and model compounds by Kowalsky and Ratner (96) gives definitive support to the structures (Fig. 7) assigned anhydrides I and II. It is of particular interest that the less stable anhydride is formed more rapidly and in larger amounts. An anhydride linkage involving the nitrogen at N—b is an unusual feature of the structure of anhydride I. The main evidence for this is twofold: (*a*) a significant shift in the resonance position for the C-5 protons, not seen in anhydride II, and (*b*) protonation on nitrogen (revealed by area integration of spectra taken in strong acid), indicating that the imino nitrogen at N—c becomes protonated. Since neither anhydride II nor any model guanidino anhydride acquires a proton on nitrogen in trifluoroacetic acid, these compounds must be protonated on the carbonyl oxygen. Support for the free carboxyl at C-10 in structures I and II also comes from NMR data.

The structural assignments are in harmony with the chemical and physical properties. The conjugate double bonds in anhydride II correlate with the chemical stability, the presence of an absorption peak at 215 nm, and a pK value of 5.15 for the guanidino anhydride grouping. In contrast, the protonation on the imino nitrogen of anhydride I in trifluoroacetic acid and the unconjugated position of the double bonds correlate with susceptibility to oxidative cleavage, lack of an absorption maximum in the ultraviolet region, and the higher value (pK 8.18) for the dissociation of the guanidino anhydride grouping (96). The assigned structures have been confirmed recently by Lee and Politt (99, 100).

Structure of Related Guanidino Anhydrides The formation of argininosuccinate represents the primary biosynthesis of the guanidino group in all forms of life. Other guanidino compounds, including the important guanidino phos-

```
(c)  (7) O
HN     C—OH              O
 ‖     |                //
(6) C   CH—CH2—C—OH
H/ \ /(8)  (9)   (10)
(b) N   N(d)
 |      H
(5) CH2
 |
(4) CH2
 |
(3) CH2
 |  (a)
(2) CHNH2
 | O
(1) C—OH
```

```
H  H
N  N
 \\/ \            O
 C   CH—CH2—C—OH
 |    |
 N—C
 |   \\
 CH2  O
 |
 CH2
 |
 CH2
 |
 CHNH2
 | O
 C—OH          I
```

```
      O
 N—C
 ‖  |           O
 C  CH—CH2—C—OH
H/ \ /
N   N
|   H
CH2
|
CH2
|
CH2
|
CHNH2
| O
C—OH        II
```

Figure 7. Structural assignments of argininosuccinate and its anhydrides I and II. From Ratner et al. (93, 94, 96).

phogens, are formed from arginine by transamidination. Guanidinoacetic acid, creatine, and guanidinosuccinic acid and their corresponding anhydrides were therefore included in the NMR study. The spectra support five-membered ring structures III, IV, and V for guanidinosuccinic anhydride, guanidinoacetic anhydride, and creatinine (Fig. 8). In each case the double bonds are allocated as in anhydride II, rather than in the unconjugated positions usually depicted, in which the extracyclic nitrogen is imino. The assigned structures are in accord with the absorption maximum which each compound exhibits in the ultraviolet region.

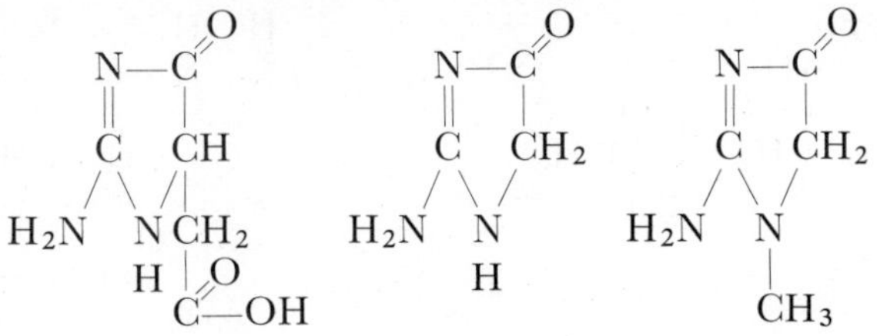

```
        O            O            O
      //           //           //
  N—C          N—C          N—C
  ‖   |         ‖   |         ‖   |
  C  CH        C  CH2       C  CH2
 / \ /|       / \ /        / \ /
H2N  N CH2  H2N  N      H2N  N
     H | O       H            |
       C—OH                  CH3
   III          IV            V
```

Figure 8. Structural assignments of guanidinosuccinic anhydride, III, guanidinoacetic anhydride, IV, and creatinine, V. From Ratner et al. (93, 94, 96).

Biological Occurrence in Ureotelic Species

In Genetic Metabolic Abnormality Argininosuccinate has been found in large amounts in infants and children afflicted with an inborn error of metabolism that is associated with the accumulation of this amino acid in plasma and cerebrospinal fluid and with high excretion in the urine (84, 95, 101). It is generally assumed that in normal individuals the only ornithine cycle intermediates present in the circulating plasma are ornithine, citrulline, and arginine. The fact that argininosuccinate has not previously been encountered in plasma is due as much to a low concentration as to its unstable behavior during conventional amino acid chromatography.

Normal Distribution Argininosuccinate levels have recently been determined (102) in the serum and liver of rats; in order to gain analytical sensitivity, the determinations were made after the animals had been subjected to a brief period of *in vivo* labeling with L-[U-^{14}C]citrulline (Table 12). The concentrations of argininosuccinate found in serum (0.112 μmole/10 ml) and liver extract (0.34 μmole/10 g wet weight) are low compared with the values for a number of other amino acids. The low tissue level should be taken as a reflection, not of the level of argininosuccinate synthetase activity, but of that of the enzymes that follow in the reaction sequence. The negatively cooperative kinetic properties of argininosuccinase, its activity relative to the synthetase, and coupling to arginase account for the low steady-state levels. The low arginine level found in liver extract is similarly explained by high arginase activity.

In accordance with the distribution of ^{14}C along the carbon chain of the administered citrulline, argininosuccinate, ornithine, arginine, and urea acquired ^{14}C during the 90 minute period of labeling. The fact that in liver or in serum the specific radioactivities of argininosuccinate, ornithine, and arginine were approximately the same and that considerable citrulline and urea were synthesized during the 90 minute period may be taken as evidence

Table 12 ^{14}**C-Labeled Amino Acids in Rat Serum and Liver after Injection of L-[U-^{14}C] Citrulline**[a]

	Serum			Liver		
Amino acid	Specific Radioactivity (cpm/μmole)	Total Radioactivity (cpm)	Micromoles (per 10 ml)	Specific Radioactivity (cpm/μmole)	Total Radioactivity (cpm)	Micromoles (per 10 g)
Argininosuccinate	6,100	690	0.112	4,030	1,370	0.34
Glutamate		120		230	2,780	12.09
Glutamine		970			5,100	22.17
Ornithine	7,400	7,000	0.95	4,400	27,500	6.25
Arginine	6,900	18,300	2.65	3,360	800	0.24
Urea	102	12,600	123	114	9,800	86.0
Proline	680	1,160	1.71	546	3,000	5.49
Citrulline	78,800	130,000	1.65		8,900	

[a] Each rat received 1.85 μmoles of L-[U-^{14}C]citrulline (specific radioactivity 4.12×10^6 cpm/μmole) per 275 g by intracardiac injection 90 minutes before exsanguination. Argininosuccinate was determined in 10 ml of serum or deproteinized extract from 10 g wet weight of liver by fractional elution from Dowex-1 (139) and further identified by conversion to the anhydrides and rechromatography. Neutral and basic amino acids were determined in aliquots taken from the first throughput by chromatography on IR-120. In addition to ^{14}C and ninhydrin analyses, ornithine, citrulline, and arginine were determined by specific colorimetric procedures. All manipulative procedures were monitored by ^{14}C. It was calculated from the specific radioactivity and size of the urea pool that at least 18% of the injected ^{14}C located in the ureido carbon had been converted to urea. From Ratner (102).

of rapid turnover of the ornithine cycle. The levels of citrulline in the serum and of ornithine in the liver are somewhat higher than reported values (103, 104), presumably because of citrulline administration.

It was of interest to find, keeping the respective pool sizes in mind, that the extent of ^{14}C-labeling in glutamic acid and proline indicates appreciable conversion of ornithine to these compounds (Table 12). It seems very probably, in confirmation of many earlier studies, that these conversions occurred through glutamic semialdehyde as the common intermediate (105, 107). The activity of ornithine δ-aminotransaminase in rat liver (108) is approximately 10% of the argininosuccinase activity (109). The formation of ornithine from proline is relatively slow in liver tissue (110). Replenishment of the ornithine level may become impaired in conditions in which the arginine pool is reduced because of malnutrition or abnormal biosynthesis. A reduction in the steady-state ornithine level would then adversely affect ornithine cycle.

REFERENCES

1. Krebs, H. A., and Henseleit, K. (1932). *Z. Physiol. Chem.* **210**, 33–66.

2. Krebs, H. A. (1952). In J. B. Sumner and K. Myrback, Eds., *The Enzymes*, **2**, Part 2, Academic, New York, pp. 866–885.

3. Gornall, A. G., and Hunter, A. (1943). *J. Biol. Chem.* **147**, 593–615.

4. Borsook, H., and Dubnoff, J. W. (1941). *J. Biol. Chem.* **138**, 389–403.

5. Cohen, P. P., and Hayano, M. (1946). *J. Biol. Chem.* **166**, 239–250, 251–259.

5*a*. Cohen, P. P., and Hayano, M. (1948). *J. Biol. Chem.* **172**, 405–415.

6. Ratner, S. (1947). *J. Biol. Chem.* **170**, 761–762.

7. Ratner, S., and Pappas, A. (1949). *J. Biol. Chem.* **179**, 1183–1198, 1199–1212.

8. Ratner, S. (1949). *Federation Proceedings* **8**, 603–609.

9. Krebs, H. A., and Eggleston, L. V. (1948). *Acta Biochim. Biophys.* **2**, 319–328.

10. Ratner, S. (1954). In F. F. Nord, Ed., *Advances in Enzymology* **15**, Interscience, New York, pp. 319–387.

11. De Duve, C., Wattiaux, R., and Baudhuin, P. (1962). In F. F. Nord, Ed., *Advances in Enzymology* **24**, Interscience, New York, pp. 291–358.

12. Wixom, R. L., Reddy, M. K., and Cohen, P. P. (1972). *J. Biol. Chem.* **247**, 3684–3692.

13. Ratner, S., and Petrack, B. (1953). *J. Biol. Chem.* **200**, 175–185.

14. Walker, J. B. (1958). *Proc. Soc. Exptl. Biol. Med.* **98**, 7–9.

15. Ratner, S., Morell, H., and Carvalho, E. (1960). *Arch. Biochem. Biophys.* **91**, 280–289.

16. Jones, M. E., Anderson, A. D., Anderson, C., and Hodes, S. (1961). *Arch. Biochem. Biophys.* **95**, 499–507.

17. Tamir, H., and Ratner, S. (1963). *Arch. Biochem. Biophys.* **102**, 249–258, 259–269.

18. Snodgrass, P. J., personal communication.

19. Tomlinson, S., and Westall, R. G. (1964). *Clin. Sci.* **26**, 261–269.

20. Schimke, R. T. (1964). *J. Biol. Chem.* **239**, 136–145.

21. Shih, V. E., Littlefield, J. W., and Moser, H. W. (1969). *Biochem. Gen.* **3**, 81–83.

22. Carter, C. E., and Cohen, L. H. (1955). *J. Am. Chem. Soc.* **77**, 499–500.

23. Abrams, R., and Bentley, M. (1955). *J. Am. Chem. Soc.* **77**, 4179–4180.

24. Lieberman, I. (1956). *J. Am. Chem. Soc.* **78**, 251.

25. Lieberman, I. (1956). *J. Biol. Chem.* **223**, 327–339.

26. Sonne, J. C., Lin, I., and Buchanan, J. M. (1953). *J. Am. Chem. Soc.* **75**, 1516–1517.

27. Sonne, J. C., Lin, I., and Buchanan, J. M. (1956). *J. Biol. Chem.* **220**, 369–378.

28. Levenberg, B., Hartman, S. C., and Buchanan, J. M. (1956). *J. Biol. Chem.* **220**, 379–390.

29. Buchanan, J. M. (1973). In A. Meister, Ed., *Advances in Enzymology* **39**, Wiley-Interscience, New York, pp. 91–183.

30. Reichard, P., and Lagerkvist, U. (1953). *Acta Chem. Scand.* **7**, 1207–1217.

31. Ratner, S., and Petrack, B. (1951). *J. Biol. Chem.* **191**, 693–705.

32. Ratner, S., and Petrack, B. (1953). *J. Biol. Chem.* **200**, 161–174.

33. Petrack, B., and Ratner, S. (1958). *J. Biol. Chem.* **233**, 1494–1500.

34. Schimke, R. T. (1962). *J. Biol. Chem.* **237**, 459–468.

35. Ratner, S. (1955). *Methods Enzymol.* **II**, 356–367.

36. Schuegraf, A., Ratner, S., and Warner, R. C. (1960). *J. Biol. Chem.* **235**, 3597–3602.

37. Rochovansky, O., and Ratner, S. (1961). *J. Biol. Chem.* **236**, 2254–2260.

38. Rochovansky, O., and Ratner, S. (1967). *J. Biol. Chem.* **242**, 3839–3849.

39. Rochovansky, O., investigations carried out in author's laboratory; to be published.

40. Ratner, S. (1970). *Methods Enzymol.* **17A**, 298–303.

41. Ratner, S., and Petrack, B. (1956). *Arch. Biochem. Biophys.* **65**, 582–585.

42. Ratner, S. (1962). In P. D. Boyer, H. Lardy, and K. Myrback, Eds., *The Enzymes* **6**, 2nd ed. Academic, New York, pp. 495–513.

43. Benzinger, T. H., Kitzinger, C., Hems, R., and Burton, K. (1959). *Biochem. J.* **71**, 400–407.

44. Alberty, R. A. (1969). *J. Biol. Chem.* **244**, 3290–3302.

45. Stadtman, E. R. (1972). In P. D. Boyer, Ed., *The Enzymes*, **8**, 3rd ed., Academic, New York, pp. 1–49.

46. Rochovansky, O., and Ratner, S. (1968). *Arch. Biochem. Biophys.* **127**, 688–704.

47. Jenkins, W. T. (1961). *J. Biol. Chem.* **236**, 1121–1125.

48. Kagan, H. W., and Meister, A. (1966). *Biochemistry* **5**, 725–732.

49. Berg, P. (1956). *J. Biol. Chem.* **222**, 991–1013, 1015–1023.

50. Stulberg, M. P., and Novelli, G. D. (1962). In P. D. Boyer, H. Lardy, K. Myrback, Eds., *The Enzymes* **6**, 2nd ed., Academic, New York, pp. 401–432.

51. Cleland, W. W. (1963). *Biochim. Biophys. Acta* **67**, 104–137, 173–187, 188–196.

52. Krishnaswamy, P. R., Pamiljans, V., and Meister, A. (1962). *J. Biol. Chem.* **237**, 2932–2940.

53. Papas, T. S., and Peterkofsky, A. (1972). *Biochemistry* **11**, 4602–4608.

54. Boyer, P. D. (1959). *Arch. Biochem. Biophys.* **82**, 387–410.

55. Mitra, S. K., and Mehler, A. H. (1966). *J. Biol. Chem.* **241**, 5161–5162.

56. Mehler, A. H., and Mitra, S. K. (1967). *J. Biol. Chem.* **242**, 5495–5499.

57. Ravel, J. M., Wang, S. F., Heinemeyer, C., and Shive, W. (1965). *J. Biol. Chem.* **240**, 432–438.

58. Levitzki, A., and Koshland, D. E., Jr. (1969). *Proc. Natl. Acad. Sci. USA.* **62**, 1121–1128.

59. Ratner, S., Anslow, W. P., Jr., and Petrack, B. (1953). *J. Biol. Chem.* **204**, 115–125.

60. Ratner, S., Petrack, B., and Rochovansky, O. (1953). *J. Biol. Chem.* **204**, 95–113.

61. Havir, E. A., Tamir, H., Ratner, S., and Warner, R. C. (1965). *J. Biol. Chem.* **240**, 3079–3088.

62. Schulze, I. T., Lusty, C. J., and Ratner, S. (1970). *J. Biol. Chem.* **245**, 4534–4543.

63. Lusty, C. J., and Ratner, S. (1972). *J. Biol. Chem.* **247**, 7010–7022.

64. Bray, R. C., and Ratner, S. (1971). *Arch. Biochem. Biophys.* **146**, 531–541.

65. Hoberman, H. D., Havir, E. A., Rochovansky, O., and Ratner, S. (1964). *J. Biol. Chem.* **239**, 3818–3820.

66. Bentley, R. (1969). *Molecular Asymmetry in Biochemistry*, **II**, Academic, New York, p. 152.

67. Hanson, K. R., and Havir, E. A. (1972). In P. D. Boyer, Ed., *The Enzymes* **7**, 3rd ed., Academic, New York, pp. 75–166.

68. Ratner, S. (1972). In P. D. Boyer, Ed., *The Enzymes* **7**, 3rd ed., Academic, New York, pp. 167–197.

69. Havir, E. A., and Hanson, K. R. (1968). *Biochemistry* **7**, 1904–1914.

70. Hanson, K. R., and Havir, E. A. (1970). *Arch. Biochem. Biophys.* **141**, 1–17.

71. Givot, I. L., Smith, T. A., and Abeles, R. H. (1969). *J. Biol. Chem.* **244**, 6341–6353.

72. Givot, I. L., and Abeles, R. H. (1970). *J. Biol. Chem.* **245**, 3271–3273.

73. Klotz, I. M., Langerman, N. R., and Darnall, D. W. (1970). *Ann. Rev. Biochem.* **39**, 25–62.

74. Haschemeyer, R. H. (1970). In F. F. Nord, Ed., *Advances in Enzymology* **33**, Wiley-Interscience, New York, pp. 71–118.

75. Rochovansky, O., (1975). *J. Biol. Chem.* **250**, 7225–7230.

76. Dales, S., Schulze, I. T., and Ratner, S. (1971). *Biochim. Biophys. Acta* **229**, 771–778.

77. Moody, M. F. (1971). *Biochim. Biophys. Acta* **229**, 779–794.

78. Valentine, R. C., Shapiro, B. M., and Stadtman, E. R. (1968). *Biochemistry* **7**, 2143–2152.

79. Kiselev, N. A., Shpitzberg, C. L., and Vainshtein, B. K. (1967). *J. Mol. Biol.* **25**, 433–441.

80. Kiselev, N. A., De Rosier, D. J., and Klug, A. (1968). *J. Mol. Biol.* **35**, 561–566.

81. Josephs, R., and Borisy, G. (1972). *J. Mol. Biol.* **65**, 127–155.

82. Weisenberg, R. C. (1972). *Science* **177**, 1104–1105.

83. Borisy, G., and Olmstead, J.B. (1972). *Science* **177**, 1196–1197.

84. Cusworth, D. C., and Dent, C. E. (1960). *Biochem. J.* **74**, 550–561.

85. Tomlinson, S., and Westall, R. G. (1960). *Nature* **188**, 235–236.

86. Fernando, J., Pontremoli, S., and Horecker, B. L. (1969). *Arch. Biochem. Biophys.* **129**, 370–376.

87. Enser, M., Shapiro, S., and Horecker, B. L. (1969). *Arch. Biochem. Biophys.* **129**, 377–383.

88. Fincham, J. R. S., and Boylen, J. B. (1957). *J. Gen. Microbiol.* **16**, 438–448.

89. Newmeyer, D. (1957). *J. Gen. Microbiol.* **16**, 449–462.

90. Cohen, B. B., and Bishop, J. O. (1966). *Genetical Research, Cambridge* **8**, 243–252.

91. Srb, A. M., and Horowitz, N. H. (1944). *J. Biol. Chem.* **154**, 123–139.

92. Ratner, S. (1951). In W. D. McElroy and B. Glass, Eds., *Phosphorus Metabolism* John Hopkins, Baltimore, **1**, pp. 601–619.

93. Ratner, S., Petrack, B., and Rochovansky, O. (1953). *J. Biol. Chem.* **204**, 95–113.

94. Ratner, S., and Kunkemueller, M. (1966). *Biochemistry* **5**, 1821–1832.

95. Westall, R. G. (1960). *Biochem. J.* **77**, 135–144.

96. Kowalsky, A., and Ratner, S. (1969). *Biochemistry* **8**, 899–907.

97. Cusworth, D. C., and Westall, R. G. (1961). *Nature* **192**, 555–556.

98. Shih, V. E., Efron, M. L., and Mechanic, G. L. (1967). *Anal. Biochem.* **20**, 299–311.

99. Lee, C. R., and Politt, R. J. (1970). *Tetrahedron* **26**, 3113–3121.

100. Lee, C. R., and Politt, R. J. (1972). *Biochem. J.* **126**, 79–87.

101. Allan, J. D., Cusworth, D. C., Dent, C. E., and Wilson, V. K. (1958). *Lancet* **1**, 182–187.

102. Ratner, S. (1975). *Anal. Biochem.* **63**, 141–155.

103. Tallan, H. H., Moore, S., and Stein, W. H. (1954). *J. Biol. Chem.* **211**, 927–939.

104. Herbert, J. D., Coulson, R. A., and Hernandez, T. (1966). *Comp. Biochem. Physiol.* **17**, 583–598.

105. Stetten, M. R. (1955). In W. D. McElroy and B. Glass, Eds., *Amino Acid Metabolism*, John Hopkins, Baltimore, pp. 277–290.

106. Meister, A., Radhakrishnan, A. N., and Buckley, S. D. (1957). *J. Biol. Chem.* **229**, 789–800.

107. Strecker, H. J. (1960). *J. Biol. Chem.* **235**, 3218–3223.

108. Peraino, C., and Pitot, H. C. (1963). *Biochim. Biophys. Acta* **73**, 222–231.

109. Smith, A. D., Benziman, M., and Strecker, H. J. (1967). *Biochem. J.* **104**, 557–564.

110. Civen, M., Brown, C. B., and Trimmer, B. M. (1967). *Arch. Biochem. Biophys.* **120**, 352–358.

16

Arginase

Guillermo Soberón
and Rafael Palacios
Departmento de Biología Molecular
Instituto de Investigaciones Biomédicas
Universidad Nacional Autónoma de Mexico

Arginase (L-arginine amidinohydrolase EC 3.5.3.1) has been thoroughly studied. We will not attempt to review exhaustively all data accumulated on the enzyme. Instead, we will emphasize some facts pertinent to the points discussed in this symposium, namely, those characteristics which are of significance for the functioning of arginase in the urea cycle.

For a long time the statement "there is no arginase in the liver of uricotelic animals" was recognized as the rule of Clementi after the Italian scientist who first described such a situation (1, 2). However when more sensitive methods of enzyme assay became available, a low level of activity could be found in the liver of uricotelic animals (3). Work carried out in our laboratory (4, 5) made it clear that the arginase present in the liver of uricotelic animals differs in several respects from that of the ureotelic animals. Accordingly Clementi's rule could be reestablished by saying "there is no ureotelic arginase in the liver of uricotelic animals." As we will explain later, there are a few exceptions to this rule.

The existence of the enzymes involved in the synthesis of arginine concomitant with the presence of arginase does not necessarily mean a functional urea cycle. Indeed, Srb and Horowitz (6) claimed that *Neurospora crassa* possesses a urea cycle based on the detection of each of the enzyme activities required for the synthesis of carbamyl phosphate, ornithine, citrulline, argininosuccinic acid, arginine, and urea. Mora challenged the statement for

The authors thank Dr. J. Mora for his valuable suggestions and Dr. H. Stiglitz for her assistance in reviewing the manuscript.

the following reasons: the ammonia pool is rather elevated in *N. crassa*, and the urea cycle is a device to keep ammonia low; ammonia is not toxic to the fungi; arginine is not an indispensable amino acid for *N. crassa*; the arginine pool in Neurospora is very high in comparison to that in the liver of ureotelic animals; a mutant, unable to convert glutamic acid to glutamic semidaldehyde, does not catalyze the arginine synthesized from ornithine to give proline but instead uses and hydrolyzes exogenous arginine as a source of nitrogen; and lastly there is a high level of urease activity which hydrolyzes the urea produced to give back ammonia. Hence a functional urea cycle would be a metabolic waste, and it is well established that biological systems do not operate under such conditions (7, 8).

The above-mentioned facts suggest that a particular protein possessing the catalytic activity to hydrolyze arginine is required to intervene in the urea cycle. This situation leads to the problem of structure and function, location, regulation, and, in general, cell organization and the metabolic role of arginase and the other urea cycle enzymes.

THE PRESENCE OF ARGINASE IN BIOLOGICAL SYSTEMS

Urea is a simple molecule formed from NH_3 and CO_2; therefore it is tempting to speculate why it is necessary to build such large complex compounds as citrulline, argininosuccinic acid, and arginine to produce urea. Our reasoning is similar to that invoked by Davis (9) to explain the synthesis in some fungi of the six-carbon amino acid lysine from the seven-carbon compound precursor diamino pimelic acid, namely, that a metabolic gain can be achieved during evolution by adding a new enzyme to a multienzyme system already existing. In the case of urea formation arginase would be added to the arginine biosynthetic pathway. Again a specific protein would be required to fulfill the proposed role of forming urea as a nitrogenous excretory product.

To gain insight into the role of arginase in biological systems, let us examine different situations where arginase activity is present:

a. In organisms where the biosynthesis of arginine occurs but there is no functional urea cycle (*N. crassa*)

b. In the liver of ureotelic and ureosmotic vertebrates

c. In the intestine and other tissues of some ureotelic invertebrates

d. In the liver of "ureogenic" animals, those with a potential urea cycle which does not function due to the lack of necessity to produce urea (teleosti fishes) (10)

e. In the liver of uricotelic species which do not possess the arginine biosynthetic enzymes

Similarly, there are other organs that contain arginase and lack the capacity to synthesize arginine (kidney, brain, mammary gland, salivary gland erythrocytes). Furthermore, arginase activity has also been described in the Papiloma Shope tumor; however, it has been ascertained that the protein is specified by the host cell (11).

UREOTELIC AND URICOTELIC ARGINASES

It was well substantiated in our laboratory 10 years ago that the arginase present in the liver of ureotelic animals (ureotelic arginase) has different properties than that found in the liver of uricotelic animals (uricotelic arginase). Moreover, the arginase from *N. crassa* resembles the latter very closely (Table 1). Ureotelic arginase has a lower K_m for L-arginine, is not inhibited by *p*-chloromercurybenzoate, and is stable during dialysis. The hepatic arginase activity of ureotelic animals is high compared to that of uricotelic animals. An antibody to purified rat liver arginase cross-reacts with the liver arginase of all ureotelic species tested so far (cow, dog, cat, rabbit, mouse, human, horse, frog, toad); it does not cross-react with liver arginase of uricotelic animals (chicken, iguana) or with *N. crassa* arginase. The ratio of hydrolysis of arginine to canavanine is 1.5 for the ureotelic arginase and 7 for the uricotelic one (4, 5). The molecular weight of the former was found to be 138,000 to 142,000 (12, 5), and 276,000 for the latter (5).

SPECIFICITY AND MOLECULAR CHARACTERISTICS OF UREOTELIC ARGINASE

The enzyme has a high degree of specificity with respect to chain length, stereo arrangement, and substituents of the α-carbon, and the guanido ends of arginine. Only L-arginine and L-canavanine serve as substrates. In the latter the methylene corresponding to the delta carbon of arginine has been substituted by an oxygen atom. Longer and shorter molecules, the methyl ester, D-arginine, argininic acid, methyl-*N*-L-arginine do not function as substrates. Neither do compounds where the H atoms of the guanido grupe are replaced by methyl groups (13).

Of interest is the fact that a guanidino ureo hydrolase activity clearly due to a protein other than arginase has been found in the liver of ureotelic animals. The metabolic role of this action has not yet been elucidated (4).

Rat liver arginase has been purified and its properties defined (14). The molecular weight estimated by sedimentation velocity coefficient, diffusion coefficient, and sedimentation equilibrium was very close to 118,000 lower than the previously reported values of 138,000 to 142,000 obtained by sucrose

Table 1 Properties of Arginases from Different Sources[a]

Source of Arginase	Activity	K_m	Inhibition by Substrate	Precipitation with Antibody against Purified Rat Liver Arginase	Molecular Weight	Stability during Dialysis	Inhibition by 5×10^{-3} M p-CMB	Specificity	Ratio of Hydrolysis L-Arginine/ L-Canavanine
Rat liver	High	10^{-2} M	+	+	138,000	Stable	0	Only L-arginine and L-canavanine are substrates	1.5
Chicken liver	Low	10^{-1} M	None	–	276,000	Unstable	100%	Only L-arginine and L-canavanine are substrates	7.2
Neurospora crassa	Low (inducible)	10^{-1} M	None	–	270,000	Unstable	100%	Only L-arginine and L-canavanine are substrates	6.5

[a] Taken from Soberón et al., 1967.

gradient analysis (12, 5). It has been reported to be composed of four apparently identical polypeptide subunits (12); however, Penninckx, Simon, and Wiame (15) have claimed that rat liver arginase is a trimer.

Rat liver arginase, a Mn^{2+} requiring enzyme, is very stable and does not dissociate readily; its molecular size does not vary by removal of Mn^{2+} and replacement by Co^{2+} (14). Four moles of Mn^{2+} are bound per mole of fully activated rat liver arginase. However, the binding affinities are not equal for the four Mn^{2+} ions; two of them remain tightly attached to the enzyme with the loss of 50% of its activity. The corresponding dissociation constants for the EMn_2^{2+} and EMn_1 complexes are of the order of $3 \times 10^{-7}\ M^{-1}$ for the EMn_4^{2+} complex (16). In the absence of the divalent cation, the enzyme dissociates at low pH; however, it can reassociate in the presence of Mn^{2+}. The rate of renaturation depends on the cation concentration (17).

Rabbit liver arginase has also been purified and studied (18). It has a molecular weight of 110,000 (sedimentation constant 5.9S) and is thought to consist of a tetrameric structure composed of two different kinds of subunits, which were shown to have catalytic activity in the presence of Mn^{2+}. The immunoprecipitation pattern of the native enzyme is not altered when dissociated.

Rat liver arginase, a basic protein which is strongly absorbed by carboxyl methyl cellulose, migrates toward the cathode at pH 8.3. On the other hand rabbit liver arginase migrates toward the anode at pH 8.3 and is not retained by the carboxymethyl cellulose resin. The isoelectric point has been determined for the ureotelic arginases from several species. They were classified either as basic arginases (rat and dog, Ip 9.4); mouse is also basic, since it is retained on CMC cellulose equilibrated at pH 7.5, or as neutral or acidic arginases (rabbit, Ip 6.5). There are two arginase isozymes in pork liver (Ip 6.9 and 8.8) and in monkey liver (Ip 6.8 and 7.5) (19).

We have obtained evidence of three different molecular forms of rat liver arginase (20), which were identifiable in the supernatant of tissue extracts and were separable on carboxymethyl cellulose columns. A purification of 2500 to 5000 fold, 800 to 1000 fold, and 600 to 1000 fold, respectively, has been achieved. Significant differences in the energy of activation were found for two molecular forms studied. They are cationic at pH 5.5 and 8.8 and show different mobilities on electrophoresis. The molecular weight determination by gel filtration yielded a value of 110,000 to 115,000 for both forms and was confirmed by the use of thin-layer immunochromatography plates. Both molecular forms have similar antigenic determinants.

In addition, two molecular forms of rat liver arginase have been separated on DEAE-cellulose (21). Electrophoretic patterns of crude homogenates show two bands of activity: one that migrates toward the anode and one that migrates toward the cathode (22). Fibroblasts also contain arginase isozymes

(23). Chromatography of purified rabbit liver arginase on DEAE cellulose reveals multiple active forms (18). The metabolic role and significance of these molecular forms cannot be ascertained at the present time. It well might be that some of the molecular forms described result from interconversion by the gain or a loss of minor components from or to each of the arginase subunits and thus could give rise to three or four closely related oligomers. Traniello, Barsacchi, Magri, and Grazi (24) have demonstrated that chicken kidney contains both the ureotelic and the uricotelic types of arginase.

SPECIFICITY AND MOLECULAR CHARACTERISTICS OF URICOTELIC ARGINASE

The enzyme from chicken liver has been purified 1600-fold and its properties described (25, 26). The preparation gives a single band on disc gel electrophoresis at pH 7.0 and 9.0. It has a sedimentation coefficient of 9 on sucrose gradient analysis; the molecular weight thus derived is 220,000. Furthermore the enzyme is able to hydrolyze argininic acid in addition to arginine. For both substrates the K_m is 30 to 100 mM, and it binds 1.6 g-atoms of Mn^{2+} per molecule of enzyme.

The amino acid composition of ureotelic and uricotelic arginases is compared in Table 2. The difference between chicken liver and rat liver arginase in moles per 100,000 g of protein is as follows: lysine −24, arginine +20, aspartic acid +21, and glutamic acid +50, which accounts for the more basic character of the rat liver enzyme.

OTHER TYPES OF ARGINASES

Arginase from human erythrocytes was purified 4200-fold. The molecular weight determination is 120,000; the sedimentation coefficient was found to be 6.4S by ultracentrifugation; Mn^{2+} is an activator of the enzyme activity. The K_m value for L-arginine is 1.56×10^{-3} M. The isoelectric point of the enzyme is 7.5 (27).

The structure and properties of the intestine arginase from the marine polychaete Annelid *Pista pacifica* have also been identified (28). It is an oligomer with a molecular weight of 205,000 composed of subunits which are enzymatically active and have a molecular weight of 34,000. The only usable substrates are arginine and canavanine, the K_m for the former being 155 to 160 mM; the enzyme does not exhibit substrate inhibition. The activity is enhanced by *p*-chloromercuribenzoate; the effect of Mn^{2+} as an activator is not very noticeable, although the divalent cation is necessary to maintain quaternary structure. There is no indication of a functional urea cycle in this annelid, but since it is aquatic, it is probably ammonotelic.

Table 2 Amino Acid Composition of the Uricotelic Chicken Liver Arginase as Compared with Rat Liver Arginase[a]

Amino Acid	Composition (moles/100,000 g of arginase)	
	Chicken Liver	Rat Liver
Lysine	49.6	73.9
Histidine	15.5	19.5
Arginine	47.1	27.5
Aspartic acid	94.7	73.7
Threonine	41.6	57.7
Serine	50.1	50.8
Glutamic acid	123.5	73.7
Proline	47.0	60.2
Glycine	74.0	76.2
Alanine	74.1	54.2
Half-cystine	17.7	6.1
Valine	66.6	73.6
Methionine	15.5	10.8
Isoleucine	38.6	43.2
Leucine	80.1	60.2
Tyrosine	23.1	20.3
Phenylalanine	30.0	26.2
Tryptophan	10.6	10.3

[a] Reproduced from Grazi and Magri, 1972.

CORRELATION BETWEEN FUNCTIONAL UREA CYCLE AND THE MOLECULAR WEIGHT OF ARGINASE

The original description of two different types of arginases (5), has been further extended to include proteins with the same catalytic activity and a wider spectrum of molecular weights. The rat liver arginase subunit has a molecular weight of 30 to 37.5 $\times$ 10^3. It is possible to arrange all arginases whose molecular weights have been determined belonging to proteins with high (220 to 270 $\times$ 10^3), medium (105 to 150 $\times$ 10^3), and low (27 to 70 $\times$ 10^3) molecular weights (Fig. 1).

Very likely all high- and medium-molecular-weight arginases have the same weight, and the differences found could be explained by the different procedures employed for the determinations.

With respect to ureotelism it can be established that all liver arginases of ureotelic animals fit in the medium-molecular-weight type. On the other hand there are invertebrates in which a functional urea cycle has been described

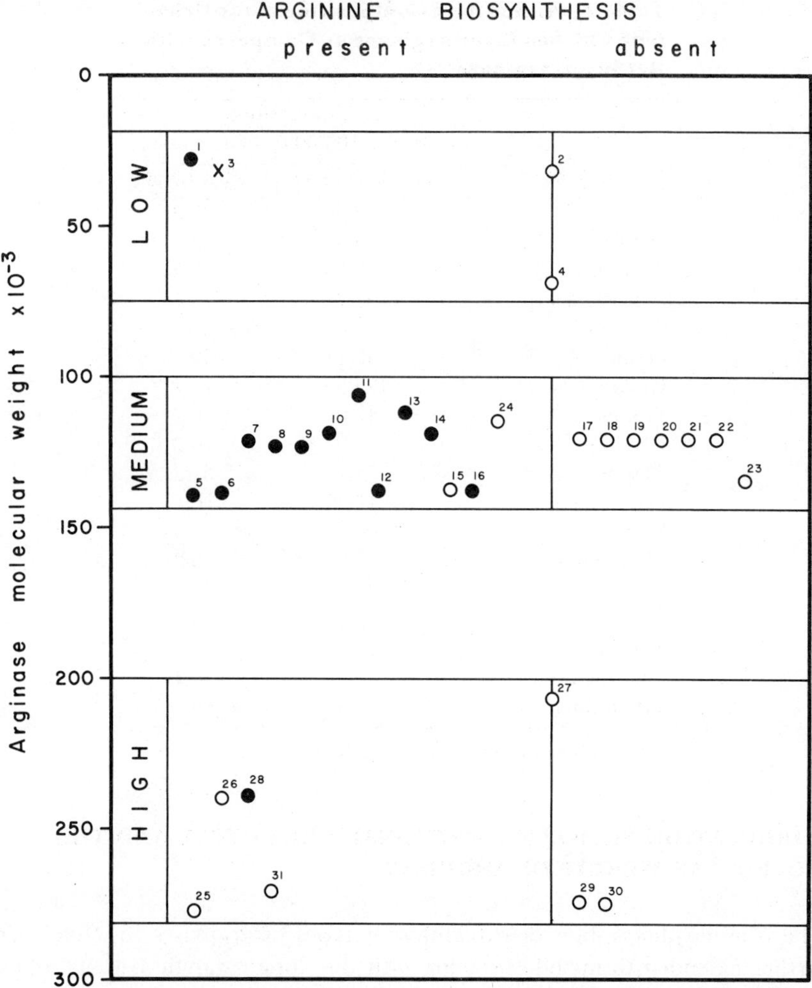

Figure 1. Correlation between ureotelism and molecular weight of arginase. The full circle indicates evidence of a functional urea cycle and the open circle lack of such evidence. X applies to the special case of an active low molecular weight arginase in the rat liver (subunit produced by dissociation?). The numbers indicate the source of arginase as follows: 1. *Earth worm* (29); 2. *Silkmoth* (57); 3. *Rat liver* (58); 4. *Cockroach* (57); 5. *Rat liver* (12); 6. *Rat liver* (4); 7. *Rat liver* (14); 8. *Rat liver* (58); 9. *Rat liver* (27); 10. *Rat liver* (46); 11. *Rat liver* (15); 12. *Horse liver* (54); 13. *Rabbit liver* (18); 14. *Human liver* (53); 15. *Axolotl liver* (38); 16. *Axolotl liver* (38); 17. *Human erythrocyte* (27); 18. *Rat kidney* (21); 19. *Rat submaxilary gland* (21); 20. *Rat brain* (21); 21. *Ox brain* (21); 22. *Chicken liver* (33), (34); 23. *Sea gull liver* (29); 24. *Yeast* (epiarginase) (36); 25. *Neurospora* (47); 26. *Snail* (29); 27. *Silkmoth* (57); 28. *Land planaria* (29); 29. *Chicken liver* (4,5); 30. *Iguana liver* (4,5); 31. *Axolotl liver* (38).

whose arginases are of other types (Fig. 1). Thus, land planaria, which has been proposed to be the first aquatic living organism to conquer terrestrial life through the advent of ureotelism, has a 238,000 molecular weight arginase (29). The earthworm, which becomes ureotelic under conditions of starvation (30), has an arginase of 27,000 molecular weight. However, it is not clear if the study of these arginases was carried out under conditions of ureotelic habit; likewise the possibility exists that the arginase was dissociated through manipulation, as has been demonstrated to be the case with beef enzyme (19). There is also the counterpart situation, medium-molecular-weight arginases without a functional urea cycle in some tissues where the activities of arginine biosynthetic enzymes cannot be detected, as in kidney, submaxillary salivary gland, and brain of the rat (21), ox brain (31), and the liver of uricotelic animals that also lack arginine biosynthetic enzymes: the sea gull liver (32, 29) and the terrestrial turtle liver (5). The existence of ureotelic arginase in the liver of uricotelic animals has been interpreted as a remnant from a ureotelic past. It is of great interest that starvation, the administration of cortisol, or insulin causes the appearance of ureotelic arginase in the mitochondrial matrix of the chicken liver (33, 34, 35).

In the yeast *Saccharomices cerevisiae*, the cycle does not function in spite of the concomitancy of the arginine biosynthetic pathway and a medium-molecular-weight arginase (36). This situation is of particular interest, because it was shown that arginase plays a regulatory role by interacting, in the presence of ornithine and arginine, with ornithine transcarbamylase, which it inhibits. Messengu and Wiame (36) propose to call this type of arginase epiarginase because of its regulatory action.

Incidentally, Penninckx, Simon, and Wiame (15), using SDS acrylamide gel electrophoresis, have reported that *S. cerevisiae* arginase (mol. wt. 114,000) and rat liver arginase (mol. wt. 105,000) are trimers composed of subunits having the molecular weight of 37,000 and 35,000 respectively. It is tempting to speculate that the interaction of arginase with ornithine transcarbamylase, an enzyme of the arginine biosynthetic pathway, might be the first step toward the functional integration of the cycle, since a further mutation of arginase or of ornithine transcarbamylase which would prevent the inhibition of the latter enzyme would ensure the continuous functioning of the cycle and facilitate handling of its metabolites in a closed circuit.

We wonder if the association of the subunits to give just dimers and tetramers could reconcile Wiame's results, which suggest a trimer structure, with previous reports that indicate a tetrameric one (12, 14, 18). Indeed, molecular weights of 30,000, 60,000, and 120,000 for the monomer, dimer, and tetramer determined by ultracentrifugation could be close enough to Wiame's three bands detected on SDS acrylamide gels.

The claim that ureogenic species (10) maintain a potentially functional urea cycle cannot be framed in the context of the issue now discussed, because it is not known what type of arginase is involved.

FUNCTIONAL INTEGRATION OF THE UREOTELIC ARGINASE WITH THE ARGININE BIOSYNTHESIS PATH

Rat liver arginase reveals itself to be distributed among nuclei, mitochondria, microsomes (37, 4). However, it is released by I *M* KCl; thus, it well might be that it is nonspecifically adsorbed during the process (4). We had suggested that it could be originally located in the mitochondria, because carbamyl phosphate synthetase and ornithine transcarbamylase are located there, thinking that argininosuccinic synthetase and arginino succinase would also leak out of these particles by the handling procedure (38). More recently Lehninger has described that carbamyl phosphate synthetase and ornithine transcarbamylase are present in mitochondria and the three other enzymes are in the cytoplasm. He has also reported a nonspecific transport system for citrulline and a highly specific transport system for ornithine, which requires the functioning of the electron transport system and is coupled to respiration-dependent inward transport of phosphate (39, 40).

Even though part of the urea cycle could take place in mitochondria and part in the cytoplasm, it appears that the medium-molecular-weight arginase is the one involved in the functioning of the cycle.

If a particular type of arginase is required for ureotelism, it can be inferred that such a protein should establish "functional links" with the arginine formation sites in order to derive arginine toward the formation of urea and to reshuffle ornithine into the arginine biosynthetic pathway.

Our studies on the Mexican axolotl have thrown light on the problem of functional links between arginase and the arginine formation sites. The axolotl, a neotenic species which does not metamorphose spontaneously, reproduces itself in the aquatic larval stage. When thyroid hormones are administered, it goes into the terrestrial adult stage. While living in water, the axolotl is ammonotelic; when metamorphosis is induced, it becomes ureotelic. Contrary to what is observed in the advent of ureotelism on natural or induced metamorphosis of the anura, where it corresponds to an 8- to 30-fold increase of the enzyme activities of the urea cycle, including *de novo* synthesis of carbamyl phosphate synthetase, the metamorphosis of the Mexican axolotl does not show any significant increase in the activity of the urea-cycle enzymes which are already high in the ammonotelic larval stage (43), just a moderate elevation of CPS.

It was detected that liver arginase from the nonmetamorphosed Mexican axolotl was not able to hydrolyze the arginine formed in a system containing citrulline, aspartic acid, ATP, and liver tissue homogenate. Incubation of such a system at pH 7.0 leads to the accumulation of arginine and the production of a small amount of urea. Liver tissue homogenate from a metamorphosed axolotl is able to hydrolyze more efficiently the arginine originated from citrulline, aspartic acid, and ATP in spite of the fact that the level of arginase activity was the same in the liver of the nonmetamorphosed animal; activity was measured by the splitting of exogenous arginine either under so-called optimal conditions (pH 9.0, Mn^{2+} added) or under conditions leading to the formation of endogenous arginine (pH 7.0, no Mn^{2+} added). The rat liver homogenate is very efficient in converting endogenous arginine to urea and ornithine (4, 43).

The addition of a small amount of purified rat liver arginase (10 u) to the above-described system prepared with liver tissue homogenate from nonmetamorphosed axolotl leads to the splitting of the endogenous arginine produced; on the other hand, addition of purified *N. crassa* arginase fails to achieve this effect even when more activity was added (200 u) (43).

It seems that certain characteristics of arginase which are present in rat liver arginase but not in *N. crassa* arginase are mandatory to enable the enzyme to hydrolyze endogenous arginine. These characteristics would appear on metamorphosis of the Mexican axolotl.

A sucrose gradient analysis of the nonmetamorphosed axolotl liver revealed the existence of two peaks of arginase activity, one corresponding to a medium-molecular-weight type and the other to a high-molecular-weight type. After metamorphosis, the profile is shifted toward the medium-molecular-weight type (43).

Axolotl liver arginase is very unstable and decays rapidly under the conditions in which arginine is formed from citrulline, aspartic acid, and ATP and in which it is converted to ornithine and urea. The instability of the enzyme, however, is the same for the nonmetamorphosed and the metamorphosed axolotl. In the latter the capacity to hydrolyze endogenous arginine is kept to a certain time if the system is functioning and is completely lost by preincubation of the system. This suggests either that the presence of citrulline, aspartic acid, ATP, or Mg^{2+} is necessary to maintain such capacity or that the flow of conversion of the metabolites to urea preserves the arginase functionally linked (44). We have also demonstrated that some divalent cations render axolotl liver arginase completely stable and enable the enzyme to hydrolyze endogenous arginine (45). The cations that are effective (Mn, Fe, Co, Ni, and Zn) have atomic numbers between 25 and 30, ionic radii between 0.74 and 0.80 Å, and a $4s^2$ electronic structure. Cu^{2+}, unable to produce the same effect, has an atomic number of 29, 0.96 Å atomic radius, and a $4s^1$ electronic configuration.

It was possible to distinguish the effect of the divalent cations on the activity, stability, and capacity to hydrolyze endogenous arginine by using different concentrations of the metal ions. The activity of arginase in the presence of the metal ions was studied with purified preparations of the enzyme. It was shown that it is possible to obtain different enzyme-metal complexes (made of identical protein moieties but different metal species and with varying metal-protein ratio) which are completely stable, having the same level of arginase activity but different capacities for hydrolyzing endogenous arginine. This strongly suggests that some conformational states of arginase are more suitable to be functionally linked with the arginine formation sites (45).

The conclusions reached with axolotl liver arginase were supported by experiments carried out with rat liver homogenate, indicating that rat liver arginase preferentially hydrolyzes endogenous arginine as opposed to that generated by the carboxypeptidase B hydrolysis of hyppuryl-L-arginine (46).

In addition, it should be considered the low pool of arginine in the liver of ureotelic animals and the high K_m of ureotelic arginase; moreover, the regulatory role of Wiame's epiarginase, which requires its interaction with the mitochondrial enzyme ornithine transcarbamylase, and the fact that the ureotelic arginase induced on starvation in certain types of chicks appears specifically in the mitochondrial matrix, supports the idea that the function of arginase inside the cell depends on appropriate links with other proteins.

POSSIBLE ROLES OF ARGINASE OTHER THAN ITS FUNCTION IN UREOTELISM

The high-molecular-weight arginase has been thoroughly studied in *N. crassa*. It has been established that, through its participation in arginine catabolism, arginine can be used as a source of N and also as an intermediate in an alternate pathway for the biosynthesis of proline (7, 47, 48, 49).

It is worth mentioning a few facts about the regulation of *N. crassa* arginase, since this will not be dealt with in the corresponding section of this symposium. *N. crassa* arginase is an inducible enzyme. Its control is the result of the interaction of arginine, glutamine, and ammonia. The analysis of the experimental data available has made Mora and coworkers (59) advance the following hypothesis (*a*) glutamine prevents the synthesis of a protein required for the induction of arginase by arginine; (*b*) this protein becomes inactive or is transformed into a repressor of arginase in the presence of ammonia and glutamine; (*c*) glutamine synthetase appears to be necessary for the induction of arginase by arginine (50).

Uricotelic animals have the high-molecular-weight type of arginase; during evolution they have lost the enzymes involved in the biosynthesis of arginine. It is possible that this arginase does not have another role, as does its possible predecessor, *N. crassa* arginase, that is, its participation in the catabolism of arginine. The case of mammary gland arginase should be mentioned, because the enzyme activity has been assigned the role of converting arginine to proline (51). There is no information on the molecular weight of mammary gland arginase, although it has been shown that it does not cross-react with an antibody to purified rat liver arginase and that these arginases differ in solubility and stability properties (temperature, pH, and treatment with various agents) (52).

Grazi (26) has suggested that the ureotelic-like arginase activity appearing in the chicken liver on starvation might have something to do with control of protein synthesis.

FINAL COMMENT

The characterization of the arginases present in biological systems and the definition of their functional role illustrate how the emerging of a catalytic activity to be added to a multienzyme system already existing determines the advent of a new way of nitrogen excretion which is essential for the establishment of terrestrial life.

A complete survey of all arginases so far detected in order to relate their structure to possible function is in order. For this purpose the purification of arginase by affinity chromatography and the determination of molecular weight and immunological properties seem to be the most appropriate criteria.

REFERENCES

1. Clementi, A. (1914). *R. C. Acad. Lincei*, **23**, 612.
2. Clementi, A. (1919). *R. C. Acad. Lincei*. **27**, 299.
3. Tamir, H., and Ratner, S. (1963). *Arch. Biochem. Biophys.* **102**, 249.
4. Mora, J., Martuscelli, J., Ortiz-Pineda, J., and Soberón, G. (1965). *Biochem. J.* **96**, 28.
5. Mora, J., Tarrab, R., Martuscelli, J., and Soberón, G. (1965). *Biochem. J.* **96**, 588.
6. Srb, A. M., and Horowitz, N. H. (1944). *J. Biol. Chem.* **154**, 129.
7. Castañeda, M., Martuscelli, J., and Mora, J. (1967). *Biochim. Biophys. Acta*. **141**, 276.
8. Cañedo, L., Martuscelli, J., and Mora, J. (1967). *Monogr. Nat. Cancer Inst.* **27**, 273.
9. Davis, B. D. (1961). *Cold Spring Harbor Symp. Quant. Biol.* **26**, 1.
10. Huggins, A. K., Skutsch, G., and Baldwing, E. (1969). *Comp. Biochem. Physiol.* **28**, 587.
11. Orth, G., Jibard, N., and Vielle-Breitburds, F. (1971). *C. R. Acad. Sci. Paris*. **273**, 1171.

12. Schimke, R. T. (1962). *J. Biol. Chem.* **237**, 459.

13. Hunter, A., and Downs, C. F. (1945). *J. Biol. Chem.* **157**, 427.

14. Hirsch-Kolb, M., and Greenberg, D. M. (1968). *J. Biol. Chem.* **243**, 6123.

15. Penninckx, M., Simon, J. P., and Wiame, J. M. (1974). *Eur. J. Biochem.* **49**, 429.

16. Hirsch-Kolb, H., Kolb, H. J., and Greenberg, D. M. (1971). *J. Biol. Chem.* **246**, 395.

17. Hosoyama, Y., (1972). *Eur. J. Biochem.* **27**, 48.

18. Vielle-Breitburd, F., and Orth, G. (1972). *J. Biol. Chem.* **247**, 1227.

19. Hirsch-Kolb, H., Heine, J. P., Kolb, H. J., and Greenberg, D. M. (1970). *Comp. Biochem. Physiol.* **37**, 345.

20. Tarrab, R., Rodriguez, J., Huitrón, C., Palacios, R., and Soberón, G. (1974). *Eur. J. Biochem.* **49**, 457.

21. Gasiorowska, I., Porembska, Z., Jachiomowicz, J., and Mochnacka, I. (1970). *Acta Biochimica Pol.* **17**, 19.

22. Porembska, Z., Jachimowicz, J., and Gasiorowska, I. (1971). *Bull. Acad. Pol. Sci. Ser. Sci. Biol.* **19**, 27.

23. Vanelsen, A. F., and Leroy, J. A. (1975). *Biochem. Biophys. Res. Comp.* **62**, 191.

24. Tranello, S., Barsachi, R., Magri, E., and Grazi, E. (1975). *Biochem. J.* **145**, 153.

25. Grazi, E., and Magri, E. (1972). *Biochem. J.* **126**, 667.

26. Grazi, E. (1973). *Acta Vitam. Enzymol. Milano.* **27**, 37.

27. Nishibe, H. (1973). *Physiol. Chem. Phys.* **5**, 453.

28. O'Malley, K. L., and Terwilliger, R. C. (1974). *Biochem. J.* **143**, 591.

29. Reddy, S. R., and Campbell, J. W. (1970). *Comp. Biochem. Physiol.* **32**, 499.

30. Bishop, S. H., and Campbell, J. W. (1965). *Comp. Biochem. Physiol.* **15**, 51.

31. Gasiorowska, I., Porembska, Z., and Mocknacka, I. (1969). *Acta Biochim. Pol.* **16**, 175.

32. Brown, G. W. Jr. (1966). *Arch. Biochem. Biophys.* **114**, 184.

33. Rossi, N. and Grazi, E. (1969). *Eur. J. Biochem.* **7**, 348.

34. Grazi, E., Rossi, N., and Sangiorg, G. (1969). *Ital. J. Biochem.* **18**, 426.

35. Sandri, G., Sottocasa, G. L., Ponfili, E., Soranzo, M. R., Trianiello, S., and Grazi, E. (1974). *Ital. J. Biochem.* **23**, 165.

36. Messengu, F., and Wiame, J. M. (1969). *FEBS Letters* **3**, 47.

37. Cohen, P. P., and Hayano, M. (1948). *J. Biol. Chem.* **172**, 405.

38. Soberón, G., Tarrab, R., and Palacios, R. (1969). *Bol. Estud. Med. Biol. Méx.* **26**, 15.

39. Lehninger, A. L. (1971). *Biomembranes* **2**. L. A. Manson, Ed., Plenum Press, New York-London, p. 147.

40. Gable, J. G. and Lehninger, A. L. (1973). *J. Biol. Chem.* **248**, (2), 610.

41. Brown, G. W., Jr., Brown, W. R., and Cohen, P. P. (1959). *J. Biol. Chem.* **234**, 1775.

42. Cohen, P. P. (1966). *The Harvey Lectures, Series 60*, Academic Press, New York and London.

43. Soberón, G., Ortiz-Pineda, J., and Tarrab, R. (1967). *National Cancer Institute Monograph.* **27**, M. P. Stalberg, Ed., Bethesda, Maryland.

44. Palacios, R., Tarrab, R., and Soberón, G. (1968). *Biochem. J.* **110**, 425.

45. Palacios, R., Huitrón, C., and Soberón, G., (1969). *Biochem. J.* **114**, 449.

46. Palacios, R., Huitrón, C., and Soberón, G. (1970). *Biochem. Biophys. Res. Commun.* **38**, 438.

47. Mora, J., Tarrab, R., and Bojali, L. F. (1966). *Biochim. Biophys. Acta.* **118**, 206.

48. Davis, R. H., and Mora, J. (1968). *J. Bact.* **96**, 383.

49. Davis, R. H., Lawless, M. B., and Port, L. A. (1970). *J. Bact.* **102**, 299.

50. Espin, G., and Mora, J. *Personal Communication.*

51. Yip, M. C. M. and Knox, W. E. (1972). *Biochem. J.* **127**, 893.

52. Glass, R. D., and Knox, W. E. (1973). *J. Biol. Chem.* **248**, (16), 5785.

53. Carvajal, N., Venegas, A., Ostreicher, G., and Plaza, M. P. (1971). *Biochim. Biophys. Acta* **250**, 437.

54. Greenberg, D. M., Bagot, A. E., and Roholt, O. A., Jr. (1956). *Arch. Biochem. Biophys.* **62**, 446.

55. Porembska, Z., and Kedra, M. (1971). *Bull. Acad. Pol. Sci. Biochem.* **19**, 633.

56. Reddy, S. R., and Campbell, J. W. (1968). *Biochim. Biophys. Acta* **159**, 557.

57. Reddy, S. R., and Campbell, J. W. (1969). *Comp. Biochem. Physiol.* **28**, 515.

58. Sorof, S., and Kirsch, V. M. (1969). *Cancer Res.* **29**, 261.

59. Vaca, G., González, A., Espin, G., Mora, Y., and Mora, J. (1974). In *Los Perfiles de la Bioquimica en México.* J. Mora, S. Estrada, and J. Martuscelli, Eds. UNAM, México, pp. 191–210.

Discussion

Dr. Cedrangolo: I wonder whether the Ratner-Pappas reaction has been experimentally blocked in these last years *in vivo* and what happened to the animals thereafter, what kind of disturbance did they suffer, did they eventually die? This could provide clear-cut evidence in favor of the exclusive physiological role of this very important reaction. Maybe Dr. Ratner recalls the experiments carried out by Cedrangolo and coworkers in 1961 and 1962. We succeeded in completely blocking this reaction in rat liver treated with DL-methyl-aspartate (MA), but, surprisingly, the treated animals didn't show any disturbance and excreted the same amount of urea as nontreated animals.

I couldn't reproduce these results, nor could Dr. Ratner, Dr. Crockaert, or Dr. Karlinski in Buenos Aires. I cannot explain why. Maybe we used a different stock of MA or, maybe more likely, different rat strains. Anyway Dr. Meister objected, for MA is unable to pass through liver cell membranes and is absorbed on their external surface. This would have explained the inhibition obtained *in vitro* with the liver homogenization following the *in vivo* treatment, but experiments carried out in 1964 by Salvatore et al. answered this objection. MA inhibits the reaction *in vivo* between citrulline and aspartate. In brief, the experiments consisted in treating *in vivo* three groups of rats in the following way:

1. The first group received a 50% NH_3 lethal dose.
2. A second group received the 50% lethal dose of NH_3 together with an ornithine-aspartate mixture. All the animals were protected, and none died.
3. A third group received NH_3 ornithine-aspartate mixture, and MA. The mortality was 50%, exactly as in the first group.

Finally, I recall a 1971 report from Bruce, Rowe, and Miller in *Proc. Soc. Exp. Biol. Med.*, who, in part, confirmed, with perfused rat liver, our results of 1961; MA added to the perfusing liquid did not inhibit urea biosynthesis, but the compound did penetrate liver cells, since the compound sometimes showed the paradoxical effect of stimulating urea biosynthesis.

Dr. Ratner: Well, I hope that Dr. Cedrangolo has seen a paper that we published a few years ago investigating your results. We repeated his experiments. We did not follow the excretion of urea or excretion of the inhibitor, but we did find an absence of inhibition, and when we took the treated animals that showed no inhibition but had been given the inhibitor and removed the liver and added the analogs to the homogenate, we did find inhibition. So that either because the analogs have been excreted too rapidly to be effective or had failed to penetrate I would say for whatever reason two people now have confirmed the failure for inhibition to occur by the injected analogs. I think it would be nice to have an inhibitor, not so much for total block, but for partial blocks to study the effect of partial blocks of enzymes in the ornithine cycle, for I think it would be very useful for kinetic reasons to study a multi-enzyme system where you could successively block each of the enzymes, and in my dreams I have thought it would be very nice to have a shelf of selected inhibitors. However, to use an inhibitor to prove a pathway would not lie within my dreams, because I think it's an unstable basis on which a pathway should be proven. You can't be sure where the site of inhibition is; there may be multiple sites of inhibition, and I think proof of the pathway should come in a positive direction. It is unfortunate that for permeability reasons no one will ever be able to feed the substrates to a cell population and get a direct precursor-product relationship. Argininosuccinate is one of the poorest penetrators one could imagine. It has three carboxyl groups and two basic groups and is highly polar. So, I think that the evidence has to come from other directions, and it has come from the amount of enzyme present, from the activity in homogenates; certainly it is hard to decide which of the two synthetases is rate-limiting. But I have often thought about the use of blocking agents as tools, and I think penetration will always be a problem in the choice of an analog.

Dr. Illiano: Dr. Cedrangolo agrees with Dr. Ratner, and he thinks that a more effective inhibitor would be useful in elucidating this metabolic step.

Dr. Soberón: Before going into the question that Prof. Cedrangolo has posed, I would like to comment on Dr. Ratner's remarks. I think there is another factor which one should not forget, in trying to extrapolate *in vivo* to *in vitro* results, and this is cell organization. Even though a given substance can penetrate the cell, we do not know the conformation inside the cell, and in your enzyme there is clearly proof of allosteric effects. So, one might visualize a conformation which will not permit the inhibitor to fit properly the enzymatic site as it might be subjected to another environment *in vitro*. Therefore, I think the enzyme in its natural location might have a conformation determined by the surrounding protein and metabolites which could prevent the inhibitor from coming into play, thus interfering with a possible *in vivo* inhibitory role already proven *in vitro*. Now in relation to the possibility that the uricotelic arginase might be an aggregation of the ureotelic one, of course this was a very tempting possibility, being twice the molecular weight. However, there was one thing against it, namely, the immunological properties. The uricotelic arginase did not cross-react with the antibody to the ureotelic enzyme although we thought that the aggregation might change the conformation of the antigenic determinants. It was later proved that subunits cross-react with the antibody to the rat arginase. Lastly, the amino acid composition worked out in the laboratories of Hirsch-Kolb and Greenberg for the rat liver enzyme and of Professor Grazi's with the chicken liver arginase show that they are different proteins. [Dr. Soberón's discussion is entirely in line with Citri's ideas (*Adv. in Enzymology*, **37**, 397, 1973).—Editor]

Dr. Aebi: This is a question to Dr. Ratner. I was much impressed by your presentation of the properties of argininosuccinate lyase being another enzyme tetramer. Almost all that was said matches exactly the situation in catalase. Catalase is a tetramer too, with slightly higher molecular weight (240,000); besides this, even the SH titration results resemble those obtained for catalase. Considering these similarities, I would like to ask you whether you tried to make specific antitetramer and antidimer antibodies? So far it was not possible with catalase to make monospecific antimonomer antibodies. However, by using antitetramer and antidimer IgG, you can demonstrate hidden determinants, and, on the other hand, on dissociation, determinants are lost. By comparing them, you may get additional evidence on how the four subunits are put together.

Dr. Ratner: No, we did not do that, and I can't even say that we know what the antigen is with respect to molecular weight. It is ground up in Freund's adjuvant and injected subcutaneously, and it takes a month to develop antibody, and I don't know what traveled to the antibody sites. But I think it is very important to know that, and I thought about it very much. I would like to know, as you evidently do, whether the antigenic sites of the monomer, subunit, were the same as the tetramer. We could do what you suggest with the monomer, because there is no chance of getting reassociation, or further dissociation, and it's a very nice idea, but we have not done it. But may I ask you, I wasn't clear what your results were with catalase with regard to their antigenicity.

Dr. Aebi: There are a number of determinants common to the tetramer and the dimer, but there are also determinants specific for the tetramer as well as for the dimer. I just wanted to make this suggestion. The technical details have, in part, been published (*Eur. J. Biochem*, 1974).

Dr. Grisolia: It is interesting that at least two main enzymes of the urea cycle are cold labile, CPS and, as just discussed by Dr. Ratner, argininosuccinase.

Dr. Tatibana: I have a question for Dr. Soberón. I am going to discuss this afternoon the concentration of arginine in liver cells. After feeding a 60% casein diet the concentration becomes quite high, 0.15 mM, 150 nmoles/g. This concentration is impossible if the arginine is functioning "normally". I would like to know the K_m values of arginase for arginine, also effects of salt, intermediates, and other factors on the K_m values.

Dr. Soberón: The K_m for arginine, as originally determined, was of the order of 100 mM for the uricotelic and 10 mM, a 10-fold difference, for the ureotelic. There is a wide spectrum of K_m's in the many species that have been studied. However, the order of magnitude is 10 to 200 mM. This is quite high.

Dr. Tatibana: This is quite high. Do some salts and intermediates affect the K_m values?

Dr. Soberón: This I don't know. But this is very important. We have worried that, in spite of high activity of arginase in the liver, with this high K_m and the low pool of arginine that you measure, it will be difficult for the cycle to operate. Perhaps the product of the previous reaction comes closer to the arginase. It is very difficult to understand how an enzyme with such a high K_m and a low pool of arginine could manage.

Dr. Cohen: It is even worse, considering the optimum pH of arginase, around 9, and you are probably trying to operate on the order of 7. I think Dr. Colombo wanted to make a comment.

Dr. Colombo: I have a short question for Dr. Ratner. As I recall from your slides, you have shown in the overall reaction that bovine argininosuccinate synthetase has 10% activity with homocitrulline as substrate. Is that correct? Have you ever demonstrated or isolated homoargininosuccinate or homoarginine in these reactions? Do you think that this bypass reaction will play a biological role?

Dr. Ratner: No, we haven't tried it; the synthetase has always been in short supply. It is a hard enzyme to make, and at the time those studies were done the preparation was very difficult; we had to go through an acetone powder. Now we have eliminated that step; so perhaps we can. But I know why you asked the question, because in citrullinemic patients I do believe that there is some evidence, let's see, I think there is a lysinaemia and there is some evidence that homocitrulline has been converted to lysine, I believe. This would require the participation of two enzymes, and I don't know whether the normal enzyme can affect this reaction. But I know that in citrullinaemia patients there is also an accumulation of homocitrulline, if I'm not mistaken, and how that comes about is not concerned with this enzyme. The only other analog that we have made enzymatically is the one starting with beta-methylaspartate to give the methylargininosuccinate analog.

Dr. Cohen: I think because of the late hour, and I'm sure people are getting anxious to indulge in what has been a common activity around here, namely, overeating, perhaps the time has come to release you for lunch. I call this meeting adjourned, and I thank you for your participation.

Regulation

17

Enzyme and Reactant Concentrations and the Regulation of Urea Synthesis

Luisa Raijman
Department of Biochemistry
University of Southern California
School of Medicine
Los Angeles, California

The capacity of mammals to increase or decrease the amount of urea they produce per day in response to their protein intake has been known for at least 70 years. In 1905, Otto Folin (1) published some studies in humans in which he correlated the amount of urea excreted with the protein intake and established many important facts regarding the composition of the urine. Figure 1 shows Folin's pattern of urea excretion when on a normal and a low-protein diet. His starch and cream diet (low-protein) consisted of 400 g of pure arrowroot starch, 300 cm^3 of cream containing 15 to 20% fat, and 1500 cm^3 of water. The data in Fig. 1 show that rapidly, within a day, pronounced changes in the amount of urea excreted took place, in the same direction as the changes in his protein intake. An increase in urea excretion greater than tenfold occurred on the fourteenth day, the day on which Folin returned to the normal diet. This and other results allowed him to state that urea is "the only nitrogenous substance which suffers a relative as well as an absolute diminution with a diminution in total protein-metabolism." Incidentally, this remarkable paper by Folin (1) also contains data on what may be one of the earliest recorded cases of protein intolerance by a normal adult.

In more recent times, much of the information available on the regulation of the urea cycle was contained in Schimke's work on the adaptive characteristics of the urea-cycle enzymes in rat liver. Schimke studied the effects of the

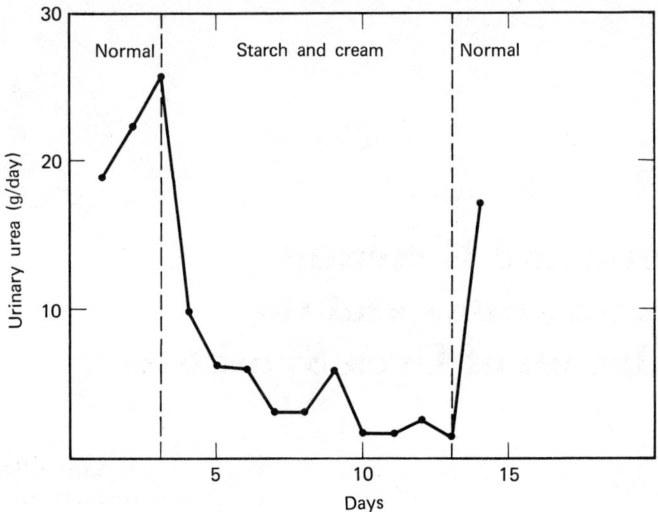

Figure 1. Dietary protein and the excretion of urea. Data compiled from Table II of (1).

composition of the diet, of starvation, of adrenal function and exogenous corticosteroids, and of certain amino acids on the activity of urea cycle and some cycle-related enzymes in rat liver and in human cells in culture (2–5). The nutritional range he studied was broad, from caloric sufficiency with protein starvation, through conditions under which caloric requirements are increasingly met by proteins (30% and 60% protein diets), to starvation which should rapidly lead to a complete dependence of the animal on endogenous proteins for all its caloric and biosynthetic needs. In this manner, Schimke coaxed the mammalian liver into manifesting fully its ability to adapt to different protein loads.

In Table 1, the activities of the enzymes of the urea cycle are expressed relative to those found in the livers of rats fed a 15% protein diet, which are assigned a value of 100% and are referred to as normal. All these data were obtained using whole liver homogenates. They show that all the enzymes of the urea cycle undergo similar adaptive changes; overall, the activities span a remarkable 10- to 20-fold range, and only arginase seems to vary within a more limited range.

Schimke also showed that adaptive increases in enzyme activity are essentially complete after 4 days on a certain diet; in starved rats, however, further large increases take place between the fourth and seventh days (3). Rats do not survive starvation for much longer than 7 days, at which time morphological signs of severe mitochondrial damage can be observed by electron microscopy; it is remarkable that large increases in the enzymes of urea synthesis

Table 1 Effect of Dietary Protein and Starvation on the Activity of Urea-Cycle Enzymes in Rat Liver Homogenates[a]

Enzymatic activities found in animals fed a 15% protein diet (normal) are regarded as 100%; activities shown in the table are expressed as percent of normal.

Diet	Carbamyl Phosphate Synthetase I	Ornithine Transcarbamylase	Argininosuccinate Synthetase	Argininosuccinase	Arginase
			(Percent of normal)		
0% Protein (1 week)	45	27	27	27	60
30% Protein (2 weeks)	170	200	170	180	130
60% Protein (2 weeks)	280	330	270	340	170
Fasting (1 week)	530	450	340	530	190

[a] Compiled from data in (2) and (3).

continue to occur even when body proteins are being extensively degraded. In the case of ornithine transcarbamylase and arginase, the measured increments are actually due to increases in the absolute amounts of enzyme proteins (2). Glutamate-alanine transaminase and glutamate-aspartate transaminase, two enzymes closely related to the pathway of urea synthesis, also increased with increasing dietary protein, although to a lesser extent than the urea cycle enzymes; no changes were detected in the activity of glutamate dehydrogenase (2), which was surprising, since this enzyme is thought to be one of the main sources of ammonia for carbamyl phosphate synthesis. These slow adaptive processes have been shown to occur as well in two species of macaques, and limited evidence suggests that they also occur in man (6).

Schimke measured the liver content of ornithine, citrulline, and arginine under different dietary conditions but was unable to demonstrate any changes; in fact, the values he obtained are extremely low if compared with other reported values (7–9), probably due to the methods used for the preparation of liver extracts. Schimke suggested that the rate of urea synthesis is altered by changes in the liver content of enzymes rather than of substrates (2, 3), despite his demonstration that 4 days were required for the adaptations to be essentially completed; this is much too long to explain effects such as those observed by Folin (1).

Recently, McGivan *et al.* (10) have studied the effect of high protein and protein-free diets on the citrulline synthesizing capacity of isolated rat liver mitochondria and on the activity of some mitochondrial enzymes. Certain of their findings are shown in Table 2. After 3 days on the experimental diet, the capacity of intact mitochondria to synthesize citrulline changes, as one

Table 2 Effect of Dietary Protein on Citrulline Synthesis by Rat Liver Mitochondria and by Mitochondrial Extracts[a]

	Standard Diet	High-Protein Diet (nmole/minute/mg)	Protein-Free Diet
1. Citrulline synthesis[b]			
From NH_4Cl	7.3 ± 1.8	19.8 ± 4.3	0.74 ± 0.30
From glutamate	2.5 ± 0.2	4.8 ± 0.5	0.69 ± 0.12
2. Activity of extracts[c]			
Carbamyl phosphate synthetase I	27.8 ± 5.5	33.8 ± 6.4	21.4 ± 1.3
Ornithine transcarbamylase	690 ± 150	1220 ± 238	697 ± 49

[a] Compiled from (10).
[b] Whole mitochondria used.
[c] Mitochondrial sonicate used.

would have predicted on the basis of Schimke's findings: it increases with high-protein and decreases with low-protein diets. However, minimal or no changes were found in the carbamyl phosphate synthetase I activity of mitochondrial extracts, and small changes in ornithine transcarbamylase were found only in one case (high-protein diet). This is unexpected, since in homogenates, and after 4 days, changes due to high-protein diets are essentially complete and those due to protein-free diets are very pronounced (2, 3).

The liver content of acetylglutamate, an essential cofactor for carbamyl phosphate synthetase I activity, also responds to the protein intake: it increases in the livers of animals fed a 60% protein diet and decreases with protein-free diets (11). The overall changes found by Shigesada and Tatibana were only about twofold; however, we will see later that decreases in the concentration of acetyl glutamate may have significant effects on the rate of carbamyl phosphate synthesis by mitochondria.

These enzyme adaptations are beautifully coherent, but one wonders, perhaps simplemindedly, why they need occur as they do. In Table 3 are shown the activities of the urea-cycle enzymes in the livers of young adult rats. Ornithine transcarbamylase activity is 40- to 70-fold higher than the synthetases and argininosuccinase, and arginase activity is two orders of magnitude higher.

Table 3 Activity of the Urea-Cycle Enzymes in Rat Liver Homogenates[a]

The enzymes were assayed under optimal conditions or near-optimal conditions.

Enzyme	Activity (μmole/hour/g)
Carbamyl phosphate synthetase	340 ± 30
Ornithine transcarbamylase	13,400 ± 300
Argininosuccinate synthetase	175 ± 30
Argininosuccinase	269 ± 31
Arginase	35,000 ± 1,500

[a] From (3).

Why should ornithine transcarbamylase, present in such a large excess over carbamyl phosphate synthetase I, increase sometimes even more than the latter when rats are starved or fed a high-protein diet (see Table 1)? Why should arginase increase at all under those conditions? We know that the velocities of the urea-cycle enzymes *in vivo* are much lower than their respective maximal velocities, but for the adaptive increases in carbamyl phosphate synthetase I, for example, to demand an equal or greater increase in ornithine transcarbamylase in order to maintain the flow of carbamyl phosphate to citrulline, the ratios *v in vivo*/V_{max} of the synthetase and the transcarbamylase must differ by about two orders of magnitude. In other words, ornithine transcarbamylase *in vivo* may be much more inhibited than carbamyl phosphate synthetase I.

There are other indications that the obvious candidates for rate-limiting functions, carbamyl phosphate synthetase I and argininosuccinate synthetase, may not be the real ones. Briefly, since Dr. Colombo will discuss this subject, hyperammonemia and a decreased capacity to synthesize urea are common to all the syndromes produced by congenital defects of the urea-cycle enzymes, including those of arginase (12). It is not evident to me why a reduction of ornithine transcarbamylase to 10 or 20% of its normal activity (with no change in K_m values) should cause such severe hyperammonemia and such impairment in the citrulline-synthesizing capacity of the liver, as is suggested by the decrease in the amount of urea excreted and the increase in orotic acid excretion (12), nor is it evident why a partial arginase deficiency should cause a "back-up" all the way to ammonia.

The pattern of adaptation of the enzymes of the urea cycle and the biochemical findings in patients with congenital defects of these enzymes lead me to think that no one step in the pathway is always rate-limiting; rather, that the concentrations of different reactants of the cycle may be limiting under different metabolic conditions and may determine which of the enzymes is rate-limiting at that particular metabolic time. Activators or inhibitors of the enzymes could, of course, affect the rates significantly.

I intended to begin studying this possibility in two ways: (*a*) by measuring the concentration of urea-cycle and related intermediates, first in normal rat liver, then under conditions which should affect the flow of ammonia through the urea cycle; (*b*) by studying the properties of carbamyl phosphate synthetase I and ornithine transcarbamylase under conditions approaching the physiological, including enzyme interactions with potential effectors.

The liver content of urea-cycle intermediates is shown in Table 4. The values obtained in three laboratories differ somewhat for reasons that are unclear. Dr. Ratner tagged the intermediates *in vivo* by injecting rats with 1.85 μmoles of [U$-^{14}$C] citrulline of high specific radioactivity; this may have caused the values she obtained for ornithine to be higher than previously reported (8). Our data were obtained from deproteinized extracts of freeze-clamped livers (13). The citrulline value probably represents an upper limit (14). Except for ornithine, the values under columns A and C are in acceptable agreement; I cannot explain the discrepancy between the value for arginine in column B (7) and the others.

The first unexpected finding is the relatively high amount of carbamyl phosphate, especially because it is probably located mostly in the mito-

Table 4 Liver Content of Some Urea-Cycle Intermediates

Intermediate	(nmole/g liver)		
	A[a]	B[b]	C[c]
Carbamyl phosphate			110
Ornithine	625	74	150
Citrulline			< 50
Argininosuccinate	34		
Arginine	24	2	43
Urea	8600		6500

[a] From (8).
[b] From (7).
[c] From (9) and (14).

chondria. Assuming that 1 g of liver is equivalent to 1 ml and that mitochondria represent 20% of the normal wet weight of liver, the approximate concentration of carbamyl phosphate in mitochondria is 0.55 m*M*. Ornithine, on the other hand, must be distributed between mitochondria and cytoplasm (15), so that its concentration in the former must be less than 0.75 m*M*. This is deceptively high; the true substrate of ornithine transcarbamylase is the ornithine zwitterion resulting from the deprotonation of a group with a p*K* of 8.7 (16), which is only about 2% of the total ornithine at pH 7.1. The K_m of rat liver ornithine transcarbamylase for total ornithine is 1.8 m*M* at pH 7.5; that for carbamyl phosphate is 22 μM (9). The normal *in vivo* concentration of ornithine, therefore, may be limiting to ornithine transcarbamylase; that of carbamyl phosphate probably is not.

Citrulline and arginine are present at fairly low concentrations, less than 0.05 m*M* if it is assumed that they are evenly distributed throughout the cell or mostly in the cytoplasmic compartment.

From these data, it would appear that ornithine transcarbamylase may be rate-limiting, owing to the limiting concentration of ornithine; it is not clear whether the low levels of argininosuccinate are due to slow synthesis, or efficient use, or both. Data on the levels of these metabolites under different ammonia loads are unfortunately still unavailable. Our methods do not lend themselves to multiple determinations, but we expect this will change in the near future, and we will continue this aspect of our work, which should give valuable information.

As for our work with purified rat liver carbamyl phosphate synthetase I, we were able to stabilize it in mixtures of glycerol and dithiothreitol and to purify it to apparent homogeneity by a modification of the method of Guthöhrlein and Knappe (17). The enzyme consists of two subunits of identical molecular weight, about 120,000 (14). The molecular weight of the active enzyme estimated by density gradient centrifugation is 232,000 ± 10,000 (14, 17).

Assuming that the pure enzyme has a specific activity of 30, we calculated from its molecular weight and from the maximal velocity of carbamyl phosphate synthesis by liver homogenates (140 μmole/15 minutes at 37°) that 1 g of liver contains 20 nmole of carbamyl phosphate synthetase I (14). Once again, assuming that 1 g of liver is equivalent to 1 ml, that mitochondria represent 20% of normal liver volume, and that the matrix represents about one quarter of the total mitochondrial volume (18), the concentration of this synthetase in the matrix is calculated to be 0.4 to 0.5 m*M* (the enzyme is in the matrix in normal liver). That is a very high concentration, not least of all questionable because it means that carbamyl phosphate synthetase I would be about 20% of the solid matter of the matrix, and this is improbably high; however, it is not impossibly high.

There are obvious possible sources of error in this calculation. Our preparation could be grossly impure, but this is not very likely. The specific activity of the pure enzyme may be higher than 30: this is possible; the mammalian synthetase is a labile protein, and it could happen that inactive forms produced early during purification copurify with the active enzyme. I intend to answer this by following the homospecific activity (activity per milligram of enzyme) of preparations during purification (19) and by direct measurement of the amount of carbamyl phosphate synthetase by radioimmunoassay. In the meantime, let us assume that our calculations may be in error by half an order of magnitude.

Is it worthwhile to invest much time determining the actual amount of carbamyl phosphate synthetase I in rat liver? I think it is. Schimke's work (2) shows that this enzymatic activity can increase three fold or more; if nothing else changed, the enzyme would then represent 60% or more of the dry matter of the matrix, which probably is impossibly high. One can hypothesize that the subcellular location of the enzyme changes during the adaptation; the enzymatically active protein is most likely synthesized *de novo* outside the mitochondria and is normally transported into them (20). These processes could be altered under extreme adaptation conditions. Translocation of enzymes apparently occurs, among other states, in acute uremia (21). More likely, the mitochondrial matrix may change under those conditions, its protein content and its volume increase, or the total number of liver mitochondria may increase. The protein content of the livers changed significantly in some of Schimke's experiments (2, 3).

In addition, some characteristics of the regulation of citrulline synthesis may be very much dictated by the concentrations of carbamyl phosphate synthetase I and ornithine transcarbamylase. Puzzled by the high concentration of the former, I calculated the approximate concentration of the transcarbamylase in the matrix; I assumed that the specific activity of the pure rat enzyme is similar to that of the bovine enzyme, about 780 μmole/minute/mg protein (21), and estimated a molecular weight of 105,000 (see below). The concentration of transcarbamylase in the matrix also appears to be about 0.4 m*M*. In other words, the synthetase, the transcarbamylase, carbamyl phosphate, acetylglutamate, and perhaps ornithine, all appear to be in the matrix at concentrations of the same order of magnitude.*

It is possible that the first two enzymes of urea synthesis form a complex *in vivo*; we have begun to study this by following the profiles of ornithine transcarbamylase activity obtained on density gradient centrifugation of mitochondrial extracts obtained after various preincubations. Activity was detected in the areas corresponding to a molecular weight of 102,000 to 110,000, and 331,000; the activity of the latter peak represents about 1% of the total. Preincubation of the mitochondria under proper conditions not yet established may be essential to the isolation of the hypothetical complex (22) in good yield.

The existence of a complex would have an effect on the kinetics of citrulline synthesis *in vivo*; in any case, given the relative concentrations of enzymes and substrates, one can assume that the kinetic studies of both enzymes carried out so far do not allow us to estimate the rates of carbamyl phosphate and citrulline synthesis *in vivo* (23, 24).

The probable matrix concentration of acetylglutamate is 0.2 to 0.4 m*M*; on the basis of a K_m value of 0.1 m*M* of carbamyl phosphate synthetase for this cofactor (25), one can calculate that, in normal mitochondria, from one- to two-thirds of the total enzyme would be in the active form (see 17), as shown in Table 5. (Perhaps this is one reason for the ineffectiveness of acetylglutamate when added to fresh mitochondria, since significant increases in carbamyl phosphate synthetase activity would require larger increases in the concentration of that cofactor; however, the absence of an effect may be due to lack of transport of acetylglutamate into the mitochondria (10)). Decreases in the cofactor *in vivo* should have a pronounced effect on the rate of carbamyl phosphate synthesis; increases in the amount of carbamyl phosphate synthetase I would have little effect if not accompanied by increases in acetylglutamate, given their nearly equimolar concentration in mitochondria and the K_m of the enzyme for this cofactor.

A final calculation shows that carbamyl phosphate synthetase at this concentration can efficiently bind ATP·Mg (Table 5). The binding constant used here is 4×10^{-5} *M* (25); at 1 m*M* total ATP·Mg about one-third of it

* Elsewhere in the discussions, Dr. Marshall points out that ornithine transcarbamylase is present in the mitochondrial matrix at lower concentration than the carbamyl phosphate synthetase. She is correct; the concentration of ornithine transcarbamylase appears to be around half an order of magnitude lower than that of the synthetase or slightly less.

Table 5 Calculated Concentrations of CPSase I–Bound AGA and ATP·Mg at Different Concentrations of AGA and ATP·Mg[a]

The equations used in the calculations are

$$\frac{[\text{CPSase I}]\ [\text{AGA}]}{[\text{CPSase I}\cdot\text{AGA}]} = 10^{-4}\ M \qquad (1)$$

and

$$\frac{[\text{CPSase I}]\ [\text{ATP}\cdot\text{Mg}]}{[\text{CPSase I}\cdot\text{ATP}\cdot\text{Mg}]} = 4 \times 10^{-5}\ M \qquad (2)$$

The value of equation 1 is that of the K_m of CPSase I for AGA and that of equation 2 is the dissociation constant of the CPSase I·ATP·Mg complex obtained by equilibrium dialysis (Marshall et al., 1961). The concentration of CPSase I used in the calculations was $3.8 \times 10^{-4}\ M$.

[total AGA] ($10^{-4}\ M$)	[CPS I·AGA] ($10^{-4}\ M$)	[total ATP·Mg] (M)	[CPSase I·ATP·Mg] (M)
0.5	0.379	10^{-8}	0.906×10^{-8}
1	0.735	10^{-6}	0.906×10^{-6}
2	1.35	10^{-5}	0.903×10^{-5}
4	2.31	10^{-4}	0.880×10^{-4}
		10^{-3}	0.358×10^{-3}

[a] Abbreviations: CPSase I, carbamyl phosphate synthetase I; AGA, acetylglutamate.

would be bound to the enzyme, but, at 0.1 mM total ATP·Mg or less, an essentially constant fraction of it (90%) would be bound. Many factors not quite ponderable at present must affect the binding *in vivo*, so that I will only claim that it seems likely that carbamyl phosphate synthetase I would compete favorably with other enzymes for ATP·Mg (see 26, 27).

Other than acetylglutamate, there are no known physiological effectors of the enzymes of the urea cycle. Yet effects such as the ornithine stimulation of urea synthesis described by Krebs, *et al.* (28) suggest that significant inhibitions or activations may well occur. Studies on possible physiological inhibitors or activators of urea synthesis should be done again, under conditions closer to the physiological; in the case of the first two steps, this means at high enzyme concentrations. Frieden and Colman (29) and Srere (23) have clearly pointed out that allosteric effects that occur with dilute enzymes may not take place with concentrated enzymes; there is no reason to believe that the opposite does not occur.

It appears to me that we can state that in animals on a normal diet, ornithine may severely limit the velocity of ornithine transcarbamylase *in vivo*; that the concentrations of acetylglutamate and carbamyl phosphate synthetase are such that the enzyme is about half-activated; that ATP·Mg is not likely to limit this synthetase. We are uncertain about whether HCO_3^- and ammonia are limiting; in fact, an important question is whether NH_4^+ or NH_3 binds to the synthetase, since, were NH_3 the necessary species, the velocity of the synthetase *in vivo* would be greatly restricted.

When mentioning the "concentrations" of enzymes and reactants in liver, I have done it with full awareness that they are only approximations; consequently, the conclusions drawn from using them may well contain errors. But that is not inherent to the approach; in time, I trust it will prove useful in the study of the regulation of urea synthesis, as it has in other areas, very notably in Professor Krebs's laboratory.

ACKNOWLEDGMENT

Most of the work reported here was done while I was a member of Dr. Mary Ellen Jones' research group. I am very grateful to her for her most generous support and stimulating discussions. Supported by Grant HD-06538 from the National Institutes of Health to Dr. Jones.

REFERENCES

1. Folin, O. (1905). *Am. J. Physiol.* **13**, 66–115.
2. Schimke, R. T. (1962). *J. Biol. Chem.* **237**, 459–468.
3. Schimke, R. T. (1962). *J. Biol. Chem.* **237**, 1921–1924.
4. Schimke, R. T. (1963). *J. Biol. Chem.* **238**, 1012–1018.
5. Schimke, R. T. (1964). *J. Biol. Chem.* **239**, 136–145.
6. Nuzum, C. T., and Snodgrass, P. J. (1971). *Science* **72**, 1042–1043.
7. Herbert, J. D., Coulson, R. A., and Hernandez, T. (1966). *Comp. Biochem. Physiol.* **17**, 583–598.
8. Ratner, S. (1973). *Adv. Enzymology* **39**, 1–90.
9. Raijman, L. (1974). *Biochem. J.* **138**, 225–232.
10. McGivan, J. D., Bradford, N. M., and Chappell, J. B. (1974). *Biochem. J.* **142**, 359–364.
11. Shigesada, K., and Tatibana, M. (1971). *J. Biol. Chem.* **246**, 5588–5595.
12. Levin, B. (1971). *Adv. Clin. Chem.* **14**, 65–143.
13. Wollenberger, A., Ristau, O., and Schoffa, G. (1960). *Pflügers Arch. Gesamte Physiol.* **270**, 399–412.
14. Raijman, L., and Jones, M. E., unpublished data.
15. Gamble, J. G., and Lehninger, A. (1973). *J. Biol. Chem.* **248**, 610–618.
16. Snodgrass, P. J. (1968). *Biochemistry* **7**, 3047–3051.
17. Guthöhrlein, G., and Knappe, J. (1968). *Eur. J. Biochem.* **7**, 119–127.
18. Pfaff, E., Klingenberg, M., Ritt, E., and Vogell, W. (1968). *Eur. J. Biochem.* **5**, 222–232.
19. Rush, R. A., Kindler, S. H., and Udenfriend, S. (1974). *Biochem. Biophys. Res. Commun.* **61**, 38–44.

20. Bartley, W., and Birt, L. M. (1970). In *Essays in Cell Metabolism*, W. Bartley, H. L. Kornberg, and J. R. Quayle, Eds., Wiley-Interscience, London, pp. 1–43.

21. Marshall, M., and Cohen, P. P. (1972). *J. Biol. Chem.* **247**, 1641–1653.

22. Bergquist, A., Eakin, E. A., Murali, D. K., and Wagner, R. P. (1974). *Proc. Natl. Acad. Sci., USA.* **71**, 4352–4355.

23. Srere, P. (1974). *Life Sci.* **15**, 1695–1710.

24. Cha, S. (1970). *J. Biol. Chem.* **245**, 4814–4818.

25. Marshall, M., Metzenberg, R. L., and Cohen, P. P. (1961). *J. Biol. Chem.* **236**, 2229–2237.

26. Tager, J. M., Zuurendonk, P. F., and Akerboom, T. P. M. (1973). In F. A. Hommes, C. J. Van Der Berg, Eds., *Development Biochemistry—Inborn Errors of Metabolism*, Academic, London, pp. 177–197.

27. Zuurendonk, P. F., and Tager, J. M. (1974). *Biochim. Biophys. Acta* **333**, 393–399.

28. Krebs, H. A., Hems, R., and Lund, P. (1973). In F. A. Hommes and C. J. Van Den Berg, Eds., *Developmental Biochemistry—Inborn Errors of Metabolism*, Academic, London, pp. 201–215.

29. Frieden, C., and Colman, R. F. (1967). *J. Biol. Chem.* **242**, 1705–1715.

Discussion

Dr. Grisolia: I should like to congratulate Dr. Raijman for her very interesting presentation. We all realize that what she is trying to do is go into one of the most difficult but also a most pressing problem in biochemistry, which is, as I commented on yesterday, to go from the enzyme level to what may truly be happening *in vivo*. Moreover, she is dealing with the problem of protein turnover. Although we know a great deal about protein synthesis, we know next to nothing about protein degradation and particularly the factors that govern it. For example, I do not know the half-life of rat liver carbamyl phosphate synthetase. Perhaps somebody knows that, but certainly I don't.

Probably you have already thought about that, but, in case you haven't, I thought I'd ask you, anyway, about the possibility of using antibodies against CPS to try to find out really how much enzyme there is.

Dr. Raijman: I thought I had mentioned this, but perhaps it didn't come across. We have the antibody already. If we use an antibody in a radio-immunoassay, we can determine directly the amount of CPS I protein in the mitochondria. I very much intend to do that.

Dr. Grisolia: But it's not clear to me whether or not you had more protein than that calculated from the activity for the "pure" enzyme you used for immunization.

Dr. Raijman: If I understand you correctly, that is the question that can be answered by using homospecific activities rather than specific activities. If you develop an antibody to the purified enzyme, you can determine the specific activity from the homogenates on to the purified enzyme, and you can at the same time determine the amount of the enzyme protein by means of the radioimmunoassay. You can now express activity in terms of activity per milligram enzyme. This should give you a true measurement of enzyme protein. Maybe I haven't answered your question.

Dr. Grisolia: You have except for the obvious difficulties in making sure that your antibody reflects the true protein anyway, because obviously you may have modified proteins with your "pure" enzyme and/or after injection. I should like to make a very short comment. Perhaps you may have said that, but it may not have come clear to all of us, and that is whether the concentrations that you referred to in much of your paper refer to the mitochondria or to the internal membrane, matrix, and space where much of the enzymes and some of the cofactors are. That may be very important in calculating again the activities.

Dr. Raijman: The concentrations calculated for carbamyl phosphate synthetase I and ornithine transcarbamylase which were in the slide are for the mitochondrial matrix, so that we have used two factors, one for the volume of the mitochondria as related to the total liver, and one for the volume of the matrix as related to the total volume of mitochondria. The calculations of acetylglutamate concentration also refer to the matrix; we assume that it is all in the matrix, because it's synthesized there, but we have not proof of this. The concentrations of ATP are hypothetical, but we look at them as if they were in the matrix. I don't know what the concentration of ATP in the matrix is; that's why I used that broad range from 10^{-3} to 10^{-8} M. The concentrations of ornithine and of CPS in general refer to the matrix. Anyway this should be clear in the written paper.

Dr. Cohen: I just wanted to add a few comments. One, I share Dr. Grisolia's concern that the extrapolation from certain kinetic constants obtained with purified solutions poses some problems in terms of a steady-state flux *in vivo*. It should be possible by proper labeling experiments to find out what the steady-state flow is of substrates from ammonia and CO_2 on through urea. There are data, some already available and some which could be obtained by properly designed experiments, bearing on this point. I would make two other comments. Although the situation may not be the same as in the rat, the half-lives of three mitochondrial enzymes in tadpole liver have been determined by my colleagues. GDH has a half-life of 25 hours, CPS has a half-life of 60 hours, and OTC has a half-life of 90 hours. Now these values are for an animal swimming around at a temperature of 15 to 25°C. It is highly likely that an animal living at 37° is going to have a higher turnover rate. It is obvious that the rate of synthesis and rate of degradation of these enzymes, or, in other words, their turnover rates, must be an important means of regulation. This may be of use to you in designing your experiments in rats. I would also urge, if you use the immunoassay procedure, that you consider also a pulse label for the enzyme, because it may actually turn out that the total amount of the enzyme may not appear to change. I think the experiment will be very complicated but certainly is worth the effort, and I wish you luck.

Prof. Krebs: You raised many interesting points, and there are just two on which I would like to comment. You expressed some surprise that a very large fall in the activity of ornithine transcarbamylase is required to produce protein intolerance, and on this point I have some experience, because I was associated with a study of a case of ornithine transcarbamylase deficiency. In this patient, a baby, the activity was assayed as 2% of the normal activity, and yet the child had survived, but only just without treatment. She was retarded, but, after suitable treatment, she developed quite satisfactorily.

A suitable treatment meant, of course, a low-protein diet and spreading the diet as much as possible over the 24 hour day in order to make maximum use of the enzyme capacity. The case shows that the great fall in enzyme activity is still compatible with life and even reasonably normal development. This bears out what you said. There is a second relevant point arising from this patient's case history, viz., the question whether ornithine is rate-limiting. We thought that giving this patient some ornithine might increase the capacity of the urea-synthesizing system. But the clinical tests were entirely negative. Giving ornithine or precursors as such in the very palatable form of herring roe (which contain proteins rich in arginine) made no difference to the protein tolerance of this child as tested by blood ammonia values. This suggests ornithine is not a limiting factor. The circumstances are quite different *in vivo* from those under experimental conditions when we suspend a liver slice in a relatively large volume (as we did in 1932) or more recently with the perfused liver or liver cells when the experimental material is washed with large volumes of perfusion fluid or suspension medium. This treatment causes a release of cell constituents and dilutes the intracellular concentrations, especially of amino acids. Hence one gets large effects of ornithine with isolated cells and the perfused liver, even greater than in slices.

Dr. Raijman: Actually, Professor Krebs, what I was trying to say concerning the transcarbamylase deficiency was exactly the opposite. I was surprised, because, as you know, there are some cases in which the transcarbamylase activity only drops to maybe 30% of normal. And still in those cases there is a marked hyperammonemia under normal protein intake. What I was surprised at was that there should be hyperammonemia at all. It seems to me, and admittedly this may be very simpleminded, that 30% should be more than enough to cope with the needs of citrulline synthesis if the ratio of carbamyl phosphate synthetase to ornithine transcarbamylase *in vivo* has anything to do with the ratio *in vitro*.

Prof. Krebs: The assay of enzyme activity in cases of congenital disorders is only part of the story. It is highly probable that often the enzyme we assay is not a normal one. Humans who survive are, of course, people where the defect of the ornithine cycle is not due to a gene deletion. In that case there would be no enzyme, and this would not be compatible with survival. When there is an enzyme, it may well be an isoenzyme which either has a lower capacity or a low activity under physiological conditions, because the kinetic constants differ from those of the normal enzymes. In our case the biopsy material was insufficient for measuring kinetic properties of the enzyme. Unless one has detailed information on the kinetic characteristics of the enzyme, one cannot go into depth about the nature of the genetic abnormality.

Dr. Aebi: You raised a very interesting question, Sir Hans. There are two ways how a structural mutation may lead to an enzyme deficiency. One possibility is that this is due to an enzyme variant of low specific activity, however, approximately normal stability; the other possibility is that it might be an enzyme variant with approximately normal specific activity but of low stability. For example, the Japanese and the Swiss type of acatalesemia represent either case. The practical problem is that by using a standard technique for the quantitative determination of enzyme activity the result obtained with enzyme variant may be misleading. So, if a variant is unusually unstable, the resulting value of, say, 1 or 2% of residual activity may be too low (see H. Aebi et al., *Biochem. Genet.*, **2**, 245–251, 1968). The problem always remains the same. How can I get enough liver tissue of such a case in order to make an appropriate study? If this is not possible, the next question then is: Are there cellular elements in blood (e.g., white cells or platelets) which may possibly serve as a substitute?

Dr. Cohen: I just want to mention one other aspect to this discussion in case you want to pursue the experiment you proposed, which I hope you will. We are currently finishing up some work with tadpoles in which we are able to demonstrate that the synthesis of CPS, in the thyroxine-induced transition from the premetamorphic to the fully metamorphosed animal, is associated with the appearance of a new population of mitochondria. It is conceivable that the lag period you alluded to in both feeding and starvation experiments reflects the time necessary to develop a new population of mitochondria. We know very little about mitochondriagenesis and what factors regulate it. When designing your experiments, you might keep that in mind in the case of the rat.

Dr. Raijman: I agree; I think I stated that more likely more mitochondria or bigger mitochondria develop. This is known to happen in muscle, and I am told, though I cannot give a reference, that it happens also in liver. I think that is very much a possibility. There was something you said that I wanted to comment on, for others who might want to take up the problem: that if it is true that there is so much CPS in mitochondria, this enzyme may turn out to be a very good marker to use in studies of mitochondrial biogenesis in mammals. One does not often find 10^{-4} M markers.

Dr. Tatibana: I have many problems to discuss concerning Dr. Raijman's paper, but I'd like to discuss them after my presentation. As far as OTC deficiency is concerned, I had a patient whose enzyme activity was 30% of normal. Still she developed hyperammonemia. I think that the direct toxic substance may not be ammonia. Of course hyperammonemia could develop, but the direct toxic agent might be carbamyl phosphate, I think. As far as I studied, the carbamyl phosphate level *in vivo* is very low, 10^{-6} M. If this is elevated 5 or 10 times, some disturbances will occur. I will present data later.

Dr. Aebi: Final comment, Dr. Raijman?

Dr. Raijman: Some on Dr. Tatibana's statement, but maybe we can leave them until after his talk.

Dr. Aebi: Very many thanks, Dr. Raijman; your contribution is indeed very stimulating.

18

Developmental Changes of Urea-Cycle Enzymes in Mammalian Liver

Niels C. R. Räihä
Departments of Obstetrics and Gynecology and Medical Chemistry
University of Helsinki, Helsinki, Finland

In the mammal the process of birth with its change from the intrauterine to the extrauterine environment must be accompanied by many biochemical adaptations which are necessary for maintaining metabolic homeostasis after the placental circulation has stopped. A fundamental mechanism in this adaptation is that of gene expression and its relation to the synthesis and appearance of specific enzyme proteins. The liver plays a central role in this process and must assume many functions necessary for extrauterine existence which were performed by the placenta or the maternal organism during fetal life. One such function of the mature liver is urea biosynthesis.

During intrauterine life the aquatic type of environment allows a rapid exchange of water and metabolites between the fetus and the mother, and this could permit the fetus to excrete its nitrogenous waste products as ammonia. In preparation for birth a more efficient biosynthetic mechanism for ammonia detoxication is needed. Needham (1) in 1931 first suggested that the embryo excretes ammonia and eventually develops a mechanism for nitrogen excretion characteristic for the species.

These studies have been supported by the Sigrid Juselius Stiftelse and Signe and Ane Gyllenbergs Stiftelse.

In the following presentation I will discuss the changes taking place in the urea-synthesizing enzymes and in urea synthesis during fetal and neonatal development of the mammal.

UREA PRODUCTION BY THE FETUS

Interest in the ontogenesis of urea production dates back to the nineteenth century, but, before the discovery of the urea cycle and its enzymes, studies in the mammalian fetus were limited to measurements of urea and ammonia in tissues and body fluids on both sides of the placenta. In a study of maternal and fetal serum concentrations of nonprotein nitrogen and urea at the time of birth Slemons and Morris in 1916 (2) found no difference in the human but suggested that in animals with a thicker placental membrane the fetal serum should have a higher concentration. Hammett in 1918 (3) determined the urea and ammonia content in human placental tissue at birth and found a high urea concentration especially in the placentas from toxemic patients. Friedberg in 1955 (4) found an accumulation of urea in the amniotic fluid after the fifth month of human pregnancy, and recent studies by Teoh and coworkers (5) have also found a steady increase in the urea concentration of amniotic fluid with increasing gestational age of the fetus.

The first report concerning direct measurements of urea production by fetal liver is that of Manderscheid in 1933 (6). Using the tissue slice technique employed by Krebs and Henseleit (7), he found urea production in the liver of 3 to 4 month old human fetuses. More recently we have studied the overall capacity of human fetal liver slices to synthesize urea from ammonia and bicarbonate in the presence of optimal concentrations of ornithine (8). Urea was produced under these conditions by 16 to 20 week fetal liver at a rate of about 4 μmole/g liver/hour. Kekomäki and coworkers (9) have shown that urea is released from isolated perfused human fetal livers at the end of the first trimester, and addition of alanine to the perfusion medium did not increase the rate of urea production, suggesting that the human fetus has very little capacity to expand its urea production at this stage of development. After birth the rate of urea synthesis as reflected by plasma urea concentrations is a direct correlate to the amount of protein in the diet, also in very small premature infants (10). These observations indicate that the capacity for urea synthesis develops relatively early during fetal life in the human.

In the rat urea excretion is low during the suckling period (first 2 weeks of life) and rises sharply during weaning, when the protein content of the diet increases (11). Urea production from ammonia and bicarbonate in rat liver slices is negligible before the twentieth day of gestation and increases gradually towards birth. Figure 1 shows that there is a rapid rise in the urea-producing capacity of rat liver slices after birth (12).

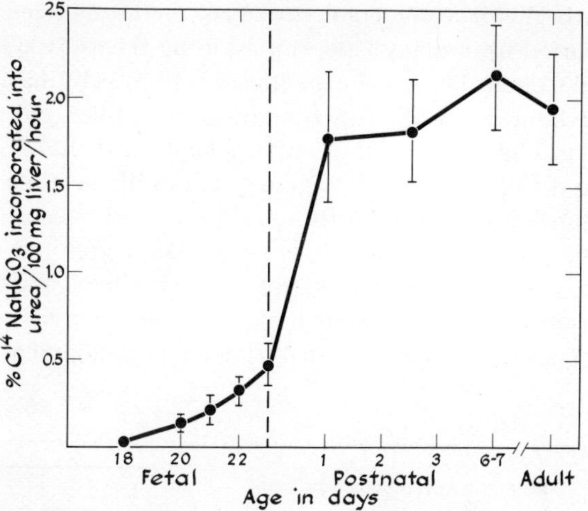

Figure 1. Urea forming capacity in rat liver slices during development.

DEVELOPMENT OF UREA-CYCLE ENZYMES

In Experimental Animals

Arginase activity was studied during postnatal life in rat liver by Lightbody as early as 1938 (13). He found a marked rise during the first 3 to 4 days of life. Subsequently the arginase activity decreased and increased again after the sixteenth postnatal day, when weaning had started. The availability of methods (14) for assaying all the individual urea-cycle enzymes in cell-free extracts of liver made it possible to study developmental aspects of these enzymes. The first report on the development of all urea-cycle enzymes in the liver of the rat and the pig was published in 1959 by Kennan and Cohen (15). The development of carbamyl phosphate synthetase, ornithine transcarbamylase, the arginine synthetase system (the condensing and the cleavage enzymes measured as a single overall system), and arginase was found to be asynchronous in rat liver, and urea did not appear to be synthesized at any significant rate until late fetal life, due to very low activity of the arginine synthetase system before birth. In the pig, on the other hand, all four enzymes were found to be present at significant levels in the liver of the youngest embryo studied at 28 days of a 112 to 116 day gestation. The marked species difference in the commencement of the urea-cycle enzyme activities was discussed by these authors in relation to the fetal membranes, the intrauterine maturation of the fetus, and the developmental differences of the excretory mechanisms of the two animals.

Subsequently the developmental pattern of the urea-cycle enzymes in rat liver was reported from many laboratories, using the assay system described by Brown and Cohen (14) or some modifications of it—Räihä and Suihkonen (16, 17), Charbonneau et al. (18), Illnerova (19), Miller and Chu (20), and Schwartz (21). The results of these studies confirmed the previous findings by Kennan and Cohen (15) and suggested that in the rat the urea cycle does not function until just prior to birth, since the activity of arginine synthetase system is either not measurable or very low. The activity of the other four urea-cycle enzymes is also relatively low in the rat fetus and increases after birth. The arginine synthetase system has the lowest activity and is thus the rate-limiting enzyme of the cycle also during fetal development (17). Figure 2

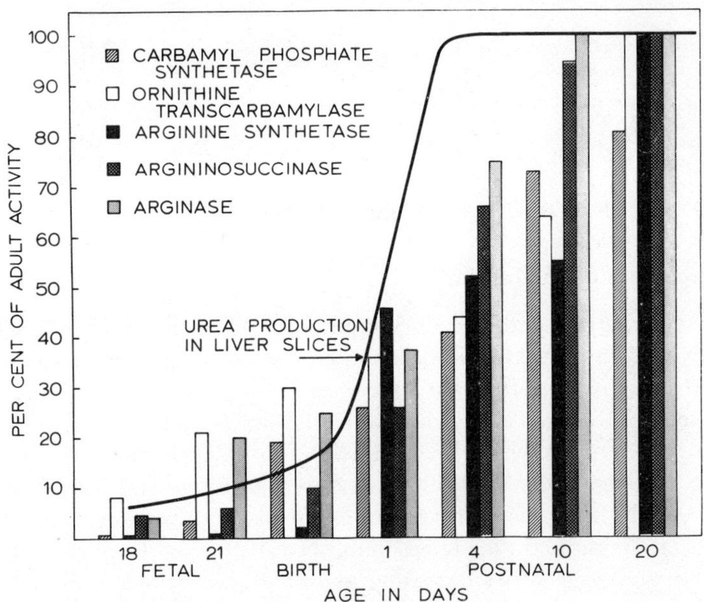

Figure 2. Changes in urea-synthesizing enzymes in rat liver during fetal development.

shows our results on the development of the urea-cycle enzymes in rat liver in relation to the overall urea-producing capacity of liver slices. Two definite periods of activity increase are seen for the arginine synthetase system, one occurring immediately after birth and the other between the tenth and twentieth day. Both the rapid postnatal increase in the capacity for urea production by liver slices and the parallel increase in the activity of the arginine synthetase system are inhibited by actinomycin D and by puromycin.

These findings are compatible with the view that the postnatal rise in the activity of the arginine synthetase system is due to formation of new enzyme protein (17). The second rise in the activity of the arginine synthetase system occurs in connection with weaning and might be a dietary adaptation which will be discussed later (11, 17).

In the Human Fetus

In the human ammonia detoxication starts early during fetal life, as mentioned previously. Kennan and Cohen (22) assayed the urea-cycle enzymes in homogenates of three human fetal livers at 12, 20, and 40 weeks of gestation and from two adults. The arginine synthetase system had the lowest relative activity and was present in the youngest fetus studied at 12 weeks. The activity was about 25% of that found in the adult livers. When calculations were made for the amount of urea which could be formed *in vitro* on the basis of the rate-limiting step (the arginine synthetase system), it was found to be as much as 3 g daily for the human fetus around term. Subsequent studies from our laboratory (8) and by Colombo and Richterer (23) have confirmed the presence of all urea-cycle enzymes in human fetal liver at an early stage of fetal development (Fig. 3).

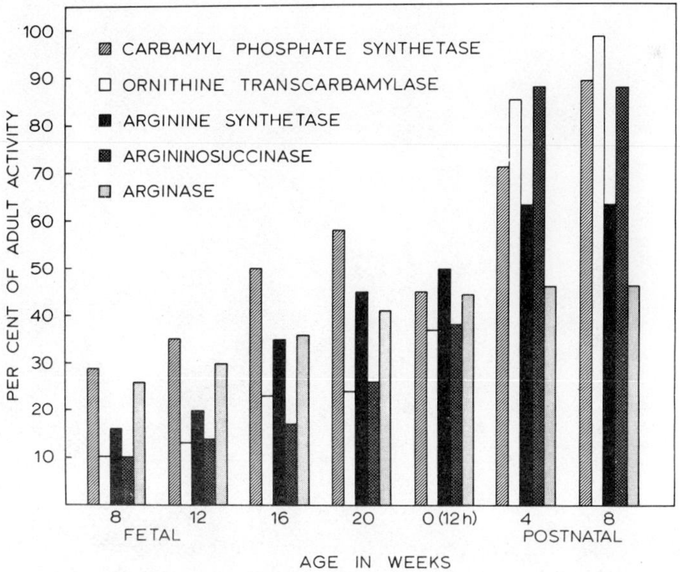

Figure 3. Developmental changes of the urea cycle enzymes in human liver.

REGULATION OF ARGININE SYNTHETASE SYSTEM ACTIVITY DURING DEVELOPMENT (IN VIVO STUDIES)

In adult rats it has been shown by Schimke (24) that the total liver content of all the urea-cycle enzymes is directly proportional to the daily consumption of protein of the animal. The urea-cycle enzyme activities can also be altered in adult rats by corticosteroids (25, 26) and by glucagon and insulin (27). Very little is, however, known about the factors which influence the activity of the urea-cycle enzymes during development.

Effect of Delivery

The effects of premature and postmature delivery on the postnatal increase of the arginine synthetase system activity was studied in our laboratory (17) (Fig. 4). Rats delivered 24 hours before normal term did not exhibit the usual postnatal increase in enzyme activity, and activity 24 hours after premature delivery by cesarean section was equal to that of control animals at the moment of normal birth. When pregnancy was prolonged for 24 to 48 hours by injections of progesterone, the enzyme activity at the moment of delayed birth was similar to that found in control animals exhibiting the normal postnatal rise. Injections of progesterone to pregnant rats between the fifteenth and twenty-first day of pregnancy did not effect the arginine synthe-

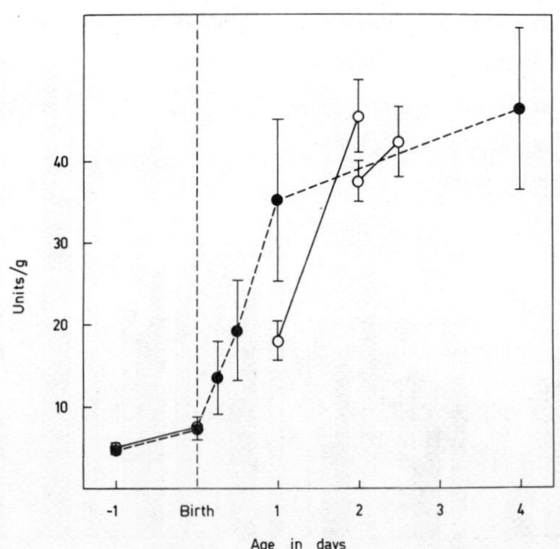

Figure 4. Development of the activity of the arginine synthetase system in normal rats (●), in rats delivered prematurely (○) and in rats delivered 1 day postmaturely (△) and 2 days postmaturely (□). Vertical lines represent standard deviations.

tase system activity of the fetuses found at birth. Starvation of newborn rats for up to 24 hours after birth had no effect on the postnatal increase of any of the enzymes of the urea cycle (17).

Effects of Adrenalectomy and Steroids

The role of the adrenal cortex on the postnatal increase of the arginine synthetase system in the rat was studied in our laboratory by removal of the adrenal glands immediately after birth. Adrenalectomy caused an almost complete inhibition of this increase at 24 hours; at 48 hours the inhibition was 72% (17). The administration of the synthetic corticosteroid triamcinolone to adrenalectomized newborn animals was quite effective in preventing the decrease in enzyme activity caused by the removal of the adrenal glands (Fig. 5).

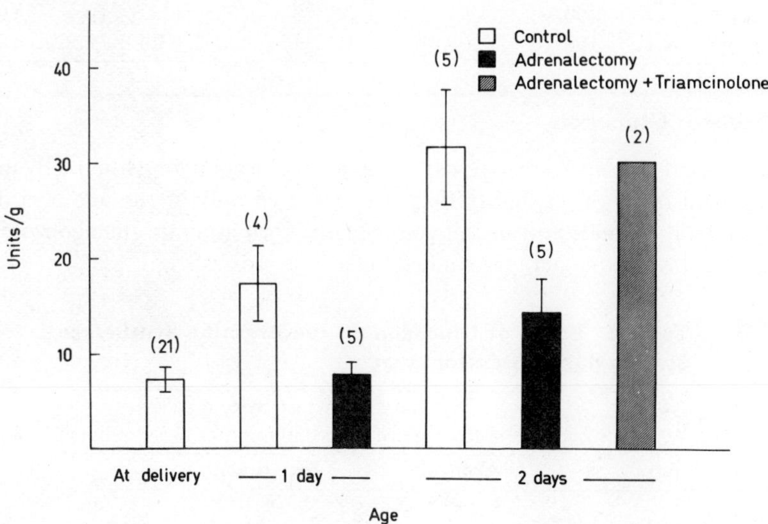

Figure 5. Effect of adrenalectomy on the development of arginine synthetase activity.

The effect of exogenous corticosteroid during different times of the perinatal period on the liver arginine synthetase system activity was further studied (21). Triamcinolone 19 to 190 μg per animal in 0.9% NaCl was injected intraperitoneally 24 hours before the study. The control animals received only saline. The fetuses in one uterine horn were injected after laparatomy performed to the mother under light anesthesia. Those in the other horn served as controls. After injection the incision was closed and the fetuses were studied after 24 hours. Table 1 shows the effect of intraperitoneally injected triamcinolone on the activity of the arginine synthetase system in rat liver. There is a statistically significant increase in the activity in all postnatal animals 24 hours after the injection. The same response although slightly less was found when hydrocortisone (cortisol) was administered. No stimulation was found in any of the fetal rats injected *in utero*.

Table 1 Effect of Triamcinolone on the Activity of the Arginine Synthetase System during Development in the Rat

Age (days)	Dose of Triamcinolone (μg/animal)	Arginine Synthetase System Activity (μmole of urea formed/30 minutes per g of liver)		
		Control	Experimental	p
−2.5	190	7.2 ± 1.1 (5)	9.6 ± 2.4 (6)	n.s.
−1.5	19	5.1 ± 0.7 (4)	5.8 ± 1.9 (6)	n.s.
+1.0	19	37.9 ± 4.1 (3)	61.4 ± 9.6 (4)	<0.005
	190	29.6 ± 3.2 (3)	73.4 ± 3.3 (4)	<0.001
+5.0	190	40.3 ± 9.1 (4)	79.4 ± 14.0 (4)	<0.005

Effects of Glucagon

When perinatal rats were exposed to 100 μg of glucagon for 24 hours by intraperitoneal injection, a slight effect was observed only at the age of 5 days. In the fetal animals and in the younger newborn animals glucagon did not increase the enzyme activity studied, as seen in Table 2.

Table 2 Effect of Glucagon on the Arginine Synthetase System during Development

Age (days)	Arginine Synthetase System Activity (μmole of urea formed/30 minutes per g of liver)		
	Control	Experimental	p
−2.5	2.4 ± 0.4 (4)	2.3 ± 0.5 (6)	n.s.
+1.0	24.7 ± 3.9 (6)	21.8 ± 4.5 (6)	n.s.
+5.0	41.2 ± 9.1 (6)	53.0 ± 4.3 (7)	<0.01

Two spurts are observed in the activity of arginine synthetase system during postnatal development of the rat, as seen in Fig. 2. One immediately after birth and the other around the tenth to twentieth day after birth. The second increase in the activity of the arginine synthetase system occurs at the time of weaning, when there is a substantial increase in the protein consumption of the animal. Many other amino acid–catabolizing enzymes show an increase in activity at this time (28, 29). The increase in the protein intake is also reflected in the urea excretion of the animal, which rises markedly during the weaning period around the fifteenth postnatal day in the rat (11).

It seems reasonable to assume that the perinatal increase in the arginine synthetase system activity is regulated by hormonal changes occuring around birth and that the second increase in activity, which coincides with the change from a low- to a high-protein diet and is accompanied by increased urea excretion, is a dietary adaptation of the enzyme.

In Vitro Studies

Liver explants from 19.5 and 21 day old rat fetuses were cultured on stainless-steel grids which supported the tissue between the culture medium and the gas phase. All explants were maintained in culture at least 24 hours before the experiments started (21, 30).

To evaluate the viability of the explants cultured in our system, the incorporation of ^{14}C-leucine and ^{3}H-orotic acid into trichloroacetic acid–precipitable material was studied. After a 24 hour preincubation the incorporation of ^{14}C-leucine (into protein) continuously increased during a 3 hour pulse period. The incorporation of ^{3}H-orotic acid into TCA-precipitable material (mainly RNA) at 30 hours in culture also showed a progressive increasing incorporation after each successive hour of pulse.

The liver explants were further examined microscopically and ultramicroscopically. The explants generally showed mitochondria of normal shape, size, and distribution, and the rough endoplasmatic reticulum also appeared normal. It was further found that tyrosine transaminase activity could rapidly be induced in the cultured fetal liver explants after addition of hydrocortisone to the medium.

Effects of Triamcinolone

The direct effect of corticosteroid on liver arginine synthetase system was examined by the use of these fetal liver explants in culture. Triamcinolone or hydrocortisone in the concentration of 20 μg/ml culture medium increased the arginine synthetase system activity threefold within 24 hours after the addition of the hormone (Table 3). A dose of 0.2 μg/ml was sufficient to produce this increase in arginine synthetase system activity in 24 hours, after a lag period of ca. 12 hours.

This rise in the activity of the arginine synthetase system was completely prevented when cycloheximide (10 μg/ml) was added to the culture medium 4 hours after the steroid. The same result was obtained when cycloheximide was added 4 hours before the steroid. Actinomycin D in a dose of 25 μg/ml added to the culture media 4 hours before or after the steroid also completely inhibited the rise of the enzyme activity.

Table 3 Effect of Triamcinolone on the Arginine Synthetase System in Fetal Rat Liver Explants in Culture

	Final Concentration of Triamcinolone (μg/ml)	Arginine Synthetase System Activity (μmole of urea formed/30 minutes per g of protein)	
		Control	Experimental
21 day fetuses	0.2	32.1 ± 3.7 (6)	89.3 ± 12.8 (6)
	20.0	28.0 ± 3.1 (15)	87.1 ± 5.7 (15)
19.5 day fetuses	20.0	34.1 ± 2.1 (6)	93.5 ± 5.1 (6)

Effect of Glucagon and Dibutyryl Cyclic AMP

Glucagon (67 μg/ml) and dibutyryl cyclic AMP (0.1 m*M*) both doubled the arginine synthetase system activity in explants of 19.5 day old fetal rat liver (Table 4). The addition of 0.5 m*M* theophylline increased slightly the response to dibutyryl cyclic AMP but not to either glucagon or triamcinolone. The maximum response to 0.1 m*M* dibutyryl cyclic AMP was the same as the maximum response to glucagon. Dibutyryl cyclic AMP together with glucagon did not have an additive effect on the arginine synthetase system activity; however, dibutyryl cyclic AMP together with triamcinolone produced an effect which was greater than that of either agent alone (Table 4).

Table 4 Effect of Combinations of Glucagon, Dibutyryl Cyclic AMP and Triamcinolone on the Arginine Synthetase System in Fetal Rat Liver Explants in Culture in the Presence of Theophylline

	Concentration	Arginine Synthetase System Activity (μmole of urea formed/30 minutes per g of protein with theophylline)
Control	—	30.4 ± 2.6 (12)
Glucagon	67.0 μg/ml	69.9 ± 4.0 (8)
Dibutyryl cyclic AMP	0.1 m*M*	78.7 ± 6.2 (12)
Triamcinolone	20 μg/ml	94.0 ± 9.1 (6)
Dibutyryl cyclic AMP + triamcinolone	0.1 m*M* + 20 μg/ml	129.1 ± 13.2 (6)

Effect of Insulin

Insulin (0.05 to 0.25 i.u./ml) had no effect on the arginine synthetase system activity of the fetal liver explants in culture.

Studies on Human Fetal Liver

The arginine synthetase system could not be stimulated in human fetal liver explants with either corticosteroid or cyclic AMP (30).

CONCLUSION

In the newborn rat corticosteroid must be present in order to obtain the normal postnatal rise in the arginine synthetase system activity, since adrenalectomy at birth abolishes this postnatal increase and it can be restored by steroid administration. Glucagon also stimulates arginine synthetase system activity in both adult and newborn animals. In fetal animals, however, neither agent has any effect on the arginine synthetase system. The lack of *in vivo* induction of arginine synthetase system activity in the fetus suggests that some suppressive factor might be present *in utero*, since the fetal liver *in vitro* in organ culture seems to be competent to respond to the external inductive stimulus exerted by the added hormonal agents. Our data on the inducibility of the arginine synthetase system in fetal liver explants are in agreement with the hypothesis that cyclic AMP mediates the action of glucagon but not of corticosteroids. Similar interpretations have been made by Wicks (32) on the hormonal induction of tyrosine transaminase. The additive actions of cyclic AMP and corticosteroid hormone on the induction of the arginine synthetase system activity suggest that the mechanism whereby the steroid produces its action is different from that of glucagon or cyclic AMP. This finding is consistent with the results by Kenney et al. (33) that the induction of tyrosine transaminase by corticosteroid is produced by a mechanism different from that of glucagon.

The arginine synthetase system activity is thus inducible in fetal rat liver maintained in organ culture by corticosteroid hormone and by glucagon and cyclic AMP, although these agents do not induce this enzyme activity *in vivo* until after birth.

In human fetal liver, in which the arginine synthetase system shows considerable activity in control liver (three- to eightfold that found in fetal rat liver), hormonal agents do not stimulate the enzyme activity in the organ culture system.

REFERENCES

1. Needham, J. (1931). *Chemical Embryology*, Macmillan, New York.
2. Slemons, J. M., and Morris, W.H. (1916). *Bull. Johns Hopkins Hosp.* **27**, 343–350.
3. Hammett, F. S. (1918). *J. Biol. Chem.* **33**, 381–385.
4. Friedberg, V. (1955). *Gynaecologia* **140**, 34–45.

5. Teoh, E. S., Lau, Y. K., Ambrose, A., and Ratman, S. S. (1973). *Acta Obst. Gyn. Scand.* **52**, 323–326.

6. Manderscheid, H. (1933). *Biochem. Z.* **263**, 245–249.

7. Krebs, H. A., and Henseleit, K. (1932). *Hoppe-Seylers Z. Physiol. Chem.* **210**, 33–41.

8. Räihä, N. C. R., and Suihkonen, J. (1968). *Acta Paediat. Scand.* **57**, 121–124.

9. Kekomäki, M., Seppälä, M., Ehnholm, C., Schwartz, A. L., and Raivio, K. (1971) *Inter. J. Cancer* **8**, 250–258.

10. Räihä, N. C. R., Heinonen, K., Rassin, D., and Gaull, G. *Pediatrics*, in press.

11. Illnerova, H. (1968). *Physiologia Bohemoslovaca* **17**, 70–76.

12. Räihä, N. C. R., and Kretchmer, N. (1965). *J. Pediat.* **67**, 950.

13. Lightbody, H. D. (1938). *J. Biol. Chem.* **124**, 169–178.

14. Brown, G. W., Jr., and Cohen, P. P. (1959). *J. Biol. Chem.* **234**, 1769–1774.

15. Kennan, A. L., and Cohen, P. P. (1959). *Dev. Biol.* **1**, 511–525.

16. Räihä, N. C. R., and Suihkonen, J. (1966). *J. Pediat.* **69**, 934.

17. Räihä, N. C. R., and Suihkonen, J. (1968). *Biochem. J.* **107**, 793–797.

18. Charbonneau, R., Roberge, A., and Berlinguet, L. (1967). *Canad. J. Biochem.* **45**, 1427–1432.

19. Illnerova, H. (1966). *Biol. Neonat.* **9**, 197–202.

20. Miller, A. L., and Chu, P. (1970). *Enzym. Biol. Clin.* **11**, 497–503.

21. Schwartz, A. L. (1972). *Biochem. J.* **126**, 89–98.

22. Kennan, A. L., and Cohen, P. P. (1961). *Proc. Soc. Exp. Biol. Med.* **106**, 170–173.

23. Colombo, J. P., and Richterer, R. (1968). *Enzym. Biol. Clin.* **9**, 68–73.

24. Schimke, R. T. (1962). *J. Biol. Chem.* **237**, 459–468.

25. Schimke, R. T. (1963). *J. Biol. Chem.* **238**, 1012–1018.

26. McClean, P., and Gurney, M. W. (1963). *Biochem. J.* **87**, 96–104.

27. McClean, P., and Novello, F. (1965). *Biochem. J.* **94**, 410–422.

28. Snell, K., and Walker, D. G. (1973). *Enzyme* **15**, 40–81.

29. Räihä, N. C. R., and Kekomäki, M. P. (1968). *Biochem. J.* **108**, 521–525.

30. Räihä, N. C. R., and Schwartz, A. L. In Hommes, F. A. and Van den Berg, C. Z., Eds., *Inborn Errors of Metabolism*, Academic, London, pp. 221–237.

31. Räihä, N. C. R., and Schwartz, A. L. (1973). *Enzyme* **15**, 330–339.

32. Wicks, W. D. (1969). *J. Biol. Chem.* **244**, 3941–3947.

33. Kenney, F. T., Reel, J. R., Hager, C. B., and Wittliff, J. L. (1968). In San Pietro, A., Lawborg, M. R. and Kenney, F. J., *Regulatory Mechanisms for Protein Synthesis in Mammalian Cells*, Academic, New York, pp. 119–142.

Discussion

Dr. Aebi: Thank you very much, Dr. Räihä. This paper is now open for discussion.

Dr. Cohen: I want to congratulate you on the very nice system that you have described. We have tried for years to find a suitable *in vitro* system using amphibian liver tissue, in the hope that we could get some information about the induction of these enzymes. We have been less successful than you, and I want to congratulate you on the success you have had with it, in particular as regards the hormone induction.

Dr. Ratner: I want to ask you if you sent out any feelers about the other enzymes of the ornithine cycle—it's just too big an experiment to do all at once—but are you guided by the assumption that they will all be going in the same direction? What is the inducing agent?

Dr. Räihä: Thank you, Dr. Ratner, for your question. We are at the moment studying the induction of all the urea-cycle enzymes in our organ culture system.

Dr. Ratner: The reason I asked the question is because Schimke's work shows that the response is a coordinated one. Sometimes arginase can change separately from the other enzymes, but certainly in response to an increase in nitrogen flux they all change together, and I think we should accept this until we are shown that it is otherwise, in response to induction, and you really have the tools, I think, to examine that.

Dr. Räihä: By using the organ culture system which I described in my paper, we will be able to examine whether the hormonal response which was observed for the arginine synthetase system will be a coordinated response which can be seen for all the enzymes in the urea cycle. It will also be possible to study *in vitro* the effects of variations in various substrate levels.

Dr. Cohen: I want to respond to Dr. Ratner's comment by pointing out that the system that is being dealt with here reveals that there is asynchrony in the emergence of the urea-biosynthetic enzymes during rat fetal development. The fact that the argininosuccinate synthetase system does not emerge until 1 day postpartum and then with a burst, while there is a progressive increase in the other enzymes, as pointed out by Dr. Räihä, implies either that some factor is operating to repress the argininosuccinate synthetase system or that there is a lack of a specific inducer. Since we don't know what factors are operating to coordinate the levels of the mitochondrial and the extramitochondrial enzymes involved in urea biosynthesis, Dr. Räihä's system offers the possibility of investigating these factors.

Dr. Ratner: I think conclusions should be withheld about what is referred to as the "Ratner system" until the assays are done over. I disclaim responsibility for the assay procedure. Unless argininosuccinase is added as a supplement in the synthetase assay, the synthetase values are seriously underestimated. We don't know anything about when the activity emerges. May I just say that if the table shown earlier by Dr. Raijman was taken from Schimke's publication, these remarks do not apply to his values for argininosuccinate synthetase, because he did supplement his assay with argininosuccinase.

19

Coordinated Changes in Enzymes of the Ornithine Cycle and Response to Dietary Conditions

H. Aebi
Medizinisch-chemisches Institut der Universität Bern
Bern, Switzerland

INTRODUCTION

The enzyme pattern of any organ or cell depends on the expression of its genetic information and on environmental factors of various kinds (1). The interdependence of both of them, as well as the sum of homeostatic regulation mechanisms, are responsible for the fact that the actual enzyme concentration measured under defined conditions and at a given moment is the resultant of a multitude of synergistic and antagonistic effects. Adaptive activity changes in enzymes forming a team often occur simultaneously. Thus, coordination in enzyme regulation is accomplished by a proportional increase (or decrease) in activity level of all enzymes participating in the reaction. The maintenance of the same proportions within a group of enzymes (e.g., in the branching-free part of the glycolytic chain) is regarded as evidence for a coordinated genetic control ("operon") (2–4).

Among the large number of environmental factors those due to nutrition are particularly striking and of considerable practical importance. There are two classical examples demonstrating such a coordinated regulation owing to a change in the diet. First, Fitch and Chaikoff (5) and Weber et al. (6) have reported that in the liver of rats fed on a diet high in fructose (60%) for 7 days there is an almost tenfold increase in activity of enzymes of the hexose-monophosphate-shunt, for example, glucose-6-phosphate-dehydrogenase and phosphogluconate dehydrogenase, as compared with that in rats fed on a standard diet containing starch as the main carbohydrate. On the other hand Schimke (7) has shown that the activity levels of all five enzymes forming the ornithine cycle remarkably well reflect the protein content of the diet. This observation, which has to be dealt with in this report in some detail, still is the most impressive example of coordinated change within a group of enzymes due to altered composition of the diet.

Many more cases are known today where dietary effects, such as quantity of food intake, composition of the diet, or rhythm of the feeding schedule, on the activity level of certain enzymes have been demonstrated. In this respect the studies of Potter (8) on the oscillation of enzyme levels are of special interest. By adapting rats to a controlled feeding schedule, for example, 40-hour fasting period and 8-hour feeding period, he has shown that under such strict standardization conditions there is a minimum of individual scattering. Furthermore, owing to a high degree of homogeneity in his data, tidal changes in enzyme levels could be observed. Significant oscillations were seen in those enzymes which are subject to induction effects in liver, such as tyrosine transaminase. The existence of a "feeding peak" in 12 different liver enzymes is an indication that adaptive alterations following feeding proceed at a relatively rapid rate and that such oscillations have to be considered as a rather general phenomenon.

DIETARY AND HORMONAL FACTORS AFFECTING THE LEVEL OF ORNITHINE-CYCLE ENZYMES

The complexity of a response to changes in environmental conditions makes it often difficult to ascribe a certain effect to one single factor exclusively. In particular there are close relations between nutritional and hormonal influences. The similarity of effects, for example, those following a high protein intake and the administration of corticosteroid hormones, justifies a common view and explains in part the complementarity of either effect as to direction and extent.

The influence of protein intake on rat liver arginase activity has first been observed by Lightbody and Kleinman in 1939 (9). The availability of specific assays for all ornithine-cycle enzymes prompted Schimke to reevaluate and to extend this finding (7, 10). The essential result of his study is shown in Fig. 1. The five diagrams clearly indicate that when varying the casein content in the diet between 15 and 60%, there is a linear correlation between protein-N intake and the activity of (1) carbamoyl-phosphate synthetase, (2) ornithine carbamoyltransferase, (3) argininosuccinate synthetase, (4) argininosuccinate

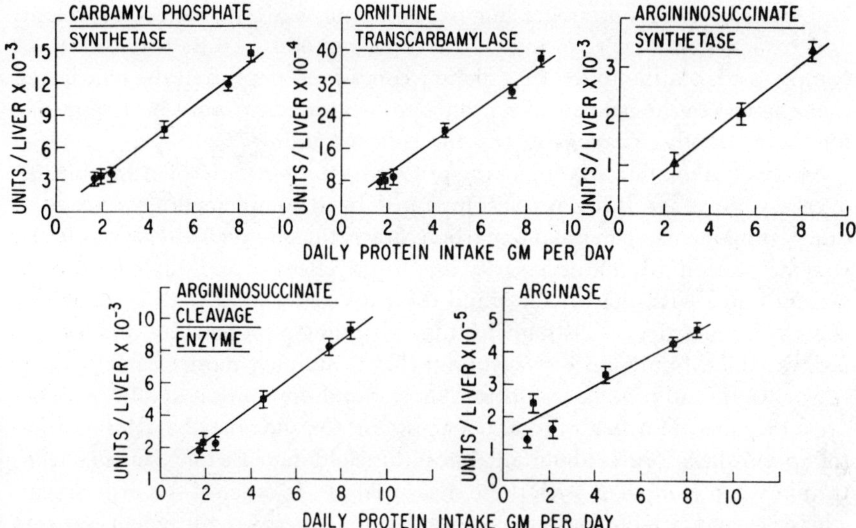

Figure 1. Relationships of total liver content of urea cycle enzymes to daily protein consumption. Protein intake in grams per day is plotted against enzyme content in units per liver. Brackets indicate ± 4 standard errors. The circles represent animals initially weighing 50–60 g each, and the squares indicate animals weighing 140–150 g each at the start of the experiment. Triangles indicate a separate group of rats weighing 140 to 150 g each, which were initially assayed only for argininosuccinate synthetase (Schimke, 1962) (7).

lyase, and (5) arginase. On the other hand, lactate dehydrogenase in rat liver showed changes opposite to those of the ornithine-cycle enzymes, whereas glucose-6-phosphate-dehydrogenase as well as aspartate- and alanineamino-transferases resembled the ornithine-cycle enzyme pattern.

These discrepancies have been commented by Schimke as follows: "Such variation in patterns of enzymes as produced by differences in dietary protein intake indicate that there is a certain specificity to such patterns, and that they are not simply a reflection of increased availability of amino acids for protein synthesis or of the storage of protein in the form of specific enzyme protein." In fact the total protein content of the rat liver as used for the determination of the nutritive value of a protein (e.g., by the method of Kosterlitz) depends much less on the level of protein intake than the ornithine-cycle enzymes (3).

Variation of the level of protein intake obviously also alters the rate of urea synthesis. The fact that a high-protein supply is not only met by an increase in argininosuccinate synthetase, the enzyme limiting the overall rate of the cycle, led Schimke to suggest the existence of a coordinating regulatory mechanism controlling the level of all five enzymes in the same way. Furthermore, the identity of enzyme preparations obtained from livers of the low-protein- and high-protein-diet group (regarding pH optimum, K_m, Q_{10}, molecular weight, and turnover number), as well as the time lag of 4 to 8 days required for complete adaptation to the new dietary conditions, permitted the conclusion that changes in the net rate of synthesis (or degradation) must be responsible for these adaptive changes in enzyme concentration (7).

Since a systematic variation of the protein content in a diet (fed ad libitum) over a wide range is always accompanied by a complementary change of other components, such as fat or carbohydrate, the observed alterations in the enzyme pattern are not necessarily due to the effect of one single food component only. With this idea in mind these dietary studies were extended by comparing the effect of fasting with that of feeding a protein-free diet for 3 to 6 days (10). Maintenance of rats on a diet containing mostly carbohydrate (dextrin) and no protein results in a more rapid diminution of all ornithine-cycle enzymes than that of total liver protein. On the other hand, starvation (of up to 7 days) brings about an almost threefold increase in enzyme concentration when compared with those measured at the onset of the experiment. This discrepancy is obviously due to the fact that energy requirements are met almost completely by dietary carbohydrate when a protein-free diet is fed, whereas in prolonged starvation energy must be derived from body fat and protein. In the former case a minimum of urea is produced ($\sim$25% of that of control rats); in the latter case protein catabolism and deamination of amino acids are leading to an increase in urea formation. During starvation urea excretion is about five times higher than that of animals maintained on a 15% casein diet. These differential effects of two forms of protein depletion indicate the existence of specific control mechanisms. This observation led Schimke to conclude that variation in urea excretion may be largely mediated by synchronous alterations in levels of the enzymes involved in urea synthesis (10).

The impact of hormonal action on the activity of the ornithine-cycle enzymes has simultaneously been studied by Schimke (11) and by McLean and Gurney (12). They have found that in rats adrenalectomy causes a reduction in the activity of all five enzymes of the ornithine cycle; however, not to the same extent. As shown in Fig. 2, this change is most pronounced in arginase and in argininosuccinate synthetase, the overall rate-limiting enzyme, both of which falling to about 40% of the pair fed control values at the fourteenth day after adrenalectomy. In these rats the activity levels can be brought to normal by prolonged administration of cortisone acetate. The same stimulating effect is seen in normal rats, where the level of those enzymes rises well above the normal range after 4 days of cortisone treatment. Contrary to cortisone, growth hormone was claimed to have a slight reducing effect of 10 to 25%. In alloxan-diabetic and in glucagon-treated rats there is a similar enzyme pattern inasmuch as mainly three out of the five enzymes forming the ornithine cycle are significantly increased (12, 13). The data in Table 1,

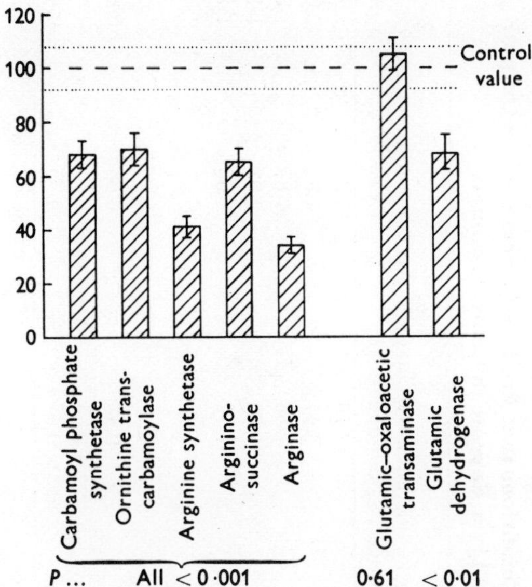

Figure 2. Effect of adrenalectomy on the total activity of enzymes of the urea cycle, glutamic dehydrogenase and aspartate amino transferase in rat liver. Results, as enzyme units in the whole liver, are expressed as a percentage of the activity in the corresponding pair-fed control groups. Vertical lines represent twice the S.E.M. The horizontal dashed line is the arbitrarily fixed control value of 100%. Fisher's P values are given at the foot of each column. Each group contains six or more animals. (McLean and Gurney 1963) (12).

which permit a comparison between various experimental conditions, indicate that—contrary to high-protein diet and cortisone treatment—only carbamoyl phosphate synthetase, argininosuccinate synthetase, and argininosuccinate lyase are above the normal range in diabetic and in glucagon-treated rats. Whereas the similarity of response in these two conditions seems reasonable, the different patterns shown in Table 1 may be taken as evidence for the specificity and the rather complex nature of the observed regulatory effects.

In order to further elucidate the primary factor or factors responsible for adaptive alterations in the activity of ornithine-cycle enzymes, various experimental approaches have been used. First, Schimke showed that urea, arginine, or any other individual amino acid added to the diet—contrary to high-protein feeding—has no enhancing effect at all; the same is true for urea-cycle intermediates, such as ornithine, citrulline, or arginine when administered parenterally (11). On the other hand, when rats are fed on an arginine-free diet, the specific activities of the four enzymes involved in arginine biosynthesis are increased up to twofold over levels predicted on the basis of urea excretion, whereas arginase activity remains unchanged. Furthermore, experimental

Table 1 Changes in the Activities of Urea-Cycle Enzymes in Some Conditions that Produce Increased Urea Output, Expressed as Ratio (Total Liver Enzyme in Treated Animals) : (Total Liver Enzyme in Control Animals)

Condition	Reference	Carbamoyl Phosphate Synthetase	Ornithine Carbamoyl Transferase	Argininosuccinate Synthetase	Argininosuccinate Lyase	Arginase
High-protein diet	Schimke (1963)	6.50	2.31	2.78	3.34	1.56
Starvation (7 days)	Schimke (1962)	5.03	4.22	3.47	4.92	2.01
Cortisone treatment	Schimke (1963)	3.15	1.42	1.38	1.85	1.26
Alloxan-diabetes	McLean and Novello (1965)	1.58	1.09	1.70	2.03	1.17
Glucagon treatment	McLean and Novello (1965)	1.24	0.88	1.42	1.58	1.11
For comparison are included:						
Adrenalectomy (situation after 14 days)	McLean and Gurney (1963)	0.68	0.70	0.43	0.68	0.36
Portocaval shunt (situation after 30 days)	Colombo et al. (1973)	0.55	0.35	0.47	0.95	0.49

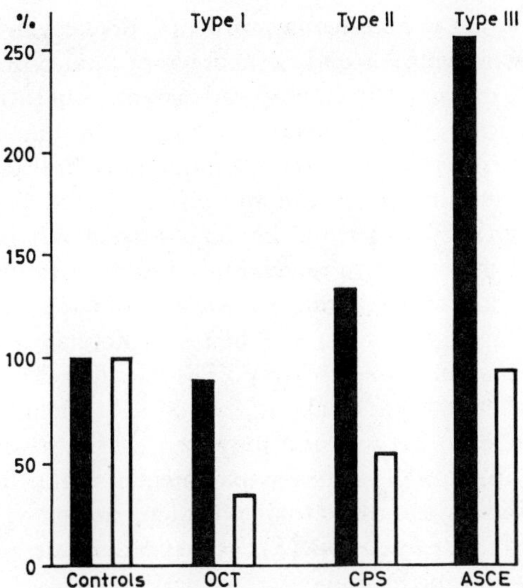

Figure 3. Different types of responses of enzyme activities of the urea cycle after portocaval shunt in the rat liver. Three typical responses are observed: type I response: normal or slightly decreased enzyme activity per gram of liver, but decreased per total liver mass (ornithine carbamoyl-transferase = OCT, argininosuccinate synthetase, arginase); type II response: increased enzyme activity per gram liver, but decreased per total liver mass (carbamoyl phosphate synthetase = CPS); type III response: increased enzyme activity per gram liver, but normal per total liver mass (argininosuccinate lyase = ASCE). The bars give mean values per gram liver wet weight (black) or per total liver mass (white), expressed in percentage of the unoperated control group. (Colombo et al., 1972/73) (15).

uremia—Brown et al., (14)—as well as portal blood diversion by portocaval shunt—Colombo et al., (15)—were used for this purpose. The characteristic alteration in the urea-cycle enzyme pattern as observed 30 days after a shunt operation is shown in Fig. 3. Three different types of responses of enzyme activity can be distinguished, the decrease in liver size being variably compensated by an increase in enzyme concentration. The authors do not offer an explanation for the preferential increase in argininosuccinate lyase; however, they consider the increase in carbamoyl phosphate synthetase, depending on ammonia as substrate, to be an adaptive consequence of systemic hyperammonemia (15).

MECHANISMS OF ENZYME REGULATION

The interplay of both enzyme formation and degradation in the control of enzyme levels has been studied in considerable detail by Schimke et al. (2, 3,

16). He has shown that in mammalian tissue, particularly in liver, there is an extensive continuous synthesis and breakdown of total protein as well as of specific enzymes. Consequently an observed enzyme concentration represents the result of both the rate of synthesis and the rate of degradation. Furthermore he has shown that either process can be altered independently by changes in the physiological and nutritional state by various factors. The determination of the turnover of an enzyme permits the calculation of its half-life time. There is a marked heterogeneity of turnover rates among enzymes present in rat liver. The half-life time of an enzyme may vary within a range of two decades: Aminolevulinicacid synthetase $t/2 \sim 1$ hour, tryptophan pyrrolase 4 hours, catalase 24 hours, and arginase 4 to 5 days. The mitochondrial stroma proteins have an even higher half-life of about 7 days (16). Therefore, the half-life time is specific for each enzyme and may vary within a wide range—even among species localized in the same compartment or organelle.

The relationship between the actual level of an enzyme (P) and its rate of synthesis (K_s) and of degradation (K_D = first-order rate constant) can be expressed by the following formula:

$$\frac{dP}{dt} = K_s - K_D \cdot P$$

Thus in the steady state the equation becomes

$$P = \frac{K_s}{K_D}$$

This highly simplified model offers an explanation why—as a rule—enzymes present in low concentration or exerting a high turnover are particularly sensitive to adaptive alterations. Rapidly regenerating enzymes such as tryptophan pyrrolase or tyrosine transaminase respond faster and more strongly to cortisone administration than is the case for arginase. The technique used for the analysis of enzyme turnover is based on the radioactive labeling of the enzyme under investigation. This is done by feeding a diet containing ^{14}C-lysine or ^{14}C-arginine until the level of specific activity in regard to ^{14}C incorporated into the liver proteins has reached a steady state (Fig. 4). This is the case in the rat after a feeding period of 3 to 4 weeks. Then the labeled amino acid is replaced in the diet by the nonlabeled compound. Total enzyme activity and the ^{14}C activity of the isolated enzyme are then followed by analyzing liver samples at appropriate intervals. By using this sophisticated procedure it has been shown that, in the case of tryptophan pyrrolase, glucocorticoid administration raises the enzyme level as a result of an increased rate of enzyme synthesis, whereas the substrate-induced accumulation of this enzyme has to be attributed to a complete stop of enzyme degradation, in the presence of continued enzyme synthesis. Accordingly total ^{14}C activity in tryptophan

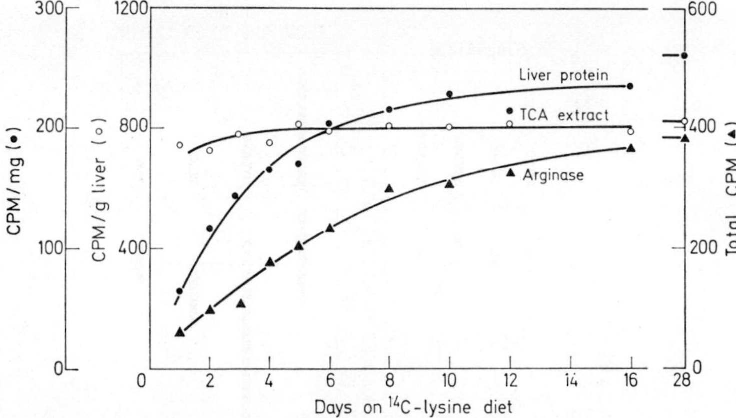

Figure 4. Incorporation of continuously administered ^{14}C-L-lysine into total protein, arginase and trichloroacetic acid (TCA) soluble extracts of rat liver. Osborne-Mendel rats maintained for 7 days on a diet consisting of 25% complete amino-acid mixture were then placed on a similar diet containing ^{14}C-L-lysine. At intervals one rat was killed. Radioactivity of the TCA-soluble fraction is expressed as counts per minute per extract from 1 g of liver, wet weight (○). Counts in total liver protein: counts per min. per mg of protein (●). Counts in the arginase: total counts precipitated (▲). Schimke, reproduced from Aebi, 1969 (3).

pyrrolase is maintained at a constant level in the presence of the inducer. However, there is a rapid fall in the control animals. This approach has shed more light on the classical concept of the "dynamic state of body constituents" demonstrating to what extent the enzyme molecules present in a given moment are subject to a continuous renewal. The example given in Fig. 4 shows that arginase has played an important role in the elucidation of enzyme protein turnover (16).

Arginase will be used again as an example in order to discuss the regulatory mechanisms and its manifestations when an adaptation to altered environmental conditions is requested. By using the same analytical procedures mentioned in the legend for Fig. 4, it has been elegantly demonstrated by Schimke (16) that drastic changes in the synthesis-degradation interplay occur when the composition of the diet is altered. Considerable changes in either rate are observed, for example, after the level of casein has been lowered suddenly from 70 to 8%. Here, the new steady state with regard to arginase activity is reached by both a decrease in the rate of synthesis and a reversible increase in the rate of degradation. However, if feeding is stopped entirely, there is an immediate and complete cessation of enzyme degradation. Here, for obvious reasons, well known as the situation in "austerity," the organism tends to save the enzyme (Fig. 5). This observation is another good example for the effectiveness and the usefulness of homeostatic mechanisms. The result of this

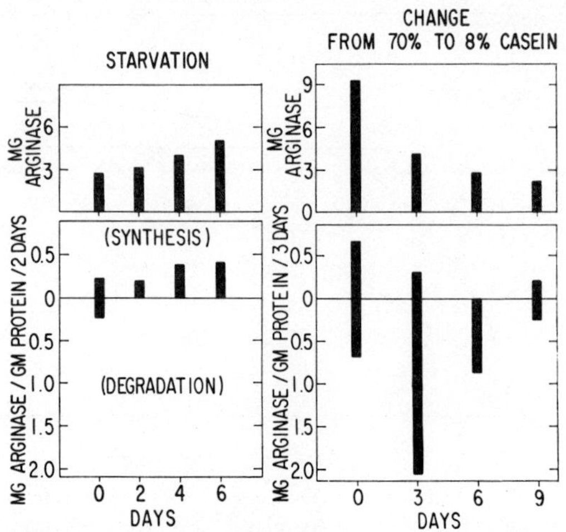

Figure 5. Rates of synthesis and degradation of rat liver arginase during fasting (left) and change from 70 to 8% casein (right). Animals were maintained on diets containing 8% (starvation) or 70% casein for 14 days prior to the experimental period. The animals were given single administrations of ^{14}C-guanidino-L-arginine 1 hour prior to change in dietary status. The loss of isotope, both total and specific activity, from arginase was followed with time. The upper set of bars indicate the total milligrams of arginase in the pooled sample of four livers (starvation) or three livers (change from 70 to 8% dietary protein). The lower set of bars show the rates of synthesis and degradation expressed as milligram of arginase synthesized and degraded per gram liver per observational period. The loss of total radioactivity in arginase is a measure of the rate of synthesis. The bases for these calculations, as well as experimental details, are given in Schimke et al. 1965 (17) and Schimke et al. 1968 (16).

investigation must be kept in mind whenever changes in enzyme levels have to be evaluated. Today, there is a good deal of knowledge what the mechanism of protein (and enzyme) synthesis and the regulation by controlling its rate is concerned. On the other hand little is known about the various systems involved in protein degradation and notably about its control. Above all, it is very difficult to see how the lysosomal proteases altogether can be activated or directed toward their targets with a high degree of specificity. Since there are such large differences in the speed of turnover of neighbouring enzymes, it is not unreasonable to assume that specific inactivating systems do exist.

More recently Das (18, 19) and Das and Waterlow (20) have accomplished a detailed investigation on the rate of adaptation of urea-cycle enzymes, aminotransferases, and glutamic dehydrogenase to changes in dietary protein intake. The adaptation from a standard diet containing 5% casein was followed at 3 to 6 hour intervals; the same procedure was followed for the study of the change in the opposite sense. It was shown that in either direction the decrease (or increase) in urinary N output and the activity of all six enzymes have reached their new steady-state levels in 30 hours (Fig. 6). The fact that the enzymes altered their activity level at practically the same rate was taken as evidence for the existence of a common control mechanism. Contrary to earlier publications an average half-life for these enzymes in the order of 7 hours was calculated. Furthermore, there was no simple relationship between the activity of the urea-cycle enzymes and the amount of N excreted. When an equal amount of gelatin was substituted for casein, the N output was doubled, but there was no change in the activity of the liver enzymes. Based on these observations it is suggested that the activity of the urea-cycle enzymes depends only in part on the amount of N available for excretion after the demands for protein synthesis have been met. Since the enzymes appear to be present in excess, an increased N load is not necessarily accompanied by an increase in enzyme activity (20).

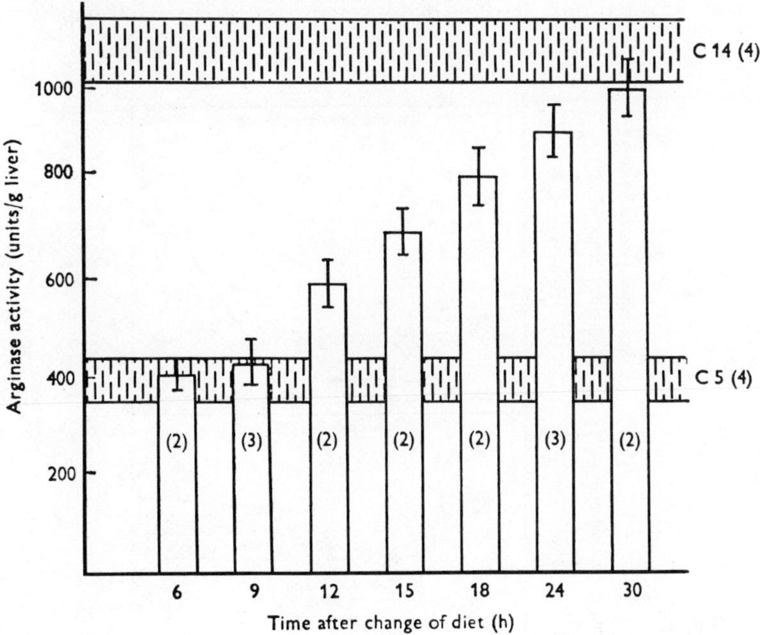

Figure 6. Effect of changing the protein content of the diet on rat liver arginase activity. Mean values and ranges (vertical bars) for arginase activity (per g liver) in rats at 6, 9, 12, 15, 18, 24, and 30 hours after transfer from a diet containing 45 g casein/kg [C_5] to one containing 135 g casein/kg [C_{14}] and in rats maintained on either C_5 or C_{14} throughout. The horizontal bands represent the range of activity in rats maintained on either C_{14} or C_5 throughout. □ = Mean values and ranges (vertical bars) of activity in rats after change in diet; number of rats in parentheses. One unit of arginase activity is the amount of enzyme catalyzing the formation of 1 μmole urea/minute at 38°. Das and Waterlow, 1974 (20).

Specificity and rate of adaptation to altered dietary conditions were investigated by analyzing 13 different rat liver enzymes by Camus et al. (21). This study is of particular interest to the situation in humans, since malnourished rats (fed on a 7% gluten diet, deficient mainly in lysine and threonine) were first starved for 24 hours and then allowed to recover for 3 days. Two groups, one fed with a 18% casein diet, the other fed with the original 7% gluten diet, were analyzed separately. Among the considerable number of findings it is of interest that, within the 3 day refeeding period, arginase activity, irrespective of the quality of the protein fed, did not vary significantly, whereas ornithine aminotransferase increased eightfold when the rats were refed with the casein diet. Here, contrary to the response of arginase, the effect of the amino acid composition of the dietary protein is evident (Fig. 7). The diversity of reaction patterns among the 13 enzymes studied may be considered additional evidence for the specificity of response in enzyme regulation. The same conclusion can be drawn from data presented by Durkin and Nishikawara (22), who studied the effect of starvation, feeding high-protein diet, and partial hepatectomy

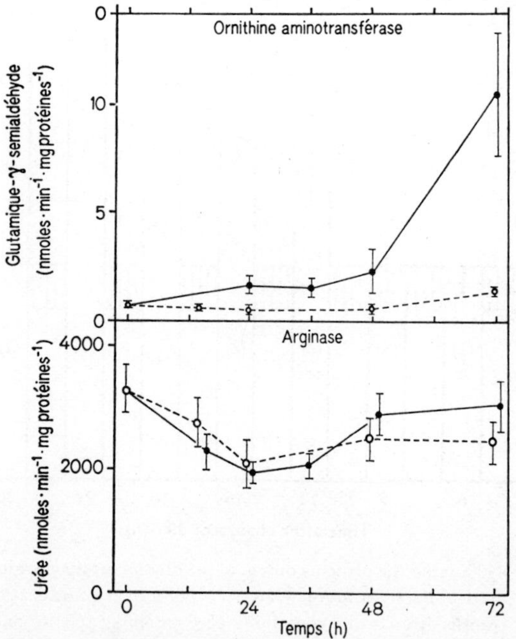

Figure 7. Ornithine aminotransferase and arginase activity in rat liver. Malnourished rats, after a fasting period of 24 hours were refed with (*a*) a diet containing 18% casein (–●–), (*b*) a diet containing 7% gluten (–○–). Each point represents the average of 8 individual values ($\overline{X} \pm s$). Camus and Vandermeers-Piret, 1969 (21).

on rat liver aspartate- and ornithine-carbamoyltransferase, as well as from those of Tatibana and Shigesada (23) on the specific roles of the two carbamoylphosphate synthetases in the control of pyrimidine and urea biosynthesis.

COMPARATIVE ASPECTS AND UREA CYCLE ENZYMES IN MAN

The activity of enzymes relevant for urea biosynthesis has been measured in a number of species under various conditions. The classical example showing that this enzyme pattern may undergo drastic changes is amphibian metamorphosis, which is paralleled by a switch from ammonotelia to urotelia. Brown et al. (24) have analyzed the activity of the urea-cycle enzymes at all developmental stages from early tadpole to the adult frog. Avian kidney arginase at different developmental stages has been studied by Dror and Nir (25). More recently urea-cycle enzymes have been measured in the rhesus monkey as well as in man (26). These data permit a direct comparison between the rat, the preferred experimental animal, the monkey, a species close to man on one hand, with the situation in the human individual. The data in Table 2 disclose that the activity levels of the single enzymes of the urea cycle vary considerably, at least when determined *in vitro* under standard conditions. In all instances arginase activity is by far highest and argininosuccinate synthetase lowest.

However, as visualized in Fig. 8 by means of a semilog plot, all three mammalian species tested exert almost exactly the same enzyme pattern. This is characterized by a group of three enzymes of approximately the same activity level, whereas arginase is present in 200 to 500 fold and ornithine carbamoyltransferase in 20 to 50 fold excess over the remainder enzyme triplet. The two series of ratios indicated in Fig. 8, intended to serve as rough indicators only, viz., arginase/ornithine carbamoyl transferase (ARG/OCT) and ornithine carbamoyl transferase/carbamoylphosphatase synthetase (OCT/CPS) are—except for the frog—of remarkable constance.

From the available data can be concluded that—at least within mammals—the urea-cycle enzymes behave as a constant proportion group of enzymes, just as the branching-free part of the glycolytic chain (2–4). However, there are quantitative species differences, inasmuch as the data obtained from rat and monkey coincide, although the rats used for this comparative experiment consumed 2.5 times more protein-N per kilogram than the monkeys. The extent, to which the enzyme level is altered by a tenfold increase of the protein intake is about the same in rat and monkey. In either species the activity increases 2 to 3 fold, a ratio which may also be assumed for man. For experimental purposes, for example, for the evaluation of the biological value of a protein, it is of interest that avian kidney arginase depends considerably more

Table 2 Levels of Urea-Cycle Enzymes in Liver of Various Species

Due to different methods used, the table is intended to compare the data mainly horizontally, rather than vertically. The activity data are based on 1 g of liver wet weight throughout.

Species	Reference	Carbamoyl Phosphate Synthetase	Ornithine Carbamoyl Transferase	Argininosuccinate Synthetase	Argininosuccinate Lyase	Arginase
Rana catesbiana	Brown et al.					
Tadpole stage	(1959)	55	2,320	1–2	25	1,040
Adult frog		1,040	20,000	30–50	506	28,900
Rat	Nuzum and					
Low-protein diet	Snodgrass					
(4.0 g/kg/day)	(1971)	246 ± 53	6,900 ± 1,850	142 ± 65	213 ± 57	200,000 ± 36,000
High-protein diet						
(40 g/kg/day)		841 ± 82	15,400 ± 2,900	500 ± 77	600 ± 127	330,000 ± 31,600
Rhesus monkey	Nuzum and					
Low-protein diet	Snodgrass					
(1.6 g/kg/day)	(1971)	272 ± 47	7,100 ± 470	117 ± 12	243 ± 57	91,500 ± 25,000
High-protein diet						
(16 g/kg/day)		900 ± 78	17,200 ± 3,200	359 ± 62	627 ± 94	212,000 ± 25,000
Man	Nuzum and					
protein intake	Snodgrass					
(1.3 g/kg/day)	(1971)	279 ± 65	6,600 ± 1,580	90 ± 12	220 ± 25	86,000 ± 9,300

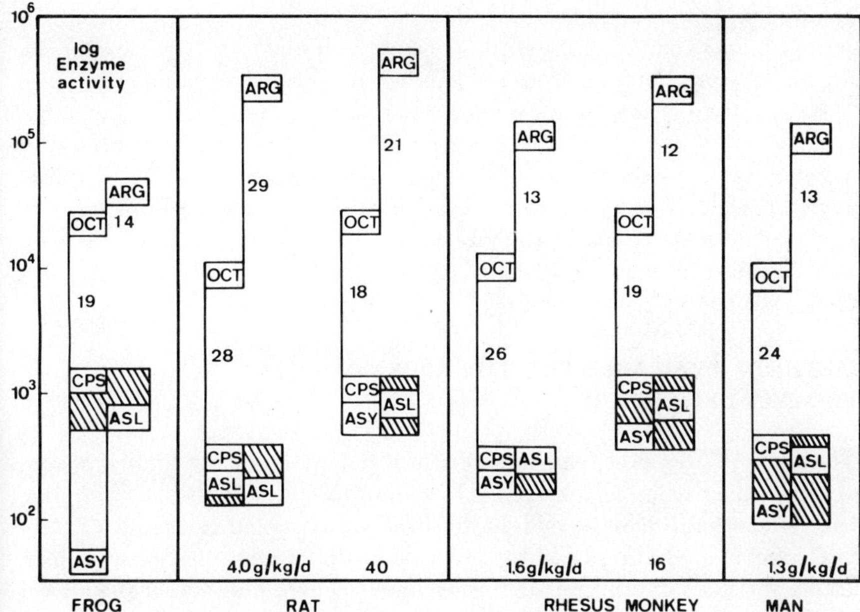

Figure 8. Pattern of urea cycle enzymes. The mean enzyme activities given in Table 2 are plotted on a log scale in order to permit a better comparison of the enzyme patterns. The abbreviations indicate: CPS = carbamoyl-phosphate synthetase; OCT = ornithine carbamoyl transferase; ASY = argininosuccinate synthetase; ASL = argininosuccinate lyase; ARG = arginase. The figures in the plots represent the activity ratio ARG/OCT and OCT/CPS respectively.

on the level of protein intake than mammalian liver arginase does. As shown by Dror and Nir (25) variation of the protein content in food between 10 and 40% causes a change in kidney arginase activity in chicks by a factor of 6 to 7. Even greater differences are observed when proteins of different biological value, such as soybean and gluten, are compared. These observations, although few in number, give ample evidence that adaptive changes in the level of urea-cycle enzymes may be considered as a general phenomenon and that the measurement of these enzymes is a valuable tool for studying the biological value of proteins.

The first proof that the level of urea-cycle enzymes in man too is affected by protein intake comes from Stephen and Waterlow (27). They have followed the recovery of 22 children suffering from malnutrition by analyzing argininosuccinate lyase activity in 5 to 15 mg liver biopsy specimens at three different stages. Despite a considerable degree of individual variation an increase in mean activity could be observed (biopsy I: 1.06; II: 1.31; III: 1.46 μmole urea/mg protein/hour). Discussing these data with respect to the difficulty in getting human material, the authors raise the question to what extent the wealth of results obtained from rats can be extrapolated to man. They state: "This is a particularly vexing problem in relation to nutrition and metabolism, since the two species differ so greatly in size and life-span, in overall metabolic rate, in the nature of their diet, and in their nutritional requirements. If general conclusions are to be drawn, direct comparisons must be made between the results in animals and man, whenever it is justifiable to do so." The more recent data presented in Table 2 and Fig. 8 meet these requirements to a considerable extent.

METHODS AVAILABLE FOR THE APPRAISAL OF PROTEIN DEFICIENCY

The study of urea-cycle enzymes in man is hampered by the limited accessability of liver tissue for analysis. Thus, from the physician's view sampling biopsy material is feasible and permissible in exceptional cases only. Therefore possible alternatives will be discussed in this paragraph. Some of them are (*a*) the study of ornithine-cycle enzymes in cells or fluids other than liver; (*b*) the use of urea production as an indirect measure for the activity of ornithine-cycle enzymes; (*c*) the availability of enzymes other than ornithine-cycle enzymes for the assessment of the nutritional status.

The search for ornithine-cycle enzymes in other tissues, such as leukocytes or thrombocytes, skin or oral mucosa specimens, even hair papilla or nail clippings, has had no success so far. More promising seems to be the use of activity measurements in serum, such as that of ornithine-carbamoyltransferase. However, there are two types of findings suggesting that data obtained from white cells or serum have to be interpreted with caution, at least what their use as indicator for the nutritional status is concerned. First, it was observed that, in rats fed at different protein levels, liver and leukocyte enzymes need not show the same dependence—e.g., glucose-6-phosphatedehydrogenase and catalase (28). However, if a significant dependence on protein intake or degree of malnutrition has been verified, such a parameter will be of considerable clinical interest (29). On the other hand the use of an enzyme activity in serum as indicator for the situation within the liver cell implies that the partition between the intra- and the extracellular compartment is constant. From observations made in other instances (e.g., creatinine phosphokinase in serum) it is evident that this partition ratio depends on a number of factors, and therefore this requirement can hardly be met (30).

Since there is a close correlation between the activity level of ornithine-cycle enzymes and the rate of urea synthesis (20), urea concentration in serum as well as the amount of urea excreted in urine per day may serve as an acceptable substitute. In fact, urea concentration in plasma or blood, provided that kidney function is normal, depends mainly on protein intake. Therefore it is widely used as an indicator for the evaluation of a given dietary situation. Variation of the daily protein intake from 0.5 to 2.5 g/kg body weight corresponds to a mean urea concentration in plasma from 8.95 ($\bar{X} \pm 2s = 6.25 - 11.6$) to 21.1 ($\bar{X} \pm 14.4 - 27.7$) mg urea-N per 100 ml (31). In field studies on protein-calorie malnutrition this parameter has repeatedly given satisfactory information (Fig. 9). The same is true for the analysis of urea-N in urine. As observed by Arroyave (32), children suffering from dietary protein deficiency have a low excretion of urea-N as determined in a fasting urine sample. When urea is related to creatinine, the collection of a timed specimen is not required. It must be pointed out, however, that a low urea nitrogen to creatinine ratio does not necessarily indicate protein-calorie malnutrition but only that in the period of the examination the child was fed a low-protein diet (33). The use of parameters related to the rate of urea synthesis in liver is further complicated by the fact that urea may be subject to alternative

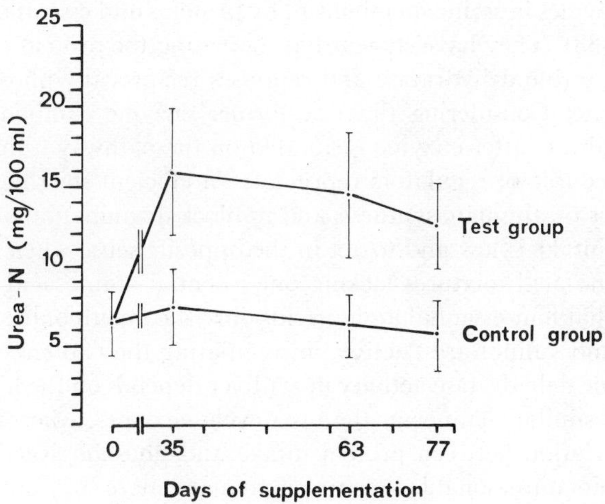

Figure 9. Level of urea-N in blood of children (4 to 8 years of age) from northern Ivory Coast. Two groups of individuals (40 each) were compared, the control group eating its usual diet (containing approximately 0.5 g protein/kg body weight), the test group consuming as a supplement a commercial casein preparation corresponding to an additional protein intake of 2 g/kg body weight. In the graph the mean values ± standard deviation are shown. The data are taken from a preliminary report (A field study on the effect of malnutrition on the immune response) submitted to the Nestlé Foundation, Lausanne, Switzerland by Drs. K. Schopfer and L. Matter.

metabolic pathways. Whereas reutilization of urea-N for synthetic processes (34–36) as well as ammonia production from urea hydrolysis (37) by the intestinal flora are probably negligible under normal dietary conditions, these pathways may become significant alternatives when protein intake is very low. Recent studies with ^{15}N-labeled protein or urea in malnourished children led to the conclusion that urea and ammonia as sources of nonessential nitrogen may play an important part in cases of protein-calorie malnutrition (36).

The urea-cycle enzymes are typical but by far not the only examples showing a close correlation between quantity and quality of protein consumed on one hand and enzyme activity or pattern on the other. From the impressive list given by Schimke and Doyle in a review article (2) it can be seen that more than 50 enzymes are known to be controlled *in vivo* or *in vitro* by substrates or metabolites occurring in food. Three of them, although not directly related to urea biosynthesis, will be mentioned here: xanthinoxidase, the first enzyme used for the measurement of the biological value of proteins, still is the most sensitive reagent in this respect (3). The most advanced study on enzyme regulation by varying food composition has recently been published by Mauron et al. (38). The authors investigated the reciprocal induction and repression of L-serine dehydratase and phosphoglycerate dehydrogenase, both key enzymes in serine metabolism, by proteins and essential amino acids in rat liver (38). They have shown that increasing the protein content of the diet induces serine dehydratase and represses reciprocally phosphoglycerate dehydrogenase. Considering that the former enzyme catabolizes serine to pyruvate and the latter enzyme is situated on the pathway from pyruvate to serine, this couple of regulators represents an efficient steering mechanism. Its purpose is to stimulate synthesis and to block deamination of serine when the protein intake is low and to act in the opposite sense when it is high. By feeding amino acid mixtures lacking one essential amino acid at a time, it was shown that four essential amino acids only (i.e., methionine, tryptophan, threonine, and valine) are "active" in regulating the two enzymes. In conclusion, serine dehydratase activity in rat liver depends on the level of dietary protein in a similar manner as the urea-cycle enzymes, whereas there is an inverse correlation between protein intake and phosphoglycerate dehydrogenase. It is not unreasonable to assume that many more enzymatic adaptation mechanisms of this type do exist.

CONCLUDING REMARKS

There is a multitude of responses to altered dietary conditions. They all demonstrate that nutrition is not merely a prerequisite for any form of life but is one of the environmental factors exerting a profound influence on metabolism and its regulation. The coordinated alteration of urea-cycle enzymes is a good example of how adaptation to a different situation is accomplished. Contrary to earlier belief homeostatic regulation mechanisms due to induction and repression are frequent and sometimes considerable not only in microbes but also in the mammalian organism. The extent and speed of these adaptive alterations, their specificity and reversibility, as well as their beneficial effect, are fully compatible with the assumption that the operon concept—although direct proof is still lacking—is also valid for higher organisms.

The correlation existing between the rate of protein degradation, urea synthesis, and the level of ornithine-cycle enzymes can be used in experimental and clinical medicine for the evaluation of the nutritional status, in particular for the appraisal of protein-calorie malnutrition. In fact, the test battery available today for this purpose includes the activity of ornithine carbamoyl transferase and urea elimination (Table 3). Those biochemical methods, when applied together, can give valuable information regarding the state of protein nutrition, notably in children (33). The parameters are, however, not equivalent but are linked to the different stages of protein metabolisms. Nevertheless additional techniques would be highly desirable, especially more sensitive parameters for the diagnosis of marginal cases of malnutrition (3).

Since there is direct proportionality over a rather wide range between the level of urea-cycle enzymes in rat liver and the percentage of protein in food, a few words of caution must be expressed at last. The idea that maximal enzyme activity in liver or maximal growth rate or highest N retention represents a state most beneficial to the organism and therefore should serve as a guideline for realizing optimal living conditions must be questioned for rat and even more so for man. Experimental evidence is accumulating that in rats a negative correlation between maximum growth rate and longevity does exist. Furthermore, Mauron has found that the response of some enzymes (e.g., xanthinoxidase, D-amino acid oxidase) to age is similar to their response to increased dietary protein quantity or quality (39). Although extrapolation of these findings to man is tempting, their interpretation will be restricted to the statement that the main aim is not the realization of maximal growth but rather a long life with minimal disease and maximal physical and intellectual capacity (and a better distribution of the available protein sources). With this idea in mind adaptive changes in urea-cycle enzyme activities do not simply represent an adaptive mechanism to realize maximal metabolic capacity but may as well be regarded as a homeostatic defense system to facilitate the elimination of nitrogen due to excess protein intake.

Table 3 Biochemical Methods for the Evaluation of Protein Supply[a]

	1	2	3	4	5
Process	Intake → available dietary amino acids	Protein synthesis (Growth and substitution)	Body constituents (Protein mass)	Protein catabolism (Degradation)	N excretion → main product urea
Indicator	serum amino acid pattern	OH-Proline urinary excretion	urinary creatinine and serum protein	urea-cycle enzymes in serum	urea resp. total N in urine
Method	amino acid ratio $\frac{\text{gly}+\text{ser}+\text{glu}}{\text{val}+\text{leu}+\text{Ileu}}$	hydroxyproline in 24 hours	creatinine/ height index	ornithine-carbamoyl transferase activity	urea N/ creatinine quotient

[a] Extended test battery for the diagnosis and study of protein calorie malnutrition, based on proposals submitted to the "Committee on Procedures for Appraisal of Protein Calorie Malnutrition" of the International Union of Nutritional Sciences (Lausanne Meeting 1968; see Reference: Arroyave, 1969). (33)

REFERENCES

1. Grisolia, S. (1964). The catalytic environment and its biological implications. *Physiological Rev.*, **44**, 657.

2. Schimke, R. T., and Doyle, D. (1970). Control of enzyme levels in animal tissues, in E. E. Snell, P. D. Boyer, A. Meister, and R. L. Sinsheimer, Eds., *Annual Review of Biochemistry* Annual Reviews, Inc., Palo Alto, California, pp. 929–976

3. Aebi, H. (1969). Enzymes and nutrition, in A. von Muralt, *Protein Calorie Malnutrition*, Springer-Verlag, Berlin, Heidelberg, New York, pp. 19–37.

4. Bücher, T. (1962). Enzyme unter biologischem Aspekt, in E. Linneweh, Ed., *Erbliche Stoffwechselkrankheiten*, Urban and Schwarzenberg, München, Berlin.

5. Fitch, W. M., and Chaikoff, I. L. (1960). Extent and patterns of adaptation of enzyme activities in livers of normal rats fed diets high in glucose and fructose. *J. Biol. Chem.* **235**, 554–557.

6. Weber, R., Banerjee, G., Bronstein, S. B. (1961). Role of enzymes in homeostasis III: Selective induction of increases of liver enzymes involved in carbohydrate metabolism. *J. Biol. Chem.* **236**, 3106–3111.

7. Schimke, R. T. (1962). Adaptive characteristics of urea-cycle enzymes in the rat. *J. Biol. Chem.* **237**, 459–468.

8. Potter, V. R. (1968). Systematic oscillation in metabolic functions in liver from rats adapted to controlled feeding schedules. *Fed. Proc.* **27**, 5, 1238–1245.

9. Lightbody, H. D., and Kleinman, A. (1939). Variations produced by food differences in the concentration of arginase in the livers of white rats. *J. Biol. Chem.* **129**, 71–76.

10. Schimke, R. T. (1962). Differential effects of fasting and protein-free diets on levels of urea cycle enzymes in rat liver. *J. Biol. Chem.* **237**, 1921–1924.

11. Schimke, R. T. (1963). Studies on factors affecting the levels of urea-cycle enzymes in rat liver. *J. Biol. Chem.* **238**, 1012–1018.

12. McLean, P., and Gurney, M. W. (1963). Effect of adrenalectomy and of growth hormone on enzymes concerned with urea synthesis in rat liver. *Biochem. J.* **87**, 96–104.

13. McLean, P., and Novello, F. (1965). Influence of pancreatic hormones on enzymes concerned with urea synthesis in rat liver. *Biochem. J.* **94**, 410–422.

14. Brown, C. L., Houghton, B. J., Souhami, R. L., and Richards, P. (1972). The effect of low-protein diet and uraemia upon urea-cycle enzymes and transaminases in rats. *Clin. Sci.* **43**, 371–376.

15. Colombo, J. P., Herz, R., and Bircher, J. (1972/73). Liver enzymes in the eck fistula rat: I. Urea-cycle enzymes and transaminases. *Enzyme* **14**, 353–365.

16. Schimke, R. T., Ganschow, R., Doyle, D., and Arias, I. M. (1968). Regulation of protein turnover in mammalian tissue. *Fed. Proc.* **27**, 1223–1230.

17. Schimke, R. T., Sweeney, E. W., and Berlin, C. M. (1965). Studies of the stability in vivo and in vitro of rat liver tryptophan pyrrolase. *J. Biol. Chem.* **240**, 4609–4620.

18. Das, T. K. (1971). Rate of adaptation of some of the urea-cycle enzymes to a low-protein diet. *Proc. Nutr. Soc.* **30**, 79A–80A.

19. Das, T. K. (1972). A further observation on urea-cycle enzymes in rats during alteration in dietary protein intake. *Proc. Nutr. Soc.* **31**, 78A–79A.

20. Das, T. K., and Waterlow, J. C. (1974). The rate of adaptation of urea-cycle enzymes, aminotransferases and glutamic dehydrogenase to changes in dietary protein intake. *Br. J. Nutr.* **32**, 353–373.

21. Camus, J., Vandermeers-Piret, M. C. Wodon, C. et Christophe, J. (1969). Activité de treize enzymes du métabolisme intermédiaire, dans le foie du jeune rat qui récupère d'une malnutrition protidique. *Eur. J. Biochem.* **11**, 225–233.

22. Durkin, E. J., and Nishikawara, M. T. (1971). Effect of starvation, dietary protein and partial hepatectomy on rat liver aspartate and ornithine carbamoyltransferase. *J. Nutr.* **101**, 1467–1474.

23. Tatibana, M., and Shigesada, K. (1972). Two carbamoyl phosphate synthetases of mammals: Specific roles in control of pyrimidine and urea biosynthesis, in *Adv. Enzyme Regul.* **10**, 249–271.

24. Brown, G. W., Brown, W. R., Jr., and Cohen, P. P. (1959). Comparative biochemistry of urea synthesis II; Levels of urea-cycle enzymes in metamorphosing rana catesbeiana tadpoles. *J. Biol. Chem.* **234**, 1775–1780.

25. Dror, Y., and Nir, I. (1971). Effect of protein quality on liver xanthine dehydrogenase and kidney arginase activities in chicks. *Nutr. Metabol.* **13**, 75–79.

26. Nuzum, C. T., and Snodgrass, P. J. (1971). Urea-cycle enzymes adaptation to dietary protein in primates. *Science*, **172**, 3987, 1042–1043.

27. Stephen, J. M. L., and Waterlow, J. C. (1968). Effect of malnutrition on activity of two enzymes concerned with aminoacid metabolism in human liver. *Lancet* 118–119.

28. Schoep-Chevalley, M. (1975). *Einfluss der Ernährung auf Leukozyten- und Leberenzyme bei Mensch und Ratte.* Inaugural-Dissertation, Medizinische Fakultät der Universität, Bern.

29. Munro, H. N. (1973). Leukocytes and fetal malnutrition. *Pediatrics* **51**, 926–928.

30. Aebi, H., and Richterich, R. (1963). Aktuelles zur Biologie der Enzyme. *Helv. Med. Acta* **30**, 353–390

31. Richterich, R. (1971). *Klinische Chemie; Theorie und Praxis*, 3rd ed., Verlag S. Karger, Basel, New York (see pages 286–289).

32. Arroyave, G. (1963). Biochemical signs of mild-moderate forms of protein-calorie malnutrition. Mild-moderate forms of protein-calorie malnutrition. In G. Blix, Ed., *Symposium of the Swedish Nutrition Foundation I*, Uppsala: Almqvist & Wiksells.

33. Arroyave, G. (1969). Proposed methodology for the biochemical evaluation of protein malnutrion in children, in A. von Muralt, Ed., *Protein Calorie Malnutrition*, Springer Verlag, Berlin, Heidelberg, New York, pp. 48–62.

34. Read, W. W. C., McLaren, D. S., Tchalian, M., and Nassar, S. (1969). Studies with (^{15}N)-labeled ammonia and urea in the malnourished child. *J. Clin. Invest.* **48**, 6, 1143–1149.

35. Read, W. W. C., McLaren, D. S., and Tchalian, M. (1972). Urinary excretion of nitrogen from (^{15}N) valine, (^{15}N) leucine and (^{15}N) isoleucine in malnourished and recovered children. *Clin. Sci.* **42**, 139–143.

36. Read, W. W. C., and McLaren, D. S. (1974). (^{15}N) studies of endogenous faecal nitrogen in infants. *Gut* **15**, 29–33.

37. Visek, W. J. (1972). Effects of urea hydrolysis on cell life span and metabolism. *Fed. Proc.* **31**, 1178–1193.

38. Mauron, J., Mottu, F., and Spohr, G. (1973). Reciprocal Induction and repression of serine dehydratase and phosphoglycerate dehydrogenase by proteins and dietary-essential amino acids in rat liver. *Eur. J. Biochem.* **32**, 331–342.

39. Mauron, J. (1973). Some current problems in protein nutrition, in Porter and Rolls, Eds., *Protein in Human Nutrition*, Academic, London and New York, pp. 1–9.

Discussion

Dr. Grisolia: I would like to congratulate Dr. Aebi for his elegant and stimulating presentation. The importance of the many points he has made cannot be overemphasized. I said at the beginning of this afternoon's session that we know a great deal about protein synthesis but almost nothing about degradation. It is self-evident that we have homeostatic mechanisms which control protein turnover and that we do not accumulate protein. We are sometimes misled, because we can increase very much the levels of one or a few enzymes, but, in the total of the thousand or so main enzymes in our body, that is very little; moreover, sometimes there is a concomitant decrease in other enzymes. It must be overemphasized that protein turnover must be important, because it is very expensive. Of course we all know that certain enzymes do have very short half-lives. Calculations from my laboratory, for instance, show that at least 30% of the total caloric needs for basal metabolism are used under normal circumstances for protein turnover. The point that you have made is that a great deal of protein may be more detrimental than beneficial. I think this is a very important point. Although certainly our protein turnover under normal circumstances is four to five times larger than the amount of protein that we take in normally, large protein ingestion, above certain limits, is unwarranted.

Dr. Walser: At the risk of seeming presumptuous, I would like to suggest a somewhat different point of view, which Dr. Raijman touched on at the beginning of her talk, namely, that urea production depends chiefly on substrate load and not on urea-cycle enzyme activity. As she pointed out, the rate of excretion of urea changes more rapidly than does the activity of urea-cycle enzymes following a protein load. Furthermore, a load of ammonia nitrogen or amino acid nitrogen of a given amount is going to lead to the production of a certain amount of urea, whether the activity of the urea cycle is high or low. All that will differ is the time for the process to occur. There are now ample data showing maximal urea capacity, measured by assessing amino acid load (Rudman et al., *J. Clin. Invest.*, **52**, 2241–2249, 1973). Obviously when the substrate runs out, no matter what the enzyme activity is, the reaction stops. I hope those who read the proceedings of this symposium will at least consider the role of substrate in urea formation.

Dr. Räihä: With regard to Dr. Aebi's comments in his paper concerning the adaptation of the urea-cycle enzymes during variations of the protein intake of the animal, I would like to emphasize that this adaptation might only be observed with very high concentrations of protein intake and in animals in which the rate of tissue growth is relatively small. We have some unpublished observations which indicate that in newborn rats an increased protein intake will not affect the activity of the urea biosynthetic system. This would suggest that in rapidly growing animals at least some of the increased protein intake can be used for anabolic processes and the load on the urea-forming system would be smaller. The effect of dietary protein on the urea-cycle activity would thus be different during different periods of development of the animal.

Dr. Aebi: I fully agree that this is a most complex phenomenon, and one has to be very careful in making too straightforward conclusions.

Dr. Illiano: In our laboratory we have carried out studies on the relationship between diet and enzymes of amino acid metabolism. We have found, in rats fed with a protein-rich diet, no increase in glutamic acid dehydrogenase. We found that two amino acids, among several tested, were more actively deaminated. They were threonine and histidine. Histidine was shown to be very quickly catabolized along the urocanic pathway. Among many steps at least the first one, which is due to the histidase, was induced. This, and the problem of ammonia as an intermediate in urea synthesis, I hope will be discussed tomorrow in the Round Table.

Dr. Aebi: May I add a remark for the benefit of those who may not be too familiar with this aspect? Two years ago [Mauron and coworkers] published [in the *European Journal of Biochemistry*] a study which I think is one of the most elegant approaches to the problem (*Eur. J. Biochem.*, **32**, 331–342, 1973). They studied, in rats fed on different diets, the activity levels, in liver, of serine dehydratase and of phosphoglycerate dehydrogenase. The regulation of these enzymes is directed in the opposite sense: Serine dehydratase, which degrades serine, is boosted by a high protein intake, whereas phosphoglycerate dehydrogenase involved in serine synthesis is repressed. Based on these observations they have developed a nice theory.

Dr. Ratner: I wanted to ask you to consider the consequences of abnormally low-protein diet on nitrogen metabolism. It is something we see more and more of around the world. I ask the question, because I have often thought that this will cause a double injury, that is, metabolic injury, particularly in children and in infants, who are close to a negative nitrogen balance. The dangers of malnutrition could be increased because of a second danger that the steady-state ornithine level might be lowered and thus slow up ornithine-cycle turnover more than one would be led to expect from the enzyme activity that is found to be present. I think there may be data to support this obtained, not from protein feeding experiments with infants, but from feeding amino acid hydrolysates. I wondered whether any attention has been directed to this possibility.

Dr. Aebi: Thank you, Dr. Ratner. You have raised a most important point. In fact, it is assumed today that there are vulnerable periods in brain development. There is growing evidence that the brain is the main organ that suffers from lack of amino acids to be used for all protein synthesis. In these vulnerable periods, there is not enough protein necessary for brain development, which, as a consequence, leads to irreversible changes. All of you know that many people in the third world are suffering severly from lack of protein. Their physical or mental health may be directly or indirectly affected by such damage irreversibly done in a critical phase of brain development. May I refer, for further information, to the proceedings of a joint symposium of the CIBA and the Nestle Foundations (*Lipids, Malnutrition and the Developing Brain*, Elsevier-Excerpta Medica, North Holland, Amsterdam, 1972).

Dr. Räihä: Being a pediatrician and a neonatologist, I am very glad that the important point of protein intake in relation to brain development during the perinatal period was brought up here at this meeting. Not only lack of protein and amino acids is dangerous for the developing central nervous system but an excess of protein, which produces an amino acid imbalance, might at least potentially be harmful. We know that especially the prematurely born human infant has transient enzymatic deficiencies in the liver, and the breakdown of, for example, phenylalanine, tyrosine, and methionine is deficient. By giving too much dietary protein, especially if it is of the wrong composition can produce amino acid imbalance.

20

Regulation of Urea Biosynthesis by the Acetylglutamate-Arginine System

Masamiti Tatibana
and Katsuya Shigesada*
Department of Biochemistry
Chiba University School of Medicine
Inohana, Chiba, Japan

Carbamyl phosphate synthetase I catalyzes the first step of urea biosynthesis and has been known since the discovery of Grisolia and Cohen (1) to have a unique property, that is, an absolute requirement for N-acetylglutamate as allosteric activator. To elucidate the physiological significance of this activation, we studied the mechanism of biosynthesis of acetylglutamate and its regulation. Some results were presented elsewhere (2–4). Based on our observations, we proposed the hypothesis that acetylglutamate provides a metabolic signal which coordinates the synthesis of carbamyl phosphate and amino acid metabolism.

The proposed mechanism (Fig. 1) involves two regulatory enzymes, that is, carbamyl phosphate synthetase I and acetylglutamate synthetase as well as two enzyme modifiers, arginine and acetylglutamate. Arginine and glutamate can be considered representatives of exogenous amino acids or of amino acid pool. Arginine can control the level of acetylglutamate as the modifier of the synthetase, and glutamate does so as a substrate. Acetylglutamate in turn controls the synthesis of carbamyl phosphate. Arginine is also important in supplying ornithine, which is a critical limiting substrate for the urea cycle,

* Present address, Institute for Virus Research, Kyoto University, Kyoto, Japan.

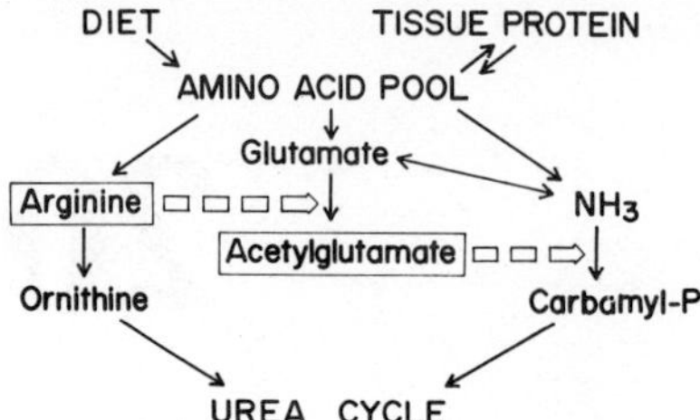

Figure 1. Proposed role of acetylglutamate in control of urea synthesis.

as originally discovered by Krebs and Henseleit in 1932 (5). This report deals with further evidence for the operation of the proposed mechanism and its significance in the control of urea synthesis *in vivo*.

ASSESSMENT OF RATE-LIMITING STEPS OF THE UREA CYCLE WITH $^{14}CO_2$ AS TRACER *IN VIVO*

Mice were given intraperitoneal injections of ^{14}C-labeled bicarbonate, and the incorporation of radioactivity into urea, citrulline, and arginine in the liver as well as in the total extrahepatic tissues was determined (Fig. 2). Since the excretion of ^{14}C-bicarbonate from the lung is rapid, its injection offers a means of pulse labeling of the intermediates. The amount of ^{14}C in urea in the liver reached a peak as early as 6 minutes after the isotope administration, and even at this early time the ^{14}C urea in the extrahepatic pool accounted

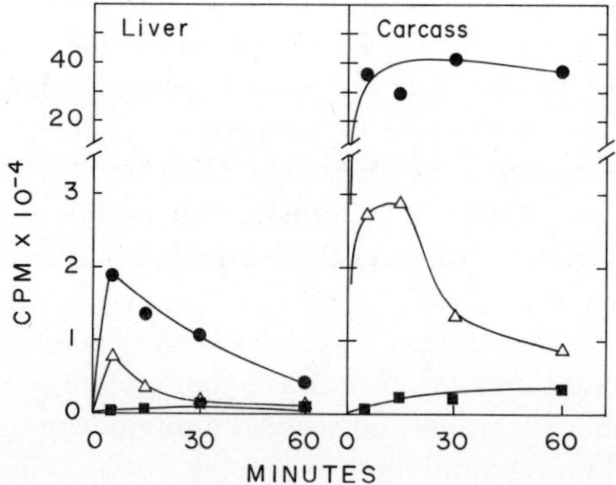

Figure 2. Time course of ^{14}C incorporation into the urea cycle intermediates. Mice previously starved overnight were given intraperitoneal injections of $NaH^{14}CO_2$ (10^7 cpm, 10 μmole) and killed at the indicated times. Trichloracetic acid extracts of liver and total extrahepatic tissues were chromatographed on Dowex 50 to isolate arginine, citrulline, and urea. The details appear elsewhere. Each point represents the mean of two animals. ■——■, arginine; △——△, citrulline; ●——●, urea.

for more than 90% of the sum of ^{14}C in the total urea-cycle members of the whole body. Citrulline in the extrahepatic pool had a small but significant radioactivity. The latter observation is in accord with the fact that the hepatic activity of argininosuccinate synthetase is relatively low. At any rate, these results indicate that under these conditions the rate-limiting step in urea synthesis of the whole body lies in the first two steps, the production or use of carbamyl phosphate.

Further evidence for the proposed mechanism, presented here, includes the following: (1) prompt variations in acetylglutamate level in response to dietary protein intake and to amino acid injections; (2) parallelism between arginine and glutamate contents in liver and in the acetylglutamate level; (3) correlation between acetylglutamate level and ureogenic capacity of liver; (4) constant low levels of carbamyl phosphate in liver.

VARIATIONS IN HEPATIC LEVEL OF ACETYLGLUTAMATE IN RESPONSE TO DIETARY PROTEIN

Male mice of dd-YF strain were fasted overnight and then, at 9 AM, fed diets of varying protein contents (Fig. 3). The acetylglutamate level rose from the initial value of 30 nmole to 60 and 120 nmole/g with 20 and 60% casein diets, respectively, within 2 hours after the start of feeding. On the contrary, with a protein-free diet, the level decreased to below 10 nmole/g. In order to elucidate factors which are involved in the observed change of the acetylglutamate level, variations in hepatic levels of enzymes and substrates were studied (Table 1).

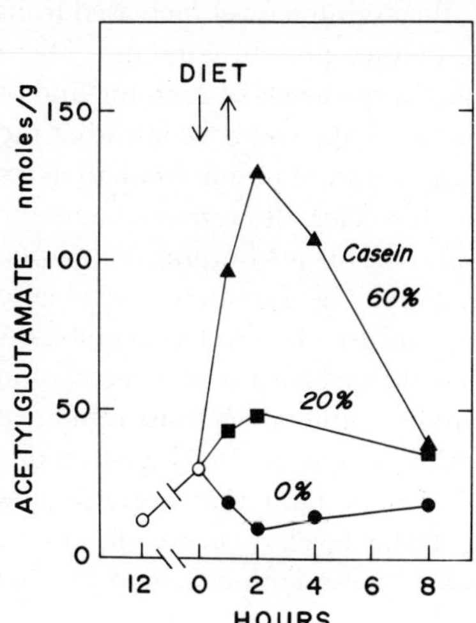

Figure 3. Effect of the protein content of the diet on the hepatic level of acetylglutamate. Three groups of mice, previously maintained on a normal laboratory chow and starved overnight (from 9:00 P.M. to 9:00 A.M.), were given synthetic diets containing 0%, 20%, and 60% casein, respectively, for a 1 hour period from 9:00 to 10:00 A.M. Each point represents the mean value of three animals. ○—○, before feeding; ●—●, fed a 0% casein diet; ■—■, fed a 20% casein diet; ▲—▲, fed a 60% casein diet.

Table 1 Effects of Dietary Protein Ingestion on Hepatic Levels of Intermediates Related to Acetylglutamate Synthesis and the Urea Cycle and of Acetylglutamate Synthetase

Mice previously starved overnight were fed synthetic diets containing 0, 20, and 60% casein in the same way as in Fig. 3, with a control group kept starved. They were killed at 2 hours after the start of refeeding and analyzed for their hepatic contents of various metabolites and acetylglutamate synthetase activity. Arginine was isolated by chromatography on Dowex 50 and assayed colorimetrically by a modified Sakagouchi reaction. Each value represents the mean of four animals. Detailed experimental procedure appears elsewhere.

Compound or Enzyme	Amount or Activity in Mouse Liver			
	Fasted	0% Casein[a]	20% Casein	60% Casein
			(μmole/g)	
Acetylglutamate	0.027	0.016	0.048	0.116
Acetyl-CoA	0.061	0.063	0.048	0.039
Glutamate	2.2	3.0	3.8	5.4
Arginine	0.045	0.026	0.048	0.170
Ornithine	0.56	0.25	0.68	2.7
			(μmole/g/hour)	
Acetylglutamate synthetase	0.044	—	—	0.048

[a] Dietary protein.

In these experiments livers were quickly isolated, freeze-clamped, and further processed for analysis. As shown in Table 1, the arginine level increased from 10 to 170 nmole/g with the increase in the dietary protein contents. There was an excellent parallelism between changes in the levels of arginine and of acetylglutamate. The results on the arginine level apparently contradict the general view that the level is extremely low and constant in mammalian livers under a variety of nutritional conditions (6). This discrepancy may be ascribable to differences in animal conditioning, the tissue sampling technique, and the procedure for arginine determination. The glutamate level also increased with protein intake, as shown. In contrast, the level of acetyl-CoA underwent only minor changes, which even showed an inverse correlation with the dietary protein content. Acetyglutamate synthetase activity remained unchanged during such a short experimental period even after ingestion of a high-protein diet. It is likely on the basis of these data that arginine and glutamate are critical factors in control of the level of acetylglutamate, although intramitochondrial concentrations of these amino acids were not known.

EFFECTS OF AMINO ACID ADMINISTRATION ON THE LEVEL OF ACETYLGLUTAMATE

A single injection of 150 μmole of arginine per mouse (30 g) could increase the acetylglutamate level from 30 to 60 nmole/g liver in 30 minutes. Furthermore, it was shown that the increase in the level of acetylglutamate by arginine was due to the increase in the rate of biosynthesis: when [^{14}C] glutamate was injected intraperitoneally to mice, ^{14}C in hepatic acetylglutamate reached a peak value as early as at 10 minutes after the injection and then declined rapidly. The peak value was increased about threefold by arginine administration. The increase in ^{14}C incorporation could be taken to indicate the increase in the synthetic rate, since general metabolism of glutamate, including entry into the liver, did not appear to be significantly affected by arginine. The half-life (decrease in radioactivity) of acetylglutamate was less than 20 minutes in either group, indicating that arginine did not inhibit the degradation of the compound.

The effect of arginine *in vivo* was fairly specific, as shown in Table 2, which summarizes the effects of various amino acids when injected alone or with

Table 2 Effect of Intraperitoneal Administration of Arginine and Other Amino Acids on the Acetylglutamate Level

Mice previously starved overnight, as in Fig. 3, were given single intraperitoneal injections of indicated amino acid with or without arginine each at a dose of 150 μmole, as 1.0 ml water solution, per 30 g of body weight. Livers were excised at 30 minutes after the injection and analyzed for acetylglutamate content (2). Each value represents the mean of four animals.

Amino Acid	Acetylglutamate Level	
	None	+ Arginine
	nmole/g liver	
Control (NaCl)	25.1	40.1
Glutamate	34.4	42.9
Glutamine	22.4	108
Aspartate	14.3	64.6
Alanine	20.6	128
Proline	21.1	62.2
Histidine	22.8	51.4
Methionine	39.2	90.2
Leucine	20.2	111
Phenylalanine	24.3	50.1
Lysine	28.0	73.3

arginine. In addition to arginine, administration of glutamate and methionine had some effect. Other amino acids could not raise the level of acetylglutamate, at a dose of 150 μmole/30 g of body weight. However, even these amino acids could increase the effect of arginine when injected together. In this respect, alanine, glutamine, leucine, and methionine were especially effective. Since these amino acids do not affect the acetylglutamate synthetase activity *in vitro*, the effects shown here might be attributable, at least in part, to their possible influence on the metabolism of glutamate and arginine.

RELATIONSHIP BETWEEN ACETYLGLUTAMATE LEVEL AND UREOGENIC CAPACITY OF LIVER

In the experiments shown in Table 3 four groups of mice were starved overnight; then three of them were fed on diets containing either 0, 20, or 60% casein, while the remaining group was fasted. Two hours after the start of refeeding, livers were isolated and cut into slices for determination of ureogenic capacity in the original system of Krebs and Henseleit (5). Acetylglutamate contents and activities of some urea-cycle enzymes were assayed. Carbamyl phosphate synthetase I and arginase did not change during such a short period. The level of acetylglutamate changed considerably, as already discussed, depending on the dietary protein. Urea synthesis proceeded linearly with time for at least 30 minutes, depending on addition of NH_4Cl; it was stimulated several fold by exogenous ornithine, as initially discovered by Krebs and Henseleit (5). The rate of urea synthesis varied with dietary conditions, being lowest in the 0% casein group and highest in the 60% casein group. It is notable that there is an approximately proportional relationship between the acetylglutamate level and the urea synthesis in the presence of added ornithine. When the amount of urea produced was plotted against the level of acetylglutamate in the slices, a linear relationship was demonstrated (not shown here). The observations give strong support for the proposal that acetylglutamate, in addition to ornithine, is an important limiting factor for urea synthesis.

CONSTANT LOW LEVELS OF CARBAMYL PHOSPHATE IN LIVER

Acetylglutamate regulates the production of carbamyl phosphate, as has been discussed, and, on the other hand, ornithine has been known to regulate its use. It is interesting to see how the intracellular level of carbamyl phosphate is affected by changes in the levels of the two compounds.

Table 3 Effects of Prior Dietary Ingestions on Hepatic Urea Synthesis in Vitro and Its Relationship with Levels of Acetylglutamate and Urea-Cycle Enzymes

Groups of four mice were fasted and refed on diets of varying casein contents in the same way as in Fig. 3. Livers were excised at 2 hours after the start of refeeding and quickly cut to slices. Urea synthesis was measured on each batch of slices at 37° for 30 minutes with or without NH_4Cl as nitrogen source (5). The remaining slices were used for determination of acetylglutamate and enzyme activities. The value for urea production represents the mean of four animals.

Dietary Condition	Urea Produced			Acetylglutamate Level	Carbamyl Phosphate Synthetase I	Arginase
	None	+ NH_4Cl	NH_4Cl + Ornithine			
	(μmole/g dry wt/minute)			(nmole/g)	(μmole/g/min)	
Fasted	0.18	0.72	2.72	36.4	2.77	680
0% casein	0.05	0.37	1.67	14.8		
20% casein	0.13	0.75	4.00	36.6		
60% casein	0.42	2.97	7.10	77.4	2.40	670

In Table 4 are shown the measured amounts of carbamyl phosphate in several tissues of mice under postabsorptive conditions. The carbamyl phosphate level was highest in the liver, followed by small intestine, and the level was much lower in other tissues. Since the liver has two carbamyl phosphate synthesases, the measured level of carbamyl phosphate in the liver represents two pools associated with the two enzymes. The size of the pool associated with carbamyl phosphate synthetase II may be relatively small in comparison with that associated with the synthetase I. In the assay of carbamyl phosphate, isolated tissues were frozen as quickly as possible by freeze-clamping or a similar method and then analyzed by a method similar to that

Table 4 Carbamyl Phosphate Level in Various Tissues of Mice

Tissues were isolated from mice previously starved overnight and quickly freeze-clamped. Carbamyl phosphate was determined by a modification of the method of Williams et al (7). Each value represents the mean of four animals.

Tissue	Carbamyl Phosphate Level (nmole/g)
Liver	1.23
Small intestine	0.41
Spleen	0.17
Brain	0.15
Testes	0.08
Kidney	0.06
Muscle	0.03

used for *Neurospora* cells by Williams, Bernhardt, and Davis (7). Powdered frozen tissues were extracted with perchloric acid, and the extracts were promptly treated with Dowex 50 and then neutralized. Carbamyl phosphate was quantitatively converted to [^{14}C]citrulline in the presence of [^{14}C]ornithine and ornithine transcarbamylase purified from bovine liver. [^{14}C]citrulline was isolated and counted. Removal of endogenous ornithine with Dowex 50, the use of freshly purified [^{14}C]ornithine, and the use of purified ornithine transcarbamylase were essential for accurate assay of carbamyl phosphate in tissues. The generally low levels of carbamyl phosphate reported here are not in accord with the values (0.1 to 0.12 μmole/g) recently presented by Raijman (8). Difference in the analytical methods used appears to be responsible for this discrepancy.

In Table 5 are shown the effects of a variety of dietary conditions on the hepatic levels of carbamyl phosphate in mice. Even when mice were fasted or fed on different amounts of casein, the level did not show remarkable changes, suggesting that the rates of production and use of carbamyl phosphate were closely coordinated under physiological conditions. However, it is possible to make an imbalance between its production and use to induce a change in the level of carbamyl phosphate by means of some treatments. The treatments included injections of NH_4Cl, ornithine, norvaline, or acetylglutamate dialkylester. Norvaline can compete with ornithine and thus inhibit ornithine transcarbamylase reaction, as found by Marshall and Cohen (9). The ester is expected to easily penetrate into the liver cells and is then hydrolyzed by esterase to give free acetylglutamate.

Table 5 Effects of Dietary Protein Ingestion on the Hepatic Carbamyl Phosphate Level

Mice were starved overnight and refed diets of varying casein contents in the same way as in Fig. 3. Livers were isolated at indicated times after the start of refeeding. Each value represents the mean of four animals.

Treatment of Animal	Time after Refeeding (hours)	Carbamyl Phosphate Level (nmole/g)
Fasted		1.23
Fed, 0% casein	2	0.52
Fed, 20% casein	2	1.02
	1	0.93
Fed, 60% casein	2	0.87
	4	1.09

In the experiments shown in Table 6, the mice were injected intraperitoneally with the compounds listed in the left-hand side column and then injected with NH_4Cl at a dose of 150 μmole/30 g body weight. Changes in the carbamyl phosphate level in the liver were followed. With NH_4Cl alone, the level abruptly rose to a peak value of nearly 3 nmole/g within 8 minutes and then declined. One hundred micromoles of L-ornithine, by itself, increased the level but inhibited the rise in carbamyl phosphate induced by NH_4Cl. Acetylglutamate diethyl ester slightly increased the level by itself and with NH_4Cl caused a significant elevation in the carbamyl phosphate level. The effect was partly counteracted by ornithine. With DL-norvaline and NH_4Cl the level was increased as high as over 5 nmole/g. The effect of norvaline was completely counteracted by ornithine.

Table 6 Changes in the Hepatic Carbamyl Phosphate Level after Intraperitoneal Administration of Ammonia and Various Agents Affecting Production and Use of the Compound

Mice were starved overnight, as in Fig. 3, and then given intraperitoneal injections of the specified doses of various agents and further of 150 μmole of NH_4Cl, as a 1.0 ml water solution, 30 minutes after the first injection. Each value represents the mean of four animals.

Compound[a] injected prior to NH_4Cl	Carbamyl Phosphate Level			
	Time after NH_4Cl injection (minutes)			
	−15	0	8	16
	(nmole/g)			
NaCl (150)	0.40	0.34	2.69	2.00
Ornithine (100)	1.30	1.84	2.43	1.94
AGE[b] (20)	1.05	0.41	5.34	2.30
AGE[b] (20) + Ornithine (100)	1.10	1.68	3.02	1.74
DL-Norvaline (50)		1.54	5.33	
DL-Norvaline (50) + Ornithine (100)		0.81	1.26	
Arginine (150)		1.04	1.84	1.43
Arginine (150) + Alanine (150)		1.32	1.27	1.01

[a] The dose of each compound (μmole/30 g body weight) was indicated in parentheses.
[b] AGE stands for *N*-acetyl L-glutamic acid diethyl ester.

The results indicate that the level of carbamyl phosphate is under independent control by at least the three factors, that is, ammonia, acetylglutamate, and ornithine. On the other hand, there was a remarkable constancy in the level of carbamyl phosphate, as shown above, even when the rate of urea synthesis widely varied after ingestion of diet of different protein content. The observations could be taken to indicate that the synthesis of carbamyl phosphate generally limits the rate of citrulline synthesis under physiological conditions and also offers an important question concerning the control of carbamyl phosphate metabolism.

The question is how the low and constant level of carbamyl phosphate is maintained. Is this maintained merely by the kinetic balance, that is, by the mass action law between the carbamyl phosphate synthetase and ornithine transcarbamylase reactions? Or are there some specific mechanisms which match the rates of carbamyl phosphate production and its conversion to citrulline? Although no definite answer is available, evidence appears to favor the latter possibility. To make the question simpler, let us examine an interesting observation often pointed out by Krebs et al. (5, 10) and other investigators (11). When ornithine was added to an *in vitro* system for assay of urea production, such as liver slices, isolated hepatocytes, and perfused liver, it induced almost invariably a prompt and definite increase, generally about 10 to 30%, in the rate of urea synthesis. Although the effect is variable in intact animals under physiological conditions, as far as studied, the following five possible mechanisms could be offered for explanation of the effect: (1) presence of surplus carbamyl phosphate; (2) release of carbamyl phosphate synthetase I from product inhibition; (3) activation of carbamyl phosphate synthetase I by ornithine; (4) involvement of the arginine-acetylglutamate system; (5) involvement of membrane function.

Possibility 1 could be excluded by the constantly low levels of carbamyl phosphate (Table 5), though it is possible that the level of carbamyl phosphate becomes relatively high under certain other conditions *in vitro*. For possibility 2, product inhibition of carbamyl phosphate synthetase is unlikely, since the K_i for carbamyl phosphate is high, as reported by Elliott (12). For possibility 3, it is not known that ornithine can affect the rate of the enzyme reaction. In regard to 4, the arginine-acetylglutamate system may be responsible for the effect of ornithine from 10 to 20 minutes after its administration, but it is hard to explain the immediate response by this mechanism only. For possibility 5, data are not available at the present. Thus, although one cannot answer the question after all, it is pertinent to reconsider possibility 1, since one may assume that a small part of carbamyl phosphate synthesized in the mitochondria usually leaks out to the cytosol, where it may be degraded or consumed in some reactions (13–15) other than citrulline synthesis, as summarized in Fig. 4. However, no available evidence supports the idea that

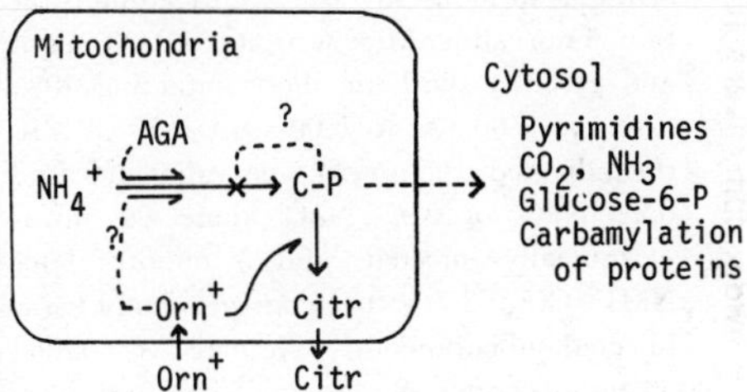

Figure 4. Possible fate of carbamyl phosphate synthesized in liver mitochondria.

leakage of carbamyl phosphate out of mitochondria takes place to any significant extent *in vivo* under physiological conditions. Indeed, evidence for this was obtained in the following experiments, where we measured pyrimidine synthesis as a marker for the leak of carbamyl phosphate.

Mice were given various amounts of DL-norvaline about 3 hours after feeding 40% casein diets. Pyrimidine synthesis in the liver as well as production of urea were followed with $^{14}CO_2$ as tracer (Fig. 5). As shown, with 40 μmole of norvaline or more, ^{14}C incorporation into uridine nucleotides in the liver increased very sharply. Under selected conditions, the ^{14}C incorporation gives a good measure for the synthetic rate of uridine nucleotides. The effect of norvaline was completely inhibited by ornithine. These observations, along with the data on assay of carbamyl phosphate content shown above, indicated that by this treatment carbamyl phosphate accumulated and leaked out from the mitochondria, and then it was used for pyrimidine biosynthesis in the cytosol. A notable observation here is that 40 μmole of norvaline markedly increased pyrimidine synthesis but did not inhibit urea synthesis to a measurable extent, although in the figure the scale for urea synthesis is 100 times as large as that for pyrimidine synthesis. This shows that the pyrimidine synthesis in the liver may be a sensitive marker for the presence of surplus carbamyl phosphate that can leak out from the mitochondria. In spite of this, the pyrimidine synthesis in normal liver receiving no norvaline is generally not very active in comparison with the urea synthesis (Fig. 5). This may indicate that leakage of carbamyl phosphate from mitochondria is not significant under normal conditions. Carbamyl phosphate synthetase II in the cytosol of liver cells (16, 17) may primarily provide carbamyl phosphate for pyrimidine synthesis. The view is consistent with the recent report by Pausch et al. (18).

In the foregoing discussion, recent observations on the role of acetylglutamate in control of urea synthesis are reviewed. Although there are still many questions that remain to be answered, the behavior of arginine in the

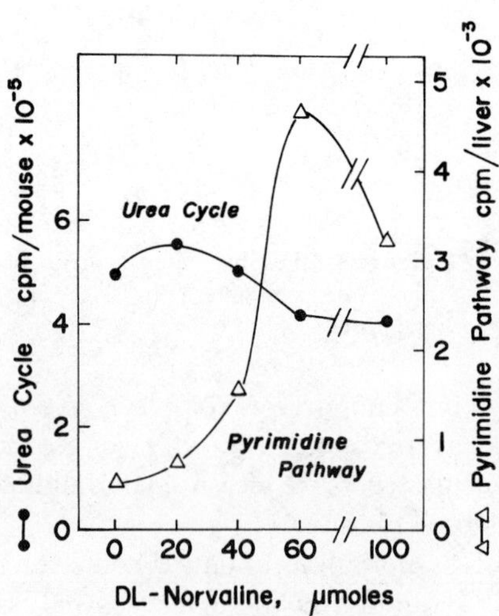

Figure 5. Stimulation of pyrimidine biosynthesis in mouse liver *in vivo* by administration of norvaline. Mice were starved overnight and given a synthetic diet containing 40% casein at 5:00 A.M. to 7:00 A.M.; at 10:30 A.M. the indicated amount of DL-norvaline in 0.5 ml 0.9% NaCl or 0.9% NaCl alone was intraperitoneally injected, and 5 minutes later $NaH^{14}CO_3$ (3.7×10^7 cpm per mouse) was injected subcutaneously. Animals were killed 30 minutes after the last injection, and livers and total extrahepatic tissues were extracted with trichloracetic acid. Liver extracts were analyzed for the ^{14}C content in total uridine nucleotides and also in urea, extracts of extrahepatic tissues were analyzed for the ^{14}C content in urea. Values are expressed as cpm per mouse and each point represents the mean of two animals.

liver cells, particularly its transport through the mitochondrial membrane, and the mechanism for possible coupling of the two initial reactions, carbamyl phosphate synthetase I and ornithine transcarbamylase reactions, would be important for further understanding of the proposed mechanism for control of the urea cycle.

REFERENCES

1. Grisolia, S., and Cohen, P. P. (1953). *J. Biol. Chem.* **204**, 753–757.
2. Shigesada, K., and Tatibana, M. (1971). *J. Biol. Chem.* **246**, 5585–5595.
3. Shigesada, K., and Tatibana, M. (1971). *Biochem. Biophys. Res. Commun.* **44**, 1117–1124.
4. Tatibana, M., Shigesada, K., and Mori, M., another paper presented in this Symposium.
5. Krebs, H. A., and Henseleit, K. (1932). *Z. Physiol. Chem.* **210**, 33–66.
6. Wannemacher, R. W., Jr., and Allison, J. B. (1968). In J. H. Leathem, Ed., *Protein Nutrition and Free Amino Acid Pattern*, Rutgers University Press, New Brunswick, pp. 206–227.
7. Williams, L. G., Bernhardt, S. A., and Davis, R. H. (1971). *J. Biol. Chem.* **246**, 973–978.
8. Raijman, L. (1974). *Biochem. J.* **138**, 225–232.
9. Cohen, P. P., and Marshall, M. (1973). *J. Biol. Chem.* **247**, 1654–1668.
10. Krebs, H. A., Hems, R., and Lund, P. (1973). *Adv. Enzyme Regul.* **11**, 361–377.
11. Saheki, T., and Katunuma, N. (1975). *J. Biochem.* (Tokyo) **77**, 659–669.
12. Elliott, K. R. F., and Tipton, K. F. (1974). *Biochem. J.* **141**, 817–824.
13. Ramponi, G., and Grisolia, S. (1970). *Biochem. Biophys. Res. Commun.* **38**, 1056–1063.
14. Natale, P. J., and Tremblay, G. C. (1969). *Biochem. Biophys. Res. Commun.* **37**, 512–517.
15. Lueck, J. D., and Nordlie, R. C. (1970). *Biochem. Biophys. Res. Commun.* **39**, 190–196.
16. Nakanishi, S., Ito, K., and Tatibana, M. (1968). *Biochem. Biophys. Res. Commun.* **33**, 774–781.
17. Mori, M., Ishida, H., and Tatibana, M. (1975). *Biochemistry* **14**, 2622–2630.
18. Pausch, J., Wilkening, J., Nowack, J., and Decker, K. (1975). *Eur. J. Biochem.* **53**, 349–356.

Discussion

Dr. Cohen: I believe, Dr. Tatibana, that before this meeting is over we might be able to convince the "substrate" people that regulation of enzyme activity, specifically CPS, is a very important factor in urea biosynthesis, and your studies, in particular, provide the evidence. A paper has recently appeared reporting that 4-pentenoic acid inhibits urea biosynthesis with a consequent elevation of plasma ammonia (see A. M. Glasgow and H. P. Chase, *Biochem. Biophys. Res. Comm.*, **62**, 362–366, 1975). Apparently 4-pentenoic acid ties up acetyl-CoA and has no direct effect on CPS and OTC. A compound of this kind could be used to study the turnover of acetylglutamate and its effect on urea biosynthesis *in vivo*. Second, I want to take the liberty, if the Chairman will permit me, to say that Dr. Marshall has some comments she would like to make regarding the OTC figures given by Dr. Raijman and perhaps will comment on the role of lysine in the OTC reaction.

Dr. Marshall: My comment was that according to my calculations the OTC levels on a molar basis are about one-tenth of those you quoted relative to CPS in the mitochondria. Also, there is an experiment in the literature bearing on the levels of CPS, namely, that Packer and his group have done SDS gel electrophoresis of fractions of mitochondria, and in the matrix fraction they find that 30% of the protein has a molecular weight of 130,000 and I think that is probably large enough to be unique for CPS and that therefore this is additional support for the high concentration independent of the other evidence. Insofar as the lysine levels are concerned, all I know is that lysine is a very poor substrate with a high K_m for transcarbamylase, so that I don't think it would be a very effective competitive inhibitor, would it?

Dr. Raijman: Dr. Marshall, even though it might be a very poor competitive inhibitor, if the intramitochondrial concentration of ornithine is low enough, couldn't it be, in effect, a fairly good competitive inhibitor?

Dr. Marshall: It's hard to judge. . . .

Dr. Bryla: We have found that the K_m for ornithine for citrulline synthesis is about 0.3 mM, so that it seems to me that it can limit the reaction and lysine can affect it.

Dr. Ratner: I don't think what I am about to say will explain the mitochondrial results, but lysine is a strong competitive inhibitor of arginase, so that it might be acting that way. I wondered if you had considered that possibility.

Dr. Bryla: Since lysine inhibits citrulline formation in the mitochondria, it should be taken into consideration that inhibition of urea formation was due to only inhibition of the arginase.

Dr. Grisolia: Dr. Tatibana, I noticed in your second slide, I believe, that when you gave no casein, the concentration of acetyl glutamic acid was at a much lower level than on fasting. Is that right? I think this is very important. I would like to make two other comments. As first shown by Dr. Raijman (*Adv. Chem. Ser.*, **44**, 128, 1964), the activation of CPS by acetyl glutamic acid is time-dependent, and, as shown by Dr. Chabas (*Eur. J. Biochem.* **29**, 333, 1972), so is the deactivation. There is quite a time lapse of the order of minutes from the maximal activity obtained after activation with acetylglutamate to negligible activity after removal of acetylglutamate. That worries me a great deal, and I think it is very important insofar as the significance of activation *in vivo* of CPS by acetylglutamate. Lastly, some time ago we injected ^{14}C-acetate and formate to rats kept under normal feeding conditions. We found very little of either acetate or formate incorporation into acetyl glutamic acid. The levels were similar to the ones you reported before, Dr. Tatibana, but much less than the ones you report now. So it seems to me that the role of levels of acetylglutamate as part of a fine control of CPS needs more clarification.

Dr. Räihä: I would like to ask Dr. Tatibana a very short question. In one of your slides you showed that, by increasing the protein content in the diet of the animal, there was no increase in the acetylglutamate synthetase activity.

Dr. Tatibana: No. Only during a 2 hour experiment, there was no increase, no change in the enzyme level.

Dr. Räihä: Do you have any information concerning changes of the acetylglutamate synthetase activity during longer experiments with increased protein in the diet? Can the acetylglutamate synthetase activity be altered by hormonal agents in rat or mouse liver?

Dr. Tatibana: Yes, I did some experiments on mice which showed that the synthetase activity increased 1.5 to twofold after 60% casein diet for 3 or 4 days.

Dr. Raijman: I apologize for taking this much time, but I think I almost have to, since there is such a large discrepancy between Dr. Tatibana's values and my own for carbamyl phosphate in liver. They differ by between 10- and 100-fold, and that is not something one can overlook. All I can say is that I don't know in great detail Dr. Tatibana's methods; he doesn't know mine. Carbamyl phosphate is a touchy compound, and I would like to point out only one possibility, for example. If you look at the rate of decomposition of carbamyl phosphate with pH, you will find (and since we have an expert here, I may fall back on you, Dr. Jones) that between pH about 2.5 and 8.5 there is a plateau at which the rate of decomposition is essentially constant. If you go below pH 2.5 or above 8.5 and you don't have to go much below, then the rate of hydrolysis or decomposition increases enormously. Being aware of this, and I don't mean to imply that Dr. Tatibana wasn't, we added perchloric acid in an amount sufficient to deproteinize the tissues but always trying to keep the pH above 2.5. We are not so fast as he is, and it takes us 15 to 20 minutes to have the CP incubating to convert it into stable citrulline. But this comment, just to give you an idea of the possible sources of differences; I don't know what is the reason for the discrepancy. I won't take too much time. You have some indications in several of your experiments that ornithine may stimulate carbamyl phosphate synthesis. I wanted to tell you that we have tested the effect of ornithine on CPS I *in vitro* with the purified enzyme, and also with mitochondria, measuring citrulline synthesis, and find that ornithine activates the synthetase at most by 10 to 15% in our hands. Once again, perhaps at higher enzyme concentrations, the effect is different. And finally, I don't really understand Dr. Cohen's point about convincing the substrate people that enzymes are important, because we don't need any convincing. The question here is, as a matter of fact, the interaction between substrates and enzymes may be most important, due to the fact that they are at such similar concentrations. At least that's the way I looked at it. Thank you.

Dr. Tatibana: First, regarding assay of carbamyl phosphate, I am quite sure my assays are good enough. I used perchloric acid to extract the tissues but always had control tubes to which I added carbamyl phosphate, 5 nmole, or something like that, at the step of homogenization, and the recovery was 60 to 80%. We used a final concentration of perchloric acid of 0.5 *M*, and exposure of carbamyl phosphate to the acid was just 3 to 5 minutes. For the second question, in referring to the first two enzymes of the urea cycle, I mean carbamyl phosphate synthetase and ornithine transcarbamylase.

Dr. Raijman: I did not mean to imply that your method is wrong.

Dr. Aebi: Now, at the end of this afternoon session, there will be a round table discussion on "Consensus of Best Available Methods for Detection and Preparation of Intermediates and Enzymes." May I ask Dr. Raijman and Dr. Kennedy to take the floor.

Discussion

The Best Available Methods for Detecting and Preparing Intermediates and Enzymes

Dr. Kennedy: The purpose of this round table discussion is to try to develop a consensus on the best available methods for detection and preparation of both intermediates and enzymes involved in urea synthesis. We can dispense quickly with one part of this, and that is preparation of intermediates, since all the urea-cycle intermediates are available in reasonably pure form commercially. I think that is equally true of compounds which are labeled with radioactive material, although I am not sure about ^{14}C-argininosuccinate. Is that available commercially?

Dr. Ratner: I think so.

Dr. Kennedy: One caution, some of these labeled intermediates have to be cleaned up a little before they are used. In regard to purification and preparation of enzymes, would you agree, Dr. Raijman, that we have homogeneous preparations of all five enzymes from one or more mammalian sources?

Dr. Raijman: Yes, I think so.

Dr. Kennedy: Does anybody wish to comment on this?

Dr. Raijman: Would most of you like to have these enzymes available commercially? Which ones in particular? Everybody is for carbamyl phosphate synthetase. But something, for example, like ornithine transcarbamylase, would you find it very useful? Okay. What about Dr. Ratner's enzymes? Argininosuccinase, of course; that's a good coupling enzyme and would be handy for many of us. Good arginase.

Dr. Ratner: As far as I know, the activity of commercial arginase is really not high enough for coupling. I have not tested any recently.

Dr. Aebi: I have a question, but I hope Dr. Grisolia will get it. Might it be possible to include in the monograph to be published a list giving this information that you just gave now? So that everybody will have free access to this kind of information.

Dr. Grisolia: Yes. In fact, Dr. Ratner very kindly informed me this morning of some very important work by Snodgrass and Nuzum and raised the possibility of inviting them to contribute it to the monograph. It relates to the use of needle biopsies, which I think are going to become more frequently used in assessing disorders of nitrogen metabolism in man. That is to say, many of these methods are being used at the clinical level. In my opinion, it will be very important to incorporate the methods developed by Snodgrass and Nuzum and we will invite them to do so.

Dr. Cohen: I think one has to make a decision as to the purpose of this discussion. Perhaps I am anticipating somewhat. I think before these diagnostic reagents are developed commercially, it would be much more desirable to get a group of interested people together who are competent to set up a standard protocol and establish an arrangement for careful and reliable studies of the very valuable clinical cases which emerge periodically all over the world. Unless this is done, the literature will record uncertain and even erroneous data as a result of random, unprepared assays of liver biopsies. I think it is nothing short of criminal to expose a patient, particularly an infant, to an unwarranted biopsy. As I have previously mentioned, there are other diagnostic means, particularly stable isotopes, which should be considered in cases of hyperammonemia before enzyme assays on a liver biopsy are undertaken. The instability of some of these enzymes and the limited use for them raise a question as to whether commercial production would be worthwhile.

Dr. Ratner: If this is the time to discuss methods, I agree with everything that Dr. Cohen said about what our goals should be. However, I think measuring urea formation with isotopes may not give us all the answers, because some of the mutations don't show great changes in urea formation, and yet there is a partial block. Some other procedures could be devised, possibly with isotopes, but they have yet to be developed and correlated with reliable enzymatic assays. I would like to describe very briefly the methods I mentioned to Dr. Grisolia. They are really based on the Schimke procedures scaled down to handle the amount of material you would get with a needle biopsy. The two synthetase assays are both supplemented with ATP-generating systems and excess coupling enzymes. The procedures require that everything be prepared in advance and kept frozen for the times for which they are known to be stable and that normal rat liver from an animal on a standard diet be run along with all the assays every single time. To carry out all this does require a trained person. Perhaps if such a laboratory could be set up to service large areas in a country, or one laboratory set up per country, a central facility might be the solution. I might add that the tissue samples obtained from the cases studied by Dr. Rosenberg that have the OTC X chromosome–linked mutation were analyzed by these procedures. Many of the assays were carried out by Nuzum and Snodgrass. The full details have not yet been published. Improvements in the needle biopsy assay techniques would get around the problem of the loss of enzymatic activity during storage and the general unreliability of assays performed on autopsy sample due to loss of activity.

Dr. Kennedy: Dr. Ratner, how many milligrams of tissue are required to do all five enzymes?

Dr. Ratner: Maybe 100 mg, whatever you get in a needle puncture.

Dr. Shih: I fully agree with Dr. Cohen and Dr. Ratner in their concerns over methods used by clinicians. I think that one of the most important contributions of this symposium is to bring the biochemists together with the clinicians and discuss the best available method for clinical use. At least for myself, I would be more than happy to have a place where we could send the tissue to have the assays done correctly. I find it very difficult to interpret all the enzyme information in the literature. Sometimes one cannot help wondering whether the data are valid. So I think this will be a very important thing to discuss.

Dr. Krebs: In Britain, a beginning has been made to centralize enzyme assays in inborn errors of metabolism, because this is the only safe way of getting comparable and reliable results. Biopsy material from cases of glycogen storage diseases are now handled at Charing Cross Hospital, London, by Professor Brenda Ryman. Something similar, perhaps on an international basis, is required, considering the large number of assays that are necessary or will become necessary in the future.

Dr. Cohen: The dietary history of the infant with hyperammonemia is often unrecorded in publications. I wanted to ask Dr. Räihä whether he knows if the infant's diet has any effect on the level of the urea-biosynthetic enzymes. These are sick children of course and usually they are not being fed a normal diet. Thus the assay values obtained may be a reflection of the dietary restriction rather than a primary genetic effect on enzyme level.

Dr. Kennedy: Very briefly we would like to discuss perhaps the best method for assaying each of the five enzymes, not from the standpoint of using clinical material, but in the laboratory, and, Dr. Raijman, would you take CPS, and I would like to call on Dr. Marshall for OTC, Dr. Ratner for her enzymes, and if Dr. Soberón is here, for arginase.

Dr. Raijman: We use basically the assays designed by Guthöhrlein. One can use, for example, a spectrophotometric method which has the advantage that it gives a continuous record of the reaction, which in the case of CPS I think is very important, because of lags, falling off of activity, the enzyme doesn't like to be very linear. If one uses a colorimetric method, one must have good coupling, so that we need a good source of OTC, active and uncontaminated. In the case of the spectrophotometric method, again it is important to have very good coupling, to make certain that one does not have impurities such as AMP which will give false readings, and to keep in mind that one must have enough of the coupling enzymes (but not too much) in the right proportions.

Dr. Cohen: I think we have just gotten perhaps the worst advice we could possibly get. In the first place, you have at most 100 mg of tissue, and you are suggesting a colorimetric method.

Dr. Raijman: I wasn't talking about that, Dr. Cohen.

Dr. Cohen: What you need to use in the case of CPS is a standardized isotope procedure and the use of a specific antibody. Conditions for such an assay have been described by Marshall and Cohen (*J. Biol. Chem.*, **236**, 718–724, 1961). If you really want to get the information that I think the geneticists need, you have to do this thing scientifically. It can't be a routine procedure.

Dr. Raijman: One of the things is that we need to agree on what we are talking about. I thought we were talking about the assay of CPS, perhaps of bovine origin. At least in my mind, I was no longer talking about congenital defects of the urea-cycle enzymes, but the plain assay of purified CPS.

Dr. Kennedy: Dr. Marshall, would you like to make any comments about the assay of purified OTC? from animal sources? Do you feel that the routine determination of citrulline by the colorimetric method is the best?

Dr. Marshall: The main problem is that the more sensitive colorimetric methods do not discriminate between urea and citrulline. The old-fashioned Archibald procedure, which is not nearly so sensitive, at least gives a color that is at least relatively specific for citrulline.

Dr. Kennedy: What assay do you use for the purified enzyme?

Dr. Marshall: I use the method of Cerriotti and Gazzaniga, modified to a slightly smaller scale that works in the range of 10 to 50 nmole. The problem with it is that carbamyl phosphate always contains a little urea, so that, unless you reprecipitate it every few months, the blanks get very high.

Dr. Kennedy: Dr. Ratner, would you comment very briefly on the synthetase and the argininosuccinase.

Dr. Ratner: For crude homogenates, the urea color can be measured for enzymes under carefully selected, controlled assay conditions. In both cases, the agent to produce urea is one that gives a small amount of color with citrulline and a much larger intensity with urea. But, because in the argininosuccinase synthetase assay citrulline is there as a substrate, you make a special set of standards in which a blank contains high citrulline and no urea, and every successive standard contains both: citrulline decreases and urea goes up. You run the standard every time, developing the color in the dark under carefully controlled conditions. Because some oxidative step is involved in color production, even the size and shape of the tubes affect the color intensity. We also mix the reagents under standard conditions because of that variable. There are ways of getting around this problem, but we haven't added other reagents to stabilize the color. We do run a separate set of standards for each enzyme. The synthetase assay is supplemented with an ATP-generating system, with excess argininosuccinase and with excess arginase. The argininosuccinase assay is also colorimetric for crude enzyme preparations; it is supplemented with excess arginase, even though this enzyme may be present. The composition of the standards is somewhat simpler. Always assays are run at two enzyme concentrations, and they are expected to agree. These details are similar to those incorporated in the micro procedures for assay of homogenates I referred to before. Always a sufficient number of blanks are run, including both an incubated blank and an unincubated blank and sometimes an additional one, to supply all the necessary controls. Blanks are of greatest importance and may lead to considerable error if omitted.

Dr. Kennedy: Thank you. Dr. Soberón, do you use the same assay system in both crude and purified preparations for arginase?

Dr. Soberón: Well, it has been more than 2 years since I moved from biochemistry to the fire department, so that I have not been in a position to review the subject recently. However, we always had trouble with the high blanks in the arginase assay. In one experiment I referred to this morning, namely, the conversion of ^{14}C-citrulline in the ureido group to arginine and then to urea, we started working on a procedure based on the hydrolysis of urea with urease and trapping the CO_2 with hyamine. This was started with promising results, but the student that was doing this got married, and we couldn't finish it. But I think this method could be very sensitive and rather inexpensive.

Dr. Kennedy: Before closing, I would like to emphasize that different methods are necessary for assay of enzymes of the urea cycle, depending on the source material. The methods alluded to in the latter part of this discussion are adequate for assay of purified enzymes and in some cases crude homogenates from nonhuman sources where a relatively large amount of material is available. However, these methods, for the most part, cannot be adapted for human studies because of the limited amount of tissue that can be obtained by needle biopsy. The procedures of Nuzum and Snodgrass seem to hold the most promise for investigating disorders of urea synthesis in man. Finally, I think there is general agreement here, with Dr. Cohen's suggestion, that a standard protocol should be devised for assay of urea-cycle enzymes in human tissues and that such assays be done only at a very limited number of centers or laboratories which have the necessary experience and expertise in this specific area.

Multiple Assays

21

Multiple Assays of the Five Urea-Cycle Enzymes in Human Liver Homogenates

C. Thomas Nuzum
and Philip J. Snodgrass
Department of Medicine
University of North Carolina School of Medicine
Chapel Hill, North Carolina
and Veterans Administration Hospital
and Department of Medicine
Indiana University School of Medicine
Indianapolis, Indiana

INTRODUCTION

Assays for the human enzymes of urea synthesis are needed to test for adaptive responses of this pathway (1, 2) and to identify causes of hyperammonemia (3, 4). In order to provide a reliable procedure for clinical studies, published assays (1, 5–11) were modified, scaled down, and coordinated to measure, under optimal conditions, the activities of all five enzymes in a human liver biopsy. The principal reactions are outlined in Fig. 1.

These studies, performed in the Department of Medicine, Peter Bent Brigham Hospital, Harvard Medical School, and the Department of Medicine, University of Kentucky College of Medicine, were supported by United States Public Health Service Grants AM-14838 and RR-05374, a Daland Fellowship of The American Philosophical Society, and a grant from the W. L. Lyons Brown Foundation.

I

Carbamyl phosphate synthetase

$$NH_4^+ + HCO_3^- + 2ATP \xrightarrow[\textit{N}\text{-acetyl-L-glutamate}]{Mg^{2+}} \text{carbamyl phosphate} + 2ADP + P_i$$

II

Ornithine transcarbamylase

$$\text{Carbamyl phosphate} + \text{ornithine} \rightarrow \text{citrulline} + P_i$$

III

Argininosuccinate synthetase

$$\text{Citrulline} + \text{aspartate} + ATP \xrightarrow{Mg^{2+}} \text{argininosuccinate} + AMP + PP_i$$

IV

Argininosuccinase

$$\text{Argininosuccinate} \rightarrow \text{arginine} + \text{fumarate}$$

V

Arginase

$$\text{Arginine} \xrightarrow{Mn^{2+}} \text{urea} + \text{ornithine}$$

Figure 1. Enzyme reactions of the urea cycle. I. The formation of carbamyl phosphate from ammonia, bicarbonate, and ATP is measured as citrulline produced in a coupled assay with ornithine and ornithine transcarbamylase provided in excess. II. Ornithine transcarbamylase activity is measured as citrulline formed from ornithine and carbamyl phosphate. III. Argininosuccinate, formed by condensation of citrulline and aspartate in the presence of ATP, is measured as urea produced in a coupled assay in which argininosuccinase and arginase are provided in excess. IV. Arginine, produced by cleavage of argininosuccinate is measured as urea formed with arginase provided in excess. V. Hydrolysis of arginine is measured as the rate of urea formation. ATP generating systems are provided for reactions I and III.

MATERIALS AND METHODS

Equipment

Homogenizers consisted of glass tubes and close-fitting Teflon plungers. Calibrated Levy-Lang micropipettes were used in most transfers up to 0.5 ml. Conical graduated 12 ml centrifuge tubes held the incubation mixtures in a circulating water bath at 37 $\pm$0.1°C. (A second bath was set at 55° for the preincubation of arginase.) Color was developed in 13 × 100 mm test tubes with capillary-vented neoprene stoppers. Covered metal canning pots served as boiling and cooling vessels. Absorbance was read in a Beckman DU spectrophotometer using glass cuvettes with a 1 cm light path.

Preparation of Tissue Samples

Assay conditions were studied with fresh human liver biopsies and autopsy specimens obtained soon after death. Fresh liver from 200 g male Sprague-Dawley rats, adapted to 23% protein chow, was used in control assays. Specimens were placed in cold saline. After blotting and weighing, 50 to 500 mg of tissue was homogenized in 19 vol of cold distilled water for 45 seconds (by eight passes at 450 rpm) in an ice bath. Further dilutions were made shortly before starting the incubations. The amounts of homogenate noted below were suitable for assaying normal human liver, but greater dilutions were necessary in the assays of rat liver. The protein content of the homogenate, diluted 1:161, was determined by the Lowry method (12) with bovine serum albumin standards. In studies of needle biopsies, DNA was measured by the fluorometric method of Kissane and Robins (13) with salmon sperm standards. The preparation of needle biopsy samples is described in a later section.

Reagents

Phosphoenolpyruvate (trisodium salt), ATP (disodium, from equine muscle), and carbamyl phosphate (dilithium salt) were purchased from the Sigma Chemical Company; l-ornithine HCl, l-citrulline, l-aspartic acid, l-argininosuccinic acid (barium salt), and l-arginine HCl from Calbiochem; *N*-acetyl-L-glutamic acid from Mann Research Laboratories, Inc.; triethanolamine, diacetyl monoxime, and α-isonitrosopropiophenone from Eastman Kodak Co.

Solutions of reagents were made up in metal-free distilled water. In preparing buffers and the stock solutions of ATP, *N*-acetyl-l-glutamate, ornithine, and aspartate, the pH was adjusted at 37° using KOH and metal-free HCl. The pH of arginine, glycine, and maleate solutions was adjusted with NaOH. Potassium argininosuccinate was prepared from a solution of the barium salt by precipitating barium with K_2SO_4 at 1.25 times the argininosuccinic acid concentration; the supernatant was adjusted to 50 m*M* argininosuccinate and pH 8.0. A stock solution of 4 m*M* urea (Mallinkrodt, analytical reagent) in 100 m*M* H_2SO_4 was used to prepare the standards.

Supplementary Enzymes

Pyruvate kinase (Sigma Type II) and phosphoenolpyruvate were used as an ATP-generating system. Arginase (70 Archibald u/mg) was purchased from Boehringer. A solution of arginase (30 mg/10 ml) was dialyzed in 10 m*M* maleate, pH 7.5, for 3 hours with two changes of dialyzate (1 liter). Aliquots of the dialyzed enzyme, stored at −20°, retained original activity for several months. Two indispensable coupling enzymes, not available commercially, were partially purified in the laboratory. A preparation of rat liver ornithine transcarbamylase containing 5 mg protein (15,000 u) per ml was obtained by the method of Marshall and Cohen (14) carried through the heat step. Steer liver argininosuccinase 50 mg protein (6000 u) per ml was prepared by the method of Ratner (15) carried through step 4. Both were stored at −20°. Each supplementary enzyme preparation proved free of the activities of the other urea-cycle enzymes and noninhibitory in the assays in which it was employed.

Enzyme Assays

The assay of a sample for each enzyme entails four measurements, namely determination of activity at two enzyme concentrations, with corresponding blanks. Each assay comprises two stages: (1) an incubation at 37° of suitably diluted liver homogenate with appropriate substrates, buffers, cofactors, and supplementary enzymes, terminated by the addition of cold trichloroacetic acid; and (2) colorimetric determination of the citrulline or urea produced. The arginase assay also required preincubation of the homogenate with $MnCl_2$ at 55°.

Table 1 presents the constituents and composition of the *incubation mixtures*. In order to reduce the number of pipettings, and to conserve expensive compounds and supplementary enzymes, biochemical "cocktails" were devised for the coupled assays. *Reagent pools* consisting of substrates, cofactors, and buffers (Table 2*A*) were prepared in advance, subdivided into the volumes used to make the assay pools, and stored at −20°C. Just before incubation, supplementary enzymes were added to complete the *assay pools* (Table 2*B*), and aliquots were placed in tubes in an ice bath.

The colorimetric methods were based on those of Archibald (16, 17). Essentially, samples of assay supernatant and standards were mixed with the color development reagents in acid solutions, heated, cooled, and read by spectrophotometer. The citrulline or urea content was determined from the appropriate standard curve, and the enzyme activity was calculated. A unit of enzyme catalyzes the formation of 1 μmole of product per hour at 37°C. Activity is expressed as units per milligram liver protein, or per gram wet weight, or per milligram DNA.

The five assay methods are presented together because, for clinical purposes, it is highly desireable to assay all five, rather than one or two, enzymes of the urea cycle. This format emphasizes features shared by more than one assay. The incubation conditions summarized in Table 1 may be traced from individual assay descriptions and from Tables 2*A* and 2*B*. Color development is described in terms of common procedures modified by individual assay requirements. The alternative conditions which were studied are indicated in the Results section. The Appendix is a specific guide to performing the assays simultaneously.

Table 1 Composition of Incubation Mixtures

The amount of any constituent in an assay can be found by multiplying its concentration by 0.3 ml.

Constituents	Carbamyl Phosphate Synthetase (pH 7.0)	Ornithine Transcarbamylase (pH 7.7)	Argininosuccinate Synthetase (pH 7.5)	Argininosuccinase (pH 7.5)	Arginase (pH 9.8)
Substrates, *Cofactors*, and *Buffers* (μmole/ml)					
NH_4HCO_3	50				
N-Acetyl-L-glutamate	5				
L-Ornithine	5	2.5			
Carbamyl phosphate		5			
L-Citrulline			6.7[a]		
L-Aspartate			6.7		
L-Argininosuccinate				20	
L-Arginine					250
ATP	12		4		
$MgSO_4$	24		6		
PEP	20		25		
$MnCl_2$[c]					4
Glycine[c]					40
Triethanolamine		270			
K_2HPO_4/KH_2PO_4 (4:1)			50	50	
CO_2	[b]				
Supplementary Enzymes (u/ml)					
Ornithine transcarbamylase	160				
Pyruvate kinase	870		1020		
Argininosuccinase			310		
Arginase			1830	3000	
Liver (wet weight, mg/ml)[d]	2.38	0.27	16.67	16.67	0.18

[a] Blanks made without citrulline.
[b] Pool gassed with CO_2.
[c] Constituents carried over from activation medium.
[d] Duplicate tubes contained 1/2 this concentration. (Approximately 20% of wet weight was protein.) The arginase incubation mixture also contained 0.33 mg bovine serum albumin per ml, carried over from activation medium.

Table 2*A* Composition of Reagent Pools in 100 × the Amounts Contained in a Single Incubation Mixture

Each reagent pool provides 6 or 7 working pools.

Stock Solutions of Reagents	Milliliters in Each Pool			
	Carbamyl Phosphate Synthetase	Argininosuccinate Synthetase		Argininosuccinase
NH_4HCO_3, 1.0 *M*	1.5			
N-Acetyl-L-glutamate, 0.1 *M*, pH 7.2	1.5			
L-Ornithine, 0.1 *M*	1.5			
ATP, 0.1 *M*, pH 7.2	3.6	1.2		
$MgSO_4$, 0.2 *M*	3.6	0.9		
PEP, 0.1 *M*	6.0	7.5		
L-Aspartate, 0.1 *M*, pH 7.5		2.0		
K_2HPO_4/KH_2PO_4 (4:1), 1.0 *M*, pH 7.5		1.5		1.5
L-Argininosuccinate, 0.05 *M*, pH 8.0				12.0
		(6.55) [a]	(6.55)	
L-Citrulline, 0.1 *M*		1.0		
H_2O	1.8		1.0	1.5
(total volume)	19.5	7.55	7.55	15.0

[a] The common argininosuccinate synthetase pool is divided into a "test" pool with citrulline and a "blank" pool without that substrate.

Table 2*B* Composition of Assay Pools

A pool is sufficient for three assays.

	Milliliters in Each Pool			
Components	Carbamyl Phosphate Synthetase	Argininosuccinate Synthetase *Test*	Argininosuccinate Synthetase *Blank*	Argininosuccinase
Working Pool	3.0[a]	1.0	1.0	2.0
Supplementary Enzymes[b]				
Ornithine transcarbamylase (15,000)	0.05			
Pyruvate kinase (200,000)	0.02	0.01	0.01	
Argininosuccinase (6,000)		0.10	0.10	
Arginase (18,000)		0.20	0.20	0.66
(final volume)	3.07	1.31	1.31	2.66

[a] CO_2 is bubbled into the pool through a capillary pipette.
[b] The numbers given in parenthesis represent units per ml of supplementary enzyme solution. Units are defined as micromoles of product formed per hour at 37°C under optimal assay conditions.

Carbamyl Phosphate Synthetase*

Before supplementary enzymes were added, the pool was bubbled with CO_2 to obtain H_2CO_3/HCO_3^- buffering at pH 7. The concentrations of constituents of the assay pool were NH_4HCO_3 75 m*M*, *N*-acetyl-L-glutamate 7.5 m*M*, ornithine 7.5 m*M*, ATP 18 m*M*, $MgSO_4$ 36 m*M*, phosphoenolpyruvate 30 m*M*, pyruvate kinase 1300 u/ml, and rat liver ornithine transcarbamylase 240 u/ml. Addition of 0.10 or 0.05 ml* of homogenate (further diluted 1:7 in H_2O) to 0.20 ml assay pool started the 15 minute incubations. Zero-time blanks were employed. Termination of the incubation and subsequent treatment were the same as for ornithine transcarbamylase.

Ornithine Transcarbamylase

Both substrates were prepared in 270 m*M* triethanolamine-HCl, pH 7.7. The homogenate was diluted (1:61) in the same buffer, and 0.10 or 0.05 ml† was added to 0.10 ml of 7.5 m*M* ornithine in buffer. Addition of 0.10 ml of a freshly prepared solution of 15 m*M* carbamyl phosphate in buffer started the 10 minute incubations. Incubated blanks contained heat-inactivated enzyme.

The carbamyl phosphate synthetase and ornithine transcarbamylase reactions were stopped by addition of 0.5 ml of 8% trichloroacetic acid. After centrifuging, 0.5 ml supernatant was added to 3.0 ml of the solution used for color development. The latter was prepared for each set of assays from 2 parts of the acid reagent (3.7 g antipyrine, 2.5 g ferric ammonium sulfate, 450 ml H_2O, 250 ml concd. H_2SO_4, and 250 ml concd. H_3PO_4, made up to 1 liter) and 1 part of an 0.4% solution of diacetyl monoxime in 7.5% NaCl, kept in a dark bottle. After the contents were mixed, tubes for both assays were placed together in boiling water for exactly 15 minutes and then transferred to water at room temperature. The solutions were protected from light during boiling and cooling and until readings were taken.

Two standards (0.10 m*M* and 0.05 m*M* citrulline in 5% trichloroacetic acid) were always included. The relationship of the concentration of citrulline and light absorbance obeys Beer's law, with a millimolar extinction coefficient of 37.8, read against the reagent blank in a 1 cm light path at 464 mμ. Activity was calculated as follows:

$$\frac{\Delta\text{O.D.} \times 0.8 \times 3.5 \times (4 \text{ or } 6)}{37.8 \times 0.5 \times \text{mg protein, or g liver, per assay}}$$

* Carbamyl phosphate synthetase herein refers to CPS I, the hepatic mitochondrial enzyme which requires ammonia as substrate and an *N*-acetyl-L-glutamate cofactor.

† To provide a 0.30 ml volume for all incubations, 0.05 ml H_2O or buffer was added in advance to those tubes which would receive only 0.05 ml homogenate.

where 37.8 is the extinction coefficient of citrulline, 0.5 ml is the sample taken for color development, 0.8 ml is the final volume of the assay, 3.5 ml is the final volume of the color development mixture, and 4 or 6 the factor correcting the 15 or 10 minute incubations to 1 hour.

Argininosuccinate Synthetase

The concentrations of constituents of the assay pool were aspartate 10 mM, citrulline 10 mM, ATP 6 mM, $MgSO_4$ 9 mM, phosphoenolpyruvate 37.5 mM, pyruvate kinase 1530 u/ml, arginase 2740 u/ml, beef liver argininosuccinase 450 u/ml, and potassium phosphate buffer 75 mM, pH 7.5. Addition of 0.10 or 0.05 ml* of undiluted homogenate to 0.20 ml assay pool started the 30 minute incubations. Incubated blanks contained no citrulline.

Argininosuccinase

The concentrations of constituents of the assay pool were argininosuccinate 30 mM, arginase 4500 u/ml, and potassium phosphate buffer 75 mM, pH 7.5. Addition of 0.10 or 0.05 ml* of undiluted homogenate to 0.20 ml assay pool started the 30 minute incubations. Zero-time blanks were used.

Arginase

The activation medium was prepared shortly before use by mixing equal parts of 30 mM $MnCl_2$ and 300 mM 1-glycine, pH 9.8. The liver homogenate was initially diluted 1:21 and 1:42 in 0.5% bovine serum albumin; from each suspension 0.5 ml was mixed with 2.0 ml activation medium and placed at 55° to start preincubation. After 20 minutes, a 0.10 ml sample of each mixture was added to 0.20 ml aliquots of 375 mM 1-arginine, pH 9.8, starting the 10 minute incubations at 37°. Zero-time blanks were used.

Addition of 0.10 ml of 20% trichloroacetic acid stopped the argininosuccinate synthetase, argininosuccinase, and arginase incubations. A single support rack was used for simultaneous color development of all urea assays and standards. Supernatant or standard, 0.20 ml, was added to 2.0 ml acid reagent, followed by 0.10 ml of 3% α-isonitrosopropiophenone in 95% ethanol. The acid mixture consisted of 90 ml concd. H_2SO_4 (Mallinkrodt, technical grade), 270 ml concd. H_3PO_4 (reagent grade), and 1 ml 0.1 M $FeCl_3$, added to 600 ml water and made up to 1 liter. After mechanical shaking, the stoppered tubes were held in boiling water for exactly 1 hour. They were transferred to another covered container of water at room temperature for 10

* To provide a 0.30 ml volume for all incubations, 0.05 ml H_2O or buffer was added in advance to those tubes which would receive only 0.05 ml homogenate.

minutes cooling. With minimal light exposure, the solutions were poured into cuvettes. Their absorbance was read at 540 mμ.

Ratner (8) found that, because citrulline yields some color by this method and also suppresses the urea color, estimation of argininosuccinate synthetase activity requires standards "compensated" with amounts of citrulline which maintain the same sum of urea plus citrulline as in the incubation mixtures (Table 3). Urea standards without citrulline were used to provide a calibration curve for argininosuccinate synthetase blanks and all argininosuccinase measurements. Both compensated and uncompensated standards, as well as

Table 3 Composition of Compensated Standard Solutions Used in Estimating Argininosuccinate Synthetase Activity[a]

In addition to the indicated volumes of urea and citrulline solutions, each included 10 ml of 50% trichloroacetic acid and was made to 100 ml with H_2O.

ml/100 ml Standard Solution		μm/0.2 ml of Standard Solution	
Urea 4 mM	Citrulline 10 mM	Urea	Citrulline
0	50	0	1.00
2.5	49	0.02	0.98
10	46	0.08	0.92
20	42	0.16	0.84
30	38	0.24	0.76
40	34	0.32	0.68
50	30	0.40	0.60

[a] The uncompensated urea standards, prepared in the same way but without citrulline, were used to estimate urea in argininosuccinate synthetase "blanks" and the argininosuccinase assays. The calibration curve for arginase assays employed uncompensated urea standards containing arginine, 187 mM, and $MnCl_2$, 3 mM.

the assay tests and blanks, were read against a zero-urea, zero-citrulline reagent blank. Standards for the arginase assays and their reagent blank contained arginine and $MnCl_2$ in concentrations corresponding to the assay supernatant.

The calibration curve fitting each set of standards was determined by the method of least squares (18). Activity was calculated as follows:

$$\frac{\Delta U \times 0.4 \times (2 \text{ or } 6)}{0.2 \times \text{mg protein, or g liver, per assay}}$$

where ΔU represents the difference in urea content between test and blank samples; 0.2 ml is the sample; 0.4 ml is the final assay volume; and 2 or 6 the factor correcting the 30 or 10 minute incubations to 1 hour.

Protocol

The Appendix presents the incubation timetable used in performing the assays together. Avoiding delay by simultaneous incubations soon after tissue excision minimized losses of enzyme activity. A single worker followed this procedure, assisted only in obtaining and homogenizing the liver samples. Usually a control and two experimental specimens were assayed. Reducing the pipetting interval from 30 to 15 seconds made it possible to assay the five enzymes in each of six specimens.

Quality Control

Each assay related activity to two concentrations of enzyme. Fresh liver, from a rat of the same strain, sex, size, and diet used to define normal activity levels, was assayed with each experiment. When the pools were replenished, a rat liver was assayed with samples from old and new pools. Certain commercial reagents (carbamyl phosphate, ATP, PEP, argininosuccinate) were assayed for purity. Supplementary enzymes were assayed periodically to ensure that a sufficient excess of coupling enzyme activity was included in the incubation mixtures. Calibration curves were reviewed for anomalous readings or change of slope.

RESULTS

Ratner's recent review (19) is an authoritative account of the biochemistry of urea synthesis, with an extensive bibliography and valuable discussion of clinical, regulatory, and comparative aspects. The following observations derive from a search for optimal conditions for assaying the enzymes in human liver.

Carbamyl Phosphate Synthetase

Ammonium bicarbonate concentrations from 25 to 100 mM gave maximal activity which varied little with pH from 7.0 to 7.8. Triethanolamine 200 mM, pH 7.2, or TES 200 mM, pH 7.7, did not enhance activity compared with CO_2/HCO_3^- alone. Maleate, imidazole, and phosphate buffers appeared inhibitory. Activity was negligible when ammonia was replaced by glutamine. *N*-Acetyl-L-glutamate, an activator of carbamyl phosphate synthetase I, gave optimal activity at 2.5 to 10 mM concentrations, but preincubation of the homogenate in *N*-acetyl-L-glutamate did not increase activity.

Blank values were the same with trichloroacetic acid added either at zero-time or after incubation with preheated (90°) homogenate. When *N*-acetyl-L-glutamate was omitted, as in the incubated blank of Wixom et al.

(20), a low level of time-dependent citrulline formation occurred in undiluted homogenates. Zero-time blanks were adopted on the assumption that this minimal activity represents neither carbamyl phosphate synthetase II nor performed carbamyl phosphate, but rather carbamyl phosphate synthetase I in the presence of *N*-acetyl-L-glutamate in the homogenate.

For both human and rat liver homogenates, approximately 50% greater activity was obtained with $2Mg^{2+}:1ATP$ than with equimolar concentrations. Kerson and Appel (21) also observed that Mg^{2+} in excess of ATP enhanced activity. Using 24 m*M* Mg and 12 m*M* ATP, the addition of an ATP-generating system (phosphoenolpyruvate and pyruvate kinase) increased activity at least 20%. Despite the ATP-generating system, a reduction of the Mg^{2+} concentration to 12 m*M* and of ATP to 6 m*M* decreased the activity by almost 50%. Varying K^+ concentrations from the usual 40 m*M* (the result of KOH used in preparing reagents) to 100 m*M* did not seem to affect catalysis; higher concentrations suppressed activity.

Table 4 illustrates the role of supplementary ornithine transcarbamylase in this assay. Without added ornithine transcarbamylase, an assay of the liver from an infant with genetic ornithine transcarbamylase deficiency (22) gave the false impression of carbamyl phosphate synthetase deficiency. Properly assayed, the carbamyl phosphate synthetase activity of this liver was high. In the rat liver homogenate, carbamyl phosphate synthetase activity was reduced 20% when coupling enzyme was not included in the assay pool, despite an intrinsic 27-fold excess of ornithine transcarbamylase.

Table 4 Effect of Supplementary Ornithine Transcarbamylase (OTC) in the Carbamyl Phosphate Synthetase (CPS) Assay

	Enzyme Activity in Units per Gram Wet Weight		
Source of Liver	CPS Assayed without Supplementary OTC	CPS Assayed with Supplementary OTC	Intrinsic OTC Activity of Homogenate
Rat	322	417	11,300
Hyperammonemic infant[a]	0	310	0

[a] A male infant appeared normal at delivery but lapsed into coma following protein feeding and died with refractory hyperammonemia 4 1/2 days after birth. Activities of carbamyl phosphate synthetase, argininosuccinate synthetase, argininosuccinase, and arginase were similar to those obtained in autopsy specimens of normal infant liver, but no ornithine transcarbamylase activity could be obtained in undiluted 5% homogenate.

Ornithine Transcarbamylase

The assay has been described previously (11). At pH 7.7 ornithine concentrations greater than 2.5 m*M* inhibit human ornithine transcarbamylase. No higher activity was found between pH 8.5 and 6.0 with ornithine concentrations ranging from 1 to 25 m*M* (11). Marshall and Cohen, using purified bovine enzyme, found similar substrate inhibition as a function of increasing pH (23).

Carbamyl phosphate was used to start the incubation, because it reacts spontaneously with ornithine (24). Blanks containing preheated homogenate corrected for this nonenzymatic citrulline formation (less than 8 nmole per assay). Because carbamyl phosphate decomposes rapidly in solution (25), it was dissolved immediately before use. The optimal concentration did not vary with pH (see ref. 11 and 24 for further discussion of the ornithine transcarbamylase assay).

Diluting human liver homogenate in water instead of buffer did not alter activity consistently in incubation mixtures having final triethanolamine concentrations of either 180 or 270 m*M*. However, the specific activity of partially purified rat liver ornithine transcarbamylase was reduced one-third by dilution in water. Dilution in the carbamyl phosphate synthetase assay pool did not affect activity of the purified enzyme.

Argininosuccinate Synthetase

An ATP-regenerating system was essential to the assay. Reducing the phosphoenolpyruvate concentration below 20 m*M* decreased activity, regardless of the initial ATP concentration. Unlike the carbamyl phosphate synthetase assay, varying the Mg^{2+}:ATP ratio from equimolar to 2:1 had no effect. Increasing citrulline and aspartate concentrations impaired the method's precision (through the effects of citrulline on color development), without enhancing activity. Reducing the concentrations to 5 m*M* did not alter activity, but the higher level assured that less than half of each substrate was consumed.

Figure 2 shows that supplementary argininosuccinase must be added to demonstrate normal argininosuccinate synthetase activity in a liver homogenate. Only 60% of the synthetase activity is elicited without supplementary argininosuccinase. A 10-fold excess of the cleavage enzyme did not fully couple the reactions. Table 5 shows that assaying the "arginine-synthesizing system" without the supplementary enzymes underestimates the synthetase activity. Ratner noted (19) "the overall procedure was originally used to ascertain the *presence* of a two step system and . . . is not a rate-limiting assay." In applying the method to crude tissue preparations the addition of an excess of cleavage enzyme was prescribed (8).

Other microassays were tried. In attempts to follow the formation of AMP by coupled oxidation of DPNH, the high ATPase activity of homogenates produced an excessive blank. Measuring the disappearance of citrulline was simpler but less sensitive and less precise than the coupled assay of urea formation.

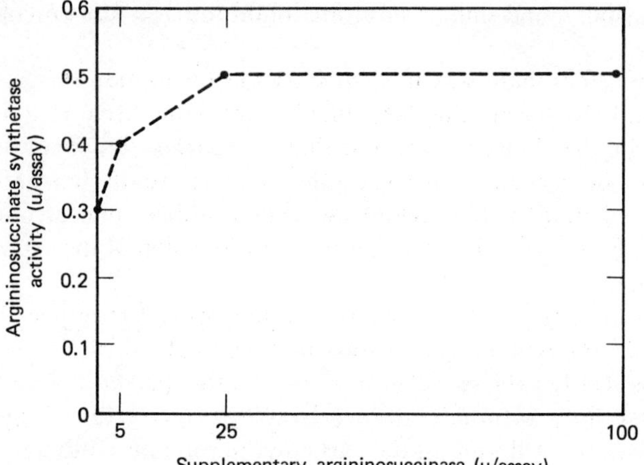

Figure 2. The effect of purified beef liver argininosuccinase in the assay of argininosuccinate synthetase in normal rat liver. The same homogenate, supplemented with arginase in large excess, was used in each assay. The intrinsic argininosuccinase activity of this homogenate was 0.83 u/assay.

Table 5 Effects of Supplementary Arginase and Argininosuccinase in the Argininosuccinate Synthetase Assay, and of Supplementary Arginase in the Argininosuccinase Assay

Enzyme Assayed	Omissions	Activity (u/g wet weight)[a]	
		Host liver	*Tumor*
Argininosuccinate synthetase	None	119	20
	Arginase	113	4.8
	Argininosuccinase	61	1.6
	Arginase and argininosuccinase	56	0
Argininosuccinase	None	189	21
	Arginase	193	8
Arginase		85,700	2050

[a] Fresh specimens were taken from a 360 g male Charles River rat's liver and a subcutaneously implanted Morris Hepatoma #7795.

Argininosuccinase

The substrate, argininosuccinate, appeared limiting at concentrations below 10 m*M*. The findings were consistent with Ratner's observation that activity is proportional to enzyme concentration if less than 40% of the substrate is consumed (26). The amount of supplementary arginase required for fully coupled activity was tested with partially purified rat liver argininosuccinase; maximal activity was obtained with arginase in greater than 350-fold excess. Table 5 shows that, despite high arginase activity in normal liver, exogenous enzyme must be added to assure a large excess in assaying unknown samples. Dialysis of the commercial arginase preparation, necessary to prevent inhibition of argininosuccinate synthetase (presumably by Mn^{2+}), did not alter the amount needed in the argininosuccinase assay. Exhaustive dialysis must be avoided because it leads to loss of activity.

Arginase

Several divalent cations were tested for activation of arginase in dialyzed human liver homogenates. Mn^{2+} was the optimal cofactor, although activation occurred with cadmium, nickel, and cobalt. Activity of the Mn^{2+}-activated enzyme increased gradually from pH 9.5 to 10.0, but abrupt suppression above pH 10 led to the selection of pH 9.8 for the incubation. Activity was not altered by varying the arginine concentration from 200 to 300 m*M*. Under the conditions described, arginase activity proved higher than previously recorded for rat and human liver homogenates.

Various methods of preincubation were evaluated and a wide range of Mn^{2+} concentrations was tested with each of the alterations. For the method adopted, varying the Mn^{2+} concentration from 8 to 20 m*M* did not affect activity. A medium containing protein, as recommended by Robbins and Shields (27), allowed greater activation than an equivalent suspension without albumin. Whether the major dilution of homogenate occurred before or after preincubation did not influence activity. Activity increased with alkalinity of the preincubation medium to pH 9.8, despite deposition of a rust-colored precipitate assumed to be $Mn(OH)_2$. Full activation required at least 10 minutes at 55°C, the temperature used by Schimke (1). Similar activation was produced by preincubation periods ranging from 10 to 40 minutes and was greater than with preincubation up to 4 hours at 37°. Addition of manganese to the substrate, as recommended by Schimke (10), did not increase activity.

Color Development

As noted by Kulhanek and Vojtiskova (28) and by Ceriotti and Gazzaniga (29), the citrulline color was augmented and stabilized by antipyrine. It was not affected by trichloroacetic acid, triethanolamine, or the substrates of the carbamyl phosphate synthetase and ornithine transcarbamylase reactions. The molar extinction coefficient of 37,800 found by Snodgrass and Parry (24) was confirmed for both assays. The color intensity of standards, although light-sensitive, proved so reproducible (less than 3% variance) that calibration curves were not necessary. Two standards were run with each set of assays as controls for color development. Urea increased the color, but its effect usually was small and virtually canceled by the blanks. Assays of uremic tissue or undilute homogenate required preincubation with urease, as in the serum ornithine transcarbamylase assay (24), or calibration with standards having a urea concentration corresponding to the tissue extract (30).

Accurate estimation of urea required three sets of standards prepared and heated together with each group of assays. The colors were affected by surface area and aeration, as well as by light. When mixing produced bubbles in the solutions, color development was impaired. Scaling down the volume without altering the concentrations of the constituents reduced the resultant optical density; color was partially restored when the same mixtures were heated in narrower test tubes. Figure 3 shows that the calibration curves obtained by developing the colors in 13 × 100 mm tubes were steeper than those produced with the same volumes in 20 × 170 mm tubes.

Citrulline enhanced the color with small amounts of urea while reducing the color at high urea levels, as noted in Ratner's initial description of com-

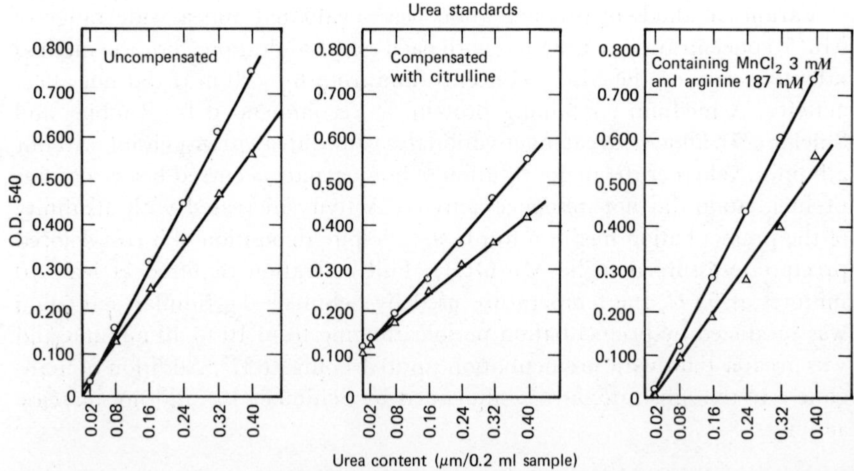

Figure 3. Urea color production in standards as affected by tube size: 0.2 ml samples were mixed and heated with 2.0 ml acid and 0.1 ml color reagent. Compare the color developed in 13 × 100 mm test tubes (○○○○) with the same solutions treated in 20 × 100 mm test tubes (△△△△).

pensated standards (8). Variability of color development in the low-urea–high-citrulline range contributes to the variance of the argininosuccinate synthetase assay results. Argininosuccinic acid, at the concentration used in the argininosuccinase assay, did not affect the color and was not included in the uncompensated standards. Archibald's observation that arginine diminished the urea color (31) was confirmed: the optical density readings of standards containing arginine in the same concentration as in the assays were one-fourth to one-third lower than the same standards without arginine. In the presence of manganese, however, arginine reduced color development very little. The standard curve was similar to the uncompensated urea curve but shifted slightly to the right of the origin. Glycine, in the concentration used, did not affect the color and was not included in the standards.

Application of the Methods

The reaction rates were proportional to amounts of liver yielding 15 to 150 nmole of citrulline for the first two assays and 80 to 800 nmole of urea for the last three. The ranges of color intensity were most accurate within these limits, although incubation conditions were adequate for higher activities. Product formation was linear with time for at least 33% longer than the incubation periods adopted. The variance of duplicate assays (single and double portions of homogenate) was 3% for ornithine transcarbamylase and argininosuccinase, 5% for carbamyl phosphate synthetase and arginase, and 10% for argininosuccinate synthetase.

Assays of human liver cell fractions indicated that, as in other animals (19, 32), carbamyl phosphate synthetase and ornithine transcarbamylase are mitochondrial enzymes, and the other three enzymes are in the cytosol. Homogenates prepared in dilute phosphate buffer or in various concentrations of cetyltrimethylammonium bromide yielded activities no higher than distilled water preparation. Three lobes of a rat liver, assayed simultaneously, showed no more variation than duplicate assays of a single homogenate. Three wedge specimens from a normal human liver also had identical levels of the enzymes.

Table 6 compares the activities of the urea-cycle enzymes in wedge specimens of normal human and rat liver assayed immediately after excision. The activities of postmortem liver were lower and less consistent than those of fresh biopsies. Biopsies immediately frozen to $-20°$ lost enzyme activities in an irregular fashion. Biopsies frozen in liquid nitrogen and then kept at $-70°$ retained initial activity levels for at least 2 weeks.

Table 7 shows that the protein content of homogenates made from needle biopsies is an inadequate reference standard for enzyme activity. Homogenate contents in mg protein per mg DNA were relatively constant among the four wedge biopsies (mean $\pm$ S.D. = 70.3 $\pm$ 4.4), but varied widely among the eight needle biopsies (102.6 $\pm$ 28.3). Based on DNA, activity values for

Table 6 Activities of the Hepatic Urea-Cycle Enzymes in Adult Humans without Liver Disease and in Control Rats

	Activities of Urea-Cycle Enzymes (mean ± standard deviation)				
Source of Liver[a]	Carbamyl Phosphate Synthetase u/g / u/mg P	Ornithine Transcarbamylase u/g / u/mg P	Argininosuccinate Synthetase u/g / u/mg P	Argininosuccinase u/g / u/mg P	Arginase u/g / u/mg P
Surgical wedge[b] biopsy $n = 8$	279 ± 65 / 1.91 ± 0.27	6,600 ± 1,580 / 44.2 ± 7.7	90 ± 12 / 0.62 ± 0.14	220 ± 25 / 1.49 ± 0.27	86,000 ± 9,300 / 579 ± 106
Autopsy (3–14 hours after death) $n = 6$	186 ± 93 / 1.39 ± 0.61	3,370 ± 976 / 25.9 ± 7.4	41 ± 26 / 0.27 ± 0.21	121 ± 67 / 0.97 ± 0.71	40,000 ± 22,300 / 298 ± 142
Rat[c] $n = 13$	743 ± 128 / 3.73 ± 0.72	14,700 ± 2,700 / 73.4 ± 13.3	280 ± 51 / 1.41 ± 0.31	343 ± 52 / 1.72 ± 0.16	148,000 ± 25,600 / 759 ± 115

Abbreviations: u = units of enzyme activity, expressed as micromoles of citrulline or urea formed per hour; g = gram wet weight of liver; mg P = milligram protein; n = number of subjects.

[a] Biopsied subjects were eating 1.0 to 1.5 g of protein per kilogram body weight per day. Rats were on 23% protein chow.

[b] Data originally presented in reference 2.

[c] Rat liver was taken immediately after the animal was sacrificed.

Table 7 Comparison of DNA and Protein as Reference Substances for Assaying Enzyme Activity in Homogenates of Hepatic Wedge and Needle Biopsies[a]

Source of Liver		Homogenate Contents mg per ml			Mean of Five Enzymes' Activity Ratios $\left[\frac{\text{needle}}{\text{wedge}}\right]$	
		DNA	Protein	Hgb	Based on DNA	Based on Protein
I	W	.106	7.9	3.1	—	—
	N_1	.081	10.0	4.6	1.07	0.62
	N_2	.108	17.4	8.7	1.00	0.46
II	W	.100	7.4	1.0	—	—
	N_1	.130	12.2	2.8	1.02	0.93
	N_2	.080	9.8	2.0	0.94	0.61
III	W	.113	7.4	2.2	—	—
	N_1	.144	11.4	4.1	0.91	0.75
	N_2	.123	8.1	3.1	1.15	1.28
IV	W	.107	7.2	2.0	—	—
	N_1	.157	11.4	5.8	1.10	1.14
	N_2	.157	16.1	7.3	1.09	0.78

[a] A wedge (W) and two needle biopsies (N_1, N_2) of normal human liver were obtained from four patients during cholecystectomy. Wedge biopsies were weighed and homogenized in 19 volumes of water, while needle biopsies were homogenized in amounts of water roughly estimated to yield 5% homogenate (0.8 to 1.0 ml). In assays of the five urea-cycle enzymes, activities were based on the homogenates' protein and DNA concentrations. Each enzyme's activity in a needle biopsy was divided by its activity in the wedge of the same liver; the mean of the five ratios is shown above. Hemoglobin was determined as cyanmethemoglobin in the supernatant after centrifugation of homogenate diluted 1:51 in Hycel reagent.

the five enzymes varied little between wedge and needle biopsies of the same liver, whereas the values based on protein were far from uniform. The variable amount of blood accompanying a needle specimen greatly affected the homogenate's protein content, but its contribution to DNA content was negligible. Attempts to rinse needle specimens with saline or to blot and weigh them resulted in lower activities than were obtained with biopsies delivered directly from the needle into the small homogenizing tube.

DISCUSSION

The exigencies underlying practices recommended here are best illustrated by the argininosuccinate synthetase assay. The enzyme's activity falls rapidly in excised tissue or on standing in homogenate. The assay system has labile constituents: an ATP-generating system and two supplementary urea-cycle enzymes. One of the substrates, citrulline, alters the color obtained with the final product, urea. The time saved by thawing a previously prepared pool, adding supplementary enzymes, and taking aliquots of the resultant pools, allows the incubation to start promptly, without postponing the incubations of the other four enzymes. A separate set of standards, compensated with citrulline, is provided specifically for argininosuccinate synthetase.

Wixom et al. (20) propose an argininosuccinate synthetase assay based on the disappearance of citrulline. Elimination of urea by urease (provided in the incubation mixture) simplifies the colorimetric determination of citrulline. Because this assay system is not supplemented with argininosuccinase, it may be affected by reaction reversal or product inhibition (19), particularly in cases of argininosuccinaciduria. A subtraction method appears subject to greater analytical error than the measurement of product formation; but, supplemented with argininosuccinase and arginase, it may be preferable for uremic tissues.

In homogenates, the apparently higher intrinsic activities of ornithine transcarbamylase (compared with carbamyl phosphate synthetase), argininosuccinase (compared with argininosuccinate synthetase), and arginase (compared with argininosuccinase) do not ensure zero-order kinetics in coupled assays. Arginase is activated by manganese and shows maximal activity near pH 10 but is used in the argininosuccinase and argininosuccinate synthetase assays at pH 7.5 without added manganese. For urea formation to be an adequate measure of argininosuccinase, the activity of arginase (as assayed at pH 9.8, after preincubation in manganese) must exceed argininosuccinase by several hundred fold. The activity of argininosuccinate synthetase is underestimated by the "arginine synthetase system" (5), which does not provide exogenous argininosuccinase. Urea formation is not linear with time (5). When arginosuccinase is added, the synthetase activity increases (Fig. 2). As Ratner emphasized (19), "To obtain a rate-limiting assay in coupled systems, supplementation with a *large* excess of purified coupling enzymes is necessary."

Several versions of the Brown and Cohen methods (33–35) use the "arginine synthetase system" and also rely on ". . . the presence of ornithine transcarbamylase in the homogenate, to form citrulline . . ." (33) in the carbamyl phosphate synthetase assay. The apparent association of carbamyl phosphate synthetase deficiency with ornithine transcarbamylase deficiency (33, 35) may be an artifact of this method (see Table 4). The patients had ornithine transcarbamylase activities less than 10 times the normal carbamyl phosphate synthetase level. Without exogenous ornithine transcarbamylase, there is uncertainty as to which enzyme is rate-limiting in the carbamyl phosphate synthetase assay. Without an ATP regenerating system, the supply of ATP may be rate-limiting, due to the high ATPase activity of crude homogenates.

Reliable kinetics require that the enzyme under assay be rate-limiting in the coupled sequence. Kerson and Appel (21) used 50 to 100 u of ornithine transcarbamylase in assays containing 0.76 u of purified carbamyl phosphate synthetase. Brown and Cohen (5), Jones et al. (36), and Schimke (1) used similar amounts of beef liver enzyme to assay carbamyl phosphate synthetase in frog or rat liver homogenates. Supplementary enzymes are essential for assays involving coupled reactions, not only to identify genetic defects, but also to analyze acquired liver disorders associated with hyperammonemia, such as Reye's syndrome, in which activities of one or both mitochondrial enzymes may be reduced (37, 38). An assay depending on more than one enzyme in the tissue lacks diagnostic specificity, particularly if deficiency of the coupling enzyme causes the same clinical syndrome as low activity of the enzyme which the assay purports to measure.

The assays described here have been applied in several studies of the urea cycle enzymes. They were used to identify severe ornithine transcarbamylase deficiency as a cause of lethal hyperammonemia in male infants (22), and to find lesser defects in their female relatives (39), supporting the hypothesis that this enzyme is X-linked (40, 41). The recent diagnosis of fatal neonatal hyperammonemia due to virtual absence of carbamyl phosphate synthetase activity (42) involved the use of these methods. They were also employed to show that the levels of the enzymes increase with dietary protein in primates (2) as they do in rats (1). The same set of assays, modified only in tissue dilutions, demonstrated the presence of the five enzymes in cultures of a clonal strain of hepatoma cells (43).

The protocol meets a current need for reliable diagnostic screening. It is designed for prompt incubation of fresh tissue. The assay conditions are specifically adapted to the activity requirements of the human liver enzymes. Khatra et al. (4) and Miller and McLean (44) reported activities in normal adults generally lower than those shown in Table 6. Standardization of assay composition, technique, and controls should facilitate comparison of values from different laboratories. The normal ranges of the five enzymes in human liver should be defined in relation to age, sex, and nutritional status. Due to the instability of the enzymes and variability of their postmortem activity losses, such normal data must be obtained from biopsies rather than autopsied liver.

The scale allows assays of all five enzymes in 40 mg of liver, but two problems limit application of the methods to needle biopsies. In genetic hyperammonemia liver architecture is usually undisturbed and sampling errors are minimal, but in cirrhosis or severe hepatitis the variable proportions of parenchyma and stroma could make a small specimen unrepresentative. As conventional baselines of enzyme activity, protein and wet weight are less reliable with needle than with wedge biopsies. The variability presumably results from the needle specimens' higher ratio of surface to mass: blood content is included in the protein determination and adherent fluid contributes to wet weight. A microassay for DNA (13) provides a standard better correlated with hepatocyte content. J. N. Davidson in the fifth edition of *The Biochemistry of the Nucleic Acids* describes the use of DNA "as a reference substance in terms of which the chemical composition of a tissue may be expressed . . ." (45).

As more abnormalities of the urea-cycle enzymes are found, other techniques will be needed to characterize them. Enzyme activity at the lower limit of colorimetric methods should be assayed by more sensitive radioisotopic methods, such as those described by Schimke (46). Studies of heat stability, electrophoretic mobility, substrate affinity, and the effects of pH on reaction kinetics may distinguish between different mutations of the same enzyme.

Appendix PROTOCOL FOR CARRYING THE ASSAYS THROUGH THE INCUBATION STAGE

The five enzymes are abbreviated as CPS, OTC, AS, Aa, and A, respectively. The dilutions are suitable for activity in normal liver from adult humans consuming 50 to 100 g protein/day. Ordinarily, a rat liver, diluted further for CPS, OTC, Aa, and A, is assayed as a control in each experiment. The CPS, AS, and Aa incubations are started with the addition of 0.10 and 0.05 ml aliquots of the appropriate liver suspensions. For OTC, the suspension is added to ornithine before the addition of carbamyl phosphate, which starts the reaction. Each incubation tube is kept on ice until, with the addition which starts the reaction, it is transferred to a 37° bath. One minute is added to the time of incubation to allow for temperature equilibration. The A incubation starts when 0.10 ml of preincubation mixture is added to arginine at 37°. Incubations are ended with 0.50 ml 8% TCA for CPS and OTC, and 0.10 ml 20% TCA for AS, Aa, and A. Zero-time "blanks" (CPS, Aa, A) are first treated with TCA and then completed with liver suspension. The schedule is given in minutes before (−min) or after (+min) preparation of the homogenate.

(−90 minutes) Arrange reagents, pools, supplementary enzymes, pipettes, and incubation tubes according to a diagram. Pools and supplementary enzymes are thawed partially at room temperature and then put in crushed ice. The circulating baths are set at 37 and 55°C, and a hot plate is brought to 80 to 90°. (−70) Weigh carbamyl phosphate 8.4 mg. (−65) Diluents for the homogenates: 0.3 ml H_2O (CPS), 3.0 ml triethanolamine (OTC), 1.0 ml and 0.5 ml 0.5% bovine serum albumin (A). (−58) To half of the incubation tubes add 0.05 ml H_2O (CPS, AS, Aa) or 0.05 ml triethanolamine (OTC). These will receive 0.05 ml of liver suspension. (−45) Dispense unpooled substrates into incubation tubes: 0.10 ml ornithine (OTC), 0.20 ml arginine (A). (−32) Saturate CPS pool with CO_2. (−30) Add supplementary enzymes to pools (Table 2*B*). (−24) Dispense the CPS, AS, and Aa assay pools (0.20 ml). (−5) Mix glycine and $MnCl_2$ for A activation medium; take 2 ml aliquots.

(0) Obtain the homogenates. (+2) Add 0.05 ml homogenate to A diluent; take 0.5 ml for further 1:1 dilution. (+7) Add 0.5 ml from each A dilution to 2 ml activation medium and place at 55° to start A preincubation. (+10) Start AS incubations (tests and blanks). (+16) Start Aa incubations. (+19) Start CPS incubations. (+24) Acidify A blanks. (+27) Put arginine (A incubation tubes) at 37°. (+28) Start A incubations. (+31) Complete A blanks. (+35) End CPS incubations. (+38) End A incubations. (+41) End AS incubations. (+47) End Aa incubations. (+50) Add 0.05 ml homogenate to OTC diluent; put 1.0 ml of this suspension in a 90° bath. (+54) Dispense cold OTC suspension into "test" tubes. (+57) Put heated OTC suspension on ice and dispense into "blanks." (+61) Add 3.0 ml triethanolamine to dissolve carbamyl phosphate. (+63) Start OTC incubations. (+70) Acidify CPS blanks. (+74) End OTC incubations. (+80) Complete CPS blanks. (+83) Acidify Aa blanks. (+86) Complete Aa blanks. Pipetting at 30 second intervals permits assay of three specimens by this protocol.

ACKNOWLEDGMENTS

We are very grateful for the encouragement, instruction, and material aid given by Dr. Sarah Ratner throughout the studies. Her contributions are essential to this set of assays. Dr. Oliver Lowry kindly taught the techniques required for measurement of DNA. Miss Carol Hood and Mrs. Connie Richardson provided valuable technical assistance. Mrs. Normal Graven cheerfully endured revisions of the manuscript.

REFERENCES

1. Schimke, R. T. (1962). *J. Biol. Chem.* **237**, 459–468.

2. Nuzum, C. T., and Snodgrass, P. J. (1971). *Science* **172**, 1042–1043.

3. Shih, V. E., and Efron, M. L. (1972). In J. B. Stanbury, J. B. Wyngaarden and D. S. Fredrickson, Eds., *The Metabolic Basis of Inherited Disease* 3rd ed., McGraw-Hill, New York, pp. 370–392.

4. Khatra, B. S., Smith, R. B. III, Millikan, W. J., Sewell, C. W., Warren, W. D., and Rudman, D. (1974). *J. Lab. Clin. Med.* **84**, 708–715.

5. Brown, G. W., Jr., and Cohen, P. P. (1959). *J. Biol. Chem.* **234**, 1769–1774.

6. Marshall, M., Metzenberg, R. L., and Cohen, P. P. (1958). *J. Biol. Chem.* **233**, 102–105.

7. Burnett, G. H., and Cohen, P. P. (1957). *J. Biol. Chem.* **229**, 337–344.

8. Ratner, S. (1955). In S. Colowick, and M. Kaplan, Eds., *Methods in Enzymology*, **II**, Academic, New York, pp. 356–367.

9. Greenberg, D. M. (1955). In S. Colowick, and M. Kaplan, Eds., *Methods in Enzymology*, **II**, Academic, New York, pp. 368–374.

10. Schimke, R. T. (1970). In H. Tabor, and C. W. Tabor, Eds., *Methods in Enzymology*, **XVIIA**, Academic, New York, pp. 313–317.

11. Snodgrass, P. J. (1968). *Biochemistry* **7**, 3047–3051.

12. Lowry, O. H., Rosebrough, N. J., Farr, A. L., and Randall, A. J. (1951). *J. Biol. Chem.* **193**, 265–275.

13. Kissane, J. M., and Robins, E. (1958). *J. Biol. Chem.* **233**, 184–188.

14. Marshall, M., and Cohen, P. P. (1972). *J. Biol. Chem.* **247**, 1641–1653.

15. Ratner, S. (1970). In H. Tabor, and C. W. Tabor, Eds., *Methods in Enzymology*, **XVIIA**, Academic, New York, pp. 304–309.

16. Archibald, R. M. (1944). *J. Biol. Chem.* **156**, 121–142.

17. Archibald, R. M. (1945). *J. Biol. Chem.* **157**, 507–518.

18. Ipsen, J., and Feigl, P. (1970). *Bancroft's Introduction to Biostatistics*, 2nd ed., Harper and Row, New York, pp. 96–97.

19. Ratner, S. (1973). *Advan. Enzymol.* **39**, 1–90.

20. Wixom, R. L., Reddy, M. K., and Cohen, P. P. (1972). *J. Biol. Chem.* **247**, 3684–3692.

21. Kerson, L. A., and Appel, S. H. (1968). *J. Biol. Chem.* **243**, 4279–4285.

22. Campbell, A. G. M., Rosenberg, L. E., Snodgrass, P. J., and Nuzum, C. T. (1973). *N. Engl. J. Med.* **288**, 1–6.

23. Marshall, M., and Cohen, P. P. (1972). *J. Biol. Chem.* **247**, 1654–1668.

24. Snodgrass, P. J., and Parry, D. J. (1969). *J. Lab. Clin. Med.* **73**, 940–950.

25. Allen, C. M., Jr., and Jones, M. E. (1964). *Biochemistry* **3**, 1238–1247.

26. Ratner, S., Anslow, W. P., Jr., and Petrack, B. (1953). *J. Biol. Chem.* **204**, 115–125.

27. Robbins, K. C., and Shields, J. (1956). *Arch. Biochem. Biophys.* **62**, 55–62.

28. Kulhanek, V., and Vojtiskova, V. (1964). *Clin. Chim. Acta* **9**, 95–96.

29. Ceriotti, G., and Gazzaniga, A. (1966). *Clin. Chim. Acta* **14**, 57–62.

30. McLean, P., Novello, F., and Gurney, M. W. (1965). *Biochem. J.* **94**, 422–426.

31. Van Slyke, D. D., and Archibald, R. M. (1946). *J. Biol. Chem.* **165**, 293–309.

32. Cohen, P. P., and Brown G. W. (1960). In M. Florkin and H. S. Mason, Eds., *Comparative Biochemistry*, **2**, pp. 161–244.

33. Levin, B. (1971). *Adv. Clin. Chem.* **14**, 65–143.

34. McLean, P., and Gurney, M. W. (1963). *Biochem. J.* **87**, 96–104.

35. Salle, B., Levin, B., Longin, B., Richard, P., Andre, M., and Gauthier, J. (1972). *Arch. Franc. Ped.* **29**, 493–504.

36. Jones, M. E., Anderson, A. D., Anderson, C., and Hodes, S. (1961). *Arch. Biochem. Biophys.* **95**, 499–507.

37. Thaler, M. M., Hoogenraad, N. J., and Boswell, M. (1974). *Lancet* **II**, 438–440.

38. Brown, T., Brown, H., Lansky, L., and Hug, G. (1974). *N. Engl. J. Med.* **291**, 797–798 (letter).

39. Short, E. M., Conn, H. O., Snodgrass, P. J., Campbell, A. G. M., and Rosenberg, L. E. (1973). *N. Engl. J. Med.* **288**, 7–12.

40. Schneider, S., Nicholson, J. F., and Chutorian, A. B. (1970). *Trans. Amer. Neurol. Assoc.* **95**, 86–90.

41. Campbell, A. G. M., Rosenberg, L. E., Snodgrass, P. J., and Nuzum, C. T. (1971). *Lancet* **II**, 217–218 (letter).

42. Gelehrter, T. D., and Snodgrass, P. J. (1974). *N. Engl. J. Med.* **290**, 430–433.

43. Richardson, U. I., Snodgrass, P. J., Nuzum, C. T., and Tashjian, A. H., Jr., (1974). *J. Cell Physiol.* **83**, 141–149.

44. Miller, A. L., and McLean, P. (1967). *Clin. Sci.* **32**, 385–393.

45. Davidson, J. N. (1965). *The Biochemistry of the Nucleic Acids*, 5th ed., Methuen, London, pp. 106–111.

46. Schimke, R. T. (1970). In H. Tabor, and C. W. Tabor, Eds. *Methods in Enzymology*, **XVIIA**, Academic, New York. pp. 324–329.

Transport and Control at the Membrane Level

Discussion

Dr. Krebs: Thank you for keeping conscientiously to the allocated time. As we have a few minutes in hand, it is perhaps appropriate to have a short discussion now. I would like to respond in particular to what you just called provocative ideas. As soon as Professor Chapell and his colleagues started to publish his revolutionary concept that glutamate dehydrogenase may not be very important in urea synthesis from alanine, we started to work in this field, because I was very skeptical. In order to clarify matters, I invited Professor Chappell and his colleagues to Oxford, and we met on May 9th, just a few weeks ago. This led to the planning of joint experiments. Some of these will be concerned with the explanation of some gross experimental discrepancies between observations at Oxford and at Bristol. We cannot confirm that leucine in isolated liver cells accelerates the synthesis of glutamine and glutamate when alanine is the added substrate. Another point concerns the possible role of the Lowenstein purine nucleotide cycle. There is a simple way of settling the matter, namely, by using N^{15}-labeled precursors. Fortunately, owing to the courtesy of Professor Whatley, at Oxford, we had at our disposal N^{15}-alanine, and we also had at our disposal, thanks to the courtesy of Professor Paton and Mr. David Harvey, of the Oxford Pharmacology Department, the use of a mass spectrometer. The crucial results came out just a few hours before we met in Oxford on May 9th. The plan of the expreiment is this: if the 6-amino group of AMP is an intermediate in the formation of urea from alanine, as suggested by Chappell, then it must become labeled with great speed. The decisive answer we obtained was that the incorporation of N^{15} into the amino group of adenine is very slow compared with the rate of urea synthesis. So far we have completed only one experiment, in which experimental conditions were not ideal. The cells were incubated for 1 hour, and the incorporation of N^{15} into the 6-amino group of alanine should have been of the order of at least 95%. In fact it was only 20%, and this rules out the Lowenstein purine adenine nucleotide cycle as a major pathway of ammonia formation from alanine. These experiments will be continued jointly with Professor Chappell's team.

Another point I wish to raise concerns the validity of experiments on isolated mitochondria. Mitochondria are, of course, a very important tool, but one must keep in mind that under natural conditions there are factors which may modify, especially quantitatively, some of the properties of mitochondria. I was struck in our own experiments that in homogenates the rates of glutamate removal by the dehydrogenase reaction, as indicated by the formation of ammonia, were greater by a factor of 5 than the rates in isolated mitochondria. Moreover, when we measured the oxygen consumption of homogenates (which contain mitochondria in the more physiological medium than saline solutions) and compared it with that of isolated mitochondria, we found that the oxygen consumption in the presence of glutamate was greater in the homogenates also by a factor of about 5. The reason is, I believe, that the supernatant fraction of homogenate acts as an ATPase and therefore increases the rate of electron transport through the respiratory chain. Hence it increases the supply of NAD in the mitochondria, and the rate of the glutamate dehydrogenase reaction *in vivo* must depend on the rate at which NAD is regenerated. These experiments bring home the fact that mitochondrial behavior, especially in respect to quantitative aspects, depends on the environment in which the mitochondria are kept.

Dr. Chappell: Now I must take objection to your statement that mitochondria are not good tools, as it were, good models. I think that the thing which has really emerged out of what I have spoken about this morning is that the effect of leucine on whole cells is mirrored exactly in the effect on the isolated mitochondria. And I think that came out very clearly. And if you won't admit that mitochondria are good objects of study, I feel very sorry, indeed, for enzymologists, who are dealing with an order of magnitude more of artifacts and that you will accept their evidence as being valid. This question of whether the mitochondria are not working at the right capacity with regard to homogenates, this we do not find. And you know that I have criticized your experiment with mitochondria on the grounds that you incubate them too hot, too long, and your response to this is that 38° is a physiological temperature. But I put it to you that 1 hour is not necessarily a physiological time. I think you have to take cognizance of the intactness of these organelles and their delicacy. We find, in fact we have winging on its way to you a letter which shows that if we do the experiments with homogenates at 38° but for a few minutes, a minute or two, then we can fully account for the oxidative activity of the homogenate in terms of the behavior of the mitochondria, and the mitochondria in the homogenate behave in exactly the same way as they do when they are washed.

Dr. Krebs: To remove any misunderstandings, I do not have the slightest doubt that mitochondria are an excellent and indispensable experimental material. But it is also essential to use mitochondria in their natural environment. I have also taken cognizance of your criticism, Brian, and since our discussion on 9th May we have carried out experiments at lower temperature and shorter periods of incubation, but the effects were essentially the same. Let me emphasize that I welcome critical comments, because criticisms and countercriticisms are absolutely essential for progress.

Dr. Lund: I would like to comment on the discrepancy between the Oxford and Bristol groups concerning the effect of leucine on glutamine synthesis from alanine. The only point on which the Chappell group and our group really disagree is on the fate of alanine. We agree that the same amounts of alanine are removed in the presence and absence of leucine, but in our experiments leucine causes the alanine amino-nitrogen not found in urea to appear as free ammonia and not as glutamate and glutamine.

Dr. Krebs: We hope to resolve the gross discrepancy of experimental observations by getting together and doing the experiments jointly. Only by paying attention to the minutiae of the experimental technique can we sort this out.

Dr. Bessman: I just want to mention something in relation to the ATP-adenine conversion to inosine. That is, that we have done a great many experiments on the ratio of inosinic acid to ATP with various conditions in muscle, which has a very large adenylicdeaminase, and there is an inverse relation between the ATP concentration and the inosinic acid concentration. It is very marked difference; the higher the ATP concentration, the lower the inosinic acid concentration *in vivo*.

22

Bicarbonate-Stimulated ATPase of Liver Mitochondria

J. Mendelson
B. Feijoo
V. Rubio
E. B. Brown
and S. Grisolia
Departments of Biochemistry and Physiology
University of Kansas Medical Center
Kansas City, Kansas

It is the purpose of this presentation to call your attention to (*a*) the presence of very active bicarbonate-dependent ATPase(s) in liver mitochondria, (*b*) the fact that bicarbonate formation, which, because it is excreted in large amounts in respiration, is still largely considered a waste product of animal metabolism, may be limiting in liver mitochondria under certain conditions, and (*c*) the possible physiological significance of the bicarbonate-dependent ATPases.

While attempting to clarify the *N*-acetyl-L-glutamate stimulation of ATPase, we found that relatively crude carbamyl phosphate synthetase preparations from rat and frog liver (1) displayed considerable bicarbonate-dependent ATPase activity in the absence of *N*-acetyl-L-glutamate (2).

It seemed to us that this bicarbonate-dependent ATPase had little to do with CPS I but that it was a contaminant possibly related or similar to the recently discovered bicarbonate-stimulated ATPase of the stomach (3), brain (4), pancreas (5), submandibular gland of rabbit and dog (6, 7), and testes (8). Interestingly, and as illustrated in Table 1, the activity of CTAB extracts of outer membrane preparations of rat and frog liver mitochondria is much higher than that described for other tissues.

Table 1 ATPase and Bicarbonate-Dependent ATPase Activities in Preparations from Stomach, Brain, and Liver

Values are given as μmoles ATP/mg protein/minute (2).

	ATPase	
Preparation	Without HCO_3^-	With HCO_3^-
Dog gastric mucosa, homogenate (9)	0.83	1.31
Dog gastric mucosa, mitochondria (3)	0.10	0.12
Dog gastric mucosa, microsome (3)	0.29	0.42
Rat brain, cerebral cortex mitochondria (4)	0.11	0.145
Rat liver, mitochondrial outer membrane (2)	1.49	3.34
Frog liver, mitochondrial outer membrane (2)	3.08	5.55

We thought that the rat and frog liver bicarbonate-stimulated ATPase might be similar to the beef heart ATPase extensively studied by Racker and coworkers (10), which is possibly involved in oxidative phosphorylation. It should be noted that Racker was first to describe in an abstract stimulation of ATPase by bicarbonate (11) but did not pursue this finding.

Although much work and marked advances have been made in the understanding of the beef heart ATPase during the last 10 years (12), much less work has been carried out with the liver enzyme. Notable exceptions are the recent reports of Catterall and Pedersen (13) and of Lambeth and Lardy (14) describing homogenous or essentially homogenous preparations of ATPase from rat liver mitochondria; interestingly, the specific activity of Lambeth and Lardy is approximately five times higher than reported by Catterall and Pedersen. It is unknown whether the extensively purified ATPase from rat liver or from beef heart mitochondria are derived entirely from the internal membrane.

As already indicated, bicarbonate-dependent ATPase activity could be detected in crude CTAB extracts of mitochondrial preparations, but the activity varied considerably from preparation to preparation. We tried another procedure. Separation of membranes from rat and from frog and chicken liver mitochondria was carried out by the use of the digitonin technique of Chan (15), which clearly demonstrated location of the bulk of ATPase activity in the outer membrane. This is illustrated in Table 2. Table 3 shows that preparations of frog liver also revealed distribution similar to that of rat liver. Although not illustrated, similar results were obtained with chicken liver. The reason for testing chicken is that these animals have no acetyl glutamate–dependent CPS.

It should be noted that, in order to compare the activities, the inner membrane and matrix fraction were also extracted with CTAB, as indicated in

Table 2 ATPase and Bicarbonate-Dependent ATPase Activities in Rat Liver Mitochondria and Fractions Thereof

Rat liver mitochondria was treated with digitonin (15). The inner membrane and matrix were then extracted with CTAB. The figures given are the mean and S.D. for eight preparations normalized to 1 g liver. Units are μmoles ATP used per minute. Protein is given in milligrams (2).

		ATPase			
		Without HCO_3^-		With HCO_3^-	
Fraction	Protein	Units	S.A.	Units	S.A.
Mitochondria	13.8	0.29	0.02 ±0.01	0.43	0.03 ±0.02
Outer membrane	1.4	2.09	1.49 ±0.14	4.68	3.34 ±0.2
Intermembrane space	3.3	0.37	0.11 ±0.17	0.55	0.17 ±0.23
Inner membrane and matrix	9.5	0.0	0.0	0.23	0.02 ±0.03

Table 3 ATPase and Bicarbonate-Dependent ATPase Activities in Frog Liver Mitochondria and Fractions Thereof

Frog liver mitochondria was extracted as per the experiments in Table 2. The figures given are the average of two preparations normalized to 1 g liver. Units are μmoles ATP used per minute. Protein is given in milligrams (2).

		ATPase			
		Without HCO_3^-		With HCO_3^-	
Fraction	Protein	Units	S.A.	Units	S.A.
Mitochondria	5.0	0.23	0.05	0.23	0.05
Outer membrane	0.4	1.23	3.08	2.22	5.55
Intermembrane space	4.4	0.60	0.14	0.60	0.14
Inner membrane and matrix	2.6	0.45	0.17	0.45	0.17

Table 2. However, when using Lubrol, there was higher total activity of the starting mitochondrial preparations and activation of the internal membrane fraction. Indeed, after Lubrol activation, the ATPase was proportionally higher in the inner membrane than in the outer membrane. (This is briefly illustrated in Table 6.) Thus, it is possible that the apparent location of ATPase shown in Tables 1 and 2 reflects contamination and artifacts due to the CTAB extraction or the existence of two ATPases. A study with a number of enzyme markers, for example, cytochrome oxidase, monamine oxidase, revealed a

small percent of possible contamination of our outer membrane preparations with inner membrane. Unfortunately, all the methods to separate membranes extract different amounts of proteins, depending on concentration and conditions. Therefore, it is not possible at this time to decide on the basis of enzyme markers whether or not we are dealing with contamination or two different ATPases.

E. Santiago (personal communication) has recently studied the bicarbonate-dependent ATPase in rat liver mitochondria. He found, as we did, that the enzyme is not a Na^+—K^+-ATPase. However, he seems to favor location of the bicarbonate-stimulated ATPase, on the basis of markers, entirely on the inner membrane. It should be noted that Dr. Santiago and coworkers used a different method than we did for the preparation of membranes.

There is evidence other than distribution that indicates differences between the ATPase preparations of Lambeth and Lardy (14) and the ATPase obtained by CTAB extraction of the outer membrane. As illustrated in Table 4, the maximum activation that we obtain with bicarbonate is 100%, but the activation noted by other workers (14, 16) is 300 to 600%. The K_m for the preparation used by Ebel, possibly mixed or largely derived from inner membrane, is 5.8 m*M*, but our value is greater than 20 m*M*. Again, as illustrated in Table 4, our preparations are not stimulated by chromate, and chromate abolishes entirely the bicarbonate stimulation of our ATPase preparations, although it has negligible effect on the bicarbonate stimulation of Lambeth and Lardy's preparation. Also, our preparations are much more labile.

Table 4 Effect of Bicarbonate and Chromate on Purified Mitochondrial and Outer Membrane Mitochondrial ATPase(s)

	Purified		Outer Membrane[c]	
Chromate Added (m*M*)	Bicarbonate Added − (% activity) +		Bicarbonate Added − (% activity) +	
	−	+	−	+
0	100	310[a], 600[b]	100	150
3	235[a]	250[a]	100	100
K_M bicarbonate	5.8 m*M*		>20 m*M*	

[a] Lambeth and Lardy (14).
[b] Ebel (16).
[c] Feijoo, Rubio, Mendelson, and Grisolia, unpublished.

Obviously, much more work is necessary to establish without a doubt whether there are ATPases in both the internal and the external membrane. Interestingly and with the exception of a brief comparative study by Lambeth and Lardy (14), little has been done with the bicarbonate stimulation of the enzyme from heart mitochondria.

Independently of the existence of one or two bicarbonate-stimulated ATPases, their existence in mitochondria is of much interest, since the bulk of CO_2 in animals arises in the mitochondria during metabolism and is subsequently exhaled. Although possibly in most tissues CO_2 is in excess, it may not be true for the ureotelic liver. As previously indicated (2) and as illustrated in Table 5, the CO_2 produced by the liver can be calculated; since man produces from ~0.3 to 1.4 mmole urea per minute, it is self-evident that CO_2 may become limiting, or nearly so, if there is also extensive CO_2 use for anaplerosis.

Table 5 Maximum Production of CO_2 and Urea by Liver

1. Liver of man uses ~15% of the O_2 needed for basal metabolism. (This figure does not essentially change with exercise.)
2. ∴ for the 70 kg average man, liver consumes 1.7 mmole O_2/minute.
3. If R.Q. = 1 (pure carbohydrate diet), then 1.7 mmole CO_2 would be formed (fat diet, 1.2 mmole).
4. Man produces ~0.3 to 1.4 mmole urea/minute.
5. ∴ CO_2 may become limiting, or nearly so, if there is also extreme CO_2 use for anaplerosis.

Although many schemes have been proposed for ATPases, particularly in transport, on the formation of protein-bound carboxyl phosphate during ATPase, and the possible relation of these ATPases to acyl phosphatases (17–22), their physiological role remains unclear.

Although it is difficult to assess the maximum activity of beef heart ATPase from the literature, it seems too low, unless extensively activated, to support maximum rates of oxidative phosphorylation. Certainly, it does not seem to be in great excess. As illustrated in Table 6, there is much lower activity for heart than for liver. However, the activity of the beef heart enzyme, to our knowledge, is not measured routinely in the presence of bicarbonate, which may, as initially shown by Racker (11), increase the activity 40 to 50%. On the other hand, the oxygen consumption of heart on a gram basis and for most animals is approximately three to five times higher than that of liver. Thus, at rest, the ATPase of heart will be sufficiently high to support its alleged role in oxidative phosphorylation, but that of liver is approximately 10 times higher than needed. Acknowledgingly, the activity is probably larger (alkaline activation, DNP, etc.) if these activities can be extrapolated to *in vivo*. On the other hand, the ATPase needs (in the context of oxidative phosphorylation)

Table 6 Comparative ATPase Activities and ATP Production by Heart and Liver

	ATPase				
	Beef Heart[a] (pH 7.4; 30°)		Rat Liver[b] (pH 8; 30°; with bicarbonate)	Rat Liver[c] (pH 7.4; 37°; with bicarbonate)	
	Heavy Mitochondria	Light Mitochondria		Outer Membrane	Inner Membrane
Activity/mg mitochondria	0.12	1.6	3.9	0.3	2
Total activity/g tissue					
Actual	2.52		42	23	
Calculated	25.2		158	92	
	O_2 [d]uptake μl/g/minute			ATP production μmole/g/minute	
Heart	98			26	
Liver	27			7	

All figures for ATPase are given in μmoles of ATP used per minute.

[a] The calculations for beef heart ATPase are based on the data of Pullman et al. (23) for heavy mitochondria and on that of Knowles and Penefsky (24) for light mitochondria. The yield is assumed to be 1 mg and 1.5 mg of mitochondrial protein per g of wet tissue for heavy and light mitochondria, respectively, according to Green and Ziegler (25). The total activity is calculated assuming that the mitochondrial protein per g of heart is 25 mg.

[b] The calculations for rat liver content are based on the data of Lambeth and Lardy (14) and on the assumption of 12.5 g per rat liver and 40 mg of mitochondrial protein per g of liver.

[c] The mitochondrial protein per g of liver is assumed to be 40 mg. The activity without Lubrol is negligible in the inner membrane. Lubrol does not affect the activity of the outer membrane. However, the latter shows ~30% more activity when extracted with CTAB.

[d] Data from R. Havel (26). The ATP production has been calculated assuming a P:O ratio of 3.

will increase nearly six times for the heart under maximum exercise but it does not affect, or barely so, the liver.

There is another important point which we would like to discuss briefly. The bicarbonate concentration in tissues is lower than that in the extracellular fluids (27), and these differences cannot be explained entirely by pH. Therefore, there must be a gradient which may involve energy. It is well recognized that intracellular HCO_3^- concentration, $[HCO_3^-]_i$, is lower than extracellular HCO_3^- concentration, $[HCO_3^-]_e$, by a factor of 2.5 to 3.0. Assuming that all the intracellular HCO_3^- is unbound, intracellular pH would be 6.9 to 7.0. This is the value found by distribution of the weak acid 5,5-dimethyl-2,4-oxazolidinedione (DMO) in skeletal muscle. Resting transmembrane electrical potential in this tissue is $\simeq$90 mV. If HCO_3^- were passively distributed, the $[HCO_3^-]_e/[HCO_3^-]_i$ would be 30 rather than 3. Therefore some active process is involved in maintaining $[HCO_3^-]_i$ at $\simeq$10 mM rather than 1 mM. It has been argued that much of the $[HCO_3^-]_i$ in skeletal muscle is bound (28), and intracellular pH is actually 5.9 to 6.0. More recently Carter and his associates (29) have obtained values in this range by direct measurement with microelectrodes. They have suggested that the value of 6.9 to 7.0 obtained by the DMO technique is a mean of values from several compartments in a multicompartment cell. Addanki (30) has reported pH values of 8.0 in maximally respiring mitochondria. If the fluid bathing mitochondria has a pH of 7.0 and an HCO_3^- of 10 mM, intramitochondrial HCO_3^- would be 100 mM if the P_{CO_2} is the same in mitochondrial fluid as it is in extracellular fluid.

Liver cells apparently have a higher $[HCO_3^-]_i$ and a lower transmembrane potential than muscle cells. Values of $\simeq$14 mM and 45 mV respectively have been reported (31). If HCO_3^- were distributed passively across liver cell membranes, intracellular concentration would be 4.5 to 5.0 mM rather than 14 mM. Therefore, although the discrepancy is not so great for liver cells as it is for muscle cells, if the values we have quoted are accurate, an active process must also be involved to account for the high $[HCO_3^-]_i$ in liver cells.

As pointed out very recently (32), the transport of bicarbonate across the mitochondrial membrane is not fully understood as yet, although it has been reported that bicarbonate may be transported actively during oxidative phosphorylation; it is generally believed to enter the membrane as dissolved CO_2 which is then presumably rehydrated in the matrix to yield the proton-bearing bicarbonate ion (33). It is then of much interest that Tavill and coworkers (34), by pulse injection of ^{14}C-carbonate, recently found evidence for the existence of two separate arginine pools indicating different bicarbonate pools.

It is known that in addition to nucleotide pumps liver mitochondria have other pumps, including an ornithine pump, which may largely control citrulline synthesis (35). Whether there are interrelations between these pumps, including the possibility of a (or more) bicarbonate-stimulated ATPase type of pump, as postulated for the stomach enzyme (3), remains to be clarified.

In conclusion, it should be noted that, until the 1940s, carbon dioxide fixation by animals had been investigated to only a very limited extent, although, in 1932, Krebs and Henseleit (36) offered proof of the participation of carbon dioxide in the formation of urea. However, the significance of this conversion was largely overlooked. It appears to us from the literature (29) and from discussion with colleagues that there are, more than likely, multiple, different, and possibly extreme pH's within the cell and probably within the mitochondria, making the understanding of the role of the bicarbonate-dependent mitochondrial ATPases very important.

ACKNOWLEDGMENT

This work was supported in part by Grant AM-01855 from the National Institutes of Health. We would like to thank Dr. R. L. Clancy, of the Department of Physiology, for his advice.

REFERENCES

1. Chabas, A., Grisolia, S., and Silverstein, R. (1972). *Eur. J. Biochem.* **29**, 333.
2. Grisolia S., and Mendelson, J. (1974). *Biochem. Biophys. Res. Commun.* **58**, 968.
3. Blum, A. L., Shah, G., St. Pierre, T., Helander, H. F., Sung, C. P., Wiebelhaus, V. D., and Sachs, G. (1971). *Biochim. Biophys. Acta* **249**, 101.
4. Kimelberg H. K., and Bourke, R. S. (1973). *J. Neurochem.* **20**, 347.
5. Simon B., and Thomas, L. (1972). *Biochim. Biophys. Acta* **288**, 434.
6. Simon, B., Kinne, R., and Knauf, H. (1972). *Pflugers Arch.* **337**, 177.
7. Izutsu K. T., and Siegel, I. A. (1972). *Biochim. Biophys. Acta,* **284**, 478.
8. Setchell, B. P., Smith, M. W., and Munn, E. A. (1972). *J. Reprod. Fert.* **28**, 413.
9. Spenney, J. G., Strych, A., Price, A. H., Helander, H. F., and Sachs, G. (1973). *Biochim. Biophys. Acta* **311**, 545.
10. Racker E. (1970). "Membranes of Mitochondria and Chloroplasts," *ACS Monograph*, **165**.
11. Racker, E. (1962). *Fed. Proc.* **21**, 54.
12. Senior, A. E. (1973). *Biochim. Biophys. Acta* **301**, 249.
13. Catterall, W. A., and Pedersen, P. L. (1971). *J. Biol. Chem.* **246**, 4987.
14. Lambeth D. O., and Lardy, H. A. (1971). *Eur. J. Biochem.* **22**, 355.
15. Chan, T. L., Greenawalt, J. W., and Pedersen, P. L. (1970). *J. Cell. Biol.* **45**, 291.
16. Ebel, R. E. (1974). *Fed. Proc.* **33**, 1399.
17. Sachs, G., Rose, J. D., and Hirschowitz, B. I. (1967). *Arch. Biochem. Biophys.* **119**, 277.
18. De Meis, I. (1969). *J. Biol. Chem.* **244**, 3733.
19. Formly, B., and Clausen, J. (1969). *Hoppe-Seylers Z. Physiol. Chem.* **350**, 973.
20. Hokin, L. E., Sastry, P. S., Galsworthy, P. R., and Yoda, H. (1965). *Proc. Natl. Acad. Sci.* **54**, 177.

21. Bader, H., and Sen, A. R. (1966). *Biochim. Biophys. Acta* **118**, 116.

22. Kahlenberg, A., Galsworthy, P. R., and Hokin, L. E. (1967). *Science* **157**, 434.

23. Pullman, M. E., Penefsky, H. S., Datta, A., and Racker, E. (1960). *J. Biol. Chem.* **235**, 3322.

24. Knowles, A. F., and Penefsky, H. S. (1972). *J. Biol. Chem.* **247**, 6617.

25. Green, D. E., and Ziegler, D. M. (1963). In S. P. Colowick and N. O. Kaplan, Eds., *Methods in Enzymology*, **6**, 416, Academic, New York.

26. Havel, R., (1970). Quoted by R. W. McGilvery, *Biochemistry: A Functional Approach*, Saunders, Philadelphia, p. 532.

27. Gambles, J. L., (1958). *Chemical Anatomy, Physiology and Pathology of Extracellular Fluid*, Harvard University Press, Cambridge, Mass.

28. Conway, E. J., and Fearon, P. J. (1944). *J. Physiol.* **103**, 274.

29. Carter, N. W., (1972). *Kidney Int.* **1**, 341.

30. Addanki, S., Cahill, F. D., and Sotos, J. F. (1968). *J. Biol. Chem.* **243**, 2337.

31. Williams, J. A., Withrow, C. D., and Woodbury, D. M. (1971). *J. Physiol.* **214**, 539.

32. Wozniak, M., Ciesielski, D., Popinigis, J., and Zydowo, M. (1973). *Physiol. Chem. Phys.* **5**, 237.

33. Elder, J. A. (1972). *Fed. Proc.* **31**, 856.

34. Tavill, A. S., East, A. G., Black, E. G., Nadkarni, D., and Hoffenberg, R. (1972). In *Protein Turnover*, Ciba Foundation Symposium, **9**, 155, Assoc. Scientific Publ., Amsterdam, 1972.

35. Gamble, J. G., and Lehninger, A. L. (1972). In G. F. Azzone et al., Eds., *Biochem. Biophys. of Mitochondrial Membranes*, Academic, New York, p. 611.

36. Krebs, H. A., and Henseleit, K. (1932). *Z. Physiol. Chem.* **210**, 33.

23

Hereditary Urea-Cycle Disorders

Vivian E. Shih
Department of Neurology
Harvard Medical School
Boston, Massachusetts

INTRODUCTION

In the past 20 years, increasing awareness of biochemical causes of mental retardation and the clinical application of chromatographic techniques for amino acids have resulted in the discovery of amino acid metabolic disorders at an exponential rate. Among these are disorders in the urea cycle. Significant events in the history of urea-cycle disorders are listed in Table 1. Argininosuccinic aciduria (ASAciduria) was the first one discovered in 1958. Four years later congenital hyperammonemia, due to ornithine transcarbamylase (OTC) deficiency, and citrullinemia were recognized. In 1964, another cause of congenital hyperammonemia, carbamyl phosphate synthetase (CPS) deficiency, was reported. Hyperargininemia was the latest one, discovered in 1969.

Inborn errors of urea-cycle metabolism are a group of fascinating disorders of clinical importance and biochemical significance. They not only cause mental retardation but also, among several other amino acid disorders, cause fulminating diseases in the neonatal period and early infancy. Recently, promising results of dietary treatment have been obtained in a limited number of patients who were detected and treated early.

From a biochemical point of view, these errors of metabolism are experiments of nature offering rare opportunities to study the urea cycle. Investigation of these patients has raised many questions concerning the regulation of urea synthesis and ammonia metabolism and the relationship of the urea cycle

Table 1 Significant Events in the History of Urea-Cycle Disorders

Carbamyl Phosphate Synthetase Deficiency

1964	First case report with the demonstration of enzyme defect by Freeman et al. (5).
1974	Recognition of the lethal neonatal form (Gelehrter and Snodgrass) (41).
1975	Promising results by treatment with α-keto analogs of essential amino acids (Batshaw et al.) (121).

Ornithine Carbamyl Transferase Deficiency

1962	First report of two patients with the demonstration of hepatic enzyme deficiency (Russell et al.) (9).
1969	Discovery of increased orotic acid production (Levin et al.) (56).
1969	Suggestive evidence for sex-linked inheritance (Levin et al.) (37).
1969	Discovery of an abnormal enzyme with altered pH dependency (Levin et al.) (137).
1970	Detailed studies of neuropathology showing severe loss of cerebral cortex and widespread formation of Alzheimer type II astrocytes (Bruton et al.) (33).
1971	Recognition of the lethal neonatal variant (Rosenberg et al.) (150).
1971	The finding of an abnormal K_m for carbamyl phosphate (Matsuda et al.) (74) and an abnormal electrolytic pattern of the enzyme (Arashima and Matsuda) (75).
1972	Detection of a mutant enzyme with a high K_m for ornithine (Cathelineau et al.) (73).
1973	Evidence for X-linked dominant inheritance (Short et al.) (13).

Argininosuccinate Synthetase Deficiency (Citrullinemia)

1962	First case report (McMurray et al.) (15).
1964	Demonstration of enzyme defect in liver (McMurray et al.) (100).
1967	Demonstration of abnormal K_m of the enzyme in cultured skin fibroblasts (Tedesco and Mellman) (64).
1971	Recognition of lethal neonatal variant (Wick et al., Van der Zee et al.) (162).
1971	Demonstration of enzyme deficiency in liver but not in kidney (Vidailhet et al.) (158).
1973	Demonstration of enzyme deficiency in cultured lymphoblasts (Spector and Bloom) (79).

Argininosuccinase Deficiency (Argininosuccinic aciduria)

1958	First report of two siblings excreting an unidentified amino acid (Allan et al.) (118).
1960	Identification of argininosuccinic acid in urine (Westall) (161).
1964	Demonstration of enzyme deficiency in erythrocytes (Tomlinson and Westall) (80).
1967	Recognition of the importance and the possible role of hyperammonemia in the pathogenesis of neurological symptoms (Moser et al.) (48).

Table 1 (*Contd.*)

Argininosuccinase Deficiency (Argininosuccinic aciduria)	
1968	Recognition of lethal neonatal variant (Baumgartner et al., Levin and Dobbs) (122, 136).
1968	Demonstration of enzyme deficiency in brain (Kint and Carton) (81).
1968	Demonstration of normal enzyme activity in the kidney of a patient with the neonatal variant (Colombo and Baumgartner) (83).
1969	Demonstration of enzyme deficiency in cultured skin fibroblasts (Shih et al.) (65).
1969	Neuropathological findings of deficit in myelin formation and the presence of Alzheimer type II astrocytes (Solitaire et al.) (34).
1972	Preliminary report of the effectiveness of early dietary treatment (Shih) (108).
1973	Prenatal diagnosis (Goodman et al.) (67).
1975	Demonstration of enzyme deficiency in liver but not in brain or kidney in the neonatal variant (Glick et al.) (55).
Arginase Deficiency (Hyperargininemia)	
1969	First report of two siblings with the demonstration of enzyme deficiency in erythrocytes (Terheggan et al.) (61).
1973	Attempted enzyme replacement therapy without success (Colombo et al.) (127).

to the metabolism of other compounds. The most intriguing findings are the normal blood urea nitrogen concentrations and the substantial amounts of urea excreted in the urine by patients with a block in the urea cycle. Hyperammonemia not only is a constant finding in urea-cycle disorders but also develops in several other enzyme deficiencies (Table 2) through yet unclarified mechanisms. The importance of ammonia in the pathogenesis of hepatic encephalopathy in adult patients with chronic liver insufficiency or portacaval shunt has long been controversial, but not until the discovery of congenital

Table 2 Hereditary Metabolic Disorders Associated with Hyperammonemia

Urea-cycle disorders
Syndrome of hyperornithinemia, hyperammonemia, and homocitrullinuria
Lysinuric protein intolerance (LPI)
Congenital lysine intolerance
Hyperlysinemia with lysine-induced crisis
Hyperammonemia and cerebroatrophic syndrome of Rett
Organic acidurias (secondary)

hyperammonemic syndromes was the role of ammonia toxicity fully recognized.

There have been two extensive reviews on the urea-cycle disorders, Colombo, (1) and Levin, (2).

Today, our discussion will be focused on the five disorders of the urea cycle. Dr. Colombo will discuss hyperargininemia and possible alternate pathways of urea synthesis. This will be followed by Dr. Walser's comments on his experiences with keto acid therapy.

INCIDENCE

Urea-cycle disorders as a group are relatively common amino acid disorders. A total of 100 cases has been reported. ASAciduria and OTC deficiency are more common than the others (Table 3). The true incidence of these disorders in the general population is not known for several reasons. Most of the screening has been done in selected populations, particularly among the mentally retarded. Any mildly affected cases may have been missed. Routine newborn screening by chromatographic studies of urine amino acids for disorders other than phenylketonuria has only been a recent practice (3). Since the congenital hyperammonemias have no specific amino acid abnormalities other than a possible increase in the excretion of glutamine and alanine, only citrullinemia, ASAciduria, and hyperargininemia are detectable by current urine screening. The incidence of ASAciduria is 1 in 90,000 newborns in Massachusetts (4). No cases of citrullinemia or hyperargininemia have been found in such a screening. Undoubtedly this incidence is too conservative; the lethal neonatal variants of these disorders cannot be detected, because the screening is performed between ages 3 and 6 weeks.

The performance of blood ammonia determinations is the only dependable method of screening for CPS and OTC deficiency. In the past few years, clinicians have become increasingly aware of the importance of hyperammonemia in the pathogenesis of severe neurological symptoms, and this is reflected in the fact that the majority of OTC deficiency cases have been reported since 1970.

CLINICAL FEATURES AND GENETICS

Urea-cycle disorders are characterized by symptoms of protein intolerance. However, there is a wide spectrum of the severity of clinical manifestations. The most severely affected patients die in the first few days of life, whereas the mildly affected ones survive to adulthood with few problems other than aversion to certain foods or occasional headaches. Episodes of vomiting and

Table 3 Relative Incidences and Outcome of Urea-Cycle Disorders

Disorder	Enzyme Defect	Total Families	Total Cases	Living	Dead	Unknown
Congenital hyperammonemia type I	CPS	6	6	3	3	
Congenital hyperammonemia type II	OTC	28	33	13	17	3
Citrullinemia	ASA Synthetase	11	12	5	7	
Argininosuccinic aciduria	ASase	34	45 (46)[a]	39	6	
Hyperargininemia	Arginase	2	4	4	0	
Total		81	100 (101)[a]	64	34	3

[a] Including one aborted fetus.

lethargy often follow infections, metabolic stresses, or an increase in protein intake. These symptoms usually respond to intravenous fluid administration and omission of protein food but may progress to irreversible coma and death. The severity of these disorders is self-evident in that over half of the patients have died (Table 3). The majority of the patients would become mentally retarded if undiagnosed and untreated.

Genetic heterogeneity in each disorder may be expected in view of the varied phenotype and the several kinds of enzyme abnormalities.

Carbamyl Phosphate Synthetase Deficiency

Clinical Features CPS deficiency, also known as congenital hyperammonemia type I, was first described by Freeman et al. (5). Subsequently five more patients with CPS deficiency have been reported. Clinically these infants all had a history of intolerance to milk feeding with early onset of poor feeding, recurrent vomiting, lethargy, stupor, and seizures (Table 4). Three patients died in infancy. One of them (Case 5) had an onset in the neonatal period with poor feeding, irritability, and increasing hypertonicity. He then developed respiratory distress and coma and died at 75 hours of age. Of the remaining three patients, one was severely retarded at 13 years of age (Case 6), one was normal at 15 months of age (Case 4), and one had cortical atrophy at 2 months of age (Case 3).

The clinical as well as laboratory findings varied considerably, so much so that it is uncertain whether all cases had a primary defect in CPS. In fact, an additional patient, first reported as CPS deficiency (50%), was later found to have a primary defect in methylmalonate metabolism ((6); Kirkman, personal communication). The metabolic acidosis, neutropenia, and mild hyperglycinemia in Case 1 are reminiscent of what has been seen in the ketotic hyperglycinemia syndrome associated with organic acidemias. Methylmalonic acid was not found in the urine of this patient, but the presence of other organic acids was not ruled out. Abnormal glucose metabolism in Cases 2 and 3 indicate complex metabolic problems in these patients.

A reduction of CPS activity has also been observed in a patient with migraine syndrome (7) and a patient with Reye's syndrome (8).

Genetics CPS deficiency has been reported in both male and female patients. The small number of cases and the lack of familial occurrence do not allow any genetic analysis of the inheritance.

Ornithine Transcarbamylase Deficiency

Clinical Features OTC deficiency, also known as congenital hyperammonemia type II, was first described in 1962 by Russell et al. (9). Thirty-three cases in 28 families have now been reported (Table 5). There is a definite

Table 4 Summary of Patients with Carbamyl Phosphate Synthetase Deficiency

Case No.	Reference	Sex	Age Onset	Presenting Symptoms	Outcome	Blood Ammonia	Other Laboratory Findings	Enzyme Defect (% of control values) Liver	Other Tissues
1	Freeman et al. (5, 39)	F	10 d	Vomiting, lethargy	Died at 5 m	↑↑	Metabolic acidosis, hyperglycinemia, neutropenia	22	—
2	Hommes et al. (134)	F	<20 d	Poor feeding, lethargy, convulsion, papilledema	Treated but died at 7-1/2 m	N	↓ blood arginine and urea; lack of insulin release after amino acid activation	40	Brain:6
3	Arashima and Matsuda (119)	F	7 d	Persistent vomiting; hypotonia	Treated and well at 15 m	↑	↓ blood urea and orotic acid; abnormal glucose tolerance	13	—
4	Odievre et al. (145)	M	6 wk	Vomiting, lethargy	Treated and IQ 97 at 15 m	↑↑	↑ blood lysine, alanine and urea; generalized amino aciduria	Not detectable	—
5	Gelehrter and Snodgrass (41)	M	2 d	Poor feeding, hypothermia, irritability	Comatous and died at 4 d	↑↑↑	Generalized hyperamino-acidemia	0	—
6	Batshaw et al. (121)	F	3 wk	Vomiting, lethargy	IQ 13 and hemiparetic at 13 yr; treated	↑↑	Normal urine, orotic acid	<15%	—

+ present, 0 absent, — not mentioned, N normal, ↑ elevated, ↓ decreased, SGOT serum glutamate oxaloacetate transaminase, EEG electro-encephalogram, p.c. postprandial.

Table 5 Summary of Patients with OTC Deficiency[a]

Genotype	Family	Case no.	Reference	Sex	Age onset (yr)	Outcome	Liver enzyme activity (% of control values) pH 7	pH 8	K_m (ORN)	K_m (CP)
X-linked dominant	1	1	Levin et al. (9)	F	5 wk	Died at 6	—	5	—	
	2	2	Levin et al. (9)	F	2	Died at 6	—	7	—	
		3 (twin)	Levin and Russell (110)	F	9	Living	—	—		
	3	4	Levin et al. (37)	F	3 wk	Severely retarded	8	20	—	
	4	5	Campbell et al. (11, 12)	M	3 d	Died on 5 d	0.4	0.4	N	
		6	Campbell et al. (11, 12)	M	2 d	Died on 9 d	—	0.1		
	5	7	Short et al. (13)	F	14 m	I.Q. normal at 5 yr	—	23		
		8	Short et al. (13)	M	2 d	Died on 2 d	—	0		
	6	9	Cathelineau et al. (163, 164)	F	1 m	Recurrent vomiting and retardation at 8 m	12	50		4x
	7	10	Salle et al. (151)	F	3 wk	Mental retardation, recurrent vomiting at 15 m	10	39		
X-linked dominant (suspected)	8	11	Palmer et al. (14)	M	Neonatal	—	0.8	30		
	9	12	Scott et al. (154)	M	Neonatal	Died on 2 d	—	0		
	10	13	Scott et al. (154)	M	Neonatal	Died in neonatal period	—	0		
	11	14	Kang et al. (115)	M	3 d	Died on 5 d	—	0.2		
		15	Kang et al. (115)	M	2 d	Died on 5 d	—	0.2		
	12	16	Palmer et al. (14); Levin (2)	F	—	—	6	16		
		17	Palmer et al. (14); Levin (2)	M	—	Died	0.7	38		
	13	18	Cathelineau et al. (164); Saudubray et al. (152)	M	2 d	Died on 11 d	—	2	100x	
	[illegible]	[illegible]	[illegible] (133)	M	7 d	Died on 25 d	<1	<1		

Variant	15	20	Levin et al. (137)	M	6 m	Normal	25	N		
	16	21	McLeod et al. (140)	M	4 wk	Died at 4 m	—	5		
	17	22	Cathelineau et al. (164); Saudubray et al. (152)	M	8	Irreversible coma and died at 8	—	10	N	
Unknown	18	23	Hopkins et al. (36)	F	7 m	Died at 18 m	—	<4		
	19	24	Sunshine et al. (111)	F	3-1/2	Mentally retarded	—	6		
	20	25	Corbell et al. (128)	F	14 d	Mentally retarded	—	35		
	21	26	Nagayama et al. (76)	F	5 m	Recurrent vomiting, mental retardation at 8 m	14	10		
	22	27	Short et al. (13)	F	Early infancy	Mildly retarded at 18 m	—	19		
	23	28	Short et al. (13)	F	Neonatal	Recurrent vomiting, normal development at 14 m	—	26		
	24	29	Matsuda et al. (74)	F	8-1/2 m	Normal development, died during a vomiting attack at 10 m	25	70		4x
	25	30	Palmer et al. (14)	F	—	Normal at 4 yr	10	11		
	26	31	Palmer et al. (14)	F	5	Died shortly after onset at 5 yr	1.7	22		
	27	32	Palmer et al. (14)	F	—	Diagnosed at 3 yr	22	33		
	28	33	Krebs et al. (135)	F	—	Protein intolerence, retardation at 13 m	—	2		

[a] See symbols and abbreviations same as Table 4.

difference in the severity of the disease between male and female. Ten of the 14 male patients had an onset of the disorder shortly after birth and died in the neonatal period. On the other hand only 5 of the 19 female patients died, none in the neonatal period. The majority of the living patients are grossly retarded in mental and physical development.

The affected male infant is normal at birth and does well for the first 1 or 2 days, when lethargy and poor feeding are observed. This is followed by grunting respiration, altered muscle tone, convulsions, and sometimes hypothermia. Rapid progression to unresponsiveness and respiratory arrest leads to death in the first week of life. These symptoms are similar to those of the neonatal variants of CPS deficiency, citrullinemia, and ASAciduria, nor are they distinguishable from the symptoms of other "acute" metabolic disorders, such as maple syrup urine disease, methylmalonic acidemia, propionic acidemia, etc. (10). Three exceptional male patients who had either a subacute course (Case 21) or a mild course (Cases 20 and 22) are considered to have variants of the disease.

There is a wide spectrum of clinical manifestations in female patients, from a mere dislike of protein food to recurrent hyperammonemic episodes resulting in death in childhood. None of the female patients had a lethal neonatal course such as described above. The onset of recurrent vomiting and feeding difficulties may be in early infancy or as late as 2 or 3 years. Mental and physical retardation or developmental regression gradually becomes noticeable.

The mothers of the affected children often have intolerance to protein as well; these symptoms may intensify during pregnancies which result in an affected infant (7).

In some cases of OTC deficiency, the intense episodic vomiting accompanied by headache, ptosis, and ataxia reminds one of the symptoms of the migraine syndrome (cyclic vomiting in childhood). Russell astutely recognized this and proceeded to test ammonia and protein tolerance in eight well-defined cases of childhood migraine with intense vomiting and lethargy justifying a description as cyclic vomiting (7). Seven were found to have abnormally high fasting blood ammonia levels. Six had abnormal ammonia and protein tolerance tests. Of four cyclic vomiters subjected to enzyme assay, three were proven to be heterozygotes for OTC deficiency.

Genetics The group at Yale was the first one to recognize an X-linked dominant inheritance after studying four kindreds, each containing one or more affected children (11, 12, 13). Subsequent reports (Cases 9 to 19) and a comprehensive review of 20 families by Palmer et al. (14) confirm their findings. The main evidence for X-linked dominant inheritance is (1) the severe nature of the disorder in males, almost invariably resulting in neonatal death; (2) almost complete absence of enzyme activity in all affected males;

(3) a wide variation in the clinical manifestations as well as in the degree of enzyme deficiency in female patients consistent with the Lyon hypothesis of random inactivation of the X chromosome; (4) history of high incidence of neonatal death, from symptoms suggestive of hyperammonemia, among the male sibs or male relatives on the maternal side; (5) the mother is frequently affected, showing either protein intolerance or partial enzyme deficiency; (6) the father usually is asymptomatic, with normal enzyme activity and an unremarkable family history. To illustrate the point, a figure giving detailed information about one family studied by Short et al. (13) is reproduced here (Fig. 1). The majority of the families reported in the literature fit into this mode of inheritance.

In those families where there is only one affected patient and the parents were either normal or not studied, the mode of inheritance cannot be determined (Cases 20 to 33). Of 14 such patients, only 3 are male. Case 20 had a mild course and was considered normal at 2 years while on dietary treatment. Case 21 had an onset of the disease in the second month of life and died in infancy. He retained about 5% of the normal OTC activity in liver and is therefore different from the lethal neonatal type. Case 22 was asymptomatic until 8 years of age, when he died in a hyperammonemic attack. It appears that all these patients are genetically distinct from the X-linked mutant.

Citrullinemia (Argininosuccinate Synthetase Deficiency)

Clinical Features Citrullinemia was first detected by McMurray et al. (15) during a chromatographic screening program of a mentally retarded population. Twelve cases have now been reported and can be tentatively divided into three types according to the clinical symptoms: the lethal neonatal type, the subacute type, and the mild type (Table 6). The symptoms of the lethal neonatal type are the same as those described for OTC deficiency. Treatment does not usually alter the course.

The onset of the subacute type is gradual, with the typical history of poor feeding, recurrent vomiting, and failure to thrive in early infancy. Four of the five patients are mentally retarded and had seizures at the time of diagnosis. Hepatomegaly has been found in two patients. Only one patient (Case 12) has been described with the mild type. He was discovered by routine screening at 3 weeks of age. He was asymptomatic at the time of diagnosis and remained so at 4 years without treatment (Wick, personal communication).

Genetics Because of the small number of patients reported, the mode of inheritance is not readily apparent. Of the 11 cases whose sex is known, there is a male predominance (M/F = 8/3). Four of the six patients with the neonatal variant are male and one was female. Parents are clinically asymptomatic. The activity of ASA synthetase in the skin fibroblasts from five

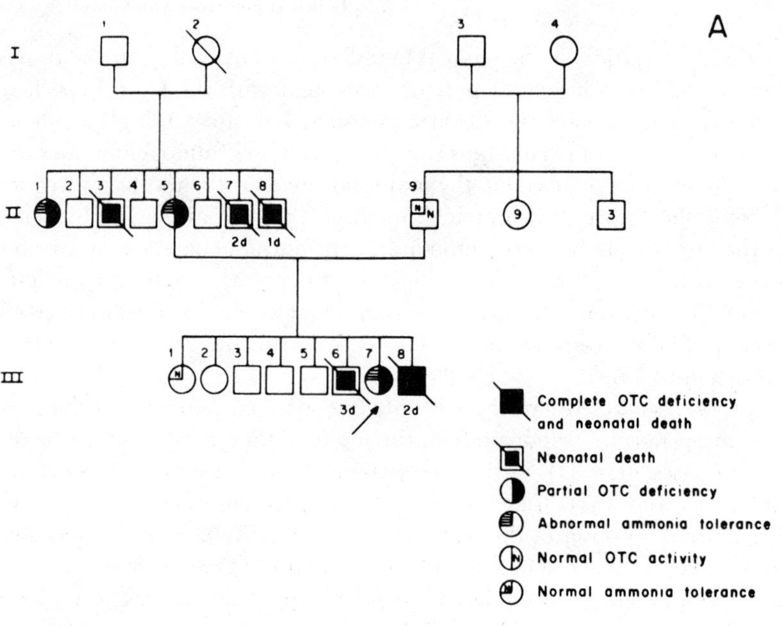

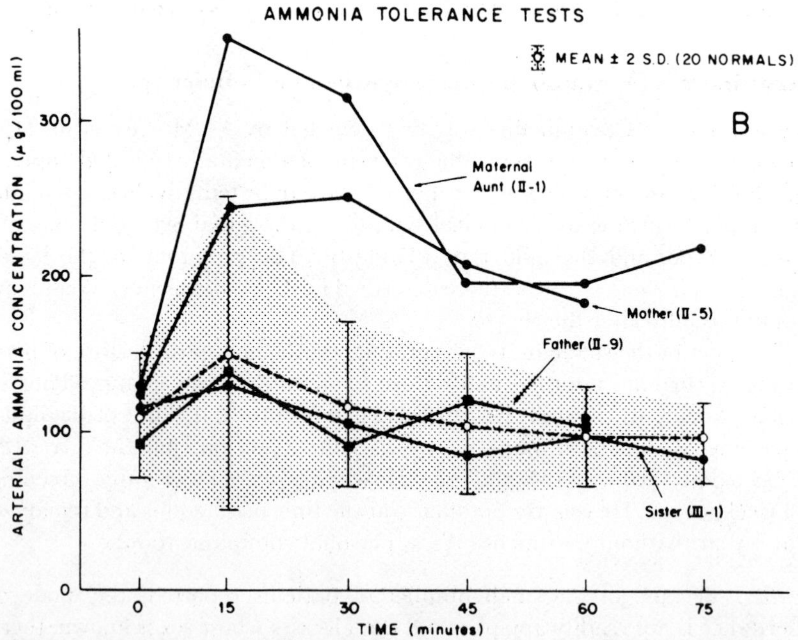

Figure 1. Pedigree of a family with OTC deficiency illustrating the X-linked dominant inheritance; OTC activity was measured in liver biopsy specimens. Oral ammonia tolerance test results (ammonium chloride, 44 mg/kg body weight; maximum dose, 3.0 g) are shown. (From Short et al., *N. Engl. J. Med.* **288**: 7, 1973.)

Table 6 Summary of Citrullinemic (Argininosuccinate Synthetase Deficiency) Patients[a]

Type	Case No.	Reference	Sex	Age Onset	Presenting Symptoms	Outcome	Blood Ammonia	Other Laboratory Findings	Enzyme Defect (% of control values) Liver	Fibroblasts
Neonatal	1	Wick et al. (17)	M	3 d	Tachypnea, refusal of feeding, muscle spasm	Coma and death on sixth day	↑↑↑	Generalized hyperamino-acidemia	20	—
	2	Wick et al. (17)	M	3 d	Same as brother (Case 1)	Coma and death on fifth day	↑↑↑	Generalized hyperamino-acidemia	14	40
	3	Van der Zee et al. (157)	F	4 d	Lethargy, poor sucking, convulsions	Death on seventh day	—	Generalized hyperamino-acidemia	0	—
	4	Ghisolfi et al. (132)	M	2 d	Hypertonicity, respiratory difficulties	Death on sixth day	N	↑↑↑ in blood glutamine, alanine, lysine ↓blood arginine ↑ in urine homocitrulline	—	—
	5	Okken et al. (165)	M	—	Coma, Opisthotonus	Treated, well at 45 days	↑↑	—	—	—
	6	Roedink et al. (84, 146)	M	67 hr	Irregular and grunting respiration, seizures	Coma and death on seventh day	N	—	0	—

Table 6 (*Contd.*)

Type	Case No.	Reference	Sex	Age Onset	Presenting Symptoms	Outcome	Blood Ammonia	Other Laboratory Findings	Enzyme Defect (% of control values) Liver	Fibroblasts
Subacute	7	McMurray et al. (91, 100)	M	9 m	Recurrent vomiting	Retarded	↑↑ (p.c.)	—	5	—
	8	Morrow et al. (102)	F	5-1/2 m	Recurrent vomiting, regression	Retarded		No orotic acid in urine	—	Abn K_m for citrulline and aspartic acid
	9	Scott-Emuakpor (78)	M	1	Vomiting, seizures	Retarded	↑ (p.c.)	↑ in homo-arginine and homocitrulline	—	Abn K_m for citrulline and aspartic acid
	10	Vidailhet et al. (158)	M	2 m	Regression of mental development, seizures	Vomiting, coma, and death at 8 m	↑↑↑	↑ in urine, homocitrulline and orotic acid	0 kidney active	— —
	11	Buist et al. (109)	F	10 d	Poor feeding, tremor, lethargy	Treated and IQ 76 at 32 m	↑	↑ homo-citrulline, orotic acid	— —	Abn K_m for citrulline and aspartic acid
	12	Wick et al. (17)	F	Routine testing	Asymptomatic	No treatment, normal at 4 yr, 2 m	N	0	—	70

[a] See symbols and abbreviations same as Table 4.

obligate heterozygotes were studied and found to be in the intermediate range (16). In one other family serum citrulline levels were slightly elevated in the parents (17). These findings suggest that at least the subacute form of citrullinemia is an autosomal recessive disorder.

Argininosuccinic Aciduria (Argininosuccinase Deficiency)

Clinical Features Analysis of the symptoms and the mode of onset of the disease in the 39 cases detected other than by routine newborn screening show that there are three types of ASAciduria: the lethal neonatal type, the subacute or infantile type, and the late-onset or chronic type. The pattern of onset seems to be recurrent in the same sibship (Table 7).

Symptoms in the neonatal variant are no different from those described previously. Five of the six infants with the neonatal onset died in the first 10 days of life. One was put on a very low-protein diet and survived to 3 months (Case 4).

The three infants with the subacute or infantile type developed recurrent vomiting, intolerance to formula feeding, and progressive enlargement of the liver in the first few months of life (Cases 7, 8, and 9). Mental and physical development was delayed.

The majority of the ASAciduric patients have the third type, with a gradual onset of the symptoms. Often the child is considered normal in the first year of life, although in retrospect, vomiting and feeding problems are common. The main clinical feature is delayed psychomotor development, which usually becomes noticeable in the second year. All except 3 of the 39 cases had some degree of mental retardation. IQ scores in the majority of patients are in the range of 30 to 60. Two patients with IQ's of 83 (Case 27) and 92 (Case 13) are considered to have low normal intelligence, and one patient reportedly was in "normal classes in school" (Case 26). At least one-third of the patients had a history of intermittent ataxia or unsteadiness and seizures. These episodes may be provoked by an infection or an increased protein intake and seem to be secondary to hyperammonemia.

Of recent interest is the effect of maternal metabolic disorders on the fetus. For instance, an offspring of a PKU mother often suffers intrauterine brain damage (18, 139), whereas the effect of maternal histidinemia is not so obvious (19, 20). Information on a case of maternal ASAciduria was kindly provided by Prof. A. M. O. Veale. This patient (Case 27), who is slightly retarded, with a short attention span, gave birth to a biochemically normal girl whose developmental age assessed at 16 to 18 months was consistent with her chronological age. This patient was not treated during pregnancy. She had no symptoms of hyperammonemia and showed an improvement in her antisocial behavior during pregnancy. The effect of the maternal ASAciduria on

Table 7 Summary of Argininosuccinic Aciduric (Argininosuccinase Deficiency) Patients[a]

Type	Case No.	Reference	Sex	Age Onset	Presenting Symptoms	Outcome	Blood Ammonia	Other Laboratory Findings	Enzyme Defect (% of control values) Liver	RBC	Other Tissues
Lethal Neonatal	1	Baumgartner et al. (122)	M	3 d	Lethargy, poor feeding, seizures, hepatomegaly	Coma, apnea, death on 10 d	—	↑ SGOT	—	—	Kidney = N
	2	Carton et al. (38)	F	4 d	Poor sucking, respiratory distress	Cardiac arrest and death on 6 d	—	Metabolic acidosis	—	—	Brain = 2.3, kidney = 0
	3	Levin and Dobbs (136, 159)	M	2 d	Lethargy, poor feeding	Coma and death on 4 d	↑↑↑	↑glutamine	0	0	—
	4	Farriaux, (131)	F	4 d	Stupor, poor feeding; had exchange transfusion and diet therapy	Failure to thrive; hematemesis; death at 3 m	↑↑↑	Anemia	"Deficient"	0	Kidney and brain = deficient
	5	Glick et al. (55)	M	40 hr	Listless, poor feeding, hepatomegaly	Tachypnea, bradycardia, death on 6 d	↑↑↑	Hypoglycemia, low BUN	0	5.1	Brain and kidney = N
	6	Murphey, W. H. (unpubl.)	F	—	—	Neonatal death	—	—	"Marked deficient"	0	Kidney = "marked deficient"

Subacute	7	Levin et al. (110); Levin (138)	M	5 d	Apathy, hypothermia, poor sucking	Retarded with seizures, hepatomegaly at 6 yr	↑↑ (p.c.)	Hair abnormality	0	0	—
	8	Schreier and Leuchte (153)	F	Early infancy	Seizures, hepatomegaly	Retarded at 9 m	↑	Hair abnormality	—	—	—
	9	Solitare et al. (34)	F	5 wk	Hepatomegaly	Seizures, coma, and death at 8 m	↑	↑SGOT, hair abnormality	17	—	—
Late Onset	10	Allan et al. (26); Westall (118)	F	Early infancy	Feeding difficulties, slow development	Severe mental retardation seizures, episodic ataxia at 13 yr	—	Hair abnormality, abnormal EEG	—	0	—
	11 (sib)	Allan et al. (26); Westall (118)	M	Second year	Slow development	Severe mental retardation at 16 yr	—	Hair abnormality, abnormal EEG	—	0	—
	12	Carson et al. (125)	F	—	—	Mental retardation at 2 yr	—	—	—	—	—
	13 (sib)	Carson et al. (125)	F	—	—	Low normal intelligence at 4 yr	—	Hair abnormality	—	—	—

Table 7 (*Contd.*)

Type	Case No.	Reference	Sex	Age Onset	Presenting Symptoms	Outcome	Blood Ammonia	Other Laboratory Findings	Enzyme Defect (% of control values) Liver	RBC	Other Tissues
	14	Wallis et al. (160)	F	Infancy	Slow development	Mental retardation and seizures at 6 yr	—	0	—	—	—
	15 (sib)	Wallis et al. (160)	M	1-1/2	Slow development	Severe mental retardation at 1-9/12 yr	—	0	—	—	—
	16	Coryell (25)	F	—	Slow development	Physical and mental retardation at 4 yr	—	Abnormal EEG	—	—	—
	17	Armstrong and Stemmerson (120)	F	—	—	Mental retardation at 18 yr	—	—	—	—	—
	18 (sib)	Armstrong and Stemmerson, (120)	F	—	—	Mental retardation at 15 yr	—	—	—	—	—
	19	Edkins and Hockey (129)	F	18 m	Abnormal hair and slow development	Physical and mental retardation	↑	Hair abnormality, abnormal EEG	—	0	—

20	Moser et al. (48)	M	18 m	Febrile seizures	Mental retardation at 23 yr	↑ (p.c.)	↑SGOT, abnormal EEG	—	0	Fibroblasts = 0
21 (sib)	Moser et al. (48)	M	Second year	Slow development	Mental retardation at 13 yr	↑ (p.c.)	0	—	0	Fibroblasts = 0
22	Miller and McLean (35); Lewis and Miller (142)	M	Infancy	Slow development	Episodic ataxia and lethargy, death at 16 yr	—	Abnormal EEG	<5	—	—
23	Moore et al. (143)	M	8 m	Slow development	Mental retardation at 7-1/2 yr	—	Abnormal EEG	—	—	—
24	Farrell et al. (130)	F	4	Slow development and seizures	Mental retardation at 20 yr	↑	Hair abnormality, Abnormal EEG	—	0	Fibroblasts = 0
25 (sib)	Farrell et al. (130)	F	5 m	Growth failure	Mental retardation, seizures at 8 yr	↑ (p.c.)	Hair abnormality, Abnormal EEG	—	0	Fibroblasts = 0
26	Porath (148)	F	—	—	Normal at 16 yr	↑	Abnormal EEG	—	—	—
27	Blackmore et al. (124)	F	—	Antisocial and aggressive behavior	IQ 83 at 16 yr; maternal ASAciduria	—	—	—	—	—

Table 7 (*Contd.*)

Type	Case No.	Reference	Sex	Age Onset	Presenting Symptoms	Outcome	Blood Ammonia	Other Laboratory Findings	Enzyme Defect (% of control values) Liver	RBC	Other Tissues
Late Onset	28 (sib)	Blackmore et al. (124)	F	—	(Family survey)	"Special class" student at 13 yr	—	—	—	—	—
	29 (sib)	Blackmore et al. (124)	M	—	(Family survey)	"Special class" student at 10 yr	—	—	—	—	—
	30	Goodman et al. (67)	F	8 wk	Growth failure	Physical and mental retardation at 2 yr	—	—	—	—	Fibroblasts = 0
	30*a* (sib)	Goodman et al. (67)	—	—	(Prenatal diagnosis)	Aborted	—	ASA in amniotic fluid and maternal urine	<3	—	Fibroblasts = 0
	31	Cederbaum et al. (103)	M	Infancy	Feeding difficulties and slow development	Seizures and treatment at 12 m, mental retardation at 21 m	↑ (p.c.)	0	—	0	—
	32	Hambraeus et al. (104)	F	8 m	Seizures	Mental retardation at 7-1/2 yr	↑	"Brittle" hair, abnormal EEG	—	—	—

33 (sib)	Hambraeus et al. (104)	F		Slow development	Mental retardation at 3-1/2 yr	↑	"brittle" hair, abnormal EEG	—	—	—
34	Hambraeus et al. (104)	F	Second yr	Slow development	Mental retardation at 5	↑	"Brittle" hair abnormal EEG	—	—	—
35	Billmeier et al. (123)	M	Gradual	Delayed speech	Mild retardation	↑ (p.c.)	"Brittle" hair, abnormal EEG	0	0	Fibroblasts = 0
36 (sib)	Billmeier et al. (123)	F	—	—	"Borderline" normal intelligence	0	Abnormal EEG	0	0	Fibroblasts = 0
37	Potter et al. (24)	F	—	—	Severe physical and mental retardation, seizures at at 4 yr	↑ (p.c.)	Hair abnormality Abnormal EEG	—	—	—
38	Hartlege et al. (117)	F	—	—	Mild mental retardation	—	Hair abnormality	—	—	—
39 (sib)	Hartlege et al. (117)	F	—	(Prenatal diagnosis)	Treated at birth and normal at 6 m	↑	Hair abnormality, ASA in amniotic fluid and maternal urine	—	0	—

[a] See symbols and abbreviations same as Table 4.

the ultimate intellectual achievement of the offspring requires long-term follow-up.

The abnormality of the hair called trichorrhexis nodosa has been observed in a little less than one-half of the patients with ASAciduria but not in those with other urea-cycle disorders. The hair tends to be short, brittle, and friable (Figs. 2 and 3). The pathogenesis is not clear. Trichorrhexis nodosa can be produced even in normal hair by frequent trauma (21, 22). It has been considered possibly a manifestation of arginine deficiency, since human hair normally contains 7.5 to 10% arginine (23). This, however, was not borne out by the study of Potter et al. (24), who found normal arginine but a reduced cystine content in the hair from their patient. Coryell et al. (25) noted that the hair of their patient became normal after institution of a good high-protein diet. However, the follow-up of Cases 10 and 11 disclosed that the general

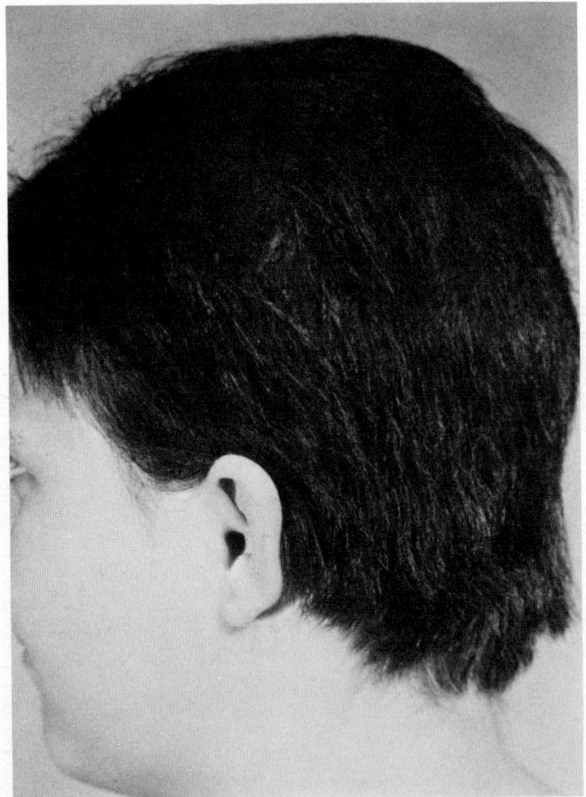

Figure 2. Patient with argininosuccinic aciduria, showing brittle and short hair. (From Farrell et al., 1969.)

Figure 3. Microscopic appearance of hair from a patient with argininosuccinic aciduria. This illustrates classical trichorrhexis nodosa, a condition in which minute nodes are formed in the shafts of the hairs, the latter splitting and breaking incompletely at these points. (From Farrell et al., 1969.)

appearance of the hair was also much improved without any therapy or appreciable change in the diet (26).

Because of the high frequency of abnormal hair in patients with ASAciduria, ASA has been sought in the urine of patients with different types of hereditary hair disease. Small amounts of ASA were found in some patients. Grosfeld (27) reported the presence of ASA in the urine of patients with monilethrix and suggested that this disease might be due to ASase deficiency in the hair follicles or skin. The same urine was studied again by Efron and Hoefnagel (28), using three different methods, including electrophoresis and paper and column chromatography. It was proved that the ninhydrin-positive compounds in the urine were not ASA. Urine specimens from three brothers in another family with the characteristic findings of monilethrix were also analyzed for ASA, but none was found. Shelley and Rawnsley (29) screened a series of 56 patients with various types of hair disorders and found two patients, both with hair loss resulting from breakage, who excreted from 10 to 20 mg ASA per day. One patient had greatly reduced ASase activity in the blood cells; the other had slightly lower than normal activity. Trace amounts of ASA were detected in occasional first-morning urine specimens of six patients with unclassified hair loss. In another study of 22 patients with hereditary hair diseases and their relatives, Winther and Bundgaard (30) found trace amounts of ASA in some patients and relatives. The relationship of ASA to the hair abnormality is not clear.

Genetics ASAciduria is the most common of the five urea-cycle disorders. To my knowledge, there are at least 45 cases. There is a female predominance. The sex of 42 patients has been reported: 12 male and 30 female. The number of patients with the neonatal and subacute types is small, but the male-to-female ratio is about equal. Of the 29 patients with the late onset type, 9 are males and 20 are females. Six children were discovered through the newborn screening program, and five of them are females (Shih and Levy, unpublished). It appears that the neonatal variant of this disorder, unlike that of OTC deficiency, is an autosomal disorder. Female preponderance in the late onset type suggests that it is sex-limited. An X-linked dominant inheritance is unlikely, because there is no clear-cut evidence of increased incidence of miscarriage or deficiency in male sibs, suggesting that this trait is lethal in males. Moreover, the parents, particularly the mothers, have been normal clinically, and intermediate levels of ASase activity in erythrocytes and fibroblasts have been found in all parents so studied (32, 65).

PATHOLOGY

Detailed pathological descriptions of urea-cycle disorders are available for only a few cases.

The general pathological findings in urea-cycle disorders are minimal and nonspecific. However, the neuropathology is interesting. The brain often shows cerebral edema which is probably the result of the terminal acute hyperammonemic episode. Cerebral atrophy seems to correlate with the severity of the disorder, particularly with the degree of hyperammonemia and the length of the illness. Bruton et al. (33) found greatly enlarged ventricles and almost total destruction of cerebral cortex and subjacent white matter in a severely affected patient with OTC deficiency who died at 6 years of age. On the other hand, the cousin of this patient, who did not have symptoms of hyperammonemia until 3 years of age, had only minimal dilation of the ventricles. The principal histological changes consist of the presence of numerous Alzheimer type II astrocytes, particularly among the neurons of the basal ganglia, pontine nuclei, and the cerebellar and the cerebral cortexes (12, 33, 34, 35). Proliferation of astrocytes with vacuolated nuclei which fit the description of Alzheimer type II astrocytes has been reported in other patients (36, 37). However, there are other patients who have not had these findings (38, 39, 40, 41).

Alzheimer types I and II cells have frequently been seen in adults with chronic hepatic encephalopathy and have also been shown to occur in experimental acute hyperammonemia (42, 43, 44, 45, 46). Delayed myelination and spongy degeneration of the brain white matter have been described in infants with urea-cycle disorders as well as with other amino acid disorders (47), and are apparently nonspecific changes.

BIOCHEMICAL FINDINGS

The prominent biochemical abnormalities in the urea-cycle disorders are hyperammonemia or the accumulation of the intermediate metabolites proximal to the enzyme block, and changes secondary to hyperammonemia such as increased glutamine and orotic acid. In general, these abnormalities are most severe in the fulminating neonatal cases.

Hyperammonemia

All five of the urea-cycle disorders are associated with varying degrees of hyperammonia. This is most severe in CPS and OTC deficiencies and less severe and less consistent in the other three disorders, where it is more likely to occur only after a protein load. The affected neonates usually have markedly increased blood ammonia.

That not all patients with CPS deficiency had significant hyperammonemia (Table 4) is a rather surprising finding, and there has been no satisfactory explanation. In OTC deficiency the severity of the clinical symptoms correlates directly with the degree of hyperammonemia. During episodic vomiting attacks, the blood ammonia concentration may be over 1000 μg/dl.

Blood ammonia levels were reported in 10 citrullinemic patients, and 5 had hyperammonemia, which is most prominent during the postprandial period (Table 6). Interestingly, three patients with the neonatal type of citrullinemia who had no measurable ASA synthetase activity had normal blood ammonia concentrations. In ASAciduria, the blood ammonia is increased only in the postprandial specimens of some patients (Table 7).

The ammonia concentration in cerebrospinal fluid is a function of that in blood. Under normal conditions, ammonia in cerebrospinal fluid is less than that in blood. In patients with hyperammonemia, the concentration of ammonia in cerebrospinal fluid approaches or may exceed that in the blood (48).

Amino Acid Abnormalities

In CPS and OTC deficiencies, there is no consistent abnormality of amino acids except for increased glutamine and alanine secondary to hyperammonemia. It is noteworthy that the plasma ornithine concentration is normal in patients with OTC or CPS deficiency with the exception of one atypical family studied by Gatfield et al. (49). In this family, CPS I deficiency was found in several patients with a syndrome of hyperornithinemia, hyperammonemia, and homocitrullinuria. Abnormally long and bizarre-shaped mitochondria in the liver were found by electron microscopy. It was speculated that the defect may be in the transport of ornithine into the mitochondria.

In ASA synthetase deficiency, citrulline is markedly increased in blood, urine, and the cerebrospinal fluid. It may be as high as 40 times normal in the blood and 100 times normal in the cerebrospinal fluid. Although the blood levels are not always directly proportional to the severity of the disease, the asymptomatic patient did have the lowest blood citrulline, only three times normal (50). The urinary excretion of citrulline is greatly increased (Fig. 4) and is in the range of several hundred milligrams to 1 g/24 hours. Normal persons excrete less than 1 mg/24 hours. Other urinary amino acid abnormalities include the presence of homocitrulline (Cases 4, 9, 10, and 11 in Table 6), homoarginine (Case 9 in Table 6), and *N*-acetylcitrulline (51).

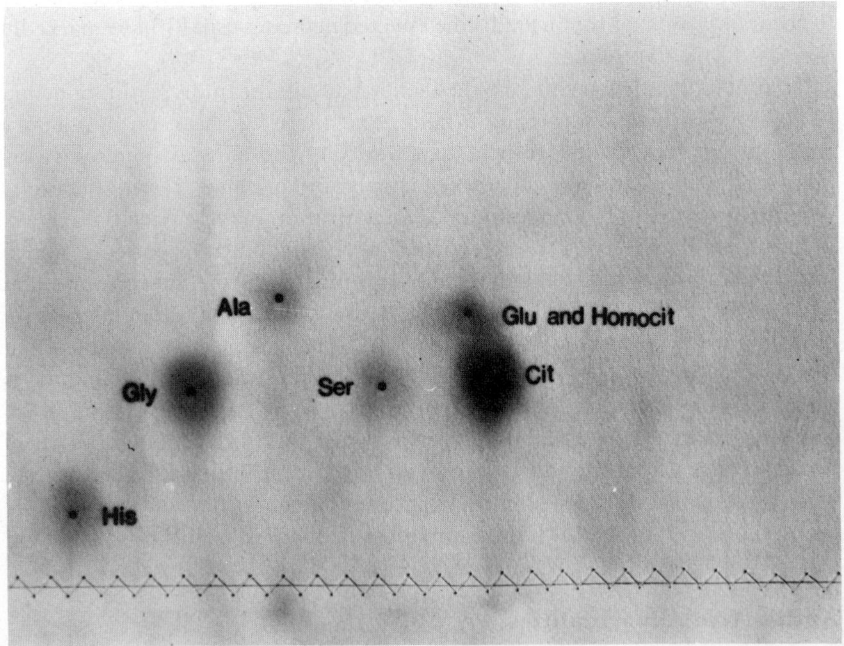

Figure 4. Photograph of a paper chromatogram showing the urinary amino acid pattern in citrullinemia. High voltage electrophoretic separation using 6% formic acid (from right to left) was followed by ascending solvent chromatography using butanol–acetic acid–water (12:3:5).

ASA, which is normally present in the urine in a very small amount up to 1.9 mg/day (52), is excreted in large quantity by patients with ASase deficiency (Fig. 5). Most of them have excreted several grams a day, and the amount varies to some degree with the protein intake. ASA is present in the urine in three forms. The major portion is in the free acid form; the anhydride I (53) and hexahydroimidazole diazepine derivatives (54) of ASA each account for less than 10%.

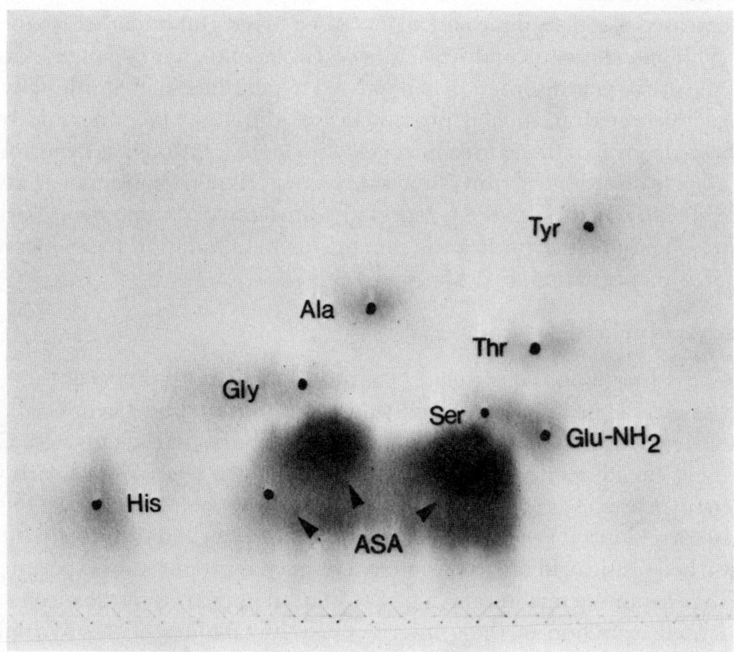

Figure 5. Photograph of a paper chromatogram showing the urinary amino acid pattern in argininosuccinic aciduria. Same conditions as described in Figure 4.

The concentrations of ASA in the blood of ASAciduric patients range from 3.5 to 4.4 mg/dl, except in newborns, who have much higher levels, up to 40 mg/dl. In almost all patients the ASA was higher in the cerebrospinal fluid than in the blood except in the two patients who excreted very little ASA in the urine and who had none detectable in the cerebrospinal fluid (Cases 14 and 15 in Table 7).

The higher concentration of ASA in the cerebrospinal fluid may be explained either by an enzyme block in the brain or by an increased synthesis. Supporting the latter hypothesis is the fact that a normal brain ASase activity has been demonstrated in one patient (55). Citrulline, the precursor of ASA, is moderately increased in the blood, urine, and cerebrospinal fluid of ASAciduric patients. With the increased transport of citrulline from the blood into the brain, ASA production could exceed its breakdown even in the presence of normal brain ASase activity. ASA is a highly polar substance, and its poor penetration from brain to blood would result in a higher accumulation in the brain and cerebrospinal fluid. Unfortunately, amino acids were not measured in the cerebrospinal fluid of this particular patient with normal ASase activity in the brain.

A decreased blood arginine and an increased blood glutamine are frequently observed in citrullinemia and ASAciduria. In the malignant neonatal form of any of the urea-cycle disorders, a generalized hyperaminoacidemia and hyperaminoaciduria with particular prominence of glutamine, alanine, and lysine have been described. Blood urea nitrogen with few exceptions has been normal in urea-cycle disorders. Serum glutamate oxaloacetate transaminase (SGOT) is often elevated in these patients especially during hyperammonemic periods. This may be due either to hepatic damage or to adaptation to an increased need for transamination.

Orotic Aciduria

Orotic acid is a minor constituent of urine. Increased excretion of this compound as well as uridine and uracil was first discovered by Levin et al. (56) in OTC deficiency. It has now been found in a number of patients with OTC deficiency, citrullinemia, and ASAciduria but not in two patients with CPS deficiency (Cases 3 and 6 in Table 4). The amount of these metabolites directly correlates with the degree of hyperammonemia and the protein intake.

Disturbed orotic acid metabolism has also been reproduced in experimental acute hyperammonemia in rats (57, 58). Thus, it appears that the mechanism of increased excretion of these intermediary metabolites of the pyrimidine pathway is a result of shunting of the mitochondrial carbamyl phosphate into the pyrimidine biosynthetic pathway.

ENZYME DEFECT

Definitive diagnosis of urea-cycle disorders depends on investigation of the enzyme defect. Table 8 shows the sources of enzymes available for diagnostic tests.

Liver is the main source of urea-cycle enzymes and is quantitatively important in the detoxification of ammonia and certainly the most satisfactory for investigation. However, because of the ethical and technical problems involved, it is not always feasible to obtain a liver biopsy. Open surgical biopsy has limitations in that it requires a laparotomy, and patients have

Table 8 Tissues Available for Biopsy Studies of Urea-Cycle Enzymes

Enzyme Defect	Liver	Jejunal Mucosa	RBC	WBC	Skin Fibroblasts	Amniotic Fluid Cells	
						Cultured	Uncultured
CPS	+[a]	+	−	+[a]	−	−	?
OTC	+[a]	+[a]	−	+[a]	−	−	+
ASA synthetase	+[a]	−	+	+	+[a]	+	?
ASase	+[a]	−	+[a]	+	+[a]	+[a]	+
Arginase	+	−	+	+	−	−	+

Enzyme defect has been demonstrated.

died following surgery (36, 39). Recent experiences with needle biopsy have proven it to be a relatively safe procedure in obtaining small amounts, 50 to 100 mg, of liver tissue for enzyme studies. It should be noted that activities of the urea-cycle enzymes have been found to be lower in needle biopsy specimens that in surgical specimens (13). In the majority of the reported cases, however, enzyme studies have been performed on postmortem tissues. It should be cautioned that long periods of storage may result in loss of enzyme activity (55) and make the interpretation of data difficult.

In studying CPS and OTC activity, leucocytes (59) and the jejunal mucosa obtained by peroral biopsy (56, 60) may be used as alternatives to liver tissue. Erythrocytes have most frequently been used for enzyme diagnosis of ASAciduria (Table 7) and hyperargininemia (61, 62). Murphey et al. (63) developed a microbiological assay for ASase activity using blood dried on filter paper which can be used for mass neonatal screening. Deficiencies of ASA synthetase activity (16, 64) and ASase activity (65) are expressed in cultured skin fibroblasts. In addition to enzyme assays, these disorders can be detected by the relatively simple technique of measuring the distribution of radioactivity in cell protein and free amino acids after incubating the fibroblasts with labelled citrulline (66, 67, 68). Although it is not known to what degree the enzyme activity in blood cells or fibroblasts reflects the hepatic enzyme level, the major advantage of being able to demonstrate the defect in blood cells is the easy accessability of blood for the confirmation of a clinical diagnosis. The demonstration of an enzyme deficiency in cultured skin fibroblasts and amniotic fluid cells makes possible the prenatal diagnosis of such disorders.

CPS Deficiency

In the six patients with CPS deficiency, the amounts of residual hepatic enzyme activity vary from 0 to 40% (Table 4). The reported CPS activities are presumably the total of CPS I and II except where indicated. A male infant who died in the neonatal period (41) represents the most severely affected case. Investigations on the liver enzyme showed less than 10% residual CPS activity which was not *N*-acetylglutamate–dependent. This finding suggests that the mitochondrial CPS I is completely absent and the soluble CPS II accounts for the residual activity.

A reduced CPS activity has been described in association with several other metabolic disorders, such as methylmalonic aciduria (Kirkman, personal communication) and OTC deficiency (14). One rather unexpected finding is the CPS I deficiency recently found in patients with hyperornithinemia, hyperammonemia, and homocitrullinuria (49). Six patients in one family were studied; three patients were shown to have no detectable CPS I activity in leucocytes, and two had approximately one-fourth of the mean control activity or 60% of the lowest control value. In one patient urea-cycle enzymes were studied in a liver biopsy specimen, and CPS I was found to be 19% of the mean control values, whereas CPS II was seven times higher than the control values. The activity of ornithine decarboxylase, which has been shown to be deficient in another patient with the same clinical disorder of hyperornithinemia, hyperammonemia, and homocitrullinuria (70), was not studied in these patients. The interrelationship of these metabolic defects and the abnormal amino acid pattern in this syndrome await further investigation.

OTC Deficiency

The severity of OTC deficiency correlates to some degree with the severity of clinical manifestations. The male infants affected with the lethal neonatal form of the disorder have had less than 2% OTC activity in the liver (Cases 5, 6, 8, 12–15, and 19 in Table 5). Those three male patients who had a less severe clinical disease and between 5 and 25% of residual enzyme activity (Cases 20, 21, 22 in Table 5) are thought to be genetically distinct from the other patients. Affected females have a deficiency of the enzyme that varies from 4 to 97% of the control values. Aside from the quantitative reduction of enzyme activity, pH dependency, K_m abnormalities, and other enzyme characteristics have been described.

In the majority of cases, OTC activity was assayed at pH 8.0 in glycylglycine buffer according to the conditions described by Brown and Cohen (71) or modifications of this method. Levin et al. (37) first observed the pH dependency of OTC activity in patients with OTC deficiency. In control liver the enzyme activity assayed at pH 7.0 using Tris buffer is approximately 86% of that at pH 8.0. In 10 patients the liver OTC activity at pH 8.0 was shown to be either moderately decreased or within normal range, but at pH 7.0 the reduction of enzyme activity was more significant. Particularly, two severely affected neonates had 30 and 38% of normal values at pH 8.0, whereas the OTC activity at pH 7.0 was negligible. The reduction of enzyme activity is too great to be explained by pH alone. If the reduction is due to inhibition by Tris buffer (72), the different behavior of the normal versus the mutant enzyme suggests that the mutant enzyme is more susceptible to Tris buffer inhibition. In four patients, such abnormal pH dependency of the enzyme activity was not observed (Cases 5, 19, 26, and 30 in Table 5).

Cathelineau et al. (60, 73) studied the enzyme characteristics of five patients and found abnormal pH-dependent variations of enzyme activity in three patients. In addition, two of these patients had an abnormal K_m value for ornithine, one 6 times and the other 100 times greater than normal. In one other patient, the K_m value for CP was 4 times normal. Matsuda et al. (74) reported an abnormal enzyme having a pH dependency and a decreased affinity to carbamyl phosphate (K_m 4 times normal). After electrolysis of the crude extract of the enzyme, two main peaks, isoelectric at pH 3.2 and 4.4, were observed in the control, whereas only one peak at pH 2.8 was found in the patient (75). Nagayama et al. (76) briefly mentioned a deficiency in the main peak of the enzyme after fractionation on a Sephadex column (Case 26 in Table 5).

OTC activity in jejunal mucosa has been studied in several patients and has, in general, shown changes similar to that observed in the liver enzyme. However, the small numbers and wide variation of the control series sometimes make it difficult to distinguish between normal and low activity.

The many different kinds of enzyme abnormalities reported in the literature requires that we investigate future cases carefully and thoroughly in order to learn whether these abnormalities are of physiological significance or whether they indeed indicate genetic heterogeneity.

Deficiency of OTC (8, 77) has recently been demonstrated in patients with Reye's syndrome, which is a disease of childhood characterized by impaired liver functions, marked hyperammonemia, acute encephalopathy, and an often fatal course. Thaler et al. (77) proposed that some patients with Reye's syndrome may have a mutant OTC which makes them susceptible to precipitating factors such as viral and toxic agents. In collaboration with Dr. Philip Snodgrass, we have also observed relative deficiency of OTC in four patients with Reye's syndrome (Snodgrass, Shih, Delong, and Glick, unpublished data). In one patient there was an initial deficiency of OTC with later improvement during convalescence, suggesting that the deficiency was transient. Investigation of additional patients and careful family studies are necessary to determine whether OTC deficiency in Reye's syndrome is an acquired condition and whether certain female patients with presumed Reye's syndrome are actually symptomatic carriers of OTC deficiency.

Argininosuccinate Synthetase Deficiency (Citrullinemia)

In citrullinemia ASA synthetase activity has been shown to be deficient in the liver. Three of the five infants with the neonatal type had virtually no detectable liver enzyme activity (Table 6). Kidney ASA synthetase was normal in one patient (Case 10). The data suggest that in certain cases of citrullinemia, analogous to ASAciduria, the enzyme may be deficient only in the liver. However, to date there has been no complete study of ASA synthetase in the liver, kidney, and brain from one single patient.

The ASA synthetase has been investigated in cultured skin fibroblasts derived from three patients (16, 64, 78). Decreased binding of citrulline and aspartate was found in two patients (Cases 8 and 11 in Table 6). In one patient (Case 9 in Table 6) the ASA synthetase activity was not detectable in both cultured skin fibroblasts and cultured lymphoblasts under standard assay conditions (Citrulline 10 m*M*, aspartate 5 m*M*), but low activities were measurable at 50 m*M* citrulline in the fibroblasts (16, 79). Wick et al. (50) reported only a mild reduction of the enzyme activity (40 and 70% of residual enzyme activity) in two unrelated patients (Cases 2 and 12), and raising the citrulline concentration in the assay medium from 2 to 10 m*M* did not increase the activity.

From the above findings, the heterogeneity of these patients is evident, and the amount of residual enzyme activity does not necessarily correlate with the severity of the clinical disease.

Argininosuccinase Deficiency (Argininosuccinic Aciduria)

Tomlinson and Westall (80) first demonstrated the absence of ASase in the blood cells of two patients with ASAciduria. Subsequently, no enzyme activity was detectable in the hemolysate of 19 additional patients (Table 7). Recently Glick et al. (55), using higher concentrations of ASA and arginase, found normal values 13 times higher than those reported by Tomlinson and Westall (80) and an 11% residual ASase activity in their newborn patient. ASase activity has also been found to be absent in fibroblasts derived from ASAciduric patients (Shih et al. (65)) (Table 6).

A marked deficiency of ASase activity has been demonstrated in the liver of eight patients (Table 7) and in the brain (81) and kidney of one patient (38). There had not been a complete study of enzyme activity in all three organs from the same individual until very recently, when Glick et al. (55) reported their careful investigation on an infant with the lethal neonatal form of ASAciduria. Their data is of particular significance in that it clearly shows that the enzyme deficiency is only present in the liver and not in the brain or kidney (Table 9). The liver ASase was below the limit of detection (less than 0.5 u/g), whereas the other urea-cycle enzymes were within or above control ranges. This well-documented case demonstrates that more than one gene either codes for the structure of ASase or regulates its activity in tissues and strengthens the hypothesis that ASase in liver and kidney are independently controlled. Data derived from Dr. Ratner's extensive investigations indicate

Table 9 Urea-Cycle Enzyme Activities[a] in Tissues of a Patient with Neonatal Type of Argininosuccinic Aciduria Compared with Six Neonatal Controls (55)

Tissue	Period of Storage at −60° (months)	CPS	OTC	ASA Synthetase	ASase	Arginase	Protein[d]
Liver							
Patient	1	137 / 1.22	6,720 / 60.0	102 / 0.91	0 / 0	25,000 / 223	112
Controls[b]	0.5–3	192 ± .68 / 1.30 ± 0.40	2,700 ± 675 / 18.1 ± 4.9	67.2 ± 10.9 / 0.49 ± 0.09	94.4 ± 28.3 / 0.64 ± 0.15	21,200 ± 8,630 / 152 ± 56	128 ± 16
Range		136–311 / 0.83–1.79	1,900–3,690 / 11.7–24.6	58.4–86.8 / 0.39–0.62	74–150 / 0.49–0.91	12,000–37,300 / 86–246	106–152
Kidney							
Patient	12	37.4 / 0.52	358 / 4.98	0 / 0	15.2 / 0.21	2,640 / 36.7	71.9
Controls[a]	2–5	28.0 ± 13.1 / 0.22 ± 0.07	10.7 / 0.09	27.7 ± 9.7 / 0.22 ± 0.09	11.3 ± 5.6 / 0.12 ± 0.04	251 / 1.97	129 ± 10
Range		14.0–52.3 / 0.12–0.35	0.71–51.6 / 0.01–0.43	9.6–35.6 / 0.07–0.34	6.4–21.6 / 0.08–0.19	93–622 / 0.61–5.13	111–132
Brain							
Patient	12	7.56 / 0.24	0.74 / 0.02	10.6 / 0.34	25.6[c] / 0.82	344 / 11.0	31.4
Controls[a]	1–4	4.94 ± 3.88 / 0.10 ± 0.05	4.50 / 0.07	104 ± 21 / 0.86 ± 0.23	13.1 ± 12.0 / 0.27 ± 0.29	373 ± 109 / 8.07 ± 2.80	54.1–9.3
Range		2.37–12.6 / 0.06–0.18	0.21–13.6 / 0.01–0.22	86–144 / 0.54–1.19	4.80–35.2 / 0.09–0.82	257–540 / 5.31–11.80	42.8–69.2

[a] Expressed as (μmole/hour/g wet tissue)/(μmole/hour/mg protein). [b] Mean ± standard deviation. [c] Assayed after 2 months at −60°.
[d] Expressed as mg/g wet weight.

that the bovine liver and kidney argininosuccinases are identical proteins (82). However, such information on the characteristics of human enzymes is not available. Thus the basic defect in ASAciduria could be a mutation of either the structure gene or the regulator gene for ASase.

Very importantly, Glick et al. (55) have studied the loss of urea-cycle enzyme activity in organs of the patient as a function of the length of storage time. They found that brain ASase activity after 12 months of storage at $-60°$ is only 6% of the activity after 2 months of storage. Kidney ASase was normal after 12 months storage at $-60°$ but decreased to 9% at 15 months. Moreover, ASase in a control liver lost 74% of the activity after 12 months of storage. Thus, one should be cautious in interpreting the low enzyme activities reported in the literature. Kint and Carton (81) reported less than 2.3% ASase activity in the brain when the assay was done after 15 months storage at $-30°$C. In the same patient, no ASase activity could be demonstrated in the kidney stored for 25 months at $-30°$ (38). The possibility of storage artifact should be considered in this case. In contrast, Colombo and Baumgartner (83) found normal activity and normal K_m of the kidney ASase from their patient with the neonatal variant. The length of storage, however, was not specified in their paper.

It may be concluded that the above findings support the concept that liver is the major site of ammonia detoxification and the major function of ASA synthetase and ASase in the kidney and brain is for the biosynthesis of arginine rather than of urea.

PATHOGENESIS OF NEUROLOGIC ABNORMALITIES

The pathogenesis of mental retardation in amino acid metabolic disease is still unknown. The accumulation of the involved amino acid and its metabolites, either acting as an inhibitor of other important enzymatic actions or interfering with the transport of other amino acids into the cells, has been considered directly or indirectly responsible for the cerebral dysfunction. On the other hand, the deficiency of an essential compound may be equally important.

In deficiencies of ASAsynthetase, ASase, and arginase, the huge amounts of citrulline, ASA, and arginine accumulated respectively may be toxic (84). Orotic acid secondary to hyperammonemia may also be a contributing factor. α-ketoglutaramate is a metabolite of glutamine recently demonstrated in the cerebrospinal fluid of humans. It was found to be increased in patients with hepatic encephalopathy (85, 86). Although α-ketoglutaramate has not been looked for in hyperammonemic patients, we may postulate that this compound is present and may be a toxic agent. A low blood arginine concentration is frequently found in patients with deficiencies of ASAsynthetase and ASase activity. This might be expected, since a block in the urea cycle would result in a decrease in the biosynthesis of arginine. The deficiency of arginine, a semiessential amino acid, may result in brain damage or failure to thrive.

A markedly increased blood ammonia level is the prominent feature in both CPS and OTC deficiency. In fact, blood ammonia elevation, although in different degrees, is a finding common to all urea-cycle disorders. In addition to the mental retardation seen in this group of patients, a characteristic symptom complex is often present, and these symptoms improve with the lowering of blood ammonia levels. These observations suggest that high blood ammonia levels either play an important role in the pathogenesis of the clinical symptoms or are at least partially responsible for them. Ammonia intoxication has also been considered a contributing factor to such neurologic abnormalities as encephalopathies in acquired liver disease (87), following a portacaval shunt (88), and in Reye's syndrome (89). The episodic stupor and coma seen in the first two conditions are similar to the attacks in urea-cycle disorders. These adult patients may have a specific neuropsychiatric syndrome (90). The neuropathologic changes in these patients are similar to those that have been found in the congenital forms of ammonia intoxication.

When the high blood ammonia level occurs early in life, it is likely to be more injurious than when it occurs later. Cerebral atrophy has been found not infrequently in patients with urea-cycle disorders (33, 36, 91). Slow mental development may be a nonspecific reaction of the developing brain to any toxic agent or injury, since it is seen in many inborn errors of metabolism.

In the brain the major mechanism for the removal of ammonia is glutamine formation. Berl et al. (92) administered ^{15}N-ammonia to cats by carotid infusion and found that most of the injected ammonia was present as free ammonia. The next largest pool in the cerebral cortex was in the amide group of glutamine, followed by the α-amino group. Glutamine is the only amino acid in the brain that is considerably increased in amount as a result of ammonia infusion; this increase occurs without a corresponding decrease in the glutamic acid concentration. It is therefore concluded that the glutamic acid of the newly formed glutamine is synthesized in the brain and is from a small and metabolically active compartment.

Little is known about the cerebral ammonia metabolism in urea-cycle disorders. With the information available on metabolic studies in hepatic encephalopathy and the data from animal experiments, a number of hypotheses concerning the mechanism of ammonia intoxication have been put forward (93, 94). Most of them explain the effect of high ammonia levels on brain function on the basis of interference with cerebral energy metabolism at some step in the citric acid cycle or a related reaction. The citric acid cycle is connected to ammonia nitrogen metabolism by a number of reactions. The former supplies oxaloacetate in the production of aspartate, which then condenses with citrulline to form ASA. α-Ketoglutarate forms glutamate by transamination with other amino acids. An increased demand for glutamate formation would result in depletion of α-ketoglutarate and in impairment of the citric acid cycle and a consequent decrease in ATP. It has also been shown experimentally that prior administration of ornithine α-ketoglutarate attenuates the rise in blood ammonia seen following the administration of ammonium salts to free ventilated dogs and also prevents the increase in anaerobic glucose consumption caused by ammonium infusion (95).

Different regions of the brain may respond differently to ammonia intoxication. Hindfelt and Siesjo (96) observed significant decreases in high-energy phosphates in the brain stem and cerebellum during acute hyperammonemic states and in the cerebral hemispheres as well during prolonged exposure to ammonia (97). The histological findings of numerous Alzheimer type II astrocytes in these regions and the clinical symptoms of ataxia are in agreement with the experimental data.

Although many observations have been made in animal experiments, both *in vivo* and *in vitro*, it should be emphasized that detailed knowledge about ammonia metabolism, the control of the urea cycle, and its relation with other metabolic pathways in human beings is still lacking. The mechanism of ammonia intoxication may be manifold.

BLOOD UREA AND UREA EXCRETION IN UREA-CYCLE DISORDERS

Studies of patients with disorders of the urea cycle have provided opportunities to investigate ammonia metabolism and the control of urea synthesis. It is of interest that patients with urea-cycle enzyme deficiencies have had normal plasma concentrations of urea and excreted substantial amounts of urea in the urine, even though there was an enzymatic block in the only known biosynthetic pathway of urea. The origin of urea in these patients is of considerable interest. Three possibilities for this anomaly have been suggested: (1) that the enzyme defect is only partial; (2) that the urea-cycle enzymes in different organs are under different genetic controls and that the biochemical defect involves one or more, but not all, organs; and (3) that there is an alternative pathway.

The Hypothesis of a Partial Enzyme Defect

A complete block in one of the steps of the urea cycle is probably incompatible with life. This view is supported by the lethality of complete CPS deficiency (41) and OTC deficiency in the male (12). It has been suggested that only one-fifth to one-fourth of the potential ability of the liver for urea synthesis is used under normal circumstances (98, 99). On the basis of this estimate, a 20% residual enzyme activity of the urea cycle would be accompanied by no significant reduction in urea excretion.

The fate of isotopic precursors of urea and the nitrogen balance have been studied in several patients with urea-cycle disorders. McMurray and co-workers (100) injected ^{14}C-citrulline intravenously to their citrullinemic patient who had a residual liver ASA synthetase activity of between 4 and 8% of the control. They found that only 11% of the injected dose was excreted unchanged in the urine and 31% was metabolized to urea in the first 24 hours. Urea formation from administered ^{15}N-labeled ammonium lactate has also been studied in a patient with ASAciduria (101). The labeled isotope excreted as urea in 3 days was 34% of the given dose, less than half the amount obtained from the control.

Nitrogen balance studies in several patients with citrullinemia (102) and ASAciduria (103, 104) have indicated that the urea-synthesizing ability of these patients is limited. Morrow et al. (102) found that the urea excretion in their citrullinemic patient was approximately 20% of the total nitrogen excretion when the patient was on a diet containing 1.5 g protein/kg/day. It increased to 44% when the protein intake was increased to 2.9 g/kg/day. However, they commented that most of the increase in urea excretion was of dietary origin and suggested that the block in the urea cycle was quite severe. Cederbaum et al. (103) observed that increasing the protein intake of an ASAciduric patient from 1.5 to 2.0 g/kg/day caused a distinct rise in the ASA nitrogen excretion from 25 to 37% that of urea nitrogen. Similarly, the three patients with ASAciduria studied by Hambraeus et al. (104) became acutely ill with neurological symptoms within 1 to 2 days after the dietary protein had been changed from 0.18 to 1.85 g/kg/day. Thus, it appears that when the protein intake is above 1.5 g/kg/day, the nitrogen disposal is met in part by increased excretion of the intermediary compounds and hyperammonemia follows. These data indicate that substantial urea production is still possible in the presence of a significant reduction of enzyme activity in the urea cycle.

Since liver is the major source of CPS and OTC activity, the hypothesis of a partial enzyme defect in patients with congenital hyperammonemia Types I and II is plausible in the absence of an alternative pathway of urea synthesis.

The Hypothesis of Different Genetic Controls for Urea-Cycle Enzyme Activities in Different Organs

In patients with ASAciduria, the concentration of ASA in the cerebrospinal fluid is higher than it is in the blood. The presence of ASA in the cerebrospinal fluid is difficult to explain, unless the synthesis of this highly polar substance takes place in the brain (105). Dent (106) speculated that the enzymatic block was present only in the brain and that the liver system was functioning normally. Ratner, on the other hand, found that the total activity of ASA synthetase in the brain is relatively small and far too low to account for the massive amounts of ASA excreted by these patients (105). ASA synthetase activity in the kidney, at least for some of the animal species investigated, is comparable to that in the liver (107). Ratner therefore proposed the possibility that ASase in the brain and kidney were inactive in the patient, whereas that in the liver functioned normally. In ASAciduria a deficiency of ASase activity has been found in several organs, including brain, liver, kidney, red blood cells, and skin fibroblasts. However, in only one patient was the ASase activity studied in all these organs. Glick et al. (55) thoroughly investigated ASase activity in the brain, liver, kidney, and erythrocytes of a neonate who died of ASAciduria (Case 5 in Table 7) and found that the enzyme activity was deficient only in the liver and erythrocytes. Their findings, as well as the demonstration of active kidney enzyme activity and deficient liver enzyme activity in a citrullinemic patient (Case 10), strengthen the hypothesis that the defect may be present in one or more but not in all organs of the body.

The Hypothesis of an Alternative Pathway for Urea Synthesis

Although alternative pathways for urea synthesis have been suggested from time to time, none has yet been proved. This will be discussed later by Dr. Colombo.

TREATMENT

The most striking clinical manifestation of the urea-cycle disorders is protein intolerance. Large protein intake results in the accumulation of blood ammonia and the respective urea-cycle intermediates. Therefore, current management of these patients is oriented toward limitation of protein intake and lowering the blood ammonia level.

Severe deficiency of urea-cycle enzymes is apparently not compatible with longevity. The course of the newborn affected with the lethal neonatal variant of any of these disorders has not been altered by any therapeutic measures, including peritoneal dialysis and exchange transfusion. Other patients improve on a restricted protein diet; if started early, normal psychomotor development is possible (108, 109). The protein requirement is the highest in the first few months of life and gradually decreases as the child grows older. A protein intake of 1.0 to 1.5 g/kg/day in the first few years of life will usually control the blood ammonia level and yet is adequate for growth. However, the protein need of each patient should be individually evaluated. The possibility of protein malnutrition and growth failure must be considered if it is necessary to keep the intake under 1 g protein/kg/day. Thus, one is faced with the problem of keeping a balance between the biochemical control of the disease and allowance of sufficient protein for growth.

As has been discussed previously, patients with urea-cycle disorders may have arginine deficiency because of decreased biosynthesis. Supplement of arginine to a low-protein diet has been used successfully in the treatment of ASAciduric infants (103, 108, unpublished observation).

Glutamine synthesis is an important mechanism for the detoxification of ammonia. In hyperammonemia the glutamine content of the blood is greatly increased. α-Ketoglutarate combines with ammonia to form, first, glutamic acid and then glutamine. The shunting of this compound from Kreb's tricarboxylic acid cycle and consequent interference with cerebral oxygen consumption has been suggested as the cause of the clinical symptoms. With this rationale, therapy with citric acid, the stable precursor of α-ketoglutaric acid, at 2 to 4 g/day, has been attempted in OTC deficiency, with varying degrees of success (7, 110, 111). Russell administered aspartate to a patient with cyclic vomiting and OTC deficiency and obtained a more dramatic effect than with citric acid therapy. When 400 mg aspartic acid was given simultaneously with a protein load, the expected abnormal rise in blood ammonia did not occur, even when the protein load was doubled (7). Although Russell's rationale for aspartate administration was to replenish the supply of α-ketoglutarate, it appears more reasonable to speculate that, in addition, aspartic acid may have removed ammonia by accelerating the pyrimidine biosynthetic pathway. Measurements of orotic acid or other pyrimidine intermediates were not mentioned in the report.

The concept of using nonessential nitrogen for protein synthesis in both healthy and uremic individuals is well established. Nutritionally essential amino acids in the diet may be replaced by their keto acid analogs. The use of these keto analogs of essential amino acids as a form of treatment will be discussed in another section by Dr. Walser.

Several measures used to lower blood ammonia in adult patients with hepatic encephalopathy such as lactulase, sterilization of the gut, and the administration of *N*-carbamyl-glutamate and L-arginine have not met with success in treating patients with congenital urea-cycle disorders (94, 112, 113). Peritoneal dialysis (114) and exchange transfusion (115) may bring about temporary improvement during acute hyperammonemic attacks.

Thyroid administration to two patients with citrullinemia resulted in no improvement in their biochemical abnormalities (91, 102).

Frequent small feedings are advisable in order to minimize the rise in blood ammonia in response to a protein load. Infection or other metabolic stress, such as surgery (36, 39), may precipitate a symptomatic hyperammonemic episode. The breakdown of tissue protein during illness may give rise to endogenous mino acids exceeding the capacity of the urea cycle. Adequate caloric intake during the illness from sources other than protein is advisable.

A recent development in preventive genetics is the increasing availability of genetic counseling and prenatal diagnosis. Activity of OTC, ASase, and arginase has been demonstrated in uncultured amniotic fluid cells (116); ASA synthetase and ASase activities are present in cultured amniotic fluid cells. Thus, prenatal diagnosis of the deficiency of these enzymes is potentially possible. In fact, fetuses deficient in ASase have been accurately diagnosed in two instances (67, 117). ASA is increased in the amniotic fluid as well as in the maternal urine. This is the only amino acid metabolic disorder in which accumulation of the metabolite occurs in utero.

REFERENCES

1. Colombo, J. P. (1971). Congenital disorders of the urea cycle and ammonia detoxication. In F. Falkner, N. Kretchmer, and E. Rossi, Eds. *Monographs in Paediatrics,* **1** S. Karger, Basel.

2. Levin, B. (1971). Hereditary metabolic disorders of the urea cycle. *Adv. Clin. Chem.* **14**, 65.

3. Levy, H. L., Madigan, P. M., and Shih, V. E. (1972). Massachusetts metabolic disorders screening program. I. Technics and results of urine screening. *Pediatrics* **49**, 825.

4. Levy, H. L. (1974). Neonatal screening for inborn errors of amino acid metabolism. *Clin. Endocr. Metab.* **3**, 153.

5. Freeman, J. M., Nicholson, J. F., Masland, W. S., Rowland, L. P. and Carter, S. (1964). Ammonia intoxication due to a congenital defect in urea synthetis. *J. Pediat.* **65**, 1039.

6. Kirkman, H. N. and Kiesel, J. L. (1969). Congenital hyperammonemia. *Pediat. Res.* **3**, 358.

7. Russell, A. (1969). A biochemical basis for migraine and cyclical vomiting. In J. D. Allen, K. S. Holt, J. T. Ireland, and R. S. Pollitt, Eds., *Enzymopenic Anaemias, Lysosomes and Other Papers,* Proceedings of the Society for the Study of Inborn Errors of Metabolism, E. & S. Livingstone, Ltd., Edinburgh.

8. Brown, T., Brown, H., Lansky, L., and Hug, G. (1974). Carbamyl phosphate synthetase and ornithine transcarbamylase in liver of Reye's syndrome. *N. Engl. J. Med.* **291**, 797.

9. Russell, A., Levin, B., Oberholzer, V. G., and Sinclair, L. (1962). Hyperammonaemia. A new instance of an inborn enzymatic defect of the biosynthesis of urea. *Lancet* **2**, 699.

10. Rosenberg, L. E. (1974). Diagnosis and management of inherited aminoacidopathies in the newborn and the unborn. *Clin. Endocr. Metabol.* **3**, 145.

11. Campbell, A. G. M., Rosenberg, L. E., Snodgrass, P. J. and Nuzum, C. T. (1971). Lethal neonatal hyperammonaemia due to complete ornithine-transcarbamylase deficiency. *Lancet* **2**, 217.

12. Campbell, A. G. M., Rosenberg, L. E., Snodgrass, P. J. and Nuzum, C. T. (1973). Ornithine transcarbamylase deficiency. A cause of lethal neonatal hyperammonemia in males. *N. Engl. J. Med.* **288**, 1.

13. Short, E. M., Conn, H. O., Snodgrass, P. J., Campbell, A. G. M., and Rosenberg, L. E. (1973). Evidence for x-linked dominant inheritance of ornithine transcarbamylase deficiency. *New Engl. J. Med.* **288**, 7.

14. Palmer, T., Oberholzer, V. G., Burgess, E. A., Butler, L. J., and Levin, B. (1974). Hyperammonaemia in 20 families. *Arch. Dis. Child.* **49**, 443.

15. McMurray, W. C., Mohyuddin, F., Rossiter, R. J., Rathbun, J. C., Valentine, G. H., Koegler, S. J., and Zarfas, D. E. (1962). Citrullinuria. A new amino-aciduria associated with mental retardation. *Lancet* **1**, 138.

16. Kennaway, N. G., Harwood, P. J., Ramberg, D. A., Koler, R. D., and Buist, N. R. M. (1975). Citrullinemia: enzymatic evidence for genetic heterogeneity. *Pediat. Res.* **9**, 554.

17. Wick, H., Brechbuhler, T., and Girard, J. (1970). Citrullinemia: elevated serum citrulline levels in healthy siblings. *Experientia* **26**, 823.

18. MacCready, R. A. and Levy, H. L. (1972). The problem of maternal phenylketonuria. *Am. J. Obstet. Gynecol.* **113**, 121.

19. Neville, B. G. R., Harris, R. F., Stern, D. J., and Stern, J. (1971). Maternal histidinaemia. *Arch. Dis. Child.* **46**, 119.

20. Lyon, I. C. T., Gardner, R. J. M., and Veale, A. M. O. (1974). Maternal histidinaemia. *Arch. Dis. Child.* **49**, 581.

21. Chernosky, M. E. and Owens, D. W. (1966). Trichorrhexis nodosa. *Arch. Derm.* **94**, 577.

22. Owens, D. W. and Chernosky, M. E. (1966). Trichorrhexis nodosa: in vitro reproduction. *Arch. Derm.* **94**, 586.

23. Rothman, S. (1954). *Physiology and Biochemistry of the Skin*, Univ. of Chicago Press, Chicago, p. 352.

24. Potter, J. L., Timmons, G. D., West, R., and Silvidi, A. A. (1974). Arginosuccinic aciduria. *Am. J. Dis. Child.* **127**, 724.

25. Coryell, M. E., Hall, W. K., Thevaos, T. G., Welter, D. A., Gatz, A. J., Horton, B. F., Sisson, B. D., Looper, J. W., Jr., and Farrow, R. T. (1964). A familial study of a human enzyme defect, argininosuccinic aciduria. *Biochem. Biophys. Res. Commun.* **14**, 307.

26. Westall, R. G. (1967). Treatment of argininosuccinic aciduria. *Amer. J. Dis. Child.* **113**, 160.

27. Grosfeld, J. C. M., Mighorst, J. A. and Moolhuysen, T. M. G. F. (1964). Argininosuccinic aciduria in monilethrix. *Lancet* **2**, 789.

28. Efron, M. L. and Hoefnagel, D. (1966). Argininosuccinic acid in monilethrix. *Lancet* **1**, 321.

29. Shelley, W. B., and Rawnsley, H. M. (1966). Aminogenic alopecia. *Trans. Assoc. Amer. Phys.* **47**, 146.

30. Winther, A., and Bundgaard, L. (1968). Argininosuccinic aciduria in hereditary hair diseases. *Acta Dermatovenéral.* **48**, 567.

31. Shih, V. E., Efron, M. L., and Moser, H. W. (1969). Hyperornithinemia, hyperammonemia and homocitrullinuria. A new disorder of amino acid metabolism associated with myoclonic seizures and mental retardation. *Amer. J. Dis. Child.* **117**, 83.

32. Shih, V. E., and Efron, M. L. (1972). Urea cycle disorders. *In* The Metabolic Basis of Inherited Disease. Third Edition. J. B. Stanbury, J. B. Wyngaarden, D. S. Frederickson, Eds., New York, McGraw Hill, Chapter 17, p. 370–392.

33. Bruton, C. J., Corsellis, J. A. N., and Russell, A. (1970). Hereditary hyperammonaemia. *Brain.* **93**, 423.

34. Solitaire, G. B., Shih, V. E., Nelligan, D. J., and Dolan, T. F., Jr. (1969). Argininosuccinic aciduria: clinical, biochemical, anatomical and neuropathological observations. *J. Ment. Defic. Res.* **13**, 153.

35. Lewis, P. D. and Miller, A. L. (1970). ASAciduria. Case report with neuropathological findings. *Brain* **93**, 413.

36. Hopkins, I. J., Connelly, J. F., Dawson, A. G., Hird, F. J. R., and Maddison, T. G. (1969). Hyperammonaemia due to ornithine transcarbamylase deficiency. *Arch. Dis. Child.* **44**, 143.

37. Levin, B., Abraham, J. M., Oberholzer, V. G., and Burgess, E. A. (1969). Hyperammonaemia: a deficiency of liver ornithine transcarbamylase. Occurrence in mother and child. *Arch. Dis. Child.* **44**, 152.

38. Carton, D., DeSchrijver, F., Kint, J., Van Durme, J., and Hooft, C. (1969). Case report. Argininosuccinic aciduria. Neonatal variant with rapid fatal course. *Acta Paediat. Scand.* **58**, 528.

39. Freeman, J. M., Nicholson, J. F., Schimke, R.T., Rowland, L. P., and Carter, S. (1970). Congenital hyperammonemia. *Arch. Neurol.* **23**, 430.

40. Ebels, E. J. (1972). Neuropathological observations in a patient with carbamylphosphate-synthetase deficiency and in two sibs. *Arch. Dis. Child.* **47**, 47.

41. Gelehrter, T. D. and Snodgrass, P. J. (1974). Lethal neonatal deficiency of carbamyl phosphate synthetase. *N. Engl. J. Med.* **290**, 430.

42. Cavanagh, J. B. and Kyu, M. H. (1969). Colchicine- like effect on astrocytes after portacaval shunt in rats. *Lancet* **2**, 620.

43. Cavanagh, J. B. and Kyu, M. H. (1971). Type II Alzheimer change experimentally produced in astrocytes in the rat. *J. Neurol. Sci.* **12**, 63.

44. Cavanagh, J. B. and Kyu, M. H. (1971). On the mechanism of type I. Alzheimer abnormality in the nuclei of astrocytes. *J. Neurol. Sci.* **12**, 241.

45. Cole, M., Rutherford, R. B. and Owen-Smith, F. (1972). Experimental encephalopathy in the primate. *Arch. Neurol.* **26**, 130.

46. Gibson, G. E., Zimber, A., and Krook, L. (1974). Brain histology and behavior of mice injected with urease. *J. Neuropathol. Exp. Neurol.* **33**, 201.

47. Donohue, W. L. (1967). Lesions in the central nervous system associated with inborn errors of amino acid metabolism. *Acta Paediat. (Uppsala)*, **56**, 116.

48. Moser, H. W., Efron, M. L., Brown, H., Diamond, R., and Neumann, C. G. (1967). Argininosuccinic aciduria. Report of two cases and demonstration of intermittent elevation of blood ammonia. *Amer. J. Med.* **42**, 9.

49. Gatfield, P. D., Taller, E., Wolfe, D. M. and Haust, M. D. (1975). Hyperornithinemia, hyperammonemia, and homocitrullinuria associated with decreased carbamyl phosphate synthetase I. activity. *Pediat. Res.* **9**, 488.

50. Wick, H., Bachmann, C., Baumgartner, R., Brechbühler, T., Colombo, J. P., Wiesmann, U., Mihatsch, M. J., and Ohnacker, H. (1973). Variants of citrullinaemia. *Arch. Dis. Child.* **48**, 636.

51. Strandholm, J. J., Buist, N. R. M., Kennaway, N. G., and Curtis, H. T. (1971). Excretion of α-N-acetylcitrulline in citrullinaemia. *Biochim. Biophys. Acta* **244**, 214.

52. Palmer, T., Oberholzer, V. G., Levin, B., and Burgess, E. A. (1973). Urinary excretion of argininosuccinic acid. *Clin. Chim. Acta* **47**, 443.

53. Ratner, S. and Kunkemueller, M. (1966). Separation and properties of argininosuccinate and its two anhydrides and their detection in biological materials. *Biochemistry* 5, 1821.

54. Lee, C. R. and Pollitt, R. J. (1972). New derivatives of argininosuccinic acid in the urine of a patient with argininosuccinic aciduria. *Biochem. J.* **126**, 79.

55. Glick, N. R., Snodgrass, P. J., and Schafer, I. A. (1976). Neonatal argininosuccinic aciduria with normal brain and kidney but absent liver argininosuccinate lyase activity. *Amer. J. Hum. Genet.*

56. Levin, B., Oberholzer, V. G., and Sinclair, L. (1969). Biochemical investigations of hyperammonaemia. *Lancet* **2**, 170.

57. Kesner, L. (1965). The effect of ammonia administration on orotic acid excretion in rats. *J. Biol. Chem.* **240**, 1722.

58. Statter, M., Russell, A., Abzug-Horowitz, S., and Pinson, A. (1974). Abnormal orotic acid metabolism associated with acute hyperammonaemia in the rat. *Biochem. Med.* **9**, 1.

59. Wolfe, D. M., and Gatfield, P. D. (1975). Leukocyte urea cycle enzymes in hyperammonemia. *Pediat. Res.* **9**, 531.

60. Cathelineau, L., Saudubray, J., and Polonovski, C. (1974). Heterogenous mutations of the structural gene of human ornithine carbamyltransferase as observed in five personal cases. *Enzyme* **18**, 103.

61. Terheggen, H. G., Schwenk, A., Lowenthal, A., van Sande, M. and Colombo, J. P. (1969). Argininaemia with arginase deficiency. *Lancet* **2**, 748.

62. Cederbaum, S. D., Shaw, K. N. F. and Valente, M. (1973). Hyperargininemia with red cell arginase deficiency. *Amer. J. Hum. Genet.* **25**, 20A.

63. Murphey, W. H., Patchen, L., and Guthrie, R. (1972). Screening tests for argininosuccinic aciduria, orotic aciduria, and other inherited enzyme deficiencies using dried blood specimens. *Biochem. Genet.* **6**, 51.

64. Tedesco, T. A., and Mellman, W. J. (1967). Argininosuccinate synthetase activity and citrulline metabolism in cells cultured from a citrullinemic subject. *Proc. Natl. Acad. Sci.* **57**, 829.

65. Shih, V. E., Littlefield, J. W., and Moser, H. W. (1969). Argininosuccinase deficiency in fibroblasts cultured from patients with argininosuccinic aciduria. *Biochem. Genet.* **3**, 81.

66. Jacoby, L. B., Littlefield, J. W., Milunsky, A., Shih, V. E. and Wilroy, R. S., Jr. (1972). A microassay for argininosuccinase in cultured cells. *Amer. J. Hum. Genet.* **24**, 321.

67. Goodman, S. I., Mace, J. W., Turner, B., and Garrett, W. J. (1973). Antenatal diagnosis of argininosuccinic aciduria. *Clin. Genet.* **4**, 236.

68. Hill, H. Z. and Goodman, S. I. (1974). Detection of inborn errors of metabolism. III. Defects in urea cycle metabolism. *Clin. Genet.* **6**, 79.

69. Palmer, T., Oberholzer, V. G., and Levin, B. (1974). Amino acid levels in patients with hyperammonaemia and argininosuccinic aciduria. *Clin. Chim. Acta* **52**, 335.

70. Shih, V. E., and Mandell, R. (1974). Metabolic defect in hyperornithinaemia. *Lancet* **2**, 1522.

71. Brown, G. W., Jr., and Cohen, P. P. (1959). Comparative biochemistry of urea synthesis. I. Methods for the quantitative assay of urea cycle enzymes in liver. *J. Biol. Chem.* **234**, 1769.

72. Snodgrass, P. J., and Parry, D. J., (1969). The kinetics of serum ornithine carbamyltransferase. *J. Lab. Clin. Med.* **73**, 940.

73. Cathelineau, L., Saudubray, J. M., and Polonovski, C. (1972). Ornithine carbamyl transferase: the effects of pH on the kinetics of a mutant human enzyme. *Clin. Clim. Acta* **41**, 305.

74. Matsuda, I., Arashima, S., Nambu, H., Takekoshi, Y., and Anakura, M. (1971). Hyperammonemia due to a mutant enzyme of ornithine transcarbamylase. *Pediatrics* **48**, 595.

75. Arashima, S. and Matsuda, I. (1971). Ornithine transcarbamylase, an isoelectric point (pH) isozyme in human liver and its deficiency. *Biochem. Biophys. Res. Commun.* **45**, 145.

76. Nagayama, E., Kitayama, T., Oguchi, H., Ogata, K., Tamura, E., and Onisawa, J. (1970). Hyperammonemia: a deficiency of liver ornithine transcarbamylase. *Paediat. Univ. Tokyo* **18**, 167.

77. Thaler, M. M., Hoogenraad, N. J., Boswell, M. (1974). Reye's syndrome due to a novel protein-tolerant variant of ornithine-transcarbamylase deficiency. *Lancet* **1**, 438.

78. Scott-Emuakpor, A., Higgins, J. V., and Kohrman, A. F. (1972). Citrullinemia: A new case, with implications concerning adaptation to defective urea synthesis. *Pediat. Res.* **6**: 626.

79. Spector, E. B., and Bloom, A. D. (1973). Citrullinemic lymphocytes in long term culture. *Pediat. Res.* **7**, 700.

80. Tomlinson, S., and Westall, R. G. (1964). Argininosuccinic aciduria: argininosuccinase and arginase in human blood cells. *Clin. Sci.* **26**, 261.

81. Kint, J. and Carton, D. (1968). Deficient argininosuccinase activity in brain in argininosuccinic aciduria. *Lancet* **2**, 635.

82. Bray, R. C. and Ratner, S. (1971). Argininosuccinase from bovine kidney: comparison of catalytic, physical, and chemical properties with the enzyme from bovine liver. *Arch. Biochem. Biophys.* **146**, 531.

83. Colombo, J. P. and Baumgartner, R. (1969). Argininosuccinate cleavage enzyme of the kidney in ASAciduria, *in Proc. 6th Symp. Soc. Study of Inborn Errors of Metabolism, Zurich, June 24–25, 1968*, E. & S. Livingstone, Ltd., Edinburgh, p. 119.

84. Roerdink, F. H., Gouw, W. L. M., Okken, A., Van der Blij, J. F., Luit-de Haan, G., and Hommes, E. A. (1973). Citrullinemia. Report of a case with studies on antenatal diagnosis. *Pediat. Res.* **7**, 863.

85. Duffy, T. E., Vergara, F., and Plum, F. (1974). α-ketoglutaramate in hepatic encephalopathy. In F. Plum, Ed., *Brain Dysfunction in Metabolic Disorders* **53**. Ravern Press, New York, pp. 39–52.

86. Vergara, F., Plum, F., Duffy, T. E. (1974). α-ketoglutaramate: increased concentrations in the cerebrospinal fluid of patients in hepatic coma. *Science* **183**, 81.

87. Stahl. J. (1963). Studies of the blood ammonia in liver disease: its diagnostic, prognostic, and therapeutic significance. *Ann. Intern. Med.* **58**, 1.

88. McDermott, W. V. and Adams, R. D. (1954). Episodic stupor associated with an Eck fistula in the human with particular reference to the metabolism of ammonia. *J. Clin. Invest.* **33**, 1.

89. Huttenlocher, P. R., Schwartz, A. D., and Klatskin, G. (1969). Reye's syndrome: ammonia intoxication as a possible factor in the encephalopathy. *Pediatrics* **43**, 443.

90. Victor, M., Adams, R. D., and Cole, M. (1965). The acquired (non-Wilsonian) type of chronic hepatocerebral degeneration. *Medicine* **44**, 345.

91. McMurray, W. C., Rathbun, J. C., Mohyuddin, F., and Koegler, S. J. (1963). Citrullinuria. *Pediatrics* **32**, 347.

92. Berl, S., Takagaki, G., Clarke, D. D. and Waelsch, H. (1962). Metabolic compartments in vivo. Ammonia and glutamic acid metabolism in brain and liver. *J. Biol. Chem.* **237**, 2562.

93. Gabuzda, G. J. (1967). Ammonium metabolism and hepatic coma. Gastroenterology, **53**, 806.

94. Schenker, S., Breen, K. J., and Hoympa, A. M. (1974). Hepatic encephalopathy: current status. *Gastroenterology* **66**, 121.

95. James, I. M., Dorf, G., Hall, S., Michel, H., Dojcinov, D., Gravagne, G., and MacDonell, L. (1972). Effect of ornithine α-ketoglutarate on disturbances of brain metabolism caused by high blood ammonia. *Gut* **13**, 551.

96. Hindfelt, B. and Siesjo, B. K. (1971). Cerebral effects of acute ammonia intoxication. II. The effect upon energy metabolism. *Scand. J. Clin. Lab. Invest.* **28**, 365.

97. Hindfelt, B. (1972). The effect of sustained hyperammonemia upon the metabolic state of the brain. *Scand. J. Clin. Lab. Invest.* **30**, 245.

98. Kennan, A. L. and Cohen, P. P. (1961). Ammonia detoxication in liver from humans. *Proc. Soc. Exp. Biol. Med.* **106**, 170.

99. Roberge, A, Dorval, G., and Charbonneau, R. (1969). Le métabolism de l'ammoniaque. IV. Effet in vivo d'injections prolongées de sulfate d'ammonium et d'arginine sur l'activité specifique des enzymes du cycle de l'ureé. *Rev. Can. Bio.* **28**, 119.

100. McMurray, W. C., Mohyuddin, F., Bayer, S. M., and Rathbun, J. C. (1964). *Citrullinuria: A disorder of amino acid metabolism associated with mental retardation. International Copenhagen Congress on the Scientific Study of Mental Retardation*, Denmark, 117, August 7–14.

101. Crane, C. W., Gay, W. M. and Jenner, F. A. (1969). Urea production from labelled ammonia in argininosuccinic aciduria. *Clin. Chim. Acta* **24**, 445.

102. Morrow, G., Barness, L. A., and Efron, M. L. (1967). Citrullinemia with defective urea production. *Pediatrics* **40**, 565.

103. Cederbaum, S. D., Shaw, K. N. F., Valente, M., and Cotton, M. E. (1973). Argininosuccinic aciduria. *Amer. J. Ment. Defic.* **77**, 395.

104. Hambraeus, L., Hardell, L. I., Westphal, O., Lorentsson, R., and Hjorth, G. (1974). Argininosuccinic aciduria. Report of three cases and the effect of high and reduced protein intake on the clinical state. *Acta Paediatr. Scand.* **63**, 525.

105. Ratner, S., Morrell, H., and Carvalho, E. (1960). Enzymes of arginine metabolism in brain. *Arch. Biochem.* **91**, 280.

106. Dent, C. E. (1959). Argininosuccinic aciduria. A new form of mental deficiency due to metabolic causes. *Proc. R. Soc. Med.* **52**, 885.

107. Ratner, S., and Petrack, B. (1953). The mechanism of arginine synthesis from citrulline in kidney. *J. Biol. Chem.* **200**, 175.

108. Shih, V. E. (1972). Early dietary management in an infant with argininosuccinase deficiency: preliminary report. *J. Pediat.* **80**, 645.

109. Buist, N. R. M., Kennaway, N. G., Hepburn, C. A., Strandholm, J. J. and Ramberg, D. A. (1974). Citrullinemia: investigation and treatment over a four-year period. *J. Pediat.* **85**, 208.

110. Levin, B. and Russell, A. (1967). Treatment of hyperammonemia. *Amer. J. Dis. Child.* **113**, 142.

111. Sunshine, P., Lindenbaum, J. E., Levy, H. L., and Freeman, J. M. (1972). Hyperammonemia due to a defect in hepatic ornithine transcarbamylase. *Pediatrics* **50**, 100.

112. Hsia, Y. Edward (1974). Inherited hyperammonemic syndromes. *Gastroenterology* **67**, 347.

113. Gelehrter, T. D. and Rosenberg, L. E., Ornithine transcarbamylase deficiency. Unsuccessful therapy of neonatal hyperammonemia with N-carbamyl-L-glutamate and L-arginine. *N. Engl. J. Med.* **292**, 351.

114. Herrin, J. T. and McCredie, D. A. (1969). Peritoneal dialysis in the reduction of blood ammonia levels in a case of hyperammonaemia. *Arch. Dis. Child.* **44**, 149.

115. Kang, E. S., Snodgrass, P. J., and Gerald, P. S. (1973). Ornithine transcarbamylase deficiency in the newborn infant. *J. Pediat.* **82**, 642.

116. Nadler, H. L. and Gerbie, A. B. (1969). Enzymes in non-cultured amniotic fluid cells. *Am. J. Obstet. Gynecol.* **103**, 710.

117. Hartlage, P. L., Coryell, M. E., Hall, W. K., and Hahn, D. A. (1974). Argininosuccinic aciduria: perinatal diagnosis and early dietary management. *J. Pediat.* **85**, 86.

118. Allan, J. D., Cusworth, D. C., Dent, C. E. and Wilson, V. K. (1958). A disease, probably hereditary, characterized by severe mental deficiency and a constant gross abnormality of amino acid metabolism. *Lancet* **1**, 182.

119. Arashima, S. and Matsuda, I. (1972). A case of carbamyl phosphate synthetase deficiency. *Tohoku J. Exp. Med.* **107**, 143.

120. Armstrong, M. D. and Stemmermann, M. G. (1964). An occurrence of argininosuccinic aciduria. *Ped.* **33**, 280.

121. Batshaw, M., Brusilow, S., and Walser, M. (1975). Treatment of carbamyl phosphate synthetase deficiency with keto-analogues of essential amino acids. *N. Engl. J. Med.* In press.

122. Baumgartner, R., Scheidegger, S., Stalder, G., and Hottinger, A. (1968). Neonatal death due to argininosuccinic aciduria. *Helv. Paediat. Acta* **23**, 77.

123. Billmeier, G. J., Molinary, S. V., Wilroy, R. S. Jr., Duenas, D. A. and Brannon, M. E. (1974). Argininosuccinic aciduria: investigation of an affected family. *J. Pediat.* **84**, 85.

124. Blackmore, R. J., Lyon, I. C. T. and Veale, A. M. O. (1972). Argininosuccinic aciduria. *Proc. Univ. Otago Med. Sch.* **50**, 4.

125. Carson, N. A. J. and Neill, D. W. (1962). Metabolic abnormalities detected in a survey of mentally backward individuals in Northern Ireland. *Arch. Dis. Child.* **37**, 505.

126. Cathelineau, L., Navarro, J., Polonovski, C., and Saudubray, J. M. (1973). X-linked transmission of structural gene mutations responsible for ornithine-transcarbamylase deficiencies. *Lancet* **1**, 261.

127. Colombo, J. P., Terheggen, H. G., Lowenthal, A., Van Sande, M., and Roggers, S. (1973). Argininaemia In F. A. Hommes and C. J. Vander Berg. Eds., Inborn Errors of Metabolism. Academic Press.

128. Corbeel, L. M., Colombo, J. P., Van Sande, M. and Weber, A. (1969). Periodic attacks of lethargy in a baby with ammonia intoxication due to a congenital defect in ureogenesis. *Arch. Dis. Child.* **44**, 681.

129. Edkins, E. and Hockey, A. (1965). A case of argininosuccinic aciduria. 4th Ann. Interstate Conf. on Mental Deficiency. p. 54, Melbourne, October.

130. Farrell, G., Rauschkolb, E. W., Moure, J., Headlee, R. E., and Moser, H. (1969). Argininosuccinic aciduria. *Tex. Med.* **65**, 90.

131. Farriaux, J. P., Pieraert, C., and Fontaine, G. (1975). Survival of infant with argininosuccinic aciduria to 3 months of age. *J. Pediat.* **86**, 639.

132. Ghisolfi, J., Augier, D., Martinez, J., Barthe, Ph., Andrieu, P., Besse, P., and Régnier, Cl. (1972). Forme neo-natale de citrullinemie a evolution mortelle rapide. *Pediatrie* **28**, 55.

133. Goldstein, A. S., Hoogenraad, N. J., Johnson, J. D., Fukanaga, K., Swierczewski, E., Cann, H. M. and Sunshine, P. (1974). Metabolic and genetic studies of a family with ornithine transcarbamylase deficiency. *Pediat. Res.* **8**, 5.

134. Hommes, F. A., De Groot, C. J., Wilmink, C. W., and Jonxis, J. H. P. (1969). Carbamylphosphate synthetase deficiency in an infant with severe cerebral damage. *Arch. Dis. Child.* **44**, 688.

135. Krebs, H. A., Hems, R., and Lund, P. (1973). Regulatory mechanisms in the synthesis of urea. In F. A. Hommes, C. J. Van Den Berg, Eds., *Inborn Errors of Metabolism*, Academic Press, London & New York, p. 201.

136. Levin, B. and Dobbs, R. H. (1968). Hereditary metabolic disorders involving the urea cycle. *Proc. R. Soc. Med.* **61**, 773.

137. Levin, B., Dobbs, R. H., Burgess, E. A., and Palmer, T. (1969). Hyperammonaemia. A variant type of deficiency of liver ornithine transcarbamylase. *Arch. Dis. Child.* **44**, 162.

138. Levin, B., Mackay, H. M. M. and Oberholzer, V. G. (1961). Argininosuccinic aciduria. An inborn error of amino acid metabolism. *Arch. Dis. Child.* **36**, 622.

139. Levy, H. L. and Shih, V. E. (1974). Maternal phenylketonuria and hyperphenylalaninemia. *Pediat. Res.* **8**, 391.

140. MacLeod, P., Mackenzie, S., and Scriver, C. R. (1972). Partial ornithine carbamyl transferase deficiency: an inborn error of the urea cycle presenting as orotic aciduria in a male infant. *Canad. Med. Assoc. J.* **107**, 405.

141. McMurray, W. C. and Mohyuddin, F. (1972). Citrullinuria, *Lancet* **2**, 352.

142. Miller, A. L. and McLean, P. (1967). Urea cycle enzymes in the liver of a patient with argininosuccinic aciduria. *Clin. Sci.* **32**, 385.

143. Moore, P. T., Martin, M. C., Coffey, V. P., and Stokes, B. M. (1968). Argininosuccinic aciduria—a case report on a rare condition, *J. Ir. Med. Assoc.* **61**, 172.

144. Nicholson, J. F. and Freeman, J. M. (1972). Metabolism of compounds labelled with ^{15}N by an infant with congenital hyperammonemia. *Pediat. Res.* **6**, 252.

145. Odievre, C., Charpentier, C., Cathelineau, L., Vedrenne, J., Delacoux des Roseaux, F. and Mercie, C. (1973). Hyperammoniemie constitutionnelle avec deficit en carbamyl-phosphate-synthetase. Evolution sous regime dietetique. *Arch. Fr. Pediat.* **30**, 5.

146. Okken, A., Vander Blij, J. F., and Hommes, F. A. (1973). Citrullinaemia and brain damage. *Pediat. Res.* **7**, 45.

147. Pollitt, R. J. (1973). Argininosuccinate lyase levels in blood, liver and cultured fibroblasts of a patient with argininosuccinic aciduria. *Clin. Chim. Acta* **46**, 33.

148. Porath, U., Liebler, G., and Schreier, K. (1969). Eine besondere verlaufsform der arginbernsteinsäure-krankheit. *Arch. Kinderheilk.* **179**, 283.

149. Ratner, S. (1973). Enzymes of arginine and urea synthesis. *Adv. Enzymol.* **31**, 1–190.

150. Rosenberg, L. E., Campbell, A. G. M., Snodgrass, P. J., and Nuzum, C. T. (1971). Complete ornithine transcarbamylase deficiency: a cause of lethal neonatal hyperammonemia. Proc. 4th International Congress of Human Genetics, Paris, p. 154.

151. Salle, B., Levin, B., Longin, B., Richard, P., Andre, M., and Gauthier, J. (1972). Hyperammoniemie congenitale par deficit en ornithine carbamyl transferase et carbamyl phosphate synthetase. *Arch. Fr. Pediatr.* **29**, 493.

152. Saudubray, J. M., Cathelineau, L., Charpentier, C., Boisse, J., Allaneau, C., Le Bont, H., and Lesage, B. (1973). Deficit hereditaire en ornithine-carbamyl-transferase avec anomalie enzymatique qualitative. *Arch. Fr. Pediatr.* **30**, 15.

153. Schreier, K., and Leuchte, G. (1965). Arginbernsteinsäure-Krankheit. *Deut. Med. Wschr.* **90**, 864.

154. Scott, C. R., Teng, C. C., Goodman, S. I., Greensher, A., and Mace, J. W. (1972). X-linked transmission of ornithine-transcarbamylase deficiency. *Lancet* **2**, 1148.

155. Siegel, N. J., and Brown, R. S. (1973). Peritoneal clearance of ammonia and creatinine in a neonate. *J. Pediatr.* **82**, 1044.

156. Snodgrass, P. J (1968). The effects of pH on the kinetics of human liver ornithine-carbamyl phosphate transferase. *Biochemistry* **7**, 3047.

157. Van der Zee, S. P. M., Trijbels, J. M. F., Monnens, L. A. H., Hommes, F. A., and Schretlen, E. D. A. M. (1971). Citrullinaemia with rapidly fatal neonatal course. *Archiv. Dis. Child.* **46**, 847.

158. Vidailhet, M., Levin, B., Dautrevaux, M., Paysant, P., Gelot, S., Badonnel, Y., Pierson, M., and Neimann, N. (1971). Citrullinemie. *Arch. Fr. Pediatr.* **28**, 521.

159. Wagstaff, T. I., Burgess, E. A., Oberholzer, V. G., and Palmer, T. (1974). Argininosuccinic aciduria: antenatal investigations in an affected family. *Amer. J. Obstet. Gynecol.* **120**, 560.

160. Wallis, K., Beer, S., and Fischl, J. (1963). A family affected by argininosuccinic aciduria. *Helv. Paediat. Acta* **4**, 339.

161. Westall, R. G. (1960). Argininosuccinic aciduria: identification and reactions of the abnormal metabolite in a newly described form of mental disease, with some preliminary metabolic studies. *Biochem. J.* **77**, 135.

162. Wick, H., Bachmann, C., Baumgartner, R., Brechbüehler, T., and Colombo, J. P. (1971). Two extreme variants of citrullinaemia. *Proc. XIII Internat. Congress of Pediat.* 243.

163. Cathelineau, L., Navarro, J., Aymard, P., Baudon, J.-J., Mondet, Y., Polonovski, C., and Laplane, R. (1972). Hyperammoniemie hereditaire par anomalie qualitative de l'ornithine-carbamyl-transferase hepatique et intestinale. *Arch. Fr. Pediatr.* **29**, 713–736.

164. Cathelineau, L., Saudubray, J. M., Navarro, J., and Polonovski, C. (1973). Transmission par le chromosome X du gène de structure de l'ornithine-carbamyl-transférase. Étude de trois familles. *Ann. Genet.* **16**, 173–182.

165. Thoene, J., Beach, B., Kulovich, S., Batshaw, M., Walser, M., Brusilow, S., and Nyhan, W. (1975). Keto acid treatment of neonatal citrullinemia. *Am. J. Hum. Genet.* **27**, 88A.

24

Argininemia

J. P. Colombo
C. Bachmann
Department of Clinical Chemistry
Inselspital
University of Berne
Berne, Switzerland

H. G. Terheggen
Department of Pediatrics
Kinder-Krankenhaus der Stadt Köln
Köln, Germany

F. Lavinha
A. Lowenthal
Department of Neurochemistry
Born-Bunge Foundation
Berchem-Antwerp, Belgium

Argininemia is due to a defect in the last step of the urea cycle. We observed a German family with six living children of whom three girls presented this condition (1–3).

This report deals with the investigations done in the third child (II/7, Fig. 1). The pertinent points of this disorder will be discussed in relation to the findings in the two other affected children.

CLINICAL FINDINGS

The mother's pregnancy, her seventh, was complicated by hypertonia and proteinuria. The girl (II/7) was born after a normal delivery at term. Despite several requests the parents were not willing to hospitalize the child immediately. Hospitalization could only be achieved at 3 weeks of age. Physical

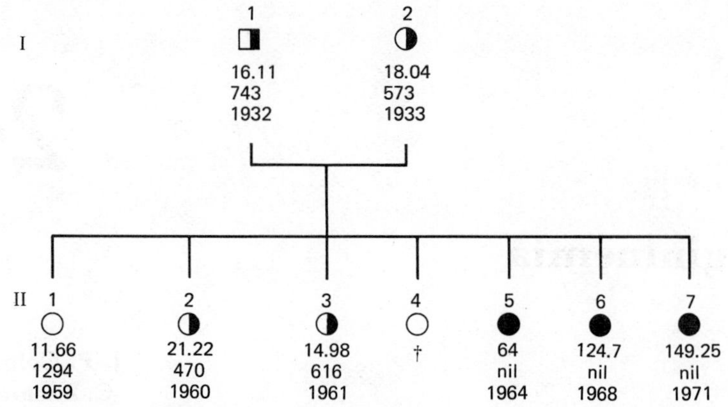

Figure 1. Arginine concentration in plasma (controls: 9.16 ± 4.5 umole/100ml) and red blood cell arginase (controls: 710 - 1330 mole/hour/g Hb) in the family with argininemia.

examination was normal except for muscular hypertonicity of the lower extremities. At 5 months of age, the girl exhibited pronounced motor retardation. In addition at 7 months the neurological examination revealed athetosis. By now the girl is 4 years old and shows marked psycho-motor retardation, athetosis, and spasticity of the lower extremities, despite low protein treatment. The EEG in contrast to that of her siblings was normal throughout this time.

BIOCHEMICAL INVESTIGATIONS

Amino Acids and Ammonia

The first examination was done on cord blood, where already elevated levels of arginine were observed (Table 1). Subsequently plasma amino acid examinations showed a persistent elevation of the methionine peak, but the concentrations of tyrosine, glycine, and ornithine were decreased. The other amino acids were present in normal concentrations. The values of arginine increased slightly in the following weeks; ornithine remained low. Some ornithine may still originate from proline (5).

In the urine which the child passed on the second day of life, already a cystinuria-lysinuria pattern was present (Table 1). This is due to an overflow of arginine, competing for the transport sites with the other basic amino acids. In the cerebrospinal fluid arginine amounted to eight times the normal value (Table 2). There was also an elevation of the glutamine concentration. This suggests an increased ammonia concentration in the central nervous system. Indeed the blood ammonia levels were particularly increased at normal and high protein intake, as shown in Table 3.

Table 1 Amino Acids in Plasma and Urine

	Reference Values	Plasma (μmoles/100 ml) 1st Day Cord Blood	7th Day	25th Day	41st Day	56th Day
Glutamine	46.93 ± 10.93	52.91	37.70	17.72	18.60	27.28
Citrulline	3.45 ± 0.92	1.17	1.24	2.01	4.72	3.39
Cystine	2.83 ± 2.17	8.78	1.85	0.87	3.38	7.74
Lysine	18.87 ± 3.83	37.77	16.08	28.83	23.56	56.90
Ornithine	10.86 ± 3.14	4.05[a]	5.73	4.18	7.20	5.57
Arginine	8.17 ± 1.81	25.20[a]	33.42	56.26	157.92	148.25
		Urine (μmoles/g creatinine) 2nd Day				
Glutamine	290 ± 88	398	189	1269	2328	
Citrulline	9 ± 8	6	n.d.	102	1401	
Cystine	83 ± 18	352	126	685	2323	
Lysine	186 ± 77	471	224	2677	10649	
Ornithine	35 ± 16	21	63	147	2356	
Arginine	33 ± 14	108	n.d.	500	17787	

[a] Cord blood, normal values according to (4):
Arginine 10.57 ± 10.0 μmoles/100 ml; ornithine 8.94 ± 2.58 μmoles/100 ml.
n.d. = not determined.

Table 2 Amino Acid and Ammonia Concentration in CSF on Normal Protein Intake (μmoles/100 ml)

	Reference Values	CSF
Glutamine	45.47 ± 12.08	130.02
Citrulline	0.21 ± 0.07	1.21
Cystine	traces	—
Lysine	1.86 ± 0.64	1.51
Ornithine	0.84 ± 0.23	0.81
Arginine	1.42 ± 0.74	11.32
NH_3	<100 μg/100 ml	237–329

Table 3 Ammonia in Blood (μg/100 ml), Urea (mg/100 ml), and Arginine (μmoles/100 ml) in Plasma in Relation to Daily Protein Supply[a]

	Protein supply (2.76–3.0 g/kg/day) 8 week period	Low-protein diet (1.5 g/kg/day) 4 week period
NH_3 Fasting	237–329	143–177
NH_3 Postprandial	356–590	280
Urea	8–39	5.2–24.1
Arginine	25–157	71–73

[a] Normal values: ammonia 45–110 μg/100 ml; urea 10–25 mg/100 ml; arginine 8.17 ± 1.81 μmoles/100 ml.

Enzyme Analysis and Loading Tests

Plasma arginine concentration and red blood cell arginase was determined in the parents and their living children (Fig. 1). Of the three clinically healthy children one had normal plasma arginine concentrations, whereas the values of the parents and the two other children exceeded the normal range. The red cell arginase activity corresponded in each case to the concentration of the plasma arginine values. In case II/7 an elevated concentration of arginine was also measured in the red blood cells (42.2 μmoles/100 ml RBC, normal control 1.05). The elevated levels of arginine in the erythrocytes point to a defective catabolism of this amino acid due to arginase deficiency. The plasma-erythrocyte ratio of arginine in this patient was 3.54. The normal ratio established by Levy and Barkin amounts to 2.16 (6). The parents and two of the children are carriers of the pathological gene in its heterozygote form. Three children are affected clinically and are homozygotes. This pattern points to an autosomal recessive inheritance of this disorder.

Following intravenous arginine administration (300 mg/kg) the disappearance rate of this amino acid from plasma was markedly delayed (Fig. 2). This delayed plasma arginine elimination makes it likely that the liver enzyme is also deficient. Unfortunately, because of refusal of the parents, liver enzymes could never be analyzed.

Urinary Excretion of Guanidino Derivatives

The following compounds were determined in the urine by column chromatography: guanidino succinic acid (GSA), guanidino acetic acid (GAA), α-*N*-acetylarginine, γ-guanidino butyric acid (GBA), argininic acid.

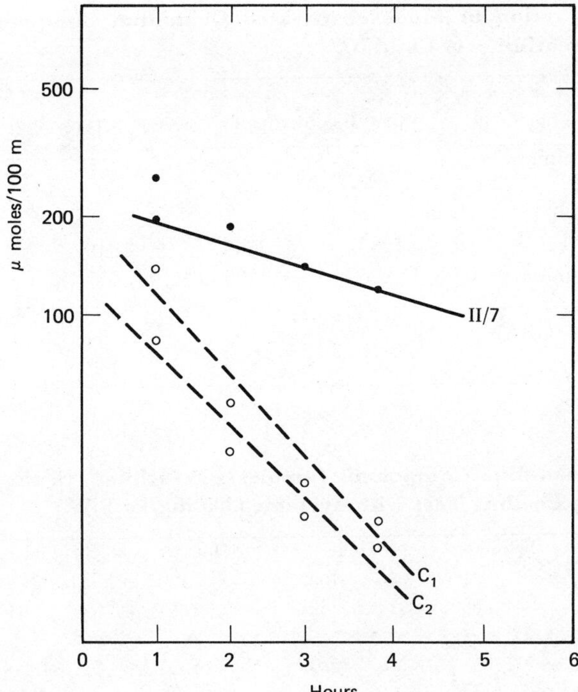

Figure 2. Disappearance rate of intravenously injected arginine in child II/7 (k = 0.28%/minute) and two control individuals C_{1+2} (k_1 = 0.92%/minute, k_2 = 0.87%/minute).

Elevated levels of GAA, α-*N*-acetylarginine, GBA, argininic acid were found in the youngest sibling already on the second day of life, compared with control subjects. Argininic acid was never measurable in normal controls. The elevation persisted throughout a 3 month observation period, most of the time of which the child was on a low-protein diet. The excretion of GSA was very low. Only traces were found, whereas normal controls excreted substantial amounts of this compound. The same observation regarding the excretion of the guanidino derivatives was done in the child II/5 (Table 4).

Loading tests in child II/7 with intravenously administered arginine (300 mg/kg B.W.) led to a marked increase of all compounds except for GSA. In the controls, only GSA, GAA, α-*N*-acetyl arginine increased after the arginine load, whereas GBA remained unchanged. (Table 5).

The excretion of these derivatives in this disorder particularly pronounced following intravenous administration of arginine points to a connection with argininemia. This connection may give raise to some speculation on the origin of these compounds.

Table 4 Excretion of Monosubstituated Guanidino Compounds in Urine (μmoles/g creatinine) of Child II/7

	2nd Day	79th Day	99th Day	Control Subjects *a*	Control Subjects *b*
Guanidinosuccinic acid (GSA)	150	—	—	88	81
Guanidino acetic acid (GAA)	1249	2668	1946	657	145
α-*N*-Acetylarginine	226	1285	657	22	31
γ-Guanidino-butyric acid (GBA)	75	95	89	26	5
Argininic acid	55	419	439	—	—

Table 5 Guanidino Compounds (μmoles/g creatinine) in the Urine after Intravenous Loading Test with Arginine (300 mg/kg B.W.)

	Before		1 Hour		2 Hours		3 Hours		4 Hours	
GSA	—	81[a]	—	108	—	173	—	145	—	80
		88		66		65		93		119
GAA	1182		1675		4031		3387		1533	
		657		1289		1220		1020		294
		145		565		338		430		351
α-*N*-Acetyl-	649		1884		7020		5264		1987	
arginine		22		597		372		273		41
		31		989		710		325		165
GBA	19		218		356		202		168	
		26		2		29		10		13
		5		57		16		10		
Argininic	61		79		350		321		178	
acid		0		0		0		0		0

[a] Control individuals.

Guanidino Succinic Acid In normal subjects the excretion of GSA in the urine is very low. It is found in increased amounts in serum and urine in states of nitrogen retention, particularly in uremic patients (7–9). Stein et al. also mention an elevated excretion after intraperitoneal administration of arginine to rats (10). The failure to demonstrate the presence of GSA in the urine of six out of seven patients with genetic defects of the urea-cycle led them to speculate that arginine, which is poorly synthetized in these patients, is a precursor in GSA synthesis (10).

Our investigations are not in agreement with this suggestion. In spite of a high arginine concentration no measurable amount of GSA was detected in the patients' urine, even after a load of arginine. Since only traces of GSA could be found in our patients, whereas measurable values were found in the control subjects, one could speculate that not arginine but a functioning arginase is necessary for GSA synthesis. A transamidination pathway could be postulated. In this case a transamidination function would have to be attributed to arginase where arginine would be the amidine donor and aspartate the acceptor leading to GSA. These possibilities, however, have so far been excluded (11, 12). The origin of GSA in normal subjects and uremic patients therefore still remains unclear.

A hypothesis has recently been forwarded by Koller et al. where GSA would be derived from canavaninosuccinic acid through reductive cleavage (13). Canavaninosuccinate would be derived from two molecules of aspartate by conversion of one of the molecules of aspartate to canaline. Condensation of canaline with carbamyl phosphate would form ureidohomoserine. Condensation of ureidohomoserine with aspartate would form canavaninosuccinate. Menyhárt et al. suggest that urea in high concentration (30 mmoles) inhibits argininosuccinate-lyase, resulting in a consecutive accumulation of argininosuccinate (ASA). Due to the elevated ASA concentration, a new enzyme is activated, converting ASA into ornithine and GSA through ammoniolysis (14).

This hypothesis would imply an arginase normally functioning, as suggested above, excluding any transamidination mechanism catalyzed by this enzyme. But it also implies an elevated urea concentration which is not present in argininemia.

Guanidino Acetic Acid Guanidino acetic acid (GAA) is a well-known intermediate in the creatine biosynthesis. It is normally excreted in human urine (13). A diminished excretion and elevated plasma levels are found in uremic patients (7, 13). The elevated excretion and its further increase in our patients after arginine loading as well as a persistently low plasma glycine concentration may be explained by an augmented transamidination between arginine and glycine. Creatine values were not measured. Creatinine was normal in the plasma.

γ-Guanidino Butyric Acid GBA excreted in the urine could arise from a transamidination reaction between arginine and γ-amino-butyric acid (GABA), yielding GBA and ornithine. This reaction is known to occur in the brain but takes place also in other organs, particularly in the kidney, where an effective synthesis of GABA has been demonstrated (15–17). The opposite direction of this transamidination reaction occurs in rabbits, where 14% of the administered guanidino-^{14}C activity of GBA is found after 24 hours in the urea excreted (18).

α-N-Acetyl Arginine This compound probably arises from an *N*-acetylation reaction with arginine and acetyl CoA, which is known to occur in mammals with other amino acids such as glycine (19).

Argininic Acid The origin of this compound is difficult to explain.

A relationship between the excretion of these guanidino compounds and the elevated arginine level in this disorder is evident. However, neither their physiological meaning nor their possible role in the pathogenesis of certain symptoms in this disease can yet be estimated.

Alternative Routes of Urea Production

Up to now several hypotheses on alternative urea production in disorders of the urea cycle have been put forward. Most of them have implied a functioning arginase. A matter of speculation in the patients with argininemia is their urea concentration in plasma which, on a regular protein intake, is either normal or near the lower limit of the normal range. This occurs despite arginase deficiency.

Several possibilities could be discussed:

The metabolic pathways for urea synthesis proposed by Cohen and coworkers suggesting GSA as a precurser of urea offers no explanation in our patients (see above).

An alternative metabolic pathway for urea synthesis was recently proposed by Scott-Emuakpor et al. (20), deduced from the observation of increased homocitrulline and homoarginine urinary excretion in certain congenital disorders of urea metabolism. In this pathway transcarbamylation of lysine to homocitrulline has been reported (21–23). The condensation of homocitrulline with aspartate to a small extent has been observed with bovine ASA synthetase (24). To our knowledge the splitting of homoargininosuccinic acid to homoarginine and fumarate has not been shown, but the reverse of the reaction has (25). The hydrolysis of homoarginine to lysine and urea, has apparently been observed recently (26). Since arginase is deficient in our patients, urea cannot arise from the splitting of homoarginine. Besides in our laboratory no activity of ornithine carbamyl transferase and arginase could be demonstrated in crude human liver homogenates, when lysine was substituted for ornithine or homoarginine for arginine respectively. Further experiments with crude and purified human liver arginase showed that homoarginine could not serve as substrate for arginase (27). Since arginase occurs in several tissues besides liver and erythrocytes, a functioning enzyme of other tissues or isoenzyme could account for a low rate of urea synthesis. From the genetic viewpoint this hypothesis involves a complex interpretation. It could not be tested by enzyme determination in our patients, but it must be considered in view of the findings of a deficient red cell arginase with a normal enzyme activity in liver tissue in a strain of macacca fascicularis (28).

A diminished substrate affinity of the mutant enzyme has also been postulated in disorders of the urea cycle and observed in citrullinemia and ornithine carbamyl transferase deficiency (29, 30). At high concentrations of substrate a normal urea synthesis could take place. This question is under investigation in the patients with argininemia at the moment.

Normal concentrations of urea in plasma could also be related to the observation of Murdaugh et al., who showed that on a low-protein intake the fraction of filtered urea excreted is diminished, leading thus to higher plasma concentrations (31).

Finally it has to be considered that urea might also be ingested with food from which a small quantity will be absorbed by the gut. Except for the last possibility an explanation for the near normal urea concentration in plasma in the patients with argininemia is still a matter of speculation.

OTHER CASES OF ARGININEMIA

To our knowledge up to now, only one other case of hyperargininemia with proven enzyme deficiency has been reported (32). Cederbaum et al. described a $5\frac{1}{2}$ year old patient with a similar clinical picture as our patients. Red blood cell arginase was less than 2% of normal. Arginine plasma values were increased. However, no hyperammonemia was observed.

Of interest is an earlier observation of Peralta Serrano, who described a 2 month old infant with argininuria, convulsions, and oligophrenia (33). This may in fact be the first observation of argininemia. The child (a girl) had convulsions and episodes of coma at the age of 2 months, with major alterations of the EEG pattern. The child exhibited a hyperaminoaciduria with a particularly elevated excretion of arginine which rose even to higher values during the neurological crises. High levels of arginine were also measured in CSF. After arginine supplement to the diet, the child's neurological condition deteriorated. No arginine was demonstrated in the urine of the parents.

SUMMARY

The biochemical findings are reported in the third child of the same family with argininemia, followed regularly shortly after birth. They agree with those previously found in the other two siblings. The unsolved aspects in this disorder are the origin of the guanidino derivatives excreted in the urine and the near normal levels of urea. Both of them are discussed.

REFERENCES

1. Terheggen, H. G., Schenk, A., Lowenthal, A., Van Sande, M., and Colombo, J. P. (1969). *Lancet* **II**, 748–749.

2. Terheggen, H. G., Schwenk, A., Lowenthal, A., Van Sande, M., and Colombo, J. P. (1970). *Z. Kinderheilk.* **107**, 298–312, 313–323.

3. Colombo, J. P., Terheggen, H. G., Lowenthal, A., Van Sande, M., and Rogers, S. (1973). *Inborn Errors of Metabolism*, F. A. Hommes and C. J. van den Berg, Eds., Academic, London-New York, 239–248.

4. Ghadimi, H., and Pecora, P. (1964). *Pediatrics* **33**, 500–506.

5. Smith, A. D., Benziman, M., and Strecker, H. J. (1967). *Biochem. J.* **104**, 557–564.

6. Levy, H. L., and Barkin E. (1971). *J. Lab. Clin. Med.* **78**, 517–523.

7. Cohen, B. D., Stein, I. M., and Bonas, J. E. (1968). *Amer. J. Med.* **45**, 63–68.

8. Dobbelstein, H., Edel, H. H., Schmidt, M., Schubert, G., and Weinzierl, M. (1971). *Klin. Wschr.* **49**, 348–357.

9. Sasaki, M., Takahara, K., and Natelson, S. (1973). *Clin. Chem.* **19**, 315–321.

10. Stein, I. M., Cohen, B. D., and Kornhauser, R. S. (1969). *N. Engl. J. Med.* **280**, 926–930.

11. Ratner, S., and Rochovansky, O. (1956). *Arch. Biochem. Biophys.* **63**, 277–295.

12. Takahara, K., Nakanishi, S., and Natelson, S. (1969). *Clin. Chem.* **15**, 397–418.

13. Koller, A., Comess, J. D., and Natelson, S. (1975). *Clin. Chem.* **21**, 235–242.

14. Menyhart, J., Grof, J., and Somogyi, J. (1975). *Abstracts, 9th Ann. Meeting Europ. Soc. Clin. Invest.*, Rotterdam.

15. Zachmann, M., Tocci, P., and Nyhan, W. L. (1966). *J. Biol. Chem.* **241**, 1355–1358.

16. Pisano, J., Abraham, D., and Udenfriend, S. (1963). *Arch. Biochem. Biophys.* **100**, 323–329.

17. Whelan, D. T., Scriver, C. R., and Mohyuddin, F. (1969). *Nature* **224**, 916–917.

18. Beeson, M. F., Buckle, A. L. J., and Jones, H. E. H. (1971). *Horm. Metab. Res.* **3**, 188–192.

19. Shigesada, K., and Tatibana, M. (1971). *J. Biol. Chem.* **246**, 5588–5595.

20. Scott-Emuakpor, A., Higgins, J. V., and Kohrman, A. F. (1972). *Pediat. Res.* **6**, 626–633.

21. Ryan, W. L., Barak, A. J., and Johnson, R. J. (1968). *Arch. Biochem. Biophys.* **123**, 294–297.

22. Cathelineau Liliane, Saudubray, J. -M., and Polonovski, C. (1974). *Enzyme* **18**, 103–113.

23. Scott-Emuakpor, A. B., and Kahrmann, A. F. (1972). *Nigerian J. Sci.* **6**, 47–56.

24. Ratner, S. (1973). *Adv. Enzymol.* **39**, 3–90.

25. Strandholm, J. J., Buist, N. R. M., and Kennaway, N. G. (1971). *Biochim. Biophys. Acta* **237**, 293–295.

26. Scott-Emuakpor, A. B. (1972). *J. West. Afr. Sci. Ass.* **17**.

27. Berüter, J., Colombo, J. P., Bachmann, C., and Peheim, E., to be published.

28. Shih, Vivian E., Jones, T. C., Levy, H. L., and Madigan, P. M. (1972). *Pediat. Res.* **6**, 548–551.

29. Tedesco, T. A., and Mellman, W. J. (1967). *Proc. Natl. Acad. Sci.* **57**, 829–834.

30. Cathelineau, L., Saudubray, J. M., Charpentier, C., and Polonovski, C. (1974). *Pediat. Res.* **8**, 857–859.

31. Murdaugh, H. V., Schmidt-Nielsen, B., Doyle, E. M., and O'Dell, R. (1958). *J. Appl. Physiol.* **13**, 263–268.

32. Cederbaum, S. D., Shaw, K. N. F., and Valente, M. (1973). *Amer. J. Hum. Genet.* **25**, 20*a* (abstracts).

33. Peralta Serrano, A. (1965). *Rev. Clin. Esp.* **97**, 176–184.

25

Ketoacid Therapy of Carbamyl Phosphate Synthetase Deficiency

Mackenzie Walser
Mark L. Batshaw
and Saul W. Brusilow
Department of Pharmacology and Experimental Therapeutics
and Department of Pediatrics
Johns Hopkins University School of Medicine
. Johns Hopkins Hospital
and John F. Kennedy Institute
Baltimore, Maryland

Present treatment of congenital hyperammonemia caused by defects in the urea-cycle enzymes is unsatisfactory, as Dr. Shih has pointed out. Protein restriction is the mainstay of therapy but does not reduce ammonia to normal and tends to prevent adequate growth. It seems clear that the only approach to the management of such children likely to afford success is to attempt to maximize nitrogen balance while at the same time minimizing urea production. In other words, the substrate load derived from protein ingestion must be somehow modified so that it provides as little stimulus as possible to urea production.

The use of essential amino acids as a sole nitrogen source was attempted in one such patient without success (1). The possibility that administration of the keto analogs of essential amino acids might accomplish this goal is worth considering, as was first pointed out to me by Professor Krebs when I worked in his laboratory. These compounds evidently suppress the formation of urea while at the same time providing essential amino acids for protein synthesis (2, 3).

This effect is most clearly demonstrated under conditions of maximal nitrogen conservation. When a mixture of five keto analogs of essential amino acids plus the four remaining essential amino acids is administered daily intravenously to patients undergoing prolonged starvation for the treatment of morbid obesity (4), as shown in Fig. 1, urea excretion falls progressively; no further fall in urea excretion is seen at this late stage of starvation in normal subjects. Of particular interest is the fact that urea excretion remains even lower during an ensuing week in those subjects who received the ketoacid infusion, suggesting that these compounds exert effects on nitrogen conservation which persist long after they have been metabolized.

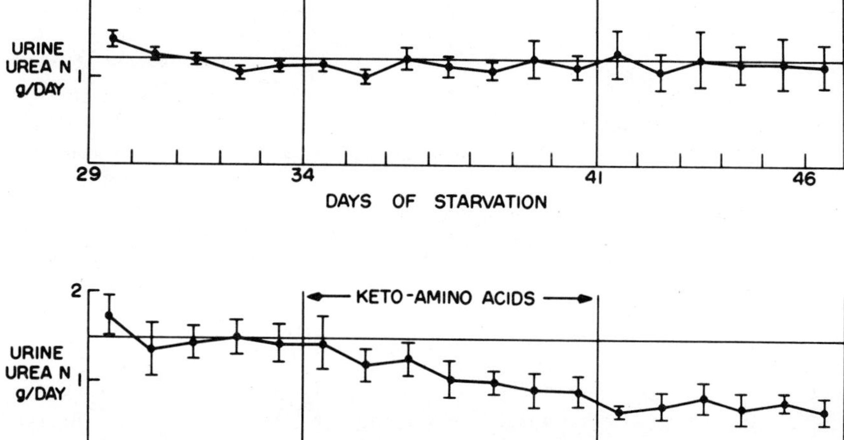

Figure 1. Urine urea nitrogen during daily ketoacid–amino acid infusions in starving obese subjects. In control subjects (above), urea excretion remained constant during this period. During infusions, urea excretion decreased and remained low in the ensuing period. (Reprinted by permission of Sapir et al.)

In portal-systemic encephalopathy, administration of various mixtures of ketoacids and amino acids has been tried, with only moderate success (5, 6). These patients differ from patients with congenital hyperammonemia in that only ammonia is elevated. Glutamine, glutamate, and alanine are not increased, suggesting that the basic defect is abnormal communication between the portal and systemic circulation, rather than an excessive accumulation of urea precursors.

We were fortunate in identifying a 13 year old girl with a partial deficiency of carbamyl phosphate synthetase. Her family history, shown in Fig. 2, includes a female sibling who was stillborn at 32 weeks of gestation and

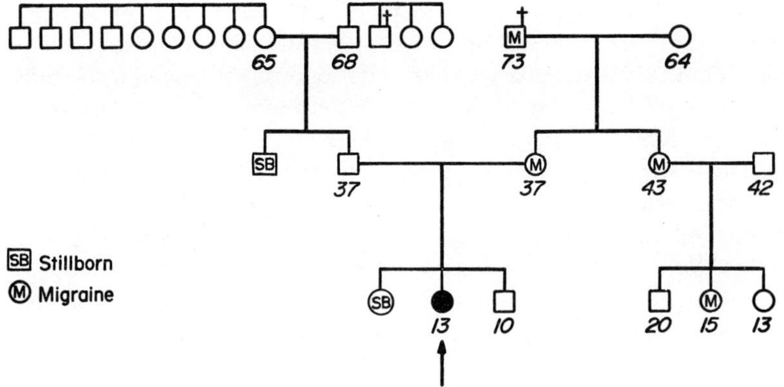

Figure 2. Family history of patient. (Reprinted by permission of Batshaw et al.)

several cases of migraine, all on the maternal side. The instances of migraine are of interest in view of the suggestion of Russell (7, 8) that some cases of migraine are the result of a partial deficiency in the urea-cycle enzymes.

Our patient was the product of an uncomplicated pregnancy, birth, and neonatal period. At 3 weeks of age, the patient first manifested vomiting and lethargy. At $2\frac{1}{2}$ years, she had a transient episode of hemiparesis. By age 3, it was apparent that she was severely mentally retarded. Later she had another episode of vomiting and lethargy associated with right hemiparesis, and during the following month she began to develop akinetic seizures which have persisted up until the present. She also continues to have monthly episodes of vomiting and lethargy, often precipitated by high-protein meals. She has voluntarily restricted herself to a low-protein diet of approximately 1.5 g/kg body weight per day. Other details of her history are reported elsewhere (9). On admission she showed pronounced growth retardation, spastic quadriplegia, bilateral Babinski signs, and diminished muscle strength on the left. Her mental age was estimated as 19 months. Blood ammonia was 3.4 to 3.6 μg/ml by the Seligson method, and serum urea nitrogen was 10 to 12 mg/100 ml. Liver function tests were normal. Methylmalonic acid was not found in the urine. A subcutaneous biopsy of the liver was performed, and activities of the urea-cycle enzymes were as shown in Table 1. Carbamyl

Table 1 Activities of Urea-Cycle Enzymes in Liver (9)

Enzyme	Patient	Normal Range
Carbamyl phosphate synthetase	23 U	180–615 U
Ornithine transcarbamylase	7299 U	3950–6550 U
Argininosuccinic acid synthetase	25 U	21–41 U

U = 1 μmole of product formed per hour per g wet weight of tissue.

phosphate synthetase was approximately 15% of normal. The last two enzymes of the cycle could not be measured because of inadequate tissue.

The child's protein intake was reduced progressively, and clinical improvement occurred. However, elevated values continued to be obtained of fasting ammonia, glutamine, and alanine in blood.

Methods were developed for measuring ammonia, glutamine, glutamate, and alanine on capillary blood samples, using a resin method for ammonia and enzymatic methods for the other substrates. In order to determine whether this child had the capacity to transaminate these keto analogs, 8.5 g of a mixture of the sodium salts of the analogs of valine, leucine, isoleucine, methionine, and phenylalanine was administered intravenously (Fig. 3).

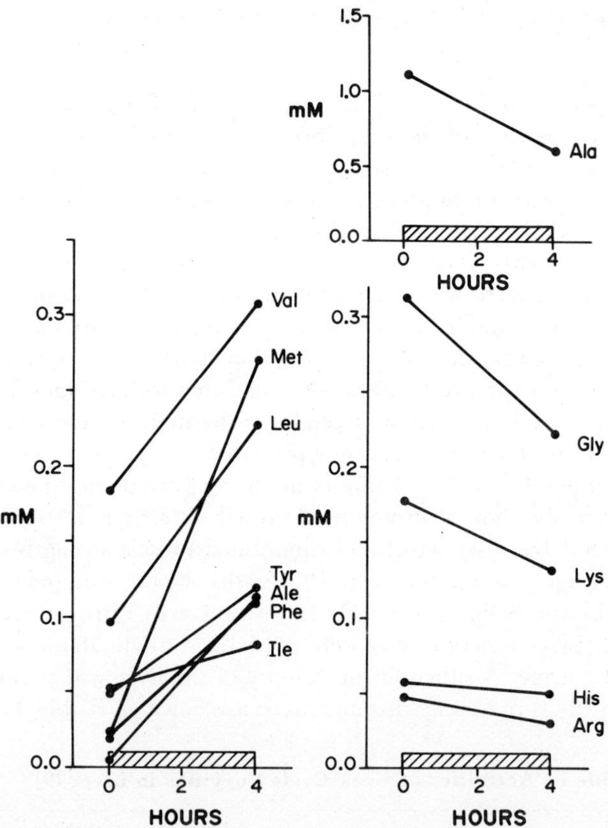

Figure 3. Changes in circulating amino acids following infusion of ketoacid mixture. The left-hand panel shows the increases in plasma amino acids corresponding to the infused ketoacids. The right-hand panel shows amino acids whose concentrations decreased. (Reprinted by permission of Batshaw et al.)

Plasma amino acids corresponding to the infused ketoacids increased sharply, including tyrosine and alloisoleucine. Alanine fell by 0.6 mM. Glycine and lysine were also decreased. Smaller decreases were noted in histidine and arginine. Glutamine and arginine were not measured in this experiment. Ammonia rose transiently immediately following the infusion but fell to normal by the following morning. Plasma ketoacids reached only 1.1 mM at the end of the infusion, indicating rapid use. During the ensuing 8 months, while on 0.75 g/kg protein, a variety of different mixtures of these same five ketoacids and essential amino acids has been administered to this child. Their efficacy is most clearly demonstrated in a recent study in which she was admitted for a period of observation during withdrawal of this dietary supplement (10). During this study, she remained on a constant protein intake of 0.75 g/kg/day. On withdrawing these supplements, ammonia rose promptly to abnormal levels (Fig. 4), and she appeared less alert. Glutamine

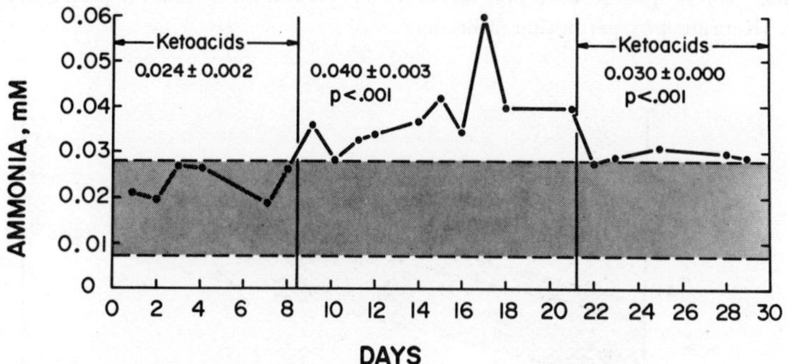

Figure 4. Effect of withdrawal of ketoacid–amino acid supplement on plasma ammonia. (Reprinted by permission of Batshaw et al.)

and alanine, as shown in Fig. 5, were also normal during therapy but became abnormal on withdrawing the supplement.

Nitrogen balance studies cannot be accurately determined in this child, because she is incontinent at night. Nevertheless, it was demonstrated on two occasions that the addition of the supplement of ketoacids and amino acids produced greater retention of dietary nitrogen, as indicated by the difference between 24 hour nitrogen intake and 12 hour urinary nitrogen excretion. On the other hand, no change has been noticed in the partition of urinary nitrogen (9, 10).

As shown in Fig. 6, she has continued to increase in height at the same rate as prior to therapy, despite restriction of protein intake and normalization of blood ammonia levels. She remains severely retarded, but seizure frequency

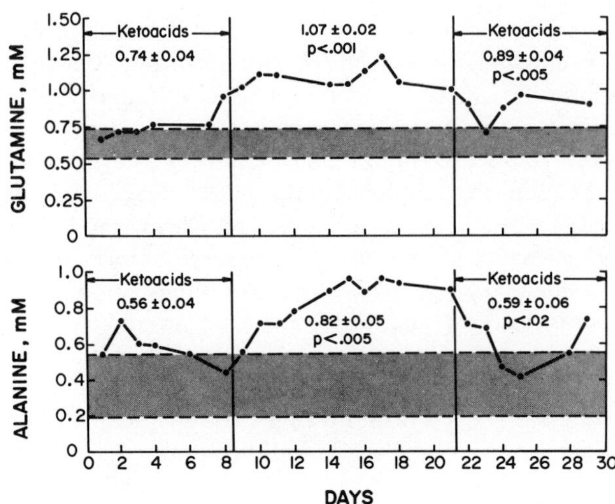

Figure 5. Effect of withdrawal of ketoacid–amino acid supplement on plasma glutamine and alanine. (Reprinted by permission of Batshaw et al.)

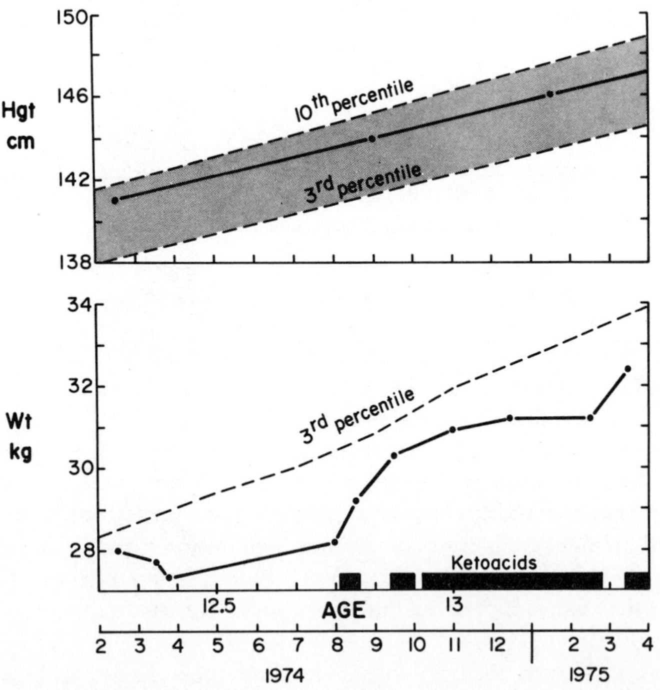

Figure 6. Effect of protein restriction and ketoacid administration on height and weight. (Reprinted by permission of Batshaw et al.)

has been substantially reduced, and no further episodes of vomiting or lethargy have occurred.

The mixture of ketoacids and amino acids we are currently using is shown in Table 2. The large dose of phenylpyruvate is provided because of the persistently low plasma level of phenylalanine. This is in part due to a low renal threshold for phenylalanine, as indicated by the fact that phenylalanine clearance reached one-third of estimated glomerular filtration rate when either phenylpyruvic acid or phenylalanine itself was administered orally. On the other hand, tyrosine is virtually absent from the urine.

Table 2 Ketoacid and Amino Acid Mixture Used in Chronic Therapy (10)

Dose of analog/24 hours (g of Ca salt)		Amino Acid/24 hours (g)	
Valine	1.44	Histidine	0.22
Leucine	1.94	Tryptophan	0.125
Isoleucine	1.32	Lysine	0.81
Phenylalanine	4.00	Arginine	0.66
Methionine	2.12	Threonine	0.48

Addition of those essential (or semi-essential) amino acids for which keto-analogs are not available to us (lysine, threonine, histidine, tryptophan, and arginine) was instituted after the first month of therapy with ketoacids alone (9), because progressive diminution in plasma concentrations of lysine, histidine, and arginine to subnormal levels occurred (10).

The only other persistent abnormality has been the presence of alloiso-leucine levels as high or higher than isoleucine levels for the entire 8 months. In order to investigate the possibility that this might lead to synthesis of abnormal proteins, plasma proteins were precipitated, washed, and hydro-lyzed with heat and acid to their constituent amino acids, using standard methods. The ratio of alloisoleucine to isoleucine in plasma protein was found to be 1–40, a value indistinguishable from that of plasma protein isolated from normal subjects and only slightly higher than that observed when pure isoleucine itself is subjected to prolonged heating with hydrochloric acid. These results indicate that, at least as far as plasma proteins are concerned, incorporation of alloisoleucine does not occur, despite persistently elevated plasma alloisoleucine. Only traces of alloisoleucine were found in the urine.

Other amino acids of the Krebs cycle are not strikingly abnormal in this child's plasma; only citrulline is slightly low. Orotic acid excretion, observed repeatedly, was normal or somewhat low.

In summary, it appears that administration of appropriate mixtures of keto analogs of essential amino acids and the remaining essential amino acids themselves can provide adequate long-term therapy for partial deficiency of carbamyl phosphate synthetase. Whether neurological damage could be prevented by the administration of such a mixture to an infant with a similar defect remains to be determined. It is also far from certain that these compounds would provide adequate therapy for a child with less capacity to synthesize urea than is present in our patient. We have not yet attempted to reduce urea excretion to even lower values by the administration of less dietary protein and greater quantities of these compounds. Nevertheless, such an attempt would appear to offer the most hopeful solution at present for the problem of severe congenital hyperammonemia in infancy.

ACKNOWLEDGMENT

This work was supported in part by a Program Project Grant from the United States Public Health Service/National Institutes of Health, 1PO1-AM-18020-01. The technical assistance of Sylvia Butler, Ellen Gordes, Valerie Hammond, and the nursing staff of the Pediatric Clinical Research Unit is gratefully acknowledged. We also owe special thanks to Dr. A. W. Coulter for provision of the keto analogs.

REFERENCES

1. Walser, M., Coulter, A. W., Dighe, S. V., and Crantz, F. R. (1973). *J. Clin. Invest.* **52**, 678–690.
2. Walser, M., Lund, P., Ruderman, N. B., and Coulter, A. W. (1973). *J. Clin. Invest.* **52**, 2865–2877.
3. Gelehrter, T. D., and Rosenberg, L. E. (1975). *N. Engl. J. Med.* **292**, 351–352.
4. Sapir, D. G., Owen, O. E., Pozefsky, T., and Walser, M. (1974). *J. Clin. Invest.* **54**, 974–980.
5. Maddrey, W. C., Chura, C. M., Coulter, A. W., and Walser, M. (1976). *Gastroenterology* in press.
6. Maddrey, W. C., Weber, F. L., Jr., and Walser, M. (1974). *Clin. Res.* **22**, 364*a*.
7. Russell, A. (1973). *Mt. Sinai Med. J.* **40**, 609–630.
8. Russell, A. (1973). *Mt. Sinai Med. J.* **40**, 723–735.
9. Batshaw, M. L., Brusilow, S., and Walser, M. (1975). *N. Engl. J. Med.* **292**, 1085–1088.
10. Batshaw, M., Brusilow, S., and Walser, M. (1976). *Pediatrics* in press.

Discussion

Dr. Krebs: We now have time for general discussion. Dr. Cohen already mentioned that he would like to raise a general point at this stage.

Dr. Cohen: Research on human subjects, and in particular fetuses and newborns, has been subjected to increased scrutiny, questioning, criticism, and even restrictive legislation in the United States and other parts of the world. In part this development has been a consequence of a few ill-considered clinical research studies which failed to include adequate informed consent on the part of the research subjects. Related to this has been the emergence of a public attitude that biomedical research has subjected many patients to some duress in requesting their permission to serve as experimental subjects in clinical studies and that certain segments of the population bear a disproportionate risk and burden as subjects to advance biomedical knowledge for the benefit of the rest of the population. There have even been questions raised as to whether the quality of medical care delivered in a given medical care facility has any necessary relation to the extent and quality of biomedical and clinical research being conducted. Recently, a U.S. Congressional Committee has raised this question with reference to the Veterans Administration Hospitals.

The purpose of these remarks is to urge those of you engaged in basic biomedical research to recognize the importance of being certain that your scientific or technical contribution as a participant in a research project involving patients or studies of tissues removed from patients under conditions involving some risk to the patient is critically considered before it is undertaken.

Hyperammonemia in the newborn or in early infancy and its possible relation to a defect in urea biosynthesis are a good example of the point I wish to make. Because not all cases of hyperammonemia are the result of a defect in urea biosynthesis, it is essential that studies other than assay of the enzymes involved in urea biosynthesis in liver biopsies be carried out beforehand. There is an opportunity now with the availability of nonradioactive isotopes, and in particular ^{13}C labeled compounds, to design procedures suitable for use in the clinic for assessing urea biosynthesis capacity. Compounds such as ^{13}C, ^{15}N labeled ammonium carbonate, ^{13}C, ^{15}N ureido labeled citrulline, and ^{13}C, ^{15}N guanido labeled arginine are available or can be made available, which should provide the means to assay *in vivo* capacity of urea biosynthesis. A urine sample collected after the labeled compound is fed or injected would then be analyzed for evidence of nonuse of labeled ammonium carbonate, citrulline, or arginine, and the urea present in the urine could be examined for isotope content by conversion to ammonia and carbon dioxide by urease treatment.

A standardized procedure should be set up for this purpose by an experienced group of investigators to establish normal values and to assess departures from normal.

If a test of the kind just described gives indication that one of the enzymes of urea biosynthesis is absent or below normal in amount, then one should consider doing a liver biopsy for the purpose of assaying the levels of enzymes involved in urea biosynthesis. However, before this is undertaken, there should be a consensus on the part of experienced investigators on the following:

a. The nature of the biopsy and the minimum sample size; standardization of procedures for handling and storage of biopsy sample

b. Standardization of all the reagents and the assay procedures; isotopic versus colorimetric procedures

c. Establishment of normal values for patients of the same age and dietary history

d. Production and purification of antibodies against each of the enzymes

Unless a concerted effort is made soon to deal with the above, valuable clinical material will be wasted. The inadequacy and even inappropriateness of many of the procedures which have been and are now being used and the uncertainty of what constitutes a "normal" level of urea-biosynthetic enzymes in a sample of liver taken from a seriously ill child, inadequately nourished because of the child's clinical condition, raise serious questions as to the validity and thus the usefulness of much of the data now in the literature.

It goes without saying that a biochemist who offers to perform enzyme assays on a liver biopsy from a sick child without recognizing the possibility that the values determined may not be diagnostically useful for the reasons mentioned above is contributing to an unethical situation.

Dr. Krebs: Go ahead. Just leave some time for others.

Dr. Cohen: I have two comments to make about Dr. Colombo's presentation. First, the matter of homocitrulline in urine. It has been demonstrated (see T. Gerritsen, J. G. Vaughn, and H. A. Waisman, *Arch. Biochem. Biophys.*, **100**, 298–301, 1963) that canned milk, in contrast to fresh milk, contains homocitrulline and that infants excreting homocitrulline when fed canned milk cease to do so when fed fresh milk. Because of the relatively high concentrations of lysine and urea in cow's milk, it seems highly likely that homocitrulline will be formed during any procedure which subjects milk to heating, concentration, evaporation to dryness, etc. It is even possible that pasteurization could bring this about under certain conditions. In view of the fact that lysine is a very poor substrate for ornithine transcarbamylase requires that an investigator who finds homocitrulline in a urine specimen rule out its dietary origin. The failure to do this has permitted the perpetuation in the literature of the experimentally unsupportable claim that homocitrulline is a significant metabolite and even a possible intermediate in an alternate pathway for urea biosynthesis.

Second, the matter of argininemia. It has been commented on earlier in this meeting, and it is well known, that arginases are widely distributed in many different cells and subcellular compartments of both ureotelic and nonureotelic animals. I suppose we must excrete at least 5 or 10% of our nitrogen in our sweat in the form of urea due to the arginine content of the sweat glands. Thus the arginase involved directly in the biosynthesis of urea in the liver is playing a special role. The presence of a high level of arginine in the blood associated with a low blood cell arginase activity and a normal blood urea level, as in the cases reported by Dr. Colombo and his colleagues, may represent cases of congenital defects in extrahepatic arginase(s) and not in the hepatic arginase associated with urea biosynthesis.

Dr. Tatibana: I have some comments relating to Dr. Chappell's work. I was very much impressed by the beautiful work he has done. I agree, partly at least, with Dr. Chappell's ideas. GDH may be important but not very much so for ammonia production. I would like to call your attention to recent findings to the fact found by Windmueller and Spaeth, of NIH and of Dr. Ishikawa, of Osata University of Japan, that glutamine is degraded in the GI tract rapidly to form ammonia. Ammonia would then be sent to the liver. This might be the largest source of ammonia for urea synthesis. Indeed, this is in agreement with important work by Duda and Handler. They injected ^{15}N labeled ammonia and showed that, after 20 minutes, almost all the ammonia was incorporated into the amide nitrogen of glutamine, since glutamine is circulating in the body, maybe the GI tract will convert it to ammonia. I think this is very important for ammonia metabolism.

Dr. Krebs: I think later this afternoon we shall hear more about the contribution of the gut to ammonia production when Patricia Lund will speak.

Dr. Colombo: I want only to answer Dr. Cohen. We know about the work of Gerritsen's group in Madison who showed that homocitrulline found in the urine may be from an exogenous source. We did not find homocitrulline or homoarginine in the urine of our cases with argininemia. These authors, who plea for a functioning homologue pathway (lysine-ornithine-pathway), which I showed, seem to have taken this objection into consideration. On a diet without milk the excretion of homocitrulline and homoarginine in a case of citrullinemia was still present. I agree with you that one may have an enzyme functioning normally in different tissues but absent in one. This was also mentioned this morning by Dr. Shih in a new case of citrullinemia. In our cases with argininemia, it might well be that there is a residual activity present also in other tissues than merely in liver and red cells which may be enough to produce urea.

Dr. Shih: I would also like to reply to Dr. Cohen's question. As Dr. Colombo indicated, we are all aware of the artifact from the milk. I think Waisman found out that babies fed canned milk excreted a fair amount of homocitrulline. But after they were put on homogenized milk, the homocitrulline excretion markedly decreased. Homogenized milk is pasteurized at a low temperature, and lysine is probably not converted to homocitrulline to any extent at that temperature. Also, in this patient with hyperornithinemia that I first described, there was a large amount of homocitrulline in the urine. I have taken the patient off milk or anything that has high lysine, and the patient still excreted significant amounts of homocitrulline, easily detectable by the chromatographic studies I showed. So I think that, even though there might not be an established known pathway from lysine to homocitrulline in the human, obviously it can be synthetized perhaps at least in abnormal conditions.

Dr. Bessman: In regard to the synthesis of glutamine, we did a number of experiments on the whole body synthesis of glutamine as alternative to urea synthesis. The enzyme concentration of glutamine synthetase in muscle is approximately one-tenth that of liver. However, there is 10 times as much gross muscle mass as liver, which means that the net enzyme in the whole organism for the synthesis of glutamine is very high. It is at least equivalent to liver. Second, it is adaptive. If you keep giving small injections of ammonia to an animal, raising its blood ammonia approximately two to threefold, you will get an increase in about 4 days in the glutamine synthetase of muscle of approximately fourfold. So the question of the partition of the source of nitrogen from one source of ammonia to urea or to glutamine depends to a great extent on the time element also. I would think that the likelihood is that after exposure for a long time to ammonia those animals would behave differently. From the normal history of ammonia encephalopathy, I believe that the same type of adaptation occurs in humans also.

Dr. Soberón: First I would like to refer to Dr. Colombo's work, in relation to the accumulation and rate of excretion of guanido compounds. I would like to mention that we described the presence of guanido ureohydrolase in liver, and it might be that only the guanido group of arginine goes to these compounds but that alterations of other enzymes might play a role. Now in relation to the comment of Dr. Tatibana, with respect to the work of Duda and Handler, we extended this work, not by following ^{15}N, but by measuring the ammonium and glutamine concentrations in different tissues, muscle, brain, liver, and on the residual carcass. We found that the muscle plays a large role in trapping ammonia. Then glutamine synthesis comes into action and then urea synthesis. That is the trapping of ammonia in muscle, and glutamine synthesis in brain and liver just serve the purpose of a reservoir, since N has to be excreted and to follow, primarily, the way of urea formation in the liver.

Now there is another point to which I would like to call your attention, and this is in reference to Dr. Chappell's presentation. I would like to ask if we would be prepared to accept without any reservation that the urea cycle functions part in the mitochondria and part in the cytosol. It seems difficult to understand how it could operate in this way. First, because of the design of the cycle itself, the ornithine moiety serves the purpose of putting on top of the molecule different groups which eventually come off to form urea, and, again, ornithine is reused. We like to present this situation to students in our biochemistry course as a streetcar named ornithine. The streetcar would collect carbamyl phosphate in the first station (ornithine transcarbamylase), forming citrulline that could be referred to as ureido-ornithine. In the next station (arginino-succinic synthetase) aspartic acid boards the streetcar to give argininosuccinate which can be envisaged as succino-guanido-ornithine; further along (argininosuccinase) fumaric acid gets off, leaving arginine or guanido ornithine; in the last station (arginase) the streetcar reaches its destination, urea stepping down, leaving ornithine available for a new trip.

So the important thing is this, arginase, which is a newcomer to the cycle, in coming together with the arginine biosynthesis path, has to play the double role of catching a metabolite and throwing the ball again. It has to catch arginine and throw ornithine back into the cycle. Second, we know that the so-called intact mitochondria, as Dr. Chappell also pointed out, have at the least, to my knowledge, disruption of the outer membrane. As isolated, even in the hands of the experts, they are not "intact" as shown by electron microscope photographs. We also know that there is a leakage of enzymes from the mitochondria and that this leakage may be prevented by the addition of phospholipids. Third, there is another piece of evidence, the finding of epiarginine, that arginase plays a regulatory role in Sacharomyces by interacting with ornithine transcarbamylase in the presence of specific effectors, ornithine, and arginine. It is very hard for me to understand how a protein in the cytosol might play this regulatory role, pick up the effectors and go into mitochondria, to come in contact with ornithine transcarbamylase. Fourth, there is another very important finding. It is the description by Professor Grazi, in Italy, that the arginase induced in the chicken liver by starvation, a ureotelic arginase, is specifically located in the mitochondrial matrix. So here you have an instance where an enzyme which has been described to be located in the cytosol is detected in the mitochondrial matrix. Fifth, we have found that an appropriate conformation of arginase is necessary to convert endogenous arginine to urea and ornithine. Sixth, in

Dr. Tatibana's proposed regulatory mechanism that he spoke about yesterday, he was worried about how arginine, activator of acetylglutamate synthetase could come into the picture, because a translocation of the metabolite would be necessary. Moreover, the low pool of arginine, the high K_m of arginase, and the optimum pH of arginase make it difficult to understand the effective action of the enzyme. Hence, I think we have to look at the operation of a cytosol/mitochondria urea cycle carefully.

Dr. Frieden: I would like to just go back to Dr. Chappell's experiments and just to reiterate some comments that I made previously and to expand a little on them. A number of years ago we surveyed glutamic dehydrogenase from a number of different sources, and we found that the animal glutamic dehydrogenases were all very similar. Those we looked at were sheep, rabbit, pig, and bovine, and kinetically they were very similar to one another. The only one which was really different was the rat liver enzyme. One of the differences, as I indicated earlier, was that under assay conditions it was very unstable. Now these assays were always done with either NADH or NADPH, using α-ketoglutarate and ammonia. One condition that stabilized the enzyme was ADP. In other words, in the presence of ADP the assays were linear. I suspect, although I don't know, because we did not do it, that leucine would do the same thing. So there is a point to wondering about the activity of the enzyme under conditions of using the sonicated enzyme or the isolated enzyme. Of course I don't know what happens in intact mitochondria. I think this difference in GDH might also apply to some of the data discussed in this meeting which indicate that what happens in the human liver may be a little bit different from what happens in rat.

Dr. Grisolia: Two very brief and relatively minor comments. Dr. Cohen's remarks about the amount of amino acids and urea which are probably put out in human sweat remind me to mention that it has been shown by Paul Hamilton to be quite considerable in the sweat of hands. Indeed, one of the amino acids which is put out in large quantities is citrulline. I often teased him to measure them in the total skin, that is, by analyzing the water after taking a bath. The other is that I remember at one time there were reports that δ-acetyl ornithine was found in some patients. Does anyone have information on that?

Dr. Krebs: I would like to ask a question addressed to Dr. Walser. What are the economic aspects of the keto acid therapy? This is of course an important matter when it is a matter of long-term treatment. I expect that if there is really a major demand, matters can be organized so that the prices of the keto acids will not become prohibitive.

Dr. Walser: To answer your question precisely, it is necessary to point out that the major demand for ketoacids is not in congenital hyperammonemia but rather in chronic renal failure. A European pharmaceutical company has now undertaken to bring these compounds to market. They don't yet know what the price will be, but it will almost certainly be less than the corresponding essential amino acids. That is the only information I can provide at the moment.

Dr. Shih: A few years ago I asked Dr. Grisolia for a small sample of δ-acetyl ornithine to see if that was one of the things present in my patient with hyperornithinemia. It turned out not to be the compound, so that I have no evidence that the patient had δ-acetyl ornithine.

Dr. Cohen: I wanted to make a couple of comments. I was curious when Dr. Shih was making her presentation as to whether anybody had examined with modern high-resolution electron microscopy the mitochondria of liver samples from patients with congenital deficiency. Certainly in the case of CPS and OTC there is some possibility that one could discern changes. I am curious to know whether, other than the report you alluded to from the study in Germany, this has been done. It is possible that you might get more information in terms of understanding what the basic lesion is in some respects by a good EM analysis of the mitochondrial picture than you might even get from the enzyme assays. The comment that Dr. Krebs made about the importance of the medium surrounding the mitochondria in maintaining the integrity of the mitochondria is particularly relevant in the case of GDH. A mitochondrion might be looked on as a particle in which there is relatively little water present as compared with the cytoplasm, which might be considered to be a large sack of water in which there are some enzymes present. GDH is tightly packed and highly structured in the mitochondrion. There must be groups in each enzyme present at high concentration, such as GDH and CPS, which interact with functional groups of other proteins, and thus the enzymes can be allosterically affected by the very nature of the way in which they are packed. I don't think you can simulate with any assurance the *in vivo* kinetics by reducing the enzyme to a homogenous, freely diffusible, and dissociable molecule, which represents the condition we employ to carry out our assays. I don't want to belabor this point; I think it is self-evident. I do think we have to recognize the biological importance of this as well as the artificiality of the assay conditions that we set up for the enzyme. I don't want to end up by sounding like a vitalist; on the other hand, I think we have to be realistic about how far we are extrapolating in our optimal assay conditions from what possibly takes place in a more intact system.

Dr. Chappell: I really must take objection to these remarks. You see, the point about the mitochondria is that, yes, it has got a mild morphological alteration during the process of isolation. But, at least to a first approximation, the enzymes are there at the concentration and in the condition in which they were in the cell. And it seems to me that this is the beauty of this situation. You have referred to the cell as a bag with some enzymes floating around in it, and that they are loose and bathing in water, and so on. I mean this is a wrong-headed way of looking at the thing. The concentration of proteins in an isolated liver cell is very high. It is 30%. It is no higher or lower than the concentration of proteins in the mitochondrial matrix. There is no difference in protein in concentration in an intact cell on one side of the mitochondrial membrane or another. They are roughly the same.

Dr. Krebs: I believe eventually these arguments will be settled in the laboratory. During an interval I have again spoken to Dr. Chappell on further collaborations which I hope will also clarify the conceptual aspects and the working hypothesis from which we start off. Any further questions or comments? We still have, according to my watch, at least 2 minutes time, and I don't know really what is planned for us between 12:30 and 1:30. I have to consult Dr. Grisolia.

Dr. Grisolia: In a few moments, as soon as we finish, they are bringing Dali's painting down here to be unveiled. They are waiting for us.

Other Aspects of Ammonia Metabolism

26

Factors Affecting Intracellular Ammonia Concentration in Liver

J. T. Brosnan
Department of Biochemistry
Memorial University of Newfoundland
St. John's, Newfoundland, (Canada)

INTRODUCTION

Ammonia exists in solution in two forms, according to the equation

$$NH_4^+ \rightleftharpoons NH_3 + H^+$$

The pK for this reaction is 9.02 (1). Thus the relative proportions of NH_3 and NH_4^+ in solution are pH-dependent. About 1 of every 100 molecules at pH 7.0 exists as the free base, the remainder as ammonium ions. In this paper the free base is designated as NH_3 and ammonium ions as NH_4^+; the sum of these is designated as ammonia.

The concentration of ammonia in liver is regulated by the rate of provision of ammonia, the rate of ammonia removal, and the affinities for ammonia (under physiological conditions) of the enzymes that remove ammonia. Under steady-state conditions the rate of removal is, of course, equal to the rate of provision. Although there are a variety of enzymes capable of removing or producing ammonia in the liver, there are a few that appear to be of the greatest quantitative importance. In Fig. 1 the cellular localization of the principal reactions of ammonia provision and removal are illustrated. Ammonia can be provided in the mitochondria by the intramatrix located glutaminase (2), is removed for urea synthesis via carbamyl phosphate

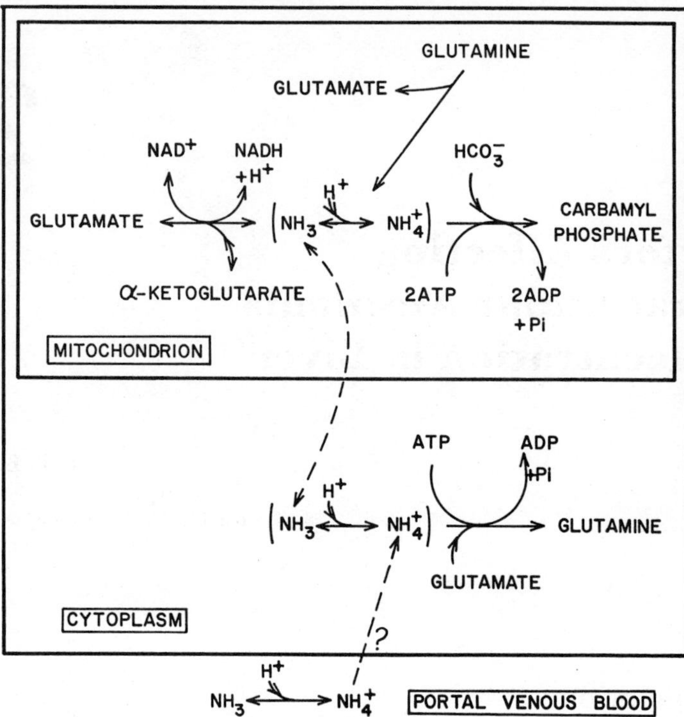

Figure 1. The subcellular distribution of the principal events in hepatic ammonia metabolism.

synthetase I (3), and may be either provided or removed by glutamate dehydrogenase, a highly active enzyme that appears to catalyze a reaction close to thermodynamic equilibrium *in vivo* (4). In the cytoplasm, glutamine synthetase may also remove ammonia. Glutamine synthetase has occasionally been characterized as a microsomal enzyme, but since it can be quantitatively removed from microsomes by physiological saline solutions (5), its membranous localization is probably artifactual and its true localization cytoplasmic. In addition to these enzymes, there is a constant inflow of ammonia in the portal blood.

Since Fig. 1 implies a compartmentation of the principal events involved in hepatic ammonia homeostasis, it seems worthwhile asking whether one can discuss hepatic ammonia homeostasis in terms of a single ammonia pool or whether one should separately consider mitochondria and cytoplasm. Jacobs (6) has pointed out that, generally, NH_3 diffuses rapidly across biological membranes. Chappell's experiments, in which the swelling of mitochondria in isotonic solutions of ammonium salts may be used to study mitochondrial permeability (7), has amply demonstrated that the inner membranes of liver mitochondria are very permeable to NH_3. Therefore, we can assume that NH_3 will equilibrate between mitochondria and cytoplasm, and, therefore, we can discuss hepatic ammonia homeostasis in terms of a single pool. This is not to say that the concentrations of NH_4^+ in both compartments will be identical but will, rather, reflect the pH of each compartment.

In this paper I will concentrate on two aspects of ammonia homeostasis. First, that the urea cycle does not reduce hepatic ammonia concentrations to near-zero but that, rather, an appreciable and constant level of ammonia is maintained. Glutamate dehydrogenase plus the alanine and aspartate aminotransferases play an important role in maintaining this constancy. Second, that the maintenance of a relatively high (0.5 to 0.7 μmole/g) tissue level of ammonia necessitates that the clearance of blood ammonia (especially portal venous ammonia) occurs against a substantial concentration gradient.

Role of Glutamate Dehydrogenase in Ammonia Metabolism

The hepatic concentration of ammonia in a variety of different experimental situations is presented in Table 1. All of the data shown were obtained by the enzymatic measurement of ammonia in extracts of freeze-clamped liver. It is clear that the concentration of ammonia is remarkably constant under these highly different conditions. One obvious explanation for this remarkable constancy is that the measured ammonia does not reflect free ammonia but, rather, reflects a large constant quantity of bound ammonia in liver. This possibility is not supported by the many experiments (4, 8, 9) that indicate that glutamate dehydrogenase enters into near-equilibrium with the measured liver ammonia concentration in all the above conditions. Thus all the liver ammonia appears to be free. That glutamate dehydrogenase catalyzes a near-equilibrium reaction implies that the liver ammonia concentration is largely determined by the mitochondrial NAD/NADH ratio and the mitochondrial glutamate and α-ketoglutarate concentrations.

Table 1 Ammonia Content of Rat Liver under Different Experimental Conditions

	μmole/g
Fed rat	0.69
Fed rat (after 5 minute ischaemia)	0.71
Starved rat	0.88
Alloxan diabetic rat	0.86
Perfused rat liver	0.61

However, there are equilibrium reactions other than glutamate dehydrogenase which may be involved in the regulation of hepatic ammonia levels. These are the alanine and aspartate aminotransferases. Our first clues that these reactions may be important in stabilizing hepatic ammonia concentrations came from experiments where we allowed rat livers to become anaerobic (8). Portions of livers were freeze-clamped immediately (i.e., within 10 seconds) after sacrifice by cervical dislocation, and, subsequently, portions were freeze-clamped after 2 minutes and 5 minutes ischaemia. Increased tissue ammonia levels have been reported in a number of tissues during anaerobiosis, but, in liver, the levels remained very constant (Table 2). However, the concentration of alanine gradually increased. Since the levels of the other amino acids measured did not increase significantly, it was apparent that the alanine increase results from metabolism rather than from proteolysis. We therefore postulated that ammonia, produced anaerobically, was being incorporated into alanine by the combined actions of glutamate dehydrogenase and alanine aminotransferase. Pyruvate may be easily provided by glycolysis. We tested this hypothesis by inhibiting alanine aminotransferase *in vivo* (by injection of L-cycloserine) and then allowing the liver to become ischaemic. In this experiment (Table 3) there was no anaerobic increase in alanine concentration. There was, however, only a modest increase in tissue ammonia content, but a definite increase in aspartate was observed. It thus appears that when alanine aminotransferase is inhibited, anaerobically generated ammonia may be incorporated into aspartate by the combined actions of glutamate dehydrogenase and aspartate aminotransferase. The source of oxaloacetate during anaerobiosis remains uncertain.

Table 2 Tissue Content of Ammonia and Some Amino Acids during Ischaemia in Rat Liver

	μmole/g		
	Duration of Ischaemia		
	0 minute	2 minutes	5 minutes
Ammonia	0.71	0.77	0.69
Alanine	0.87	1.53	2.26
Aspartate	1.19	1.01	0.92
Glutamate	2.35	2.84	2.65
Glutamine	4.68	4.57	4.26

Figure 2 illustrates the coupling of glutamate dehydrogenase to alanine aminotransferase and to aspartate aminotransferase. The coupling of aminotransferase with glutamate dehydrogenase is, of course, the same as in the

Table 3 Tissue Content of Ammonia and Some Amino Acids during Ischaemia in Livers from Rats Injected with L-Cycloserine

	μmole/g Duration of Ischaemia		
	0 second	2 seconds	5 seconds
Ammonia	0.63	0.75	0.77
Alanine	3.40	3.46	3.44
Aspartate	0.74	1.09	1.30
Glutamate	1.63	1.91	1.85

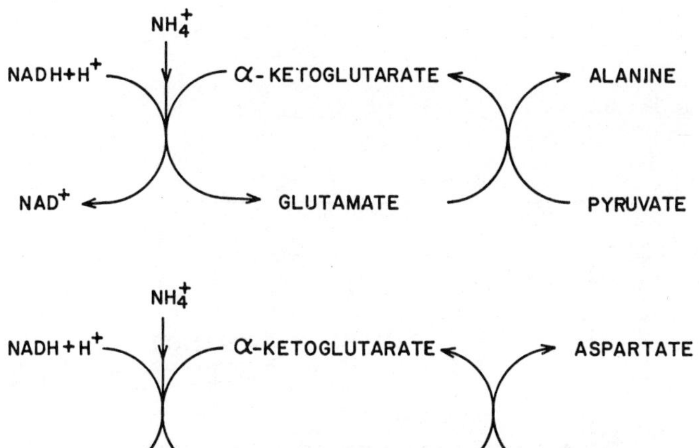

Figure 2. Enzymatic mechanisms of ammonia incorporation into alanine and aspartate.

classic scheme of Braunstein for the transdeamination of amino acids. What is especially important here is that all three enzymes catalyze reactions very close to thermodynamic equilibrium (4, 8, 10), and, thus, the system will respond to an elevated ammonia concentration by incorporating a portion of this ammonia into alanine and aspartate. If such a system can remove ammonia generated anaerobically, when the operation of the urea cycle is impaired, it may also serve to remove excess ammonia *in vivo* if the urea cycle were overextended. It is doubtful if the capacity of the urea cycle is ever normally overextended but it certainly is in patients with inborn errors of the cycle and may be experimentally overextended by injection of ammonium salts. We therefore injected rats intraperitoneally with sublethal doses of ammonium chloride and followed hepatic changes in a number of amino acids.

After an intraperitoneal ammonia load there is an immediate elevation of hepatic alanine and aspartate, reaching a peak at 5 minutes and returning back to normal after about 20 minutes (Fig. 3*a*). In the experiment depicted in Fig. 3*b* the rats were injected with L-arginine 60 minutes before the ammonia load. Such treatment has been shown to increase the rate of urea synthesis (11). It is evident that pretreatment with arginine decreased both the magnitude

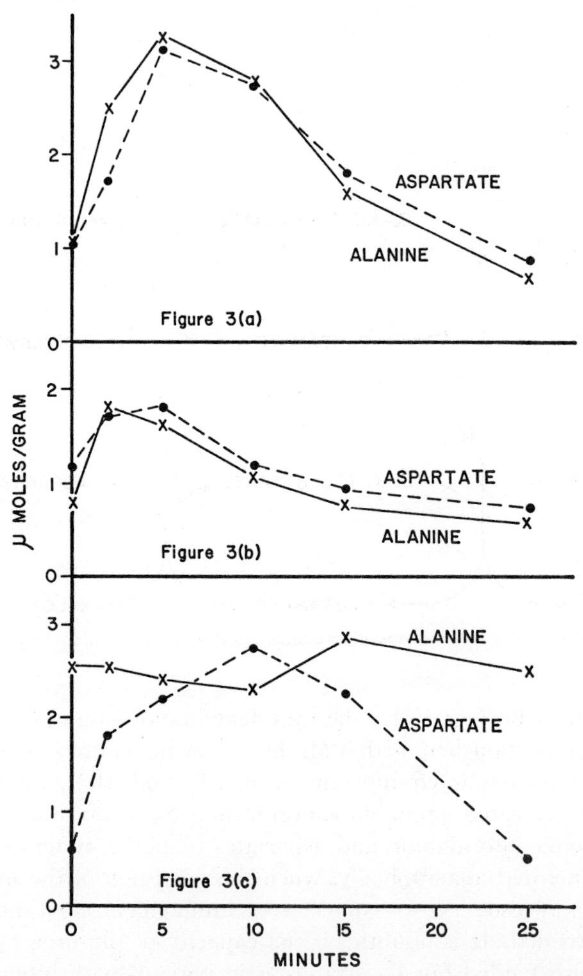

Figure 3. Alterations in the hepatic content of alanine and aspartate after intraperitoneal injection of NH_4Cl (2.5 mmole/kg body weight): (*a*) normal rats; (*b*) rats injected 60 minutes previously with L-arginine (3 mmole/kg body weight); (*c*) rats injected 60 minutes previously with L-cycloserine (25 mg/kg body weight).

and duration of the increases in alanine and aspartate observed after the ammonia load. In Fig. 3*c* is presented data from experiments where the rats were injected with L-cycloserine 60 minutes before the ammonia load. There was now no increase in alanine after the ammonia load, but a much greater increase in aspartate was evident. The result is reminiscent of the ischaemia experiment reported in Table 3. These ammonia-loading experiments demonstrate that the glutamate dehydrogenase-aminotransferase systems (Fig. 2) can serve to temporarily remove ammonia when the urea cycle is overextended. Since the reactions involved are close to thermodynamic equilibrium, the alanine and aspartate may feed ammonia into the urea cycle at the end of the "ammonia crisis."

Administration of ammonium salts is, of course, a very artificial situation. Nevertheless, it is interesting to speculate as to whether temporary storage of ammonia may ever be of importance during hyperammonemia. The more common causes of hyperammonemia are associated with cirrhosis, and it appears that the lesion responsible for hyperammonemia is not enzymatic but circulatory—the bypassing of the liver by a large fraction of ammonia-rich portal blood, due to portacaval shunts, leads to a high concentration of ammonia in the systemic circulation. According to Kirk (12), who studied the effect of oral administration of ammonium salts to patients with cirrhosis and hepatitis, such patients could metabolize ammonia perfectly well, provided that there was no shunt between the portal and peripheral venous systems. Thus the removal of ammonia by hepatic glutamate dehydrogenase and the aminotransferases can play no effective part in the amelioration of this type of hyperammonemia.

There is, however, a class of diseases in which this mechanism may be of importance in the removal of ammonia. These are the inborn metabolic defects which affect the urea cycle. Four distinct types have been reported: (*a*) hyperammonemia due to a deficiency of carbamylphosphate synthetase (13); (*b*) hyperammonemia due to a deficiency of ornithine transcarbamylase (14); (*c*) citrullinuria, probably due to a deficiency of argininosuccinic acid synthetase (15); and (*d*) argininosuccinic aciduria (16). Argininosuccinic aciduria has been fairly well studied, considering its rarity. The activities of the urea-cycle enzymes have been measured in an argininosuccinic aciduric child (17), and a huge reduction in the activity of argininosuccinase (3% of normal) has been shown. Nevertheless the activity present was sufficient to account for the production of about 30 g urea per day, which is in the normal range for a child (16 year old boy) of this age. This helps to resolve one of the puzzling features of this disease—the maintenance of normal blood urea and the daily production of a normal amount of urea. However, the argininosuccinase activity observed in the patient would only be adequate if the load presented to it was spread evenly throughout the day. This would not usually be the case, and at times (after a meal) the capacity of the defective urea cycle would be overextended. At such times hyperammonemia is observed (17). This is precisely the situation in which the removal of ammonia by the formation of alanine and aspartate could be of use in the temporary storage of ammonia. When the immediate ammonia crisis was over and the intrahepatic concentration of ammonia began to fall, the reversal of the glutamate dehydrogenase and aminotransferase reactions (by mass action) could slowly supply ammonia to the urea cycle. Some evidence that this may actually occur is to be found in the work of Moser et al. (18), who measured the plasma amino acids after an ammonia tolerance test on an argininosuccinic aciduric child. They found that the alanine concentration increased from 0.526 to 1.04 mM during the ammonia load. No other amino acid increased by so much.

An Ammonia Gradient between Liver and Blood

The intracellular mechanisms (of which the glutamate dehydrogenase-aminotransferase system is one) responsible for controlling the hepatic ammonia concentration result in a liver ammonia content of about 0.5 to 0.7 μmole/g. This is quite high relative to the concentrations found in blood. In fact appreciable ammonia gradients exist across a number of tissues (Table 4).

Table 4 Ammonia Content of Arterial Blood and in Some Rapidly Frozen Tissues of the Rat

Arterial blood	0.02 μmole/ml
Liver	0.712 μmole/g
Abdominal muscle	0.867 μmole/g
Brain	0.340 μmole/g
Thigh muscle	0.255 μmole/g
Spleen	0.204 μmole/g
Heart	0.199 μmole/g

The ammonia content of portal venous blood is known to be considerably higher than in blood from the systemic circulation. This is primarily due to the catabolic action of colonic bacteria on dietary amino acids. Administration of Neomycin greatly decreases the ammonia concentration of portal venous blood (19). Table 5 shows the concentration of ammonia in the liver and in the blood vessels serving the liver. It can be seen that the liver extracts ammonia from the portal blood against a 30-fold concentration gradient. Since the relative blood flow through the liver is approximately 75% from the portal vein and 25% from the hepatic artery, the hepatic extraction of ammonia may be calculated as being more than 80%. Seligson and Hirahara (20) have shown that in human venous blood the erythrocytes contain 2.8 times more ammonia than the plasma. If this also holds for rat blood, it would result in a plasma concentration of 0.0094 μmole/ml in hepatic venous blood, and the effective liver/plasma gradient would be more than 100-fold.

Table 5 Ammonia Content in the Liver and Its Blood Vessels

Liver	0.712 μmole/g
	= 0.975 μmole/ml tissue water
Arterial blood	0.020 μmole/ml
Portal venous blood	0.261 μmole/ml
Hepatic venous blood	0.031 μmole/ml
$\dfrac{\text{Liver ammonia}}{\text{Hepatic venous ammonia}}$	= 31.4

The study of the liver-blood ammonia gradient *in vivo* is difficult, because conditions cannot be altered readily. In view of the success in demonstrating amino acid gradients in the isolated perfused liver (21) this system was used to study the ammonia gradient also. The technique employed was that of Hems et al. (22) except that the bile duct was not cannulated. Oleate (2 m*M*) was included as a substrate, as, without fatty acid oxidation, the mitochondrial pyridine nucleotides become oxidized relative to the *in vivo* state (23). Livers from fed rats were perfused for 38 minutes, and at this time 75 μmoles of NH_4Cl was added to the medium, bringing the ammonia concentration up to about 0.5 m*M*. Samples were taken at intervals from the medium, and at the end of the experiment (100 minutes) the liver was freeze-clamped. The results of such an experiment are shown in Fig. 4.

There was a large increase in the urea concentration of the perfusion medium after 10 minutes, due to the washing out of urea from the liver. During the next 30 minutes the rate of urea production from endogenous precursors was 0.221 μmole/g/minute, which increased to 0.266 μmole/g/minute after the addition of ammonia to the medium. Toward the end of the perfusion the rate decreased to about 0.16 μmole/g/minute. The initial ammonia concentration in the medium was reduced from 0.103 to 0.060 m*M* within 10 minutes. When, at 38 minutes, 75 μmoles of NH_4Cl was added to the medium, this was rapidly removed until a steady value of about 0.07 m*M* was reached. This is approximately double the concentration found in hepatic venous blood. The rate of ammonia removal between 40 and 43 minutes was 0.51 μmole/g/minute, which is well below the maximum rate of urea synthesis—Hems et al. (22) observed a production of 0.8 μmole urea/g/minute

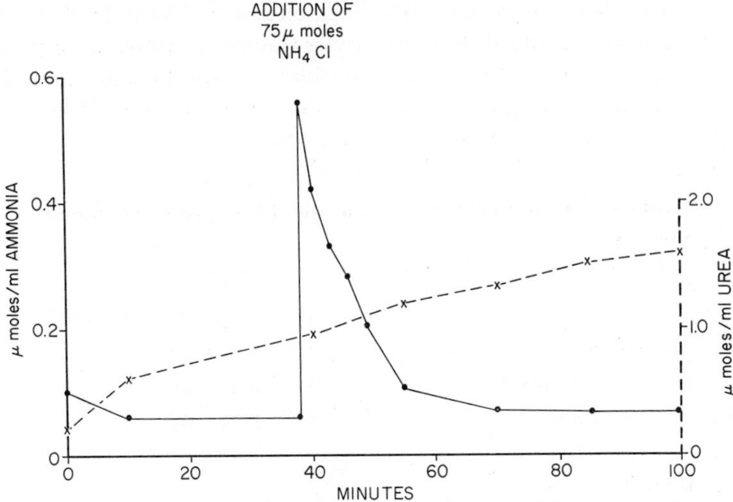

Figure 4. Removal of added ammonia by the perfused rat liver.

from 10 m*M* NH_4Cl—so that the removal of ammonia in the present experiments may be determined merely by the flow rate through the liver. This rapid removal of ammonia to a low concentration is extremely similar to that reported by Glogner et al. (24). The liver was freeze-clamped at the end of the perfusion. The hepatic concentration of ammonia was 0.62 μmole/g, so that there was a ninefold ammonia gradient between liver and medium at the end of the perfusion.

If the observed ammonia gradient is a functional one, then its maintenance may require the expenditure of energy. In anaerobiosis the gradient should, therefore, be diminished or even abolished. This can be conveniently studied in the perfusion system, where anaerobiosis is easily attained. Perfusion was therefore carried out as before (2 m*M* oleate was included in the medium so as to be comparable with the aerobic perfusion), except that:

a. The gas used was 95% N_2/5% CO_2.

b. Sodium cyanide (2 m*M* final concentration) was added to the medium just before perfusion, and an equal quantity was added every 30 minutes to replace that which may be blown off in the artificial lung.

c. No NH_4Cl was added to the perfusate.

Perfusion was continued for 70 minutes. The results of this experiment are shown in Fig. 5. There was an initial "washout" of tissue urea in the first 10 minutes of perfusion, similar to that observed in the aerobic perfusions, but

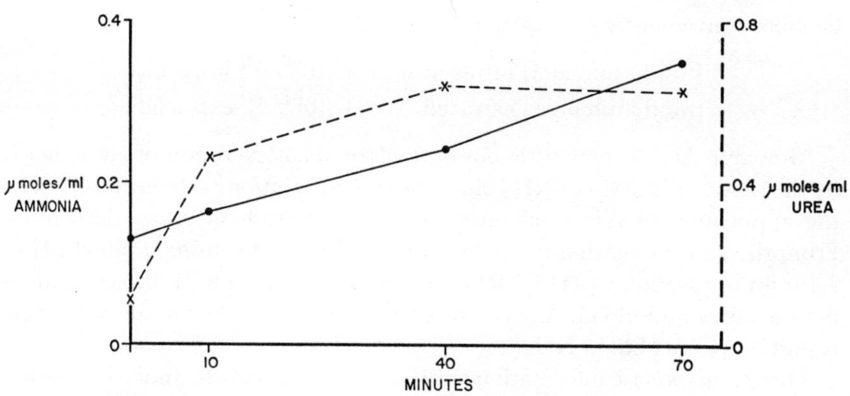

Figure 5. Ammonia production by the anaerobically perfused liver.

once this was complete (sometime between 10 and 40 minutes), there was no longer any production of urea. The ammonia concentration in the medium steadily increased at a rate equal to about 3.0 μmole/g liver/hour. The liver was freeze-clamped at the end of the perfusion. The hepatic ammonia concentration was 1.17 μmole/g, and only a threefold ammonia gradient was maintained. Other experiments showed that added ammonia was not removed by the anaerobic perfused liver. Thus, it appears that energy from oxidative metabolism is necessary for the maintenance of the normal ammonia gradient.

There appear to be three mechanisms by which ammonia could become concentrated across the liver plasma membrane:

a. Free diffusion of NH_3 into the cell in response to a lower intracellular pH
b. Distribution of NH_4^+ in response to a membrane potential
c. Active "pumping" of NH_4^+ across the plasma membrane

The first two mechanisms involve the passive movement of ammonia, although energy would have to be expended to maintain both the postulated pH gradient and the membrane potential.

Jacobs (6) has suggested that, because of its high lipid solubility and lack of charge, NH_3 readily diffuses across cell membranes, whereas NH_4^+ penetrates much more slowly because of its charge, its larger molecular diameter (due to hydration), and its low lipid solubility. Therefore NH_3 diffuses across cell membranes until its concentration becomes equal on both sides of the membrane; NH_4^+ then accumulates on the more acid side. If the membrane is absolutely impermeable to NH_4^+, then a 10-fold greater $[H^+]$ on one side would result in a 10-fold greater $[NH_4^+]$ on the same side. Since it is total ammonia that is experimentally measured (not NH_4^+), this relationship may be more conveniently expressed as follows (25):

$$\frac{[\text{total ammonia}]\ \text{intracell}}{[\text{total ammonia}]\ \text{extracell}} = \frac{1 + 10^{pK-pH}\ \text{intracell}}{1 + 10^{pK-pH}\ \text{extracell}}$$

Although there is now little doubt that the renal excretion of ammonia is by the passive diffusion of NH_3 into a more acid solution (tubular fluid) (26), the application of this mechanism to the liver gradient raises difficulties. From the above equation it can be calculated that (assuming a blood pH of 7.40) an intracellular pH of 5.89 is necessary to explain a 31.4-fold gradient between liver and blood. An ever lower pH must be invoked if the membrane is slightly permeable to NH_4^+.

There is no precise information on the pH of the liver cell. Indeed, since the cell is not a homogeneous aqueous phase but contains many different compartments, each liable to have different hydrogen ion activities, the concept of intracellular pH is not very meaningful. However, with a knowledge of the pH values and volumes of the individual compartments, it would be possible to calculate an "overall" intracellular pH. For the purposes of ammonia distribution between liver and plasma it is the overall pH that is relevant. An estimate of this overall pH can be arrived at by estimating the CO_2 and HCO_3^- concentrations of liver. Krebs (27) has estimated the overall pH of the liver on this basis and has reported a value of 7.22. The distribution of the weak acid 5,5-dimethyl-2,4-oxazolidimedione between intracellular fluid has also been applied to the problem of estimating intracellular pH. This method yields a liver pH of 7.23 (28). A liver cell pH of 5.89 is necessary to explain the *in vivo* ammonia gradient on the basis of passive diffusion of NH_3 across the plasma membrane. Hence it appears unlikely that passive diffusion is the operative mechanism.

Since the overall pH of the liver cell is not too dissimilar from the pH of blood, the ratio

$$\frac{[NH_4^+]\ \text{intracellular}}{[NH_4^+]\ \text{extracellular}}$$

will be quite similar to the ratio

$$\frac{[\text{total ammonia}]\ \text{intracellular}}{[\text{total ammonia}]\ \text{extracellular}}$$

that is, about 31.4. This gradient could be brought about by a potential across the membrane. Two conditions are necessary for this mechanism to be effective:

i. NH_4^+ must be permeable through the membrane so that the potential can influence its distribution.

ii. NH_3 must be quite impermeable through the membrane, because otherwise, since it would be at a higher concentration inside than out, it would diffuse back out of the cell, thus counteracting the effect of the membrane potential.

At equilibrium (i.e., when the tendency of the cell to accumulate cations is just balanced by their tendency to diffuse out along a concentration gradient) the situation is described by the Nernst equation:

$$\text{Potential difference (mV)} = -61 \log \frac{[NH_4{}^+]\ \text{intracellular}}{[NH_4{}^+]\ \text{extracellular}} = -91.3\ \text{mV}$$

Thus a membrane potential of -91 mV is required to explain the ammonia gradient; if NH_3 were slightly permeable, the potential would be even higher. The *in vivo* resting membrane potentials of hepatic cells are in the region of -51 mV (29). This is much too low to account for the observed ammonia gradient. A potential of -51 mV could support a 6.48-fold gradient of $NH_4{}^+$. This would mean an intracellular $[NH_4{}^+]$ of 0.205 m*M* and, therefore, an intracellular $[NH_3]$ of 0.770 m*M*. Since these proportions of $NH_3/NH_4{}^+$ require an intracellular pH of 9.6, it is clear that we cannot explain the ammonia gradient of the basis of passive distribution of $NH_4{}^+$ in response to a membrane potential.

It is conceivable that $NH_4{}^+$ could be actively pumped across the membrane. Again, a necessary precondition for this mechanism is that the diffusion of NH_3 out of the cell should be absent or very small. It is perhaps relevant that the $Na^+ + K^+$–activated ATPase is also activated by $NH_4{}^+$. $NH_4{}^+$ can replace the K^+ activation and is about equally as effective as K^+ (30, 31). That $NH_4{}^+$ can compete with K^+ for inward transport has been shown in erythrocytes by Post and Jolly (32). A similar effect has been suggested in cat cerebral cortex slices by Tower et al. (33). As rat liver possesses an extremely high activity of the $Na^+ + K^+$–activated ATPase (34), such a mechanism is, perhaps, a feasible explanation for the ammonia gradient observed *in vivo* across the rat liver.

The transport of ammonia has been studied by Mossberg (35, 36) in everted ileal sacs, and it appears that some sort of pump may be involved here also. The use of everted sacs permits the direct and accurate measurement of ammonia and pH on both sides of the permeability barrier and permits the manipulation of the composition of both fluid phases. These experimental advantages, which are not available in studies of ammonia transport into liver cells, enabled Mossberg to test the proposition that diffusion of NH_3 in response to a pH gradient is responsible for ammonia movement in the ileum. Such passive diffusion of NH_3 was disproved, since, regardless of the direction of the pH or NH_3 gradients, ammonia always moved from the mucosal to the serosal side. Mossberg also demonstrated that the movement of ammonia occurred against a concentration gradient, could not be explained by solvent drag, was abolished during anaerobic incubation, and was dependent on the presence of bicarbonate. Furthermore, Michaelis-Menten kinetics were demonstrated, possibly indicating the specific interaction of the transported species of ammonia with a carrier in the membrane. It is tempting to speculate that ammonia uptake by the liver may occur by a similar mechanism. It is especially important that Mossberg has clearly shown that NH_3 diffusion is not so rapid that it will automatically equilibrate across all membranes and, hence, that other transport systems for ammonia must be considered.

Whatever the mechanism, it is clear that the uptake of ammonia against a gradient is an important aspect of the liver's ammonia detoxification function. Together with the urea cycle it is responsible for the removal of the large quantities of ammonia supplied in the portal blood and for the maintenance of low ammonia concentrations in the systemic circulation.

SUMMARY

The operation of the urea cycle does not reduce the hepatic ammonia concentration to the low concentrations seen in blood. Concentrations of the order of 0.7 mM are normally maintained in liver. This, doubtless, reflects the fact that ammonia is not merely a waste product of nitrogen metabolism but is, of itself, an important intermediary metabolite. Due to the fact that glutamate dehydrogenase catalyzes a near-equilibrium reaction, the hepatic ammonia level is related to the mitochondrial concentrations of glutamate and of α-ketoglutarate as well as to the redox state of the mitochondrial pyridine nucleotides. Glutamine synthesis and asparagine synthesis are also important functions of hepatic ammonia. The maintenance of an appreciable hepatic ammonia concentration implies that the clearance of portal venous ammonia involves the transport of ammonia against a gradient. This may, possibly, be accomplished by the active pumping of NH_4^+.

REFERENCES

1. Jacquez, J. A., Poppell, J. W., and Jeltsch, R. (1959). *J. Appl. Physiol.* **14**, 255–263.
2. Kalra, J., and Brosnan, J. T. (1973). *FEBS Letters* **37**, 325–328.
3. Grisolia, S., and Cohen, P. P. (1953). *J. Biol. Chem.* **204**, 753–757.
4. Williamson, D. H., Lund, P., and Krebs, H. A. (1967). *Biochem. J.* **103**, 514–527.
5. Wu, C. (1963). *Biochim. Biophys. Acta* **77**, 482–493.
6. Jacobs, M. H. (1940). *Cold Spring Harbor Symp. Quant. Biol.* **8**, 30–39.
7. Chappell, J. B. and Haarhoff, K. N. (1967). In E. C. Slater, Z. Kaniuga, and L. Wojtczak, Eds., *Biochemistry of Mitochondria* Academic, London and New York, pp. 75–91.

8. Brosnan, J. T., Krebs, H. A., and Williamson, D. H. (1970). *Biochem. J.* **117**, 91–96.

9. Rawat, A. K. (1968). *Europ. J. Biochem.* **6**, 585–592.

10. Williamson, D. H., Lopes-Vieira, O., and Walker, B. (1967). *Biochem. J.* **104**, 497–502.

11. Greenstein, J. P., Winitz, M., Gullino, P., Birnbaum, S. M., and Otey, M. C. (1956). *Arch. Biochem. Biophys.* **64**, 342–353.

12. Kirk, E. (1936). *Acta. Med. Scand.* Supplement 78.

13. Russell, A., Levin, B., Oberholzer, V. G., and Sinclair, L. (1962). *Lancet* **2**, 699–700.

14. Freeman, J. M., Nicholson, J. F., Masland, W. S., Rowland, L. P., and Carter, S. (1964). *J. Pediat.* **65**, 1039–1046.

15. McMurray, W. L., Rathbun, J. C., Mohyuddin, F., and Koegler, S. J. (1963). *Pediatrics* **32**, 347–356.

16. Tomlinson, S., and Westall, R. G. (1964). *Clin. Sci.* **26**, 261–269.

17. Miller, A. L., and McLean, P. (1967). *Clin. Sci.* **32**, 385–393.

18. Moser, H. W., Efran, M. L., Brown, H., Diamond, R., and Neumann, C. G. (1967). *Am. J. Med.* **42**, 9–14.

19. Silen, W., Harper, H. A., Mawdsley, D. L., and Weirich, W. L. (1955). *Proc. Soc. Exper. Biol. Med.* **88**, 138–140.

20. Seligson, D., and Hirahara, K. (1957). *J. Lab. Clin. Med.* **49**, 962–974.

21. Schimassek, H., and Gerok, W. (1965). *Biochem. Z.* **343**, 407–419.

22. Hems, R., Ross, B. D., Berry, M. N., and Krebs, H. A. (1966). *Biochem. J.* **101**, 284–292.

23. Williamson, J. R., Browning, E. T., Scholz, R., Kreisberg, R. A., and Fritz, I. B. (1968). *Diabetes* **17**, 194–208.

24. Glogner, P., Nieth, H., Watner, H., and Gloeck, K. (1966). *Klin. Wschr.* **44**, 1040–1043.

25. Milne, M. D., Scribner, B. H., and Crawford, M. A. (1958). *Am. J. Med.* **24**, 709–729.

26. Pitts, R. F. (1964). *Am. J. Med.* **36**, 720–742.

27. Krebs, H. A. (1967). *Adv. Enz. Reg.* **5**, G. Weber, Ed., Pergamon Press, Oxford, pp. 409–434.

28. Masoro, E. J., and Siegel, M. E. (1971) *Acid-Base Regulation: Its Physiology and Pathophysiology*, W. B. Saunders, Philadelphia, pp. 35–37.

29. Schanne, O., and Caraboeuf, E. (1966). *Nature* **210**, 1390–1391.

30. Skou, J. C. (1960). *Biochim. Biophys. Acta* **42**, 6–23.

31. Rendi, R., and Uhr, M. L. (1964). *Biochim. Biophys. Acta* **89**, 520–531.

32. Post, R. L., and Jolly, P. C. (1957). *Biochim. Biophys. Acta* **25**, 118–128.

33. Tower, D. B., Wherrett, J. R., and McKhann, G. M. (1961). S. S. Kety, and J. Elkes, Eds., *Regional Neurochemistry* Pergamon Press, Oxford, p. 65.

34. Albers, R. W. (1967). *Ann. Rev. Biochem.* **36**, 727–756.

35. Mossberg, S. M., and Ross, G. (1967). *J. Clin. Invest.* **46**, 490–498.

36. Mossberg, S. M. (1967). *Am. J. Physiol.* **213**, 1327–1330.

27

The Use of Isolated Liver Cells in the Study of Hepatic Histidine Metabolism

R. Hems
Metabolic Research Laboratory
Nuffield Department of Clinical Medicine
Radcliffe Infirmary, Oxford, England

During studies on gluconeogenesis in the perfused rat liver it has been shown that, although histidine is metabolized, as was evident from the removal of histidine and the increase in formation of ammonia and urea, negligible amounts of glucose are formed; in fact the rate of glucose formation is lower than the endogenous rate (1).

Yet as early as 1913 Dakin (2) had shown in the phloridzin-treated dog that histidine is glucogenic, and in 1925 this was confirmed by Edlbacher (3) when he showed that the liver contained enzymes which can form one molecule of glutamate per molecule of histidine.

An investigation of the fate of histidine, first in the perfused liver and subsequently in isolated hepatocytes, forms the basis of this present work. It shows that in these preparations histidine gives rise to large amounts of formiminoglutamicacid within the cell and that methionine reduces this accumulation with a consequent increase in urea, ammonia, and glucose.

METHOD OF RAPID SEPARATION OF ISOLATED HEPATOCYTES FROM SUSPENSION MEDIUM

To obtain information on the distribution of metabolites between cells and suspension medium, a method of rapid separation of cells and medium, using a special design of centrifuge tube, was developed (4).

This work was carried out in collaboration with Dr. N. W. Cornell, Professor H. A. Krebs, Dr. P. Lund, and Mr. B. W. Tyler.

LIVER PREPARATION

The liver was perfused by the method of Hems et al. (5) and Krebs et al. (6). Isolated liver cells were prepared by the method of Berry and Friend (7) as modified by Cornell et al. (1) and Krebs et al. (8). All rats were fasted for 48 hours.

ANALYTICAL METHODS

Urea, glucose, lactate, and pyruvate were determined as described by Hems et al. (5). Ammonia was determined by the method of Kirsten et al. (9), histidine by the manometric method of Gale (10), with histidine decarboxylase.

Urocanate was measured by the increase in optical densities of perchloric acid filtrates at 277 mμ (11). Glutamine was determined as described by Lund (12).

Glutamic dehydrogenase was found to react quantitatively with formiminoglutamate at a rate which made it indistinguishable from glutamate when assayed according to the spectrophotometric method of Bernt and Bergmeyer (13), as modified by Cornell et al. (14). This method was therefore not suitable for the determination of glutamate in the presence of formiminoglutamate. Methods for the determination of glutamate and formiminoglutamate were based on the removal of glutamate from the sample with glutamic decarboxylase of *E. coli*, which attacks glutamate but not formiminoglutamate. Thus the sum of glutamate plus formiminoglutamate is determined in one sample and formiminoglutamate in a second sample after treatment with the glutamic decarboxylase. The value for glutamate was obtained by difference.

METABOLIC PRODUCTS FROM HISTIDINE IN THE PERFUSED LIVER

In order to establish the extent to which histidine is metabolized, the livers of starved rats were perfused with 10 m*M* histidine, and the perfusion medium was analyzed for glucose, ammonia, urea, urocanate, and glutamate. The values obtained show (Table 1) that, although substantial amounts of histidine are removed, only one equivalent of ammonia plus urea is formed. Some of the histidine removed is found as urocanic acid discharged into the perfusion medium, but there remains a deficit of 83 μmole unaccounted for by the products measured.

Progress with this problem received an impetus by using isolated hepatocytes as described by Berry and Friend (7), instead of the perfused liver. The decisive advantage of isolated hepatocytes is that the single suspension of cells

Table 1 Metabolic Products of Histidine in the Perfused Rat Liver

The concentration of histidine (added at 38 minutes) was 10 mM, and the perfusion was continued after addition of histidine for 90 minutes. The values refer to the changes in the perfusion medium. They are means of three experiments ± S.E.M. The mean weight of the liver was 6.08 g.

Metabolites	Total Change in Medium	Rate of Change (μmole/minute/g)
Histidine	−435 ± 13	−0.79
Urea	+120 ± 13	+0.22
Ammonia	+171 ± 27	+0.31
Urocanate	+272 ± 110	+0.50
Glutamate	+ 30.0 ± 5.3	+0.055
Glutamine	+ 19.5 ± 1.9	+0.036
Glucose	+ 14.0 ± 6.9	+0.026

which it provides allows for the testing of many variables, including the necessary controls.

THE EFFECT OF METHIONINE ON THE UPTAKE OF HISTIDINE BY ISOLATED HEPATOCYTES

A survey of the effect of amino acids on the fate of histidine in liver cells revealed that methionine is unique among the protein amino acids in its stimulating effect on histidine metabolism.

The effect of methionine on the time course of histidine uptake is shown in Table 2. Analysis of cell contents suggested that there was a large accumulation of glutamate in the cells, but this was not in accord with the finding of only one ammonia equivalent in the perfused liver. The discrepancy was explained by the finding that formiminoglutamate reacts with glutamate dehydrogenase, under the conditions for the assay of glutamate, at a rate which makes it indistinguishable from glutamate. Formiminoglutamate is a major product of histidine metabolism in the absence of added methionine in both the isolated liver cell and perfused liver. This had already been suggested by Tews and Harper (15). The values for glutamate plus formiminoglutamate show that these two compounds account for most of the carbon of histidine which accumulates in the cells and that the concentration approaches 20 mM. The addition of 0.1 mM methionine markedly reduced the apparent uptake of histidine and lowered the concentration of glutamate plus formiminoglutamate to one-third the level in the absence of methionine.

Table 2 The Effect of Methionine on the Time Course and Uptake of [U^{14}C]-Histidine and the Formation of Formiminoglutamate plus Glutamate

Hepatocytes (118 mg) were incubated in the presence of 2 mM-[U^{14}C]-histidine and separated from the medium by centrifugation as described in "Methods." The amount of histidine in cells and medium was calculated on the basis of the counts and the specific activity of the added histidine (29.8 × 10^3 counts/μmole). The values obtained were corrected for the amounts of formiminoglutamate plus glutamate, as measured by glutamate dehydrogenase. For further details see text.

Period of Incubation (minutes)	Counts in Cells × 10^{-4}	Ratio counts/g cells / counts/ml Supernatant	Histidine Plus Products in Cells as Histidine (μmole/g)	Glutamate Plus Formiminoglutamate in Cells (μmole/g)	Histidine in Cells (μmole/g)
With 2 mM-[U^{14}C]-histidine					
5	1.24	1.80	3.52	1.24	2.28
10	2.17	3.27	6.18	4.15	2.03
20	3.39	5.34	9.65	7.93	1.72
40	5.37	9.82	15.3	14.4	0.93
60	6.73	14.20	19.2	18.7	0.42
With 2 mM [U^{14}C]-Histidine and 0.1 mM Methionine					
5	1.14	1.62	3.25	1.55	1.70
10	1.69	2.53	4.79	2.72	2.07
20	2.08	3.29	5.92	3.75	2.17
40	2.48	4.29	7.07	5.72	1.35
60	2.90	5.38	8.25	6.60	1.65

METABOLISM OF HISTIDINE IN THE PRESENCE OF METHIONINE IN ISOLATED HEPATOCYTES

A balance of the products of histidine metabolism is shown in Table 3. Methionine has no effect on the amount of histidine removed, but a large increase in ammonia and urea occurs. The calculation of the balance between histidine removed and products formed is based on the assumption that the formation of glucose requires two histidine equivalents and causes the liberation of six ammonia equivalents, the formation of glutamate releases two ammonia equivalents, and formiminoglutamate releases one ammonia equivalent. No corrections have been made for glucose, ammonia, and urea formed in controls to which no histidine was added because of the uncertainty of the effect of addition of substrates on endogenous metabolism. The results show that a little more carbon was recovered than was expected and more ammonia equivalents were expected than found. However, the overall finding is clear, that in the absence of methionine, the sum of ammonia and urea is indicative of histidine metabolism only as far as the formation of formiminoglutamate. Addition of methionine increases the formation of ammonia, urea, and glucose, which can be accounted for by a decrease in the amounts of formiminoglutamate formed.

Table 3 Metabolism of Histidine in the Presence of Methionine in Isolated Hepatocytes

61.4 mg cells were incubated for 1 hour in 4 ml medium. The data are μmole/hour/g. Disappearance is indicated by the negative sign, appearance by the positive sign. For the calculation of the nitrogen expected and the histidine carbon recovered see text.

	Substrates Added		
Metabolite	None	Histidine (2 m*M*)	Histidine (2 m*M*) Methionine (0.1 m*M*)
Histidine	—	− 57.3	− 57.6
Glucose	+9.54	+ 11.20	+ 25.6
Formiminoglutamate		+ 42.9	+ 27.6
Glutamate	+4.25	+ 2.25	+ 8.04
Ammonia	+2.36	+ 26.8	+ 24.6
Urea (as ammonia)	+6.30	+ 78.5	+132
Carbon found (as C_3)		+ 70.4	+ 89.8
Nitrogen expected (as NH_3)		+114.6	+197
Nitrogen found (as NH_3)		+105.3	+157

SPECIFICITY OF THE METHIONINE EFFECT

In addition to testing the protein amino acids a comparison of the effect of methionine with a range of methionine analogs showed that only ethionine acts quantitatively like methionine. Selenomethionine had a small effect, as did homocysteine, but in the case of this latter compound the effect resulted probably from the formation of small amounts of methionine from homocysteine.

ACCUMULATION OF FORMIMINOGLUTAMATE IN THE PERFUSED RAT LIVER

The accumulation of formiminoglutamate in the perfused liver is shown in Table 4. Most of the formiminoglutamate is found in the tissue, the concentration in the medium being very low. The value of 108 μmoles found in the whole liver can account for the deficit of products shown in Table 1. Most of the glutamate was found in the medium, and the concentration in the tissue was of the same order as found *in vivo*.

Table 4 Formation of Glutamate and Formiminoglutamate from Histidine in the Perfused Rat Liver

Liver from a fasted rat (6.1 g) was perfused 38 minutes before the addition of 10 m*M* histidine, and the perfusion was continued for 60 minutes. At that time the liver was freeze-clamped, and a sample of the medium was deproteinized with perchloric acid.

	Total (μmole)	μmole/g or ml
Glutamate in medium	65	0.43
Formiminoglutamate in medium	7.8	0.053
Glutamate in liver	23.5	3.85
Formiminoglutamate in liver	108	17.7

RELEASE OF $^{14}CO_2$ FROM [RING 2—^{14}C] HISTIDINE

Degradation of histidine beyond the stage of formiminoglutamate by the addition of methionine raises the question of the fate of the 1-carbon fragment arising from splitting of the formimino group of formiminoglutamate.

Breakdown of formiminoglutamate with the formation of glutamate occurs when the formimino group is transferred to tetrahydrofolic acid (THF). The formiminotetrahydrofolate is then split to form N^5, N^{10}-methenyltetrahydrofolate and ammonia. In both perfused liver and isolated liver cells the rate of these reactions must be low so that the formiminoglutamate accumulates.

The stimulating effect of methionine on the formation of urea, ammonia, and glucose from histidine is indicative of an increase in the rate of the tetrahydrofolate-linked reactions. The amount of methionine removed is very small; the action of methionine is thus catalytic.

When [ring—2—^{14}C] histidine is incubated with isolated liver cells, ^{14}C appears in CO_2, and the addition of methionine results in a threefold increase of ^{14}C labeled CO_2 (Table 5).

The yield of CO_2 as a percentage of that calculated from the formation of glucose and glutamate is low, and addition of methionine causes only a small increase in the yield. The fate of the [ring—2—^{14}C] which is not found in CO_2 is still unknown.

ACCELERATION OF FORMATE OXIDATION BY METHIONINE

The oxidation level of the formimino group of formiminoglutamate corresponds to that of formate. It is of interest that ^{14}C formate added to hepatocytes yields $^{14}CO_2$ and that the rate of CO_2 production is more than doubled by

Table 5 Release of $^{14}CO_2$ from [ring—2—^{14}C] Histidine

The experiment was that recorded in Table 6, where added histidine was [ring—2—^{14}C] histidine. The amount of the CO_2 which arises from the ring carbon of histidine is calculated on the basis of the counts found in CO_2 and the specific activity of the histidine added. The expected yield of $^{14}CO_2$ was calculated on the assumption that for each molecule of glucose formed two carbon 2 were released, and for each molecule of glutamate one carbon 2 is released.

	Substrates Added			
	Histidine	Histidine, Methionine	Histidine, Oleate	Histidine, Methionine, Oleate
$^{14}CO_2$ expected from glucose plus glutamate (μmole/hour/g)	24.7	59.2	22.7	48.3
$^{14}CO_2$ found (μmole/hour/g)	6.2	17.9	6.0	17.8
Yield of $^{14}CO_2$ as percent of expected	25%	30%	26%	27%

the addition of 0.1 m*M* methionine (Table 6). It is very likely that this effect is related to that on the degradation of formiminoglutamate. Furthermore it has been shown that addition of formate reduces the formation of $^{14}CO_2$ from [ring—2—^{14}C] histidine.

Table 6 Acceleration of Formate Oxidation by Methionine in Isolated Hepatocytes

The initial concentration of ^{14}C-formate was 2 m*M*. The incubation time was 1 hour.

Expt. No.	Cell Concentration	Initial Methionine Concentration (m*M*)	CO_2 Formed (μmole/hour/g)
		0	9.4
1	89 mg in 4 ml	0.1	20.8
		0	7.9
2	92 mg in 4 ml	0.1	17.8

The formation of CO_2 from formate involves the following reactions:

$$\text{Formate} + \text{tetrahydrofolate} + \text{ATP} \rightarrow \text{formyltetrahydrofolate} + \text{ADP} \quad (1)$$

$$\text{Formyltetrahydrofolate} + \text{NADP} + H_2O \rightarrow CO_2 + \text{tetrahydrofolate} + \text{NADPH} \quad (2)$$

Reaction 1 is catalyzed by formyltetrahydrofolate synthetase (EC 6.3.4.3) (16) and reaction 2 by formyltetrahydrofolate dehydrogenase (EC 1.5.1.6) (17). It is likely that reaction 2 is common to both the formation of CO_2 from formate and the formimino group of formiminoglutamate, and this is proposed as the site of action of methionine. The pathways are shown in Scheme 1.

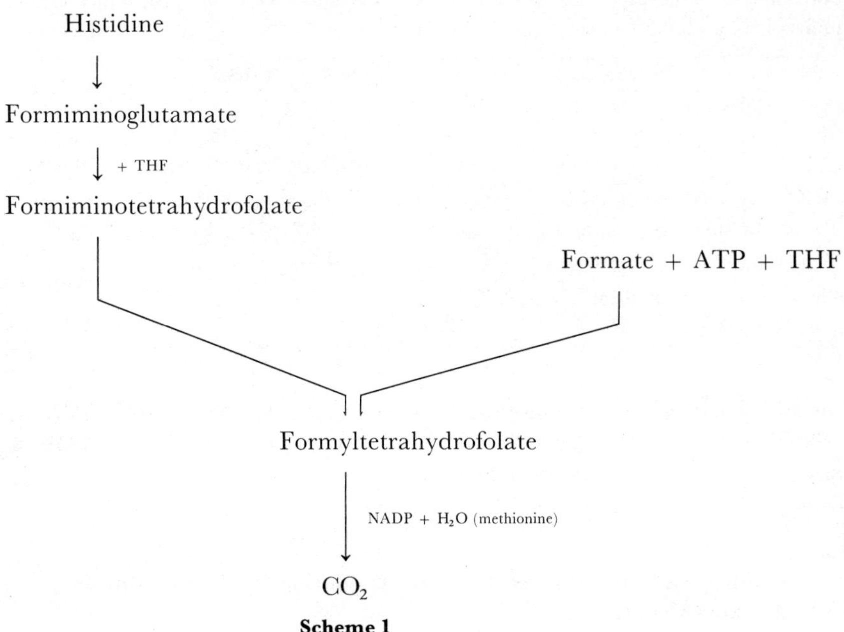

Scheme 1

SUMMARY

In the perfused liver and in the isolated liver cell histidine is metabolised with the formation of one equivalent of ammonia and the accumulation of high concentrations of formiminoglutamate within the cell. The addition of small amounts of methionine results in an increase in the amounts of the ammonia equivalents and glucose formed and a decrease in the formiminoglutamate concentration. Methionine increases the contribution of carbon 2 of the histidine ring to metabolic CO_2. The oxidation of formate to CO_2 is similarly stimulated by methionine. The results indicate that a rate-limiting reaction in the degradation of histidine beyond the stage of formiminoglutamate is accelerated by methionine. The finding of an increase in formate oxidation by methionine points to formyltetrahydrofolate dehydrogenase as the rate-limiting enzyme and the site of action of methionine.

REFERENCES

1. Cornell, N. W., Lund, P., Hems, R., and Krebs, H. A. (1973). *Biochem. J.* **134**, 671–672.
2. Dakin, H. D. (1913). *J. Biol. Chem.* **14**, 321–333.
3. Edlbacher, S. (1926). *Z. Physiol. Chem.* **157**, 106–114.
4. Hems, R., Lund, P., and Krebs, H. A. (1975). *Biochem. J.* **150**, 47–50.
5. Hems, R., Ross, B. D., Berry, M. N., and Krebs, H. A. (1966). *Biochem. J.* **101**, 284–292.
6. Krebs, H. A., Wallace, P. G., Hems, R., and Freeland, R. A. (1969). *Biochem. J.* **112**, 595–600.
7. Berry, M. N., and Friend, D. S. (1969). *J. Cell Biol.* **43**, 506–520.
8. Krebs, H. A., Cornell, N. W., Lund, P., and Hems, R. (1974). In *Alfred Benzon Symposium IV*, Regulation of hepatic metabolism, Munksgaard, Copenhagen, pp. 726–750.
9. Kirsten, E., Gerez, C., and Kirsten, R. (1963). *Biochem. Z.* **337**, 312–319.
10. Gale, E. F. (1945). *Biochem. J.* **39**, 46–52.
11. Magasanik, B., Kammskas, E., and Kimhi, Y. (1971). In H. Tabor, and C. W. Tabor, Eds., *Methods in Enzymology*, **17B** Academic, New York and London, pp. 47–50.
12. Lund, P. (1970). In H. U. Bergmeyer, Ed., *Methoden der enzymatischen Analyse* 2nd ed., Verlag Chemie, Weinheim, pp. 1670–1673.
13. Bernt, E., and Bergmeyer, H. U. (1963). In H. U. Bergmeyer, Ed., *Methods of Enzymatic Analysis*, Academic, New York and London, pp. 384–388.
14. Cornell, N. W., Lund, P., and Krebs, H. A. (1974). *Biochem. J.* **142**, 327–337.
15. Tews, J. K., and Harper, A. E. (1969). *Biochim. Biophys. Acta* **183**, 635–637.
16. Kutzbach, C., and Stokstad, E. L. R. (1968). *Biochem. Biophys. Res. Commun.* **30**, 111–117.
17. Jaenicke, L., and Brode, E. (1961). *Biochem. Z.* **334**, 108–132.

28

The Effect of Ethanol on Ammonia Metabolism

Marion Stubbs
Metabolic Research Laboratory
Nuffield Department of Clinical Medicine
Radcliffe Infirmary, Oxford, England

It is well established that there are certain reversible enzymes in the liver that are of high enough activity to establish equilibrium among the reactants of a particular enzyme system (1, 2, 3). Three of the enzymes which play an important role in urea synthesis come into this category, namely, glutamate dehydrogenase (EC 1.4.1.3), aspartate aminotransferase (EC 2.6.1.1), and alanine aminotransferase (EC 2.6.1.2). Their involvement in urea synthesis is dependent on the precursor of urea. The following examples refer to the extreme case where either an amino acid or ammonia is the precursor. In reality the precursors would be a mixture of the two. When urea is synthesized from alanine, aspartate and carbamoyl phosphate would be formed by the following reactions:

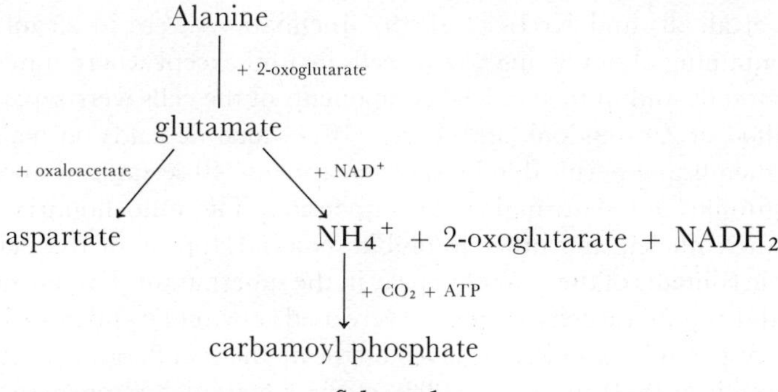

Scheme 1

When urea is synthesized from ammonia, another set of reactions takes place to form carbamoyl phosphate and aspartate:

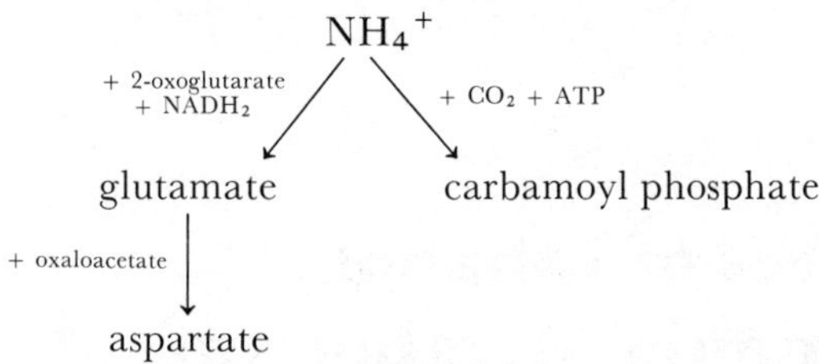

Scheme 2

The directional flow of nitrogen through the pathway, however, is determined by an essentially irreversible reaction at the argininosuccinate synthetase stage. The role of the equilibrium stages is in determining the concentrations of aspartate, glutamate, and ammonia.

It can be seen from schemes 1 and 2 that there is an expected stoichiometric requirement for urea synthesis through the ornithine cycle in that half the nitrogen must be supplied in the form of aspartate and half in the form of carbamoyl phosphate (4, 5). Krebs and coworkers (6) have shown that ethanol causes an accumulation of aspartate in the perfused liver when excess nitrogen in the form of alanine or ammonium lactate was present, thereby upsetting the expected stoichiometry. The purpose of the present study was to investigate the reasons for this effect of ethanol, and isolated hepatocytes were chosen as the experimental material.

MATERIALS AND METHODS

Hepatocytes were prepared from the livers of 48-hour starved rats, essentially as described by Berry and Friend (7), incorporating the modifications of Cornell et al. (8) and Krebs et al. (9). Incubations were in 25 ml conical flasks containing about 80 mg wet wt cells in 4 ml except where stated otherwise. Cytosolic and mitochondrial components of the cells were separated by the method of Zuurendonk and Tager (10), which depends on making the plasma membrane permeable by short treatment (40 seconds at about 5°C) with digitonin and centrifuging the suspension. The mitochondria, plasma membranes, nuclei, and other insoluble material appear in the pellet, but all soluble contents of the cytosol appear in the supernatant. Lactate dehydrogenase and glutamate dehydrogenase were used as cytosolic and mitochondrial markers respectively and were assayed by the methods of Bergmeyer et al. (11) and Schmidt (12). If more than 90% of the lactate dehydrogenase and less than 10% of the glutamate dehydrogenase were found in the cytosol fraction, adequate separation was assumed to have taken place.

Correction for Adhering Fluid in Separation Experiments

Up to 10% of lactate dehydrogenase was found in the pellet in the separation experiments. Corrections were made for this cytosolic contamination of the pellet based on the assumption that the metabolites measured shared the same cytosolic space as the lactate dehydrogenase.

Determination of Metabolites

Metabolites were determined as described by Stubbs and Krebs (13) and were measured on neutralized perchloric extracts of the cell suspensions, cells, or subcellular components.

RESULTS

Preliminary experiments indicated that the increase in aspartate formation observed in the presence of alanine and ethanol in the perfused liver could not be reproduced in the hepatocyte suspensions. A comparison of the cells and perfused organ demonstrated that glucose, lactate, and pyruvate production and alanine removal were similar but that there were differences in the alanine nitrogen metabolism (see Table 1).

Table 1 Comparison of the Effects of Ethanol on Alanine Nitrogen Metabolism in Liver Perfusion and Isolated Hepatocytes

Experimental conditions for the hepatocytes were as described in text. Results are expressed as μmole/hour/g wet wt (means ± S.D. for five observations). The values for the perfusion experiments are calculated from Table 3 in Krebs et al. (1973), expressed as μmole (tissue plus medium)/hour/g wet wt.

Metabolite	Alanine (10 m*M*)		Alanine (10 m*M*), Ethanol (10 m*M*)	
Formed	Hepatocytes	Perfused Liver	Hepatocytes	Perfused Liver
Urea	42.4 ± 10.3	17	36.4 ± 8.8	10
NH_4^+	18.2 ± 4.3	61	14.4 ± 4.8	23
Aspartate	7.2 ± 2.2	2.7	7.3 ± 2.4	15

In hepatocytes more aspartate was formed from alanine than in the perfused liver, but although ethanol increased the yield about fivefold in the perfused organ, it had no effect in hepatocytes. The main end product of alanine nitrogen metabolism in the liver cells was urea, compared with ammonia in the perfused organ. These findings raised the question of whether the aspartate accumulation was related to the production of ammonia.

When urease was added to the hepatocytes (Table 2) in order to maintain a high ammonia concentration, a threefold increase in aspartate formation on the addition of ethanol was then observed. This indicates that aspartate formation depends on the ammonia concentration. This was further tested by experiments where the alanine was replaced by lactate plus NH_4Cl.

Table 2 The Effect of Urease on Ammonia Production and Amino Acid Accumulation from Alanine in the Presence of Ethanol in Hepatocytes

Experimental conditions were as described in text. Initial concentrations of alanine and ethanol were 10 m*M*. 1.5 International Units of urease were added per flask.

	μmole Products Formed/hour/g Wet Wt in the Presence of			
Metabolite	Alanine	Alanine, Ethanol	Alanine, Urease	Alanine, Ethanol, Urease
Aspartate	7.4	10.3	8.1	23.8
NH_4^+	15.7	15.2	107	70
Urea	44.1	36.4	< 2	< 2

In the presence of lactate and NH_4Cl (see Table 3) ethanol caused a 22-fold increase in [aspartate]. Ethanol inhibited glucose synthesis by 73% and caused a highly reduced state of the cytosolic NAD couple, as indicated by the [lactate]/[pyruvate] ratio. Ethanol did not affect the removal of ammonia but it did cause more ammonia to be diverted away from urea synthesis to the formation of aspartate (and glutamate). [Alanine] and [glutamine] were not affected by ethanol. Control experiments where lactate, lactate plus ethanol, NH_4Cl alone, NH_4Cl plus ethanol, or ethanol alone were present showed no increase in the yield of aspartate, demonstrating that the extra amino acids did not arise from proteolysis. As the accumulation of aspartate only occurred when the main product of nitrogen metabolism was ammonia, ornithine might be expected to abolish the effect.

The addition of ornithine increased urea synthesis both in the absence and presence of ethanol (see Table 4). It also largely abolished the accumulation of aspartate, presumably by providing an acceptor in the form of citrulline to form argininosuccinate and subsequently urea.

As aspartate can only be formed through the aspartate aminotransferase reaction, it was thought necessary to look at the distribution of the components

Table 3 The Effect of Ethanol on Gluconeogenesis, Urea Synthesis, and Amino Acid Formation in Hepatocytes

Experimental details were as described in text. The results are expressed as μmole product formed or used per g wet wt for a 60 minute incubation (means ± S.D. for eight observations). The initial substrate concentrations were 10 mM. Values which are statistically different by Student's t test are indicated by *: $P < 0.05$.

Metabolite or Metabolite Ratio	Metabolite Changes in the Presence of: Lactate, NH_4Cl	Lactate, NH_4Cl, Ethanol
Glucose	63.4 ± 15.5	17.0 ± 10.2*
[Lactate]/[pyruvate]	11.3 ± 3.3	203 ± 35*
Urea	59.5 ± 4.1	44.4 ± 3.4*
Ammonia	−151 ± 19.9	−158 ± 19.7
Glutamate	6.92 ± 1.25	10.6 ± 1.83*
Glutamine	5.32 ± 1.26	5.98 ± 0.78
Alanine	20.7 ± 5.3	17.5 ± 7.85
Aspartate	2.77 ± 1.45	60.2 ± 21.5*

Table 4 The Effect of Ethanol on Urea Synthesis and Aspartate Formation in the Presence of Ornithine

Experimental details were as described in text. The results are expressed as μmole/g for a 60 minute incubation (means ± S.D. for three observations). The initial concentrations were 10 mM for lactate, NH_4Cl, and ethanol and 2 mM for ornithine.

Metabolite	Metabolite Changes in the Presence of: Lactate, NH_4Cl	Lactate, NH_4Cl, Ornithine	Lactate, NH_4Cl, Ethanol	Lactate, NH_4Cl, Ethanol, Ornithine
Urea	58.8 ± 12.4	183 ± 32	45.5 ± 15.6	243 ± 82
Aspartate	3.5 ± 1.9	0.71 ± 0.20	49.6 ± 10	3.21 ± 1.91

of this reaction between cytosol and mitochondria. A technique devised by Zuurendonk and Tager (10) for separation of cytosol and mitochondria in liver cells was used. As this method results in mixing cytosol contents with the medium, it was necessary to first establish the distribution of metabolites between cells and medium. This was done by the technique already described elsewhere in this book by Hems. p. 459

The major proportion of aspartate, glutamate, and oxoglutarate was intracellular (see Table 5). Malate was also shown to be intracellular, and the reason for measuring this will become apparent later.

Table 5 Distribution of Metabolites between Cells and Medium

Experimental details were as described in text. 280 mg wet wt cells were incubated for 30 minutes in 16 ml medium before separation. Initial concentrations of lactate, NH_4Cl, and ethanol were 10 m*M*.

	Lactate, NH_4Cl		Lactate, NH_4Cl, ethanol	
	Cells (μmole/g)	Medium (m*M*)	Cells (μmole/g)	Medium (m*M*)
Aspartate	3.5	0.016	34.4	0.144
Glutamate	2.02	0.042	3.88	0.055
Malate	0.88	0.01	0.90	0.011
Oxoglutarate	0.22	<0.005	0.10	<0.005

When it had been established that the components of the aspartate aminotransferase system were mostly intracellular, separation of cytosol and mitochondria was undertaken, and the components of the system determined in the two compartments (Table 6). The measurements provide information of the amounts but not of the concentration of the metabolites. Ethanol increased the amount of aspartate in the cytosol by a factor of 10, but the mitochondrial content did not change. The glutamate content increased in both compartments on the addition of ethanol. The oxoglutarate content of the cytosol was halved in the presence of ethanol, whereas the mitochondrial content was not affected. The cytosolic oxaloacetate value was calculated according to Williamson et al. (1) from the malate content, the equilibrium constant for malate dehydrogenase, and the cytosolic $[NAD^+]/[NADH]$ ratio. The calculation of the mitochondrial oxaloacetate content was not possible, as the mitochondrial malate values were too low to be reliable.

The mass action ratio for the aspartate aminotransferase reaction

$$\frac{[\text{aspartate}][\text{oxoglutarate}]}{[\text{glutamate}][\text{oxaloacetate}]}$$

in the cytosol was 11.8 in the absence of ethanol and 43.5 in its presence. Thus in the absence of ethanol the system is near equilibrium—the equilibrium constant being 6.61 at pH 7.0, 38°C, and $I = 0.25$ (3)—but ethanol causes a displacement from equilibrium.

Table 6 The Effect of Ethanol on the Cytosolic and Mitochondrial Content of Aspartate, Glutamate, and 2-Oxoglutarate Formed in the Presence of Ammonium Lactate

Experimental details were as described in the text. Initial concentrations of lactate, NH_4Cl, and ethanol were 10 mM. The data are from a representative experiment in which 424 mg wet wt cells were incubated for 10 minutes before separation. The results are expressed as μmole metabolite present in the named compartment. "Corrected" values refer to μmole metabolite present in the mitochondria corrected for adhering cytosol (for details of correction see text). *Oxaloacetate was calculated from the measured cytosolic malate content as described by Williamson et al. (1967). Calculation of mitochondrial oxaloacetate is unreliable.

Metabolite	Addition	Cytosol (μmole)	Mitochondria (μmole)	Mitochondria "Corrected" (μmole)
Aspartate	Lactate, NH_4Cl	0.36	0.40	0.36
	Lactate, NH_4Cl, ethanol	3.40	0.76	0.43
Glutamate	Lactate, NH_4Cl	0.83	0.17	0.09
	Lactate, NH_4Cl, ethanol	1.52	0.48	0.33
2-Oxoglutarate	Lactate, NH_4Cl	0.040	0.015	0.011
	Lactate, NH_4Cl, ethanol	0.019	0.013	0.012
*Oxaloacetate	Lactate, NH_4Cl	0.00151		
(calculated)	Lactate, NH_4Cl, ethanol	0.00098		

DISCUSSION

The experiments demonstrate that the accumulation of aspartate in the presence of ethanol depends on the presence of relatively high (several millimolar) concentrations of ammonia. The following facts are relevant to the interpretation of the findings:

1. The accumulated aspartate is located in the cytosol.
2. The concentration of oxoglutarate is low in the cytosol when aspartate accumulates and that of glutamate is raised in the mitochondria.
3. Ornithine abolishes the accumulation of aspartate.
4. The components of the aspartate aminotransferase system are not at equilibrium in the cytosol when the ammonia concentration is high.

The accumulation of aspartate implies that one of the major reactions by which aspartate is removed—either the cytosolic aspartate aminotransferase reaction or the argininosuccinate synthetase reaction—must be decreased in relation to the production of cytosolic aspartate. The pathways involved in aspartate accumulation in the presence of lactate, NH_4Cl, and ethanol are as follows:

$$\text{Ethanol} + \text{NAD}^+ \rightarrow \text{acetaldehyde} + \text{NADH} + \text{H}^+ \quad (\text{C})$$

$$\text{Lactate} + \text{NAD}^+ \rightarrow \text{pyruvate} + \text{NADH} + \text{H}^+ \quad (\text{C})$$

$$\text{Acetaldehyde} + \text{NAD}^+ \rightarrow \text{acetate} + \text{NADH} + \text{H}^+ \quad (\text{M})$$

$$\text{NH}_4{}^+ + \text{oxoglutarate} + \text{NADH} + \text{H}^+ \rightarrow \text{glutamate} + \text{NAD}^+ \quad (\text{M})$$

$$\text{Pyruvate} + \text{ATP} + \text{CO}_2 \rightarrow \text{oxaloacetate} + \text{ADP} + \text{P}_i \quad (\text{M})$$

$$\text{Glutamate} + \text{oxaloacetate} \rightarrow \text{aspartate} + \text{oxoglutarate} \quad (\text{M})$$

Sum:

$$\text{Ethanol} + \text{lactate} + 2\text{NAD}^+ + \text{NH}_4{}^+ + \text{ATP} + \text{CO}_2 \rightarrow \text{acetate} + 2\text{NADH} + 2\text{H}^+ + \text{ADP} + \text{P}_i + \text{aspartate}$$

(C = cytosol; M = mitochondria)

The aspartate that accumulates in the cytosol must originate from the mitochondrial aspartate aminotransferase reaction, because the two precursors—glutamate and oxaloacetate—are both generated in the mitochondrial matrix by the glutamate dehydrogenase and pyruvate carboxylase reactions respectively. As the aminotransferase is the only enzyme apart from proteolytic enzymes and asparaginase which can generate major quantities of aspartate and since the aminotransferase cannot be energy driven, the driving force forming the mitochondrial aspartate must be taken to be the tendency to establish equilibrium in the mitochondrial aspartate aminotransferase system in the face of a continuous production of glutamate and oxaloacetate. Cytosolic aspartate can only arise by translocation of mitochondrial aspartate, either by facilitated diffusion along a concentration gradient or by a mechanism involving an energy-driven pump such as that proposed by La Noue et al. (14, 15). But whatever the mechanism there is the fact that aspartate is translocated to the cytosol. Normally, in the presence of lactate, when gluconeogenesis occurs, the translocation of aspartate is accompanied by an equimolar amount of 2-oxoglutarate (for details of the aspartate shuttle, see 16, 17, 18). The finding of low concentrations of 2-oxoglutarate in the cytosol and raised concentrations of glutamate in the mitochondria is in accordance with the assumption that the stoichiometric transfer of aspartate and oxoglutarate is upset in the presence of NH_4Cl and ethanol: mitochondrial oxoglutarate, instead of being translocated, reacts with $NH_4{}^+$ and NADH to regenerate the glutamate used up in the mitochondrial aminotransferase reaction, the dehydrogeneration of ethanol and acetaldehyde providing the NADH. This can account for the excess aspartate in the cytosol which has no stoichiometric partner for the disposal reaction.

This assumption does thus explain the disequilibrium in the aspartate aminotransferase system of the cytoplasm. It is necessary to assume that at the lowered concentration of oxoglutarate the activity of the transferase is too weak to maintain equilibrium, [oxoglutarate] at 0.045 mM being far below the K_m value of 0.3 to 0.4 mM (reported for ox liver by Wada and Morino (19); there are no data for rat liver).

The nonaccumulation of aspartate in the presence of ornithine might be taken to indicate that the argininosuccinate synthetase reaction can keep pace with the rate of supply of cytosolic aspartate when ornithine promotes the rate of supply of citrulline. Thus the powerful tendency to establish equilibrium in one compartment can drive a reaction out of equilibrium in another. The experiments demonstrate the importance of equilibrium relations in the regulation of ammonia metabolism and urea synthesis and how they can be disturbed under extreme conditions.

ACKNOWLEDGMENT

This work was supported by grants from the Medical Research Council, the National Institute on Alcohol Abuse and Alcoholism (No. 1 RO1 AA00381), and Schering AG, Berlin. The author is indebted to Professor Sir Hans Krebs for invaluable discussions.

REFERENCES

1. Williamson, D. H., Lund, P., and Krebs, H. A. (1967). *Biochem. J.* **103**, 514–527.

2. Krebs, H. A., and Veech, R. L. (1969). In S. Papa, J. M. Tager, E. Quagliariello, and E. C. Slater, Eds. *The energy level and metabolic control in mitochondria* Adriatica Editrice, Bari, Italy, pp. 329–364.

3. Krebs, H. A., and Veech, R. L. (1970). In H. Sund, Ed., *Pyridine Nucleotide-Dependent Dehydrogenases* Springer-Verlag, Berlin, pp. 413–438.

4. Krebs, H. A., Hems, R., and Lund, P. (1973). In G. Weber, Ed., *Advances in Enzyme Regulation* **2** Pergamon Press, Oxford and New York, pp. 361–377.

5. Krebs, H. A., Hems, R., and Lund, P. (1973). In F. Hommes and C. J. Van den Berg, Eds., *Inborn Errors of Metabolism* Academic Press, New York and London, pp. 201–219.

6. Krebs, H. A., Hems, R., and Lund, P. (1973). *Biochem. J.* **134**, 697–705.

7. Berry, M. N., and Friend, D. S. (1969). *J. Cell. Biol.* **43**, 506–520.

8. Cornell, N. W., Lund, P., Hems, R., and Krebs, H. A. (1973). *Biochem. J.* **134**, 671–672.

9. Krebs, H. A., Cornell, N. W., Lund, P., and Hems, R. (1974). In F. Lundguist and N. Tygstrup, Eds., *Regulation of Hepatic Metabolism*, Alfred Benzon Symposium VI Munksgaard, Copenhagen, pp. 457–479.

10. Zuurendonk, P. F., and Tager, J. M. (1974). *Biochim. Biophys. Acta* **333**, 393–399.

11. Bergmeyer, H. U., Bernt, E., and Hess, B. (1963). In H. U., Bergmeyer, Ed. *Methods of Enzymatic Analysis*, Academic Press, New York and London, pp. 736–741.

12. Schmidt, E. (1963). In H. U. Bergmeyer, Ed., *Methods of Enzymatic Analysis*, Academic Press, New York and London, pp. 752–756.

13. Stubbs, M., and Krebs, H. A. (1975). *Biochem. J.* **150**, 41–46.

14. La Noue, K. F., Walajtys, E. I. and Williamson, J. R. (1973). *J. Biol. Chem.* **248**, 7171–7183.

15. La Noue, K. F., Bryla, J., and Bassett, D. J. P. (1974). *J. Biol. Chem.* **249**, 7514–7521.

16. Borst, P. (1963). In *Funktionelle und Morphologische Organisation der Zelle*, Springer-Verlag, New York, pp. 137–158.

17. Lardy, H. A., Paetkau, V., and Walter, P. (1965). *Proc. Natl. Acad. Sci. US* **53**, 1410–1415.

18. Cornell, N. W., Lund, P., and Krebs, H. A. (1974). *Biochem. J.* **142**, 327–337.

19. Wada, H., and Morino, Y. (1964). *Vitam. Horm.* **22**, 411–444.

29

Glutamine as a Precursor of Urea

Patricia Lund and Malcolm Watford
Metabolic Research Laboratory
Nuffield Department of Clinical Medicine
Radcliffe Infirmary, Oxford, England

INTRODUCTION

It seems appropriate that some discussion of glutamine metabolism should be included in a symposium honouring Sir Hans Krebs. He was the first to describe glutamine synthesis in mammalian tissues (1). In the same paper he described the tissue-specific properties of two glutaminases which only relatively recently have been separated and characterized as "phosphate-dependent" and "phosphate-independent" glutaminase isoenzymes (2, 3).

Glutamine has had a long association with the urea cycle. Its direct participation was at various times considered to be a prerequisite for urea synthesis (4, 5, 6, 7, 8, 9, 10). The details of the mechanism of urea synthesis have been fully covered by other authors in this volume. We propose to consider the quantitative aspects and to develop the argument that glutamine is an important precursor of urea under normal physiological conditions. We shall present experimental evidence that liver "glutaminase I" (L-glutamine amidohydrolase, EC 3.5.1.2)

$$\text{L-glutamine} + H_2O \rightarrow \text{L-glutamate} + NH_3$$

is nonfunctional under these conditions but that "glutaminase II" (L-glutamine-oxo-acid aminotransferase, EC 2.6.1.15)

$$\text{L-Glutamine} + \alpha\text{-ketoacid} \xrightleftharpoons{\text{aminotransferase}} \alpha\text{-ketoglutaramate} + \text{amino acid}$$

$$\alpha\text{-Ketoglutaramate} \xrightarrow{\omega\text{-amidase}} \alpha\text{-ketoglutarate} + NH_3$$

overall reaction

$$\text{L-Glutamine} + \alpha\text{-ketoacid} \rightarrow \alpha\text{-ketoglutarate} + NH_3 + \text{amino acid}$$

can be active, given the presence of a suitable ketoacid acceptor. The results support the conclusions of Matsutaka et al. (11) and of Windmueller and Spaeth (12) that hydrolysis of glutamine occurs in the intestinal mucosa and that the products are transported to the liver in the portal vein for conversion to urea. We shall also present experimental evidence that reincorporation of ammonia into glutamine does not occur to any significant extent in the liver and that the highly active glutamine synthetase measurable *in vitro* is largely inactive in the intact liver cell.

Sources of Glutamine

Glutamine is released in large amounts from skeletal muscle in the post-absorptive state in man (13), fasted rat (14, 15), dog and monkey (16). Glutamine and alanine are the chief end products of amino acid catabolism in muscle, though more emphasis tends to be placed on their importance as precursors of glucose than on their importance as precursors of urea. While the tendency to establish equilibrium in the alanine aminotransferase reaction explains the release of alanine (17), the teleological advantage of expending one molecule of ATP for each molecule of glutamine formed is harder to explain. It is generally accepted that glutamine and alanine provide a nontoxic means of transporting ammonia to the liver. The free ammonia required for glutamine synthesis may arise through the glutamate dehydrogenase reaction, although its activity is low in skeletal muscle (18). Alternatively it may be produced through operation of the purine nucleotide cycle (19). Glutamine is more important than alanine in terms of nitrogen transport. The amounts released depend on the dietary (i.e., hormonal) state (20). For example, we were unable to show net glutamine release from the hindlimbs of rats fed a normal diet supplemented with glucose (Table 1), a diet in which proteolysis is low as a result of high insulin levels. These animals were the controls in experiments designed to identify the source of the extra glutamine required by the kidney in compensating chronic metabolic acidosis. In the chronically acidotic group there was a small output of glutamine by the hindlimbs (Table 1) which was much smaller than reported for fed rats by Yamamoto et al. (21), for 24-hour starved rats by Aikawa et al. (14), or for 48-hour starved rats by Ruderman and Berger (15). These comparisons take no account of blood flow rates or the fact that some experiments refer to plasma, although erythrocytes may be involved in the peripheral exchange of glutamine, see (22).

Minor sources of glutamine *in vivo* are the brain (16, 23) and, in some species, the kidney. Rat, guinea-pig, sheep, and rabbit kidney contain

Table 1 Arteriovenous Differences for Glutamine and Glutamate across Various Organs of the Normal and Chronically Acidotic Rat

Female rats were maintained on a normal diet supplemented with 5% *w/v* glucose in the drinking water. The drinking water of the acidotic group contained 100 m*M* NH_4Cl with 5% *w/v* glucose to prevent dehydration. The acidotic rats each consumed about 5 mmoles of NH_4Cl per day for 8 to 10 days before the experiment. Blood was carefully taken under light ether anaesthesia into 1 ml syringes in which the needles were bent for ease of insertion into the vessels. A syringe was implanted and fixed in position in one vessel before blood was drawn from the other. The sample (0.5 to 1 ml) was quickly transferred to a weighed tube containing 1 ml $HClO_4$ (14% *v/v*) and reweighed. Assays were carried out on the neutralized, deproteinized supernatant. Metabolite assays were as described later in the "Experimental" section. Vessels sampled simultaneously from three different groups of normal and acidotic rats were aorta and junction of the common iliac veins, portal vein and hepatic vein, and aorta and renal vein. Values given are μmoles per milliliter whole blood, with means ± S.E.M. of the number of rats given in parentheses. The mean arteriovenous difference was calculated from the individual differences. A plus (+) sign indicates an output of glutamine from the tissue and a minus (−) an uptake. Experiments by L. G. Welt and P. Lund.

	Control		Chronic acidotic	
	Glu	Gln	Glu	Gln
Artery (aorta)	0.186 ± 0.004 (10)	0.583 ± 0.023 (9)	0.178 ± 0.004 (13)	0.532 ± 0.013 (11)
Portal vein	0.212 ± 0.004 (6)	0.449 ± 0.017 (6)	0.174 ± 0.008 (6)	0.484 ± 0.027 (6)
Hepatic vein	0.210 ± 0.008 (6)	0.534 ± 0.020 (6)	0.182 ± 0.010 (6)	0.579 ± 0.020 (6)
H.V.-P difference		+ 0.085 ± 0.016 (6)		+ 0.093 ± 0.030 (6)
Artery–renal vein difference		+ 0.023 ± 0.027 (5)		− 0.073 ± 0.021 (6)
Artery–common iliac vein difference		− 0.003 ± 0.018 (9)		+ 0.021 ± 0.015 (11)

glutamine synthetase; dog, cat, pig, pigeon, do not (1). Arteriovenous differences measured across the kidney of the rat are contradictory. Yamamoto (21) and Aikawa (14) find an uptake in both fed and starved rats. We found a small output in four of the five rats in the control group (Table 1) but an uptake in acidosis as in species lacking glutamine synthetase.

The contribution of the liver to the glutamine pool of the body is still not settled. Arteriovenous measurements are complicated by the two blood supplies of the hepatic artery and the portal vein. The Japanese group (21, 14) report glutamine uptake by livers of fed rats and glutamine output in starved. Our data (Table 1) obtained by sampling the portal vein and hepatic vein (with arterial blood obtained from a separate group of rats) show no uptake by the liver (assuming the arterial contribution to be no more than 30% of the total blood flow). On this basis there is some output of glutamine, a finding not compatible with *in vitro* experiments (see later) or with the *in vivo* experiments of Brosnan and Williamson (31). On balance the available information points to skeletal muscle as the main site of glutamine synthesis. On one important point all groups of workers are agreed: glutamine is extracted from arterial blood by the portal-drained viscera. The significance of this observation will be discussed later.

The experimental results presented in Table 1 are inconclusive in regard to identification of the site of increased glutamine production in chronic acidosis. The problem is relevant to the regulation of urea synthesis because of the switch from urea synthesis to urinary ammonia production that occurs in the metabolic acidosis of starvation. It is possible that the decreased arterial-portal vein difference for glutamine in the acidotic group of animals (Table 1) leads to a decreased supply of urea precursors to the liver.

GLUTAMINE METABOLISM IN HEPATOCYTES

Preparation of Hepatocytes

The preparative procedure was essentially as described by Berry and Friend (24) with the modifications described by Cornell et al. (25) and Krebs et al. (26). Incubations were carried out in 25 ml Erlenmeyer flasks containing hepatocytes (2 ml suspension containing about 80 mg wet wt) with 2 ml additional medium containing substrates. Rapid separation of cells from incubation medium was carried out in specially designed centrifuge tubes (27).

Determination of Metabolites

Glutamine was measured as glutamate after hydrolysis with *E. coli* glutaminase (Sigma Grade V) as described by Lund (28). Other metabolites were measured as previously described (29).

Uptake of Glutamine by Hepatocytes

At 10 m*M*, glutamine is readily converted into glucose and urea by the liver (30). In experiments designed to study factors affecting glutamine uptake from the incubation medium at near-physiological concentrations (i.e., about 0.5 m*M*) no measurable uptake of glutamine occurred. This was true for all concentrations below 1 m*M* (Table 2), although a slight output of endogenous and newly synthesized glutamine occurs which is not necessarily depressed by added glutamine. At 1 to 2 m*M* added glutamine, removal from the medium remained very low. Only at 4 m*M* and higher did glutamine degradation occur, as indicated by increased synthesis of glucose and urea. During incubation the intracellular physiological concentration of glutamine—about 4 μmoles/g in 48-hour starved rats (31)- was exceeded if the external concentration was 1 m*M* or higher. At 4 m*M* glutamine, glutamine accumulation inside the liver cells was accompanied by glutamate accumulation. This means that disposal of glutamate can become rate-limiting for glucose and urea synthesis (32), at least under conditions where glutamine is the only added substrate. Addition of ornithine does not increase urea synthesis from glutamine.

Table 2 Concentration Dependence of Glutamine Uptake by Rat Hepatocytes

Hepatocytes from 48 hour starved rat. 65 mg wet wt in 4 ml. A representative experiment. Cells were rapidly separated from medium as described by Hems et al. (27).

Glutamine added (m*M*)	Incubation time (minutes)	Found in medium (μmole/ml)				Intracellular (μmole/g wet wt)	
		Gln	Glu	Glucose	NH_3 + urea-N	Gln	Glu
None	0	0.01	<0.01	0.02	0.06	1.3	1.4
1.0	0	1.03	<0.01	0.02	0.06	1.1	1.3
None	60	0.07	0.02	0.14	0.48	0.5	0.9
0.25	60	0.25	0.03	0.15	0.57	2.3	1.2
0.50	60	0.50	0.03	0.15	0.57	3.4	1.4
1.0	60	0.92	0.04	0.16	0.63	7.3	1.4
2.0	60	1.75	0.05	0.20	0.93	13.7	2.0
4.0	60	2.71	0.06	0.48	2.28	15.1	10.0

Factors Affecting Glutamine Uptake by Hepatocytes

Earlier work from this laboratory (33) had shown that ketoacid analogs of certain essential amino acids were aminated in the perfused rat liver by endogenous glutamine. It was concluded that glutamine-oxoacid aminotransferase was the enzyme responsible because of the pattern of specificity toward the various ketoacids (34). The ketoacid most effectively aminated was keto-methionine (α-keto-γ-methiolbutyrate). When hepatocytes were incubated with low concentrations of glutamine in the presence of 2 m*M* keto-met, glutamine uptake was accelerated (Table 3), and the intracellular glutamine concentration was decreased. Increasing the concentration of keto-met did not increase glutamine uptake. Amination of the ketoacid (confirmed by amino acid analyzer) was accompanied by increments of glucose, ammonia, and urea (Table 3) which were somewhat smaller than expected on the assumption that one molecule each of α-oxoglutarate and NH_3 is produced by the aminotransferase from each molecule of glutamine. Phenylpyruvate was somewhat less effective in promoting glutamine uptake, and the physiologically important ketoacids, pyruvate and oxaloacetate, were virtually without effect.

Table 3 Effect of Keto-Met on Glutamine Uptake by Rat Hepatocytes

Experimental conditions as in Table 2. Incubation 60 minutes at 38°C. "Increment due to keto-met" refers to changes occurring in the whole suspension (72 mg hepatocytes in 4 ml) compared with the appropriate control.

Glutamine added (m*M*)	Gln found in medium (μmole/ml)		Increment due to keto-met (μmole/flask)			
	No keto-met	With 2 m*M* keto-met	Gln	Glucose	NH_3	Urea-N
0.25	0.25	0.04	−0.84	0.33	0.14	0.38
0.5	0.50	0.11	−1.6	0.35	0.15	0.38
1.0	0.93	0.09	−3.4	0.85	0.25	1.24

The amination of keto-met by glutamine was inhibited by aminooxyacetate (0.1 m*M*), a general inhibitor of pyridoxal-phosphate–dependent enzymes, see (35). This and the specificity toward ketoacid substrate are confirmation that the glutamine aminotransferase is involved. From these experiments we conclude that glutaminase is nonfunctional in the liver at physiologically occurring glutamine concentrations. On the other hand the glutamine-oxoacid aminotransferase, which has a very high affinity for glutamine in the intact liver cell, could provide a means of accelerating glutamine uptake and metabolism, given a suitable ketoacid acceptor. So far, no physiologically occurring ketoacid has proved effective.

It is of interest that an intracellular-extracellular gradient for glutamine is maintained under the conditions of our experiments. *In vivo* the normal liver to blood (or plasma) ratio is about 8 (31). When the external glutamine concentration was varied or when the rate of glutamine metabolism was accelerated by keto-methionine (Tables 2 and 3), a gradient of about 4 to 10 was maintained by the hepatocytes.

The Fate of Glutamine Synthesized by Extrahepatic Tissues

The nonuptake by the liver cell of glutamine at physiologically occurring plasma concentrations raises the question of the site of hydrolysis of the glutamine formed in extrahepatic tissues.

Arteriovenous difference measurements across the portal-drained viscera of various species show a net uptake of glutamine, especially in the post-absorptive state (36, 37, 16, 38, 14, 22, 21). By use of germ-free rats the site of glutamine degradation has recently been identified as the mucosa of the small intestine (11, 12). Windmueller and Spaeth conclude that glutamine is a major respiratory fuel of the intestinal mucosa. The nitrogenous products (citrulline, alanine, ammonia, and proline) are released into the portal plasma. The synthesis of citrulline by the small intestine is interesting in relation to urea synthesis in that it shows that the low levels of carbamoyl phosphate synthetase (39) and of ornithine carbamoyl transferase found in the small intestine (40) are physiologically important. Alanine and ammonia at low concentration, unlike glutamine, are readily taken up by the liver. The fate of citrulline is less certain. It may be that some of the citrulline is converted to arginine in the kidney (41). On the assumption that all the nitrogenous products of glutamine degradation in the intestine ultimately appear in urea it can be calculated from the data of Windmueller and Spaeth (12) that glutamine provides about 20% of the urea-nitrogen excreted by the rat.

The Disposal of Ammonia and Alanine in the Liver: Relative Contributions of Glutamine and Urea Synthesis

The capacities of carbamoyl phosphate synthetase and glutamine synthetase measured under near-optimal conditions *in vitro* are of the same order. A value of 350 μmoles carbamoyl phosphate formed per hour per g wet wt of liver from rats fed a 15% protein diet was reported by Schimke (42). The comparable value for glutamine synthetase is about 500 μmoles per hour per g (43). Precursors of urea, such as ammonia and alanine, would also be expected to be precursors of glutamine, especially as glutamate is an obligatory intermediate in the transfer of the amino group of alanine to aspartate. In fact glutamine synthesis from all precursors tested is very low in isolated liver cells (Table 4). In no case was the rate of glutamine synthesis more than threefold higher than the rate of synthesis from endogenous sources. On the other hand, the optimal rate of urea synthesis from NH_4Cl in liver cells from fed rats [about 5 μmoles/minutes/g wet wt; Krebs et al., (44)] approaches the capacity of carbamoyl phosphate synthetase and exceeds the capacity of the argininosuccinate synthetase system measured in liver extracts (42).

Low rates of glutamine synthesis from various precursors were also found in the isolated perfused liver (23). None of the substrate combinations tested

Table 4 Glutamine Synthesis from Various Precursors in Hepatocytes

Hepatocytes were prepared from 48 hour starved rats. The initial concentration of added precursor was 5 m*M*. After 60 minutes incubation at 37°C glutamine was determined in the deproteinized cell suspension. Values are means ± S.E.M. of the number of observations in parentheses, or means of two separate experiments.

Precursor added	Glutamine synthesis (μmoles/hour/g wet wt)
None (12)	2.68 ± 0.21
Alanine (6)	4.98 ± 0.42
Proline (2)	3.95
Alanine + proline (4)	6.65 ± 0.75
NH_4Cl (2)	2.1
Pyruvate + NH_4Cl (2)	5.2
Histidine (2)	4.1
Histidine + NH_4Cl (2)	3.5

elicited a rate of glutamine synthesis of more than one-fortieth the capacity of glutamine synthetase. The experiments of Brosnan and Williamson (31) also showed that glutamine synthesis does not provide an alternative to urea synthesis *in vivo* when the urea cycle is overloaded by administration of ammonia. Instead ammonia is temporarily stored as alanine and aspartate through the near-equilibrium system of glutamate dehydrogenase, alanine-, and aspartate-aminotransferases. On the basis of these findings we have concluded that glutamine synthetase is largely inactive in the intact liver cell (23). It may be that the enzyme is very much more sensitive *in situ* to the inhibitors of the purified enzyme (i.e., carbamoyl phosphate) described by Meister and coworkers (45). In fact, inhibition of liver glutamine synthetase would prevent a potentially wasteful cycle of glutamine hydrolysis in the intestinal mucosa being followed by resynthesis in the liver. Interaction between tissues will, of course, be modified by the physiological state, and net glutamine synthesis by the liver may occur under some conditions.

CONCLUDING REMARKS

The importance of glutamine as a physiological precursor of urea has emerged only within the last few years. The reasons for this rather late discovery are of some general interest. Until about 5 years ago the liver was considered to be the main site of glutamine synthesis, largely because of its high activity of glutamine synthetase. This view was held in spite of the findings in 1951 of Flock et al. (46) that plasma glutamine concentrations increased in hepatectomized dogs. Mammalian skeletal muscle was not considered a site of net glutamine synthesis, because the enzyme activity was too low to be measurable. The presence of glutamine synthetase in this tissue was established by Iqbal and Ottoway (47). In the late sixties, with the widespread use of enzymatic analysis and of improved automatic amino acid analyzers, meaningful data on arteriovenous differences for amino acids, and for glutamine in particular, were obtained. The importance of skeletal muscle as a site of glutamine synthesis was then clearly established (13). This illustrates the potential importance of a large mass of tissue possessing a very low activity of an enzyme per unit weight and shows that the presence of an enzyme at high activity in a tissue, such as glutamine synthetase in liver, is not necessarily an indication of functional activity *in vivo*. It also suggests that the glutamine synthetases of muscle and liver will be found to have different regulatory properties.

REFERENCES

1. Krebs, H. A. (1935). *Biochem. J.* **29**, 1951–1969.

2. Katunuma, N., Tomino, I., and Nishino, H. (1966). *Biochem. Biophys. Res. Commun.* **22**, 321–328.

3. Katunuma, N., Huzino, A., and Tomino, I. (1967). In G. Weber, Ed., *Advances in Enzyme Regulation*, Vol. 5, Pergamon Press, Oxford and New York, pp. 55–69.

4. Leuthardt, F. (1938) Hoppe-Seyler's *Z. Physiol. Chem.* **252**, 238–260.

5. Leuthardt, F. (1938), *Biochem. Z.* **299**, 281–306.

6. Bach, S. J. (1939). *Biochem. J.* **33**, 1833–1844.

7. Leuthardt, F., and Glasson, B. (1942). *Helv. Chim. Acta* **25**, 630–635.

8. Leuthardt, F., and Glasson, B. (1944). *Helv. Physiol. Acta* **2**, 549–567.

9. Leuthardt, F., Müller, A. F., and Nielson, H. (1949). *Helv. Chim. Actā* **32**, 744–756.

10. Bach, S. J., and Smith, M. (1956). *Biochem. J.* **64**, 417–425.

11. Matsutaka, H., Aikawa, T., Yamamoto, H., and Ishikawa, E. (1973). *J. Biochem. (Tokyo)* **74**, 1019–1029.

12. Windmueller, H. G., and Spaeth, A. E. (1974). *J. Biol. Chem.* **249**, 5070–5079.

13. Marliss, E. B., Aoki, T. T., Pozefsky, T., Most, A. S., and Cahill, G. F. (1971). *J. Clin. Invest.* **50**, 814–817.

14. Aikawa, T., Matsutaka, H., Yamamoto, H., Okuda, T., Ishikawa, E., Kawano, T., and Matsumura, E. (1973). *J. Biochem (Tokyo)* **74**, 1003–1017.

15. Ruderman, N. B., and Berger, M. (1974). *J. Biol. Chem.* **249**, 5500–5506.

16. Hills, A. G., Reid, E. L., and Kerr, W. D. (1972). *Amer. J. Physiol.* **223**, 1470–1476.

17. Krebs, H. A. (1975). In G. Weber, Ed., *Advances in Enzyme Regulation*, Vol. 13, Pergamon Press, Oxford and New York, pp. 449–472.

18. Wergedal, J. E., and Harper, A. E. (1964). *Proc. Soc. Exp. Biol. Med.* **116**, 600–604.

19. Lowenstein, J. M, (1972). *Physiol. Rev.* **52**, 382–414.

20. Blackshear, P. J., Holloway, P. A. H., and Alberti, K. G. M. M. (1974). *FEBS Lett.* **48**, 310–313.

21. Yamamoto, H., Aikawa, T., Matsutaka, H., Okuda, T., and Ishikawa, E. (1974). *Amer. J. Physiol.* **226**, 1428–1433.

22. Felig, P., Wahren, J., and Räf, L. (1973). *Proc. Natl. Acad. Sci. U.S.* **70**, 1775–1779.

23. Lund, P. (1971). *Biochem. J.* **124**, 653–660.

24. Berry, M. N., and Friend, D. S. (1969). *J. Cell. Biol.* **43**, 506–520.

25. Cornell, N. W., Lund, P., Hems, R., and Krebs, H. A. (1973). *Biochem. J.* **134**, 671–672.

26. Krebs, H. A., Cornell, N. W., Lund, P., and Hems, R. (1974). Isolated Liver Cells as Experimental Material, in F. Lundquist and N. Tygstrup, Eds, *Alfred Benzon Symposium VI* Munksgaard, Copenhagen pp. 726–750.

27. Hems, R., Lund, P., and Krebs, H. A. (1975). *Biochem. J.* **150**, 47–50.

28. Lund, P. (1974). In H. U. Bergmeyer, Ed., *Methods of Enzymatic Analysis* 2nd Ed. Academic Press Inc., New York, pp. 1719–1722.

29. Cornell, N. W., Lund, P., and Krebs, H. A. (1974). *Biochem. J.* **142**, 327–337.

30. Ross, B. D., Hems, R., and Krebs, H. A. (1967). *Biochem. J.* **102**, 942–951.

31. Brosnan, J. T. and Williamson, D. H. (1974). *Biochem. J.* **138**, 453–462.

32. McGivan, J. D., and Chappell, J. B. (1975). *FEBS Lett.* **52**, 1–7.

33. Walser, M., Lund, P., Ruderman, N. B., and Coulter, A. W. (1973). *J. Clin. Invest.* **52**, 2865–2877.

34. Cooper, A. J. L. and Meister, A. (1972). *Biochemistry* **11**, 661–671.

35. Longshaw, I. D., Bowen, N. L., and Pogson, C. I. (1972). *Eur. J. Biochem.* **25**, 366–371.

36. Addae, S. K., and Lotspeich, W. D. (1968). *Amer. J. Physiol.* **215**, 269–277.

37. Elwyn, D. H., Parikh, H. C., and Shoemaker, W. C. (1968). *Amer. J. Physiol.* **215**, 1260–1275.

38. Wolff, J. E., Bergman, E. N., and Williams, H. H. (1972). *Amer. J. Physiol.* **223**, 438–446.

39. Hall, L. M., Johnson, R. C., and Cohen, P. P. (1960). *Biochim. Biophys. Acta* **37**, 144–145.

40. Lowenstein, J. M. and Cohen, P. P. (1956). *J. Biol. Chem.*, **220**, 57–70.

41. Featherston, W. R., Rogers, Q. R., and Freedland, R. A. (1973). *Amer. J. Physiol.* **224**, 127–129.

42. Schimke, R. T. (1962). J. Biol. Chem. **237**, 1921–1924.

43. Lund, P. (1970), *Biochem. J.*, **118**, 35–39.

44. Krebs, H. A., Lund, P., and Stubbs, M. (1976). Interrelations Between Gluconeogenesis and Urea Synthesis, in M. A. Mehlman and R. W. Hanson, Eds., *Gluconeogenesis*, John Wiley & Sons Inc., New York, in press.

45. Tate, S. S., Leu, F-Y., and Meister, A. (1972). *J. Biol. Chem.* **247**, 5312–5321.

46. Flock, E. V., Mann, F. C., and Bollman, J. L. (1951). *J. Biol. Chem.* **192**, 293–300.

47. Iqbal, K., and Ottaway, J. H. (1970). *Biochem. J.* **119**, 145–156.

Use and Misuse of Urea-Cycle Intermediates in Medicine

30

Carbamyl Phosphate—Cyanate and CNS Toxicity

Robert D. Crist
University of Kansas Medical Center
Kansas City, Kansas

P. Puig Parellada
Laboratorio P.E.V.Y.A.
Barcelona, Spain

Carbamylation of biological acceptors, including proteins, by carbamyl phosphate via its decomposition product, cyanate (1), have been clearly demonstrated (1–6). Stark and Smyth (7) studied the use of cyanate for the determination of NH_2-terminal residues in proteins. Subsequently, the blocking of amino groups by carbamylation has served as a tool to study their role in the function of proteins. Kilmartin and Rossi-Bernardi (8) used carbamylation to evaluate the role of α-amino groups in CO_2 transport and the Bohr effect of horse hemoglobin.

Carbamylation of whole blood by potassium cyanate was proposed as an inhibitor of sickling (9), and the carbamylation of human hemoglobin studied by Carreras et al. (10) and Diederich (11), showed an increase in oxygen affinity of the hemoglobin as carbamylation increased and a parallel change in the Bohr effect. The alteration of the Bohr effect and effectiveness in reduction of *in vitro* sickling led to early clinical trials of cyanate to alleviate sickling (12). In some of these patients, a reversible peripheral neuropathy was observed (13, 14).

The mechanism by which cyanate causes these effects on nervous tissue is unknown. At physiologic pH, cyanate reacts irreversibly with the α-amino group of proteins. The reaction of cyanate with brain proteins has been shown

(15, 16). This reaction might be with enzymes or structural proteins and be responsible for alteration of their functional activity as well. Nervous system hypoxia, with resulting acidosis, may follow the shift in oxygen equilibrium and altered oxygen delivery by hemoglobin.

Clinical and toxicologic studies have been conducted by several groups of investigators (9, 15–32). Their findings of nervous system manifestations of cyanate administration will be discussed.

SEIZURE ACTIVITY

The LD_{50} of sodium cyanate in mice is 260 mg/kg when administered I.P. (18). After administration of a lethal dose, the animals appeared sedated, but then suddenly developed seizures which could be triggered by noise. Rats were given 245 mg/kg, I.P., daily with 70% of the animals dead by 48 hours. At 195 mg/kg, 50% of the animals had experienced seizures and death by the fifth day (22). Seizure activity was seen in dogs following sublethal oral doses (18).

The threshold for drug-induced convulsion is a standard measure of CNS excitability (31). The total dose of pentylenetetrazole necessary to induce grand mal seizures has been employed as a test of sedative potency of drugs (32) and as a search method for the cause-and-effect relationships between brain chemistry and seizures (33). Table 1 illustrates the convulsive dose of pentylenetetrazole. The amount of pentylenetetrazole necessary to induce EEG seizure activity was 16.3, 15.1, 14.8, 10.1, and 9.0 mg/kg for rats that had received 0, 25, 50, 75, and 100 mg/kg of sodium cyanate, respectively (25). Blood determinations made on these groups of animals revealed a metabolic acidosis (lowered bicarbonate, base excess, and pH). No change occurred in arterial lactate, pyruvate, or the pyruvate/lactate ratio.

Table 1 Pentylenetetrazole mg/kg Rat to Produce Grand Mal Seizures[a]

Control (NaCl)	52.29 ± 5.05
0.25 μmole CNO^-/g	48.98 ± 5.95
0.5 μmole CNO^-/g	36.99 ± 4.84

[a] The threshold for pentylenetetrazole-induced convulsion was tested as described by Payan et al. (31). Male rats (Holtzman), 22 days old, were caged in groups of six, weighed, and injected intraperitoneally with isotonic saline or sodium cyanate, 0.25 mmole/kg; 0.5 mmole/kg; and 1.0 mmole/kg, daily, Monday through Friday.

The influence of cyanate on cerebral energy metabolism could not be seen on glycogen, pyruvate, lactate, lactate/pyruvate ratio, ATP, ADP, AMP with only glucose, phosphocreatine, total CO_2, intracellular pH, and cytoplasmic $NADH/NAD^+$ showing change. The systemic metabolic and CNS intracellular acidosis can explain the increased glucose, the decreased phosphocreatine and cytoplasmic $NADH/NAD^+$ ratio, and low total CO_2.

The cyanate-induced metabolic acidosis and resulting slowing of energy flux suggest a metabolic etiology for the altered convulsive threshold.

PERIPHERAL NEUROPATHIES

Chronic administration of high doses of cyanate caused hind limb paralysis in rats (16) and spastic quadriplegia in monkeys (29). Lesions in the brain and peripheral nerves were microscopic and variable. Severe depression and polyneuropathy manifested by severe motor weakness has been observed in hemoglobin SS patients treated with oral cyanate (13). Nerve conduction velocity was reduced in these patients from 50 to 33 m/second.

Peripheral neuropathy and central motor dysfunction may be due to excessive carbamylation of brain and nerve proteins. Alternatively, the lesions may be an indirect effect of general metabolic derangement. The interruption of axoplasmic transport, through the microtubules of peripheral nerves, has been suggested as being instrumental in the pathophysiology of long-tract peripheral neuropathy (34). The extent of polymerization of purified beef brain tubulin has been examined, after incubation for 60 minutes at 4°C, pH 7.0, with sodium cyanate. Polymerization was inhibited 99% by prior incubation, 43% by addition of cyanate at the time of initiating the polymerization reaction, and 22% by the presence of NaCl. Final ionic concentration (μ) of added sodium cyanate or sodium chloride was $0.15M$ (35).

Inhibition of microtubular polymerization has been associated, in a cause-and-effect relationship, with vinca alkaloid neuropathy (36). Although a similar causal relationship for cyanate has not been shown, the similarity between carbamylation and vinca alkaloid inactivation of unpolymerized dimeric subunits is suggested.

Early acute administration of cyanate to human volunteers was associated with transient diminution in visual activity (30). This peripheral neuropathy has not been reported or reinvestigated.

CEREBRAL CORTEX FUNCTION

Experiments were designed to pharmacologically examine cyanate effect on integrated CNS function (20). (See Table 2.) Charles Rivers' male mice were

Table 2 CNS Activity Functions

Activity	↓	$P < 0.05$
Amphetamine-induced activity	↓	$P < 0.001$
Amphetamine-induced stereotypic movements	±	—
Motor coordination (rotor rod)	±	—
Exploratory activity	↓	$P < 0.001$
Barbiturate sedation	↑	$P < 0.001$

tested by the method of Svensson and Thieme (37). Cyanate (50 mg/kg) treated animals showed less spontaneous activity. The addition of 5 mg/kg amphetamine I.M. increased the activity of saline-pretreated animals three-fold but only 34% in cyanate-pretreated animals. Stereotypic movements were induced by amphetamines (38) but were not altered by pretreatment with cyanate. Likewise, the motor coordination of cyanate- and saline-pretreated mice was tested on a spinning glass rod. This rotor-rod testing showed no differences between groups.

The instinctual tendency of a rat to explore a new environment has been exploited to measure the effects of psychotropic drugs (39). Cyanate- and saline-treated rats were tested on a *planche à trous*, with a significant decrease demonstrated in animals treated with cyanate.

Sedative properties of cyanate have been described in this presentation. These were quantified by I.P. injection of 25 mg/kg sodium pentobarbital in mice. This dose does not induce sleep in control mice, but in 75% of the cyanate-treated animals it did.

Decreased spontaneous and exploratory activities as well as the antagonism of the primary action of amphetamines suggests a decrease in the levels or turnover of catecholamines (40–44). This decrease in activity is not due to motor uncoordination as demonstrated by the rotor-rod test. The lowered convulsive threshold induced by cyanate is consistent, because Jobe (45, 46) has shown that a depletion or reduction of cerebral norepinephrine and dopamine augments convulsions.

Barbiturates reduce the peripheral secretion of catecholamines (47). These studies demonstrate the capacity of cyanate to induce sleep with subhypnotic doses of barbiturates.

Specific effects of catecholamines (Table 3) were examined by measuring palpebral ptosis and rectal temperature decrease induced by reserpine (48). The animals served as controls for 1 week, followed by cyanate or cyanate + reserpine therapy. Cyanate caused an increase in ptosis and a decrease in rectal temperature. The addition of reserpine caused a significant change in both parameters, with no evidence of antagonism.

Table 3 Reserpine-Induced Changes

Rectal temperature lowering	$P < 0.001$
Palpebral ptosis	$P < 0.001$

LEARNING AND BRAIN PROTEIN CARBAMYLATION

The effects of cyanate on maze learning have been tested in rats using a Lashley III maze. Cyanate doses of 33 to 100 mg/kg/day produce significant retardation (Fig. 1). Lower doses and periods shorter than 6 weeks have failed

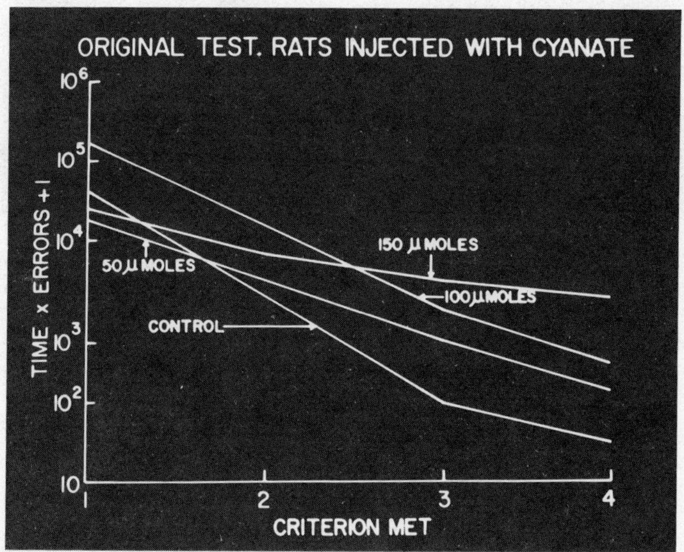

Figure 1. Relationship between dose of cyanate and maze-learning ability. Thirty-two rats were tested in a Lashley Maze III immediately after administration of sodium cyanate or isotonic saline (controls). The ordinate is the logarithmic scale of (numbers of errors + 1) × minutes × 10^2 needed to complete criterion. The abscissa indicates the criteria met. Animals were injected 5 times per week until they had received 16 injections. Least-squares analysis yields $p < 0.001$.

to show a similar result. Recovery from the retardation is seen in Fig. 2. Animals were retested 63 to 110 days after stopping cyanate.

In a parallel experiment, whole animal carbamylation was achieved by daily injection, I.P., with (^{14}C) cyanate (6.0×10^7 counts/minute) for 5 days. Serum hemoglobin and whole blood radioactivity in sephadex G25 purified proteins were followed to near base line (110 days) and sacrificed. Tissue protein radioactivity was determined and compared with similarly carbamylated animals sacrificed 6 hours after carbamylation. (See Table 4.)

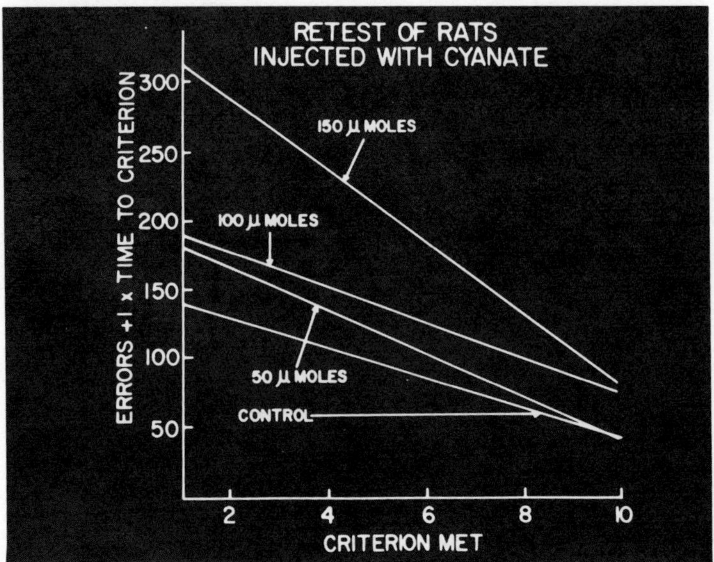

Figure 2. Relationship of dose to maze learning as tested 9 weeks after completing cyanate administration. Retesting in Lashley Maze III of animals used for the experiments in Fig. 1, 9 weeks after ceasing cyanate administration (6 weeks after maze exposure). The ordinate shows (errors + 1) × minutes × 10^2 needed to reach a criterion; the abscissa, criteria met. Least-squares analysis of variants yields $p < 0.05$.

Table 4 (^{14}C) Distribution into Tissues of Rat 110 Days after Injection of (^{14}C) Cyanate

In the right-hand column the average molecular weight was taken as 50,000. Corresponding amounts at 6 hours after injections were 440, 750, and 1800 nmole/g for brain, liver, and plasma respectively.

Tissue	Carbamylated Protein nmole/g	moles/mole
Muscle	421	0.211
Brain	198	0.099
Kidney	90	0.045
Liver	46	0.023
Plasma	6	0.003

Recent work (15) has proved that incorporation of (^{14}C) cyanate into brain proteins is via carbamylation rather than via the decomposition to CO_2 and subsequent incorporation of carbon dioxide. Both *in vitro* and *in vivo* incorporation of (^{14}C) cyanate yielded (^{14}C) hydantoins which could be hydrolyzed with loss of radio-activity.

Several intriguing possibilities are presented for the mechanism of cyanate effects on CNS. The carbamylation of enzymatic proteins may inhibit energy flux, for example, inhibition of erythrocyte glucose-6-phosphate dehydrogenase (49). Metabolic acidosis induced by hypoxia secondary to the increased oxygen affinity of carbamylated hemoglobin may likewise alter energy use. Intracellular acidosis decreases the activity of phosphofructokinase, a rate-limiting enzyme of glycolysis (50), thus reducing the overall glycolytic flux.

Catecholamine production and destruction involve complex systems of cyclic nucleotides and calcium ion control of secretory granule release. These cascading controls of catecholamine release are highly sensitive to intracellular pH. Again, the general metabolic state is of primary importance and catecholamine levels a secondary reflection of the general CNS acidosis.

The fact that carbamylated animals retain 50% of the radioactivity in brain proteins at 4 months compared with that found at 6 hours is most important. Memory and its acquisition are probably related to synaptic facilitation, which may be influenced by protein turnover and thus by elastoplastic modification (2). Carbamylation of protein, with carbamyl phosphate and cyanate used as examples of irreversible elastoplastic modifiers (2), seems to act in such a manner.

Memory is probably established by a facilitation of synaptic connections. If the synaptic junction involves protein, the turnover of these proteins could determine memory duration. Chemotrophic effectors may yield modified proteins which, in some cases, do not turn over or do so very slowly; this could explain how a relatively small number of chemotrophic agents, all reacting via a similar simple mechanism, could produce the many modified proteins involved in memory. It seems unlikely that every memory event would yield a different type of protein via protein synthesis, simply because there is no time (15); thus, if by a finite number of general mechanisms, the neuron could fix imprints and reinforce them in a repetitious manner, at least many types of memory could be explained.

REFERENCES

1. Carreras, J., Diederich, D., and Grisolia, S. (1972). *Eur. J. Biochem.* **27**, 103–108.
2. Grisolia, S., and Hood, W. (1972). In E. Kun and S. Grisolia, Eds., *Biochemical Regulatory Mechanisms in Eukaryotic Cells*, Wiley- Interscience, New York, pp. 137–203.
3. Ramponi, G., and Grisolia, S. (1970). *Biochem. Biophys. Res. Commun.* **38**, 1056–1063.

4. Carreras, J., Chabas, A, and Grisolia, S. (1971). *Biochim. Biophys. Acta* **250**, 456–457.

5. Grande, F., Grisolia, S., and Diederich, D. (1972). *Proc. Soc. Expt. Biol. Med.* **139**, 855–860.

6. Grisolia, S. (1968). *Biochem. Biophys. Res. Commun.* **32**, 56–59.

7. Stark, G. R., and Smyth, D. G. (1963). *J. Biol. Chem.* **238**, 214–226.

8. Kilmartin, J. V., and Rossi-Bernardi, L. (1969). *Nature* **222**, 1243–1246.

9. Cerami, A., and Manning, J. M. (1971). *Proc. Natl. Acad. Sci., USA* **68**, 1180–1183.

10. Diederich, D. (1972). *Biochem. Biophys. Res. Commun.* **46**, 1255–1261.

11. Gillette, P. N., Peterson, C. M., Lu, Y. S., and Cerami, A. (1974). *N. Engl. J. Med.* **290**, 654–660.

12. Peterson, C. M., Tsairis, P., Ohnishi, A., Lu, Y. S., Grady, R., Cerami, A., and Dyck, P. J. (1974). *Neurology* **24**, 365*a*.

13. Charache, S., Duffy, T. P., and Morrell, R. M. (1974). *Clin. Res.* **22**, 560*a*.

14. Crist, R. D., Grisolia, S., Bettis, C. J., and Grisolia, J. (1973). *Eur. J. Biochem.* **32**, 109–116.

15. Fando, J., and Grisolia, S. (1974). *Eur. J. Biochem.* **47**, 389–396.

16. Alter, B. P., Kan, Y. W., and Nathan, D. G. (1974). *Blood* **43**, 69–77.

17. Birch, K. M., and Schutz, F. (1946). *Brit. J. Pharmacol.* **1**, 186–193.

18. Cerami, A., Allen, T. A., Graziano, J. H., de Furia, F. G., Manning, J. M., and Gillette, P. N. (1973). *J. Pharmacol. Exp. Ther.* **185**, 653–666.

19. Crist, R., and Puig-Parellada, P. (1974). *Physiol. Chem. Physics* **6**, 371–374.

20. Crist, R. D., Puig-Parellada, P., manuscript in preparation.

21. Crist, R. D. (1973). "Carbamylation: Toxicity and Memory Modifier in the Rat," Ph.D. Thesis, Department of Biochemistry and Molecular Biology, University of Kansas.

22. Gillette, P. N., Peterson, C. M., Lu, Y. S., and Cerami, A. (1974). *N. Engl. J. Med.* **290**, 654–660.

23. Graziano, J. H., Thornton, Y. S., Leong, J. K., and Cerami, A. (1973). *J. Pharmacol. Exp. Ther.* **185**, 667–675.

24. Harkness, D. R., personal communication.

25. Leong, J. K., Grady, R. W., Herbert, J., Graziano, J. H., Cerami, A., Judge, M., and Quartermain, D. (1974). *J. Pharmacol. Exp. Ther.* **191**, 60–67.

26. May, A., Bellingham, A. J., Huehns, E. R., and Beaven, G. H. (1972). *The Lancet* **1**, 658–661.

27. Papayannopoulou, T., Stamatoyannopoulos, G., Giblett, E. R., and Anderson, J. (1973). *Life Sciences-II* **12**, 127–133.

28. Peterson, C. M., Lu, Y. S., Herbert, J. T., Cerami, A., and Gillette, P. N. (1974). *J. Pharmacol. Exp. Ther.* **189**, 577–584.

29. Shaw, C. M., Papayannopoulou, T., and Stamatoyannopoulos, G. (1974). *Pharmacology* **12**, 166–176.

30. Schutz, F. (1949). *Experientia* **5**, 133–172.

31. Payan, H., Levine, S., and Strebel, R. (1964). *Proc. Soc. Exp. Biol. N.Y.* **116**, 454–56.

32. Alexander, G. J., and Kopeloff, L. M. (1970). *Brain Res.* **22**, 231–35.

33. Lovell, R. A. (1971). In A. Lajtha, Ed., *Handbook of Neuro-chemistry*, **6**, Chapter 3, Plenum Press, New York, p. 63.

34. Samson, F. E., and Hinkley, R. E. (1972). *Anesthesiology* **36**, 417–421.

35. Crist, R. D., and Chien, S. M., unpublished observation.

36. Shelanski, M. L., and Wisniewski, H. (1969). *Arch. Neurol.* **20**, 199–206.

37. Svensson, T. H., and Thieme, G. (1969). *Psychopharmacologia* (Berlin) **14**, 157–163.

38. Simon, P., Chermat, R. (1972). *J. Pharmacol.* (Paris) **3**, 235–238.

39. Boissier, J. R., Simon, P., and Lwoff, J. M. (1964). *Therapie* **19**, 571–589.

40. Tseng, L. F., and Loh, H. H. (1974). *J. Pharmacol. Exp. Ther.* **189**, 717–724.

41. Fibiger, H. C., Fibiger, H. P., and Zis, A. P. (1973). *Br. J. Pharmac.* **47**, 683–692.

42. Smith, C. B., and Dews, P. B. (1962). *Psycho-pharmacologia* **3**, 55–59.

43. Villarreal, J. E., Guzman, M., and Smith, C. B. (1973). *J. Pharmacol. Exp. Ther.* **187**, 1–7.

44. Taylor, K. M., and Snyder, S. H. (1970). *Science* **168**, 1487–1489.

45. Jobe, P. C., Picchioni, A. L., and Chin, L. (1973). *J. Pharmacol. Exp. Ther.* **184**, 1–10.

46. Jobe, P. C., Picchioni, A. L., and Chin, L. (1973). *Life Sciences* **13**, 1–13.

47. Holmes, J. C., and Schneider, F. H. (1973). *Br. J. Pharmac.* **49**, 205–213.

48. Rubin, B., Malone, M. H., Waugh, M. H., and Burke, J. C. (1957). *J. Pharmacol. Exp. Ther.* **120**, 125–136.

49. Glader, B. E., and Conrad, M. E. (1972). *Nature* (Lond.) **273**, 336–338.

50. Lehninger, A. L. (1971). In *Bioenergetics*, 2nd ed., W. A. Benjamin, Menlo Park, Calif, pp. 38–95.

31

Physiological and Clinical Implications of Protein Carbamylation

J. Carreras
Departmento de Fisiologia y Bioquimica
Facultad de Medicina
Universidad de Barcelona (Spain)

A. Chabás
Instituto de Bioquimica Clinica "Fundacion Juan March"
Universidad Autonoma de Barcelona
Centro Coordinado del C.S.I.C. (Spain)

D. Diederich
Department of Internal Medicine
Division of Nephrology
University of Kansas Medical Center
College of Health Sciences and Hospital
Kansas City, Kansas

INTRODUCTION

In 1945 Pauling suggested that sickle cell anemia might be a "molecular disease" resulting from a genetic mutation that led to the synthesis of abnormal molecules of hemoglobin, which on deoxygenation aggregate into rods which polymerize to form elongated crystals, long enough to deform the erythrocytes (1, 2). Since then, there has been much interest in sickle cell hemoglobin (hemoglobin S) and in the molecular pathogenesis of sickling. Therapeutic attempts to prevent and treat the sickle cell crisis by decreasing the erythrocyte concentration of deoxyhemoglobin S or by stabilizing the red cell membrane were unsatisfactory and often were more dangerous than helpful (3–6).

In 1970 Nalbandian and coworkers (7, 8) reported successful reversal of sickling and of sickle cell crises with urea and attempted its use as a chemotherapeutic antisickling agent. They developed clinical protocols for the treatment of acute sickle cell crisis with intravenous urea in sugar solutions and for oral prophylactic urea therapy (9–16). Preliminary reports by this group and by others (17–21) showing improvement in symptoms were encouraging.

On the basis of the modified Murayama hypothesis for the mechanism of sickling (22, 23), two mechanisms for the desickling action of urea were proposed: alteration of the configuration of hemoglobin S molecules by hydrophobic interaction with urea, and disruption of the implicated hydrophobic bonds critical to sickling (10, 22–25). However, the possibility that the effectiveness of urea was due to the cyanate present in aqueous urea solutions prompted Cerami and Manning to test cyanate as an antisickling agent (26, 27). They found carbamylation of the terminal valine residues of hemoglobin S, inhibition of the gelling of deoxyhemoglobin S, and irreversible inhibition of sickling *in vitro*. As no interference with the physiology of the red cell was found, the use of cyanate as therapeutic agent was suggested.

Kraus and Kraus (28) reported that carbamyl phosphate was more effective than cyanate in preventing sickling *in vitro*. Moreover, they suggested that this superiority was due to the fact that when O_2 is released, carbamyl phosphate competes with 2,3-diphosphoglycerate for the phosphate binding site within the hemoglobin molecule. Then, carbamyl phosphate decomposes *in situ* to cyanate, which is now in a more favorable position to carbamylate the terminal α-amino residues of the hemoglobin chains.

To ascertain whether carbamyl phosphate was more effective than cyanate in the carbamylation of hemoglobin, particularly in the intact red cell, Grisolia and coworkers (29–32) carried out comparative experiments at or near physiological conditions, including pH and tonicity. Under all conditions tested, carbamylation of both hemoglobin in solution and in the intact erythrocyte proceeded more readily with cyanate than with carbamyl phosphate. Carbamylation prevented sickling, cyanate being as effective as carbamyl phosphate in diminishing sickling. In addition, an increase in hemoglobin oxygen affinity related to the degree of hemoglobin carbamylation and extensive carbamylation of erythrocyte membranes and of plasma proteins was noted.

Grisolia and coworkers (30–32) pointed out that, although acute toxicity was low for cyanate and lower for carbamyl phosphate, intravenous injection of cyanate resulted in extensive and relative indiscriminate carbamylation of proteins. Moreover, they suggested that carbamyl phosphate might be a more desirable carbamylating agent *in vivo* because of the fact that there is a very large amount of carbamyl phosphatase in tissues. This enzyme decomposes carbamyl phosphate to ammonia and CO_2 without formation of cyanate. This, together with the large amount of ornithine transcarbamylase in the liver, will ensure that after intravenous injection of carbamyl phosphate, any carbamyl phosphate leaving the circulatory system will be converted to citrulline or decomposed to ammonia and CO_2.

In the last few years, urea therapy for the prevention and treatment of sickle cell crises has been attempted by several groups (33–38). Although favorable results were reported in several studies (8–16, 34–35), no clinical efficacy was noted in a large-prospective controlled study (36, 37). Clinical trial with cyanate for the treatment of the sickle cell disease has been designed, and three routes for the clinical use of cyanate have been evaluated: oral administration, intravenous administration, and extracorporeal treatment of the blood with cyanate and subsequent return of the carbamylated cells to the patient (39–53). Carbamyl phosphate has been also tested for extracorporeal therapy (54–57).

Many experimental and clinical studies have been carried out to evaluate the potential usefulness and toxicity of cyanate and carbamyl phosphate. The effects of urea, cyanate, and carbamyl phosphate on sickling have been extensively studied, and the mechanism of their antisickling action has been compared (58–72). Increased survival of sickle cell erythrocytes after *in vitro* carbamylation with cyanate or with carbamyl phosphate has been shown (39–41, 44, 52–57, 83). In agreement with studies of Kilmartin and Rossi-Bernardi (73–77), who found with horse hemoglobin that carbamylation of the α-amino terminal residues increases O_2 affinity, carbamylation of hemoglobin S with cyanate or carbamyl phosphate also increases oxygen affinity (29–31, 78–88). As a result of the increased oxygen affinity, less deoxyhemoglobin S is present at a given pO_2. Since erythrocyte sickling occurs only on deoxygenation of hemoglobin S, the increased oxygen affinity of carbamylated hemoglobin S impairs sickling under hypoxic conditions. However, the inhibition of sickling is, in part, independent of the increase in oxygen affinity (66, 69, 81). Carbamylation also appears to diminish the gelling of deoxygenated hemoglobin S (26, 62–64, 69) and to lower the solubility of hemoglobin S (89).

Stark (90–97) and Smyth (98) demonstrated that cyanate readily reacts with functional groups in proteins. Several enzymes (32, 90, 99–124), hormones (30, 32, 125–128), and other proteins (30, 32, 130–137) are known to be carbamylated *in vitro* and subsequently altered in their physiologic functions by reaction with cyanate, carbamyl phosphate, and urea. Since cyanate reacts with other intracellular and membrane proteins, in addition to hemoglobin, the *in vitro* effect of cyanate on the red cell metabolism and functions have been investigated, and some metabolic alterations have been reported (82, 142–147). The possible functional significance of membrane carbamylation (29, 30) has been tested. Carbamylation of the red cell membranes significantly increases its electrophoretic mobility (148) and alters its deformability, improving *in vitro* filterability (149). The normal concentration of intracellular Na^+ and K^+, the normal osmotic fragility curves, and the normal patterns of autohemolysis suggest that cation-pumping mechanisms are not affected by carbamylation and that the integrity of membrane is maintained (82). Carbamylation of plasma proteins has been reported (32), and reduction of clottability of fibrinogen on cyanate treatment has been shown (150).

Schutz and Birch (151–153) in the 1940s carried out a series of classical studies on the pharmacology of cyanate in animals and humans. In the last few years several toxicological studies have been carried out to determine the possible effects in experimental animals of acute and chronic administration of cyanate and carbamyl phosphate (45, 62, 154–164). The fate and distribution of parenterally administered [^{14}C] cyanate and [^{14}C] carbamyl phosphate have been clarified. The effects in animals of acute and chronic cyanate or carbamyl phosphate administration on growth and development, on physiological functions of several enzymes, hormones, and organ systems have been studied. A dose-dependent peripheral neuropathy has been observed in patients receiving oral cyanate (46–48).

CHEMICAL ASPECTS OF PROTEIN CARBAMYLATION

Reaction of Cyanate with Functional Groups of Proteins

Cyanate reacts with amino, sulfhydryl, phenolic, hydroxyl, carboxyl, and imidazole groups of proteins, to yield carbamyl derivatives. However, only the reaction with amino groups results in the formation of a stable product. Products from the other reactions are relatively unstable and decompose on dilution or with a change of pH (90–98).

The reaction of cyanate with amino groups probably involves nucleophilic addition of the unprotonated amino nitrogen to neutral cyanic acid, the reactive form of cyanate (94, 98):

$$\mathrm{R{-}NH_2{:}} + \mathrm{HN{=}C{=}O} \underset{k_{-1}}{\overset{k_1}{\rightleftharpoons}} \mathrm{R{-}NH_2^+{-}C(=O)NH_2} \xrightarrow[k_2]{\ \mathrm{HO^-}\ \to\ \mathrm{H{-}O{-}H}\ } \mathrm{R{-}NH{-}C(=O)NH_2}$$

The reaction rate depends on the pK_a of the amino group and on the pH of the medium (94, 98). Logarithms of the rate constants (k_2 = rate/$[RNH_2]\cdot[HNCO]$) for the reaction of several amino acids with cyanate are related linearly to the pK_a values of the amino groups. The slope of the line fits the empirical equation $\log k_2(M^{-1}, \text{minute}^{-1}) = 7.94 - 0.71\ pK_a$. From this equation, the velocity at which an amino group is carbamylated can be predicted once its pK_a is known. As a result of their lower pK_a values, at pH 7 or below the α-amino groups of proteins can be expected to be carbamylated nearly 100 times faster than the ε-amino groups of lysine residues (94, 97). However, when pepsinogen is treated with cyanate at pH 8 to 9, nine of the ten ε-amino groups of lysine are carbamylated without reaction of the α-amino terminal leucine (120).

The carbamyl amino groups can be regenerated by hydrolysis in 0.2 *M* NaOH at 110°. In 6 *M* HCl at 110°, homocitrulline reverts slowly to lysine, whereas the α-carbamylamino compounds cyclize to hydantoins. Hydantoins can be isolated and determined then by hydrolysis to amino acids. Thus, this sequence of reactions can be used for the determination of the NH_2-terminal amino acids (91, 96).

Sulfhydryl groups react with cyanate more rapidly than do amino groups, but the reaction is readily reversible (90, 92, 97):

$$R\text{—}S^- + H\text{—}N\text{=}C\text{=}O + H\text{—}O\text{—}H \underset{k_{-1}}{\overset{k_1}{\rightleftharpoons}} R\text{—}S\text{—}C(\text{=}O)\text{—}NH_2 + H\text{—}O^-$$

The value of k_1 is constant between pH 6 and 8, but k_{-1} depends on the OH^- concentration. Thus, the position of the equilibrium constant is dependent on pH. At pH 5 and lower carbamylmercaptans are relatively stable. Between pH 6 and 8 (at 25°C) their half-life varies from ≃11 to ≃6 minutes (92).

In general, the aliphatic hydroxyl groups do not react with cyanate even under rather harsh conditions (95, 97). However, cyanate reacts, under relatively mild conditions, with the reactive serine residues of several proteases (chymotrypsin, trypsin, and subtilisin) (32, 118, 119) to give urethanes which are relatively stable (95).

The phenolic hydroxyl groups react readily with cyanate in a reversible reaction similar to the one that occurs with sulfhydryl groups. The mechanism appears to involve cyanic acid and the negatively charged phenolic anion (98):

$$R\text{—}C_6H_4\text{—}O^- + H\text{—}N\text{=}C\text{=}O + H\text{—}O\text{—}H \underset{k_{-1}}{\overset{k_1}{\rightleftharpoons}} R\text{—}C_6H_4\text{—}O\text{—}C(\text{=}O)\text{—}NH_2 + H\text{—}O^-$$

The rate of formation of *O*-carbamyltyrosine is independent of pH over the range of 5 to 9. The carbamyl derivative is stable in acidic medium but decomposes rapidly in neutral and alkaline solutions (98).

Cyanate reacts with carboxyl groups to form mixed anhydrides of carbamic acid (94, 97):

$$\mathrm{R{-}C({=}O){-}O^{-}} + \mathrm{H{-}N{=}C{=}O} + \mathrm{H{-}O{-}H} \underset{k_{-1}}{\overset{k_1}{\rightleftharpoons}} \mathrm{R{-}C({=}O){-}O{-}C({=}O){-}NH_2} + \mathrm{H{-}O^{-}}$$

The rates of both the forward and the reverse reactions vary with pH. An increase in 1 pH unit decreases 100-fold the equilibrium concentration of carbamylcarboxylate. Therefore its formation, which proceeds readily at pH 5, can be avoided entirely at pH 7 or 8 (94, 97).

The carbamylcarboxylate groups are highly reactive with many nucleophiles. Reaction with amines results in the acylation of the amino groups. Therefore a cross-link between glutamic or aspartic carboxyl residues and the ε-amino groups of lysine could be mediated by cyanate, without carbamylation of the end product (32). The carbamylcarboxylate group can also react with cyanate to yield a mixed anhydride of the carboxylic acid and isocyanic acid. Then, cyclization to an acyl- or carbamylhydantoin can occur if the carboxyl is α- to an acyl- or carbamylamino group (94, 97).

Imidazole reacts reversibly with cyanate to form carbamylimidazole (93, 97):

$$\text{R-imidazole (N, NH)} + \mathrm{H{-}N{=}C{=}O} + \mathrm{H{-}O{-}H} \underset{k_{-1}}{\overset{k_1}{\rightleftharpoons}} \text{R-imidazolium}^{+}\mathrm{{-}N{-}C({=}O){-}NH_2} + \mathrm{H{-}O^{-}}$$

At pH 7.0 carbamylimidazole is very unstable and, unlike other acylimidazoles, is not an effective acylating agent (93).

At low pH cyanate reacts with urea or amides to form biuret or a substituted urea (98). However, similar reaction with amide groups in proteins has not been found. As pointed out by Stark (97), it is very unlikely that other functional groups in proteins (guanido, indole) would react significantly with cyanate even under extreme conditions.

Tables 1 to 3 present a résumé of some of the enzymes, hormones, and other proteins which have been carbamylated *in vitro* with cyanate.

Carbamylation with Carbamyl Phosphate

Stark (94) concluded that carbamyl phosphate, like carbamyl carboxylates, is probably a very poor carbamylating agent. However, carbamylation of proteins by carbamyl phosphate has been shown to occur in a number of cases. Grisolia and coworkers (32) in experiments with some 20 proteins found that their susceptibility to carbamylation with carbamyl phosphate at or near neutral pH varies a great deal (Table 1). Ramponi and Grisolia (130) tested

Table 1 Effect of Carbamyl Phosphate on Activity and Carbamylation of Several Enzymes (32)

Enzyme	Remaining Activity, %	n*M* Carbamyl Protein
Glutamate dehydrogenase	2	190,250
Glyceraldehyde-3-phosphate dehydrogenase	29	Traces
Alcohol dehydrogenase	100	N.D.
Enolase	100	N.D.
Ribonuclease	0	235,78
Trypsin	N.D.	191
Catalase	N.D.	0
Urease	100	0

N.D.: not determined.

carbamylation of histones, both with added carbamyl phosphate and with a generating system for carbamyl phosphate. Arginine-rich and particularly lysine-rich histones were readily carbamylated. Carbamylation increased with increasing carbamyl phosphate, and acyl phosphatase markedly inhibited the reaction. Chabás and Grisolia (32, 102) studied the relative susceptibility to carbamyl phosphate of the main enzymes concerned with its metabolism in ureotelic animals, namely, glutamic dehydrogenase, glutamine synthetase, carbamyl phosphate synthetase, ornithine transcarbamylase, aspartic transcarbamylase, and carbamyl phosphatase. Glutamic dehydrogenase and, to a much less extent, carbamyl phosphate synthetase were inactivated. No loss in the enzymatic activity was observed with the other enzymes tested (Table 2). Insulin (30) and glucagon (127) also react with carbamyl phosphate. Hemoglobin (A, F, and S) and red cell membranes were rapidly carbamylated *in*

Table 2 Effect of Carbamyl Phosphate on the Stability of Various Enzymes Involved in Ammonia Utilization or Related to Urea Cycle (102)

Enzyme	Carbamyl-P added (μmole)	Incubation time (minutes)	Remaining activity (%)
Glutamate dehydrogenase	10	1	54
	25	1	26
	10	2	28
	25	2	0
Carbamyl phosphate synthetase	33	1	89
	66	1	67
Ornithine transcarbamylase	10–70	1–4	100
Aspartate transcarbamylase	10–40	1	100
Glutamine synthetase	33–50	0.5	100
Acyl phosphatase	200	24	100

vitro with carbamyl phosphate (30). Some serum proteins were very extensively carbamylated, whereas others were not (32). Table 3 presents a resume of some of the proteins which are susceptible to *in vitro* carbamylation with carbamyl phosphate.

Studies on the hydrolysis of carbamyl phosphate (165–169) showed that, whereas below pH 6 (where the monoanion predominates) ammonium ion was produced, above this pH (where the dianion predominates) cyanate was the major product. In fact, Allen and Jones (168) demonstrated separate mechanisms for carbamyl phosphate mono- and dianion decomposition in water. Decomposition of carbamyl phosphate monoanion proceeds by P—O bond cleavage with elimination of carbamic acid and formation of metaphosphate, which reacts with H_2O to form orthophosphate:

$$H_2N-\overset{\overset{O}{\|}}{C}-O-\underset{\underset{O}{\|}}{\overset{\overset{OH}{|}}{P}}-O^- \longrightarrow \begin{cases} H_2N-\overset{\overset{O}{\|}}{C}-OH \longrightarrow NH_3 + CO_2 \\ \underset{\underset{O}{|}}{\overset{\overset{O}{\|}}{P}}-O^- \xrightarrow[H_2O]{} H_2PO_4^- \end{cases}$$

Table 3 Carbamylation of Proteins and Peptides

Protein	Carbamylating Agent	Effect	Altered Group	Reference
Malate dehydrogenase (1.1.1.37)	Cyanate	Altered electrophoretic pattern		162
6-P-Gluconate dehydrogenase (1.1.1.44)	Cyanate	Altered electrophoretic pattern		162
Glucose-6-phosphate dehydrogenase (1.1.1.49)	Cyanate	Activity decreased		142
Glyceraldehyde-3-phosphate dehydrogenase (1.2.1.12)	Carbamyl-P	Activity decreased		32
	Cyanate	Altered electrophoretic pattern		162
Glutamate dehydrogenase (1.4.1.3)	Carbamyl-P	Activity decreased	Lysine residues ε-NH_2	99, 102
	Cyanate, urea	Activity decreased	groups of lysine-97 and 85; α-NH_2 group of alanine-1	100, 101
Phosphorylases a and b (2.4.1.1)	Cyanate	Activity decreased	ε-NH_2 groups of lysine residues	103, 104
Aspartate aminotransferase (2.6.1.1)	Cyanate	Altered electrophoretic pattern		162
Alanine aminotransferase (2.6.1.2)	Cyanate	Altered electrophoretic pattern		162
Tyrosine aminotransferase (2.6.1.5)	Cyanate Carbamyl-P	Interconversion of enzyme forms		105
Aminobutyrate aminotransferase (2.6.1.19)	Carbamyl-P Cyanate	Activity decreased		106, 107
Pyruvate kinase (2.7.1.40)	Cyanate	Activity decreased		78, 82

Table 3 (*continued*)

Protein	Carbamylating Agent	Effect	Altered Group	Reference
Phosphoglycerate kinase (2.7.2.3)	Cyanate	Altered electrophoretic mobility		162
Carbamyl phosphate synthetase (2.7.2.5)	Carbamyl-P	Activity decreased		102
Carbamyl phosphate synthetase (2.7.2.9)	Cyanate	Glutamine-dependent activity decreased	Glutamine binding site on light subunit	110
Phosphoglucomutase (2.7.5.1)	Urea + cyanate	Activation	Conformational change	108
	Cyanate	Altered electrophoretic pattern		162
Alkaline phosphatase (3.1.3.1)	Cyanate	Activity decreased	—SH group	111
Acid phosphatase (3.1.3.2)	Cyanate	Altered electrophoretic pattern		162
Ribonuclease (3.1.4.22)	Carbamyl-P	Activity decreased		32
	Cyanate, urea	Activity decreased	ε-NH_2 groups of lysine residues	90
Lysozyme (3.2.1.17)	Cyanate, urea	Increased-decreased activity		112
Chymotrypsinogen A	Urea	Altered tryptic activation and hydrolysis rate at low ionic strength	ε-NH_2 groups lysine and terminal α-NH_2 group	114–117
Chymotrypsin (3.4.21.1)	Cyanate	Activity decreased	Ser-195	118, 119
Trypsin (3.4.21.4)	Carbamyl-P	Carbamyl protein formation		32
Subtilisin (3.4.21.14)	Cyanate	Activity decreased	Serine	122

Papain (3.4.22.2)	Cyanate	Reversible inhibition	—SH group	113
Pepsinogen	Cyanate	Loss of potential pepsin activity	Nine ε-NH_2 groups of lysine and four tyrosine residues	120, 121
Pepsin (3.4.23.1)	Cyanate	Reduction of proteolytic activity	Five to six tyrosine residues, NH_2-terminal isoleucine	120
Adenosine deaminase (3.5.4.4)	Cyanate	Altered electrophoretic pattern		162
Enolase (4.2.1.11)	Cyanate	Altered electrophoretic pattern		162
Glucose-6-phosphate isomerase (5.3.1.9)	Cyanate	Altered electrophoretic pattern		162
Triose-phosphate isomerase (5.3.1.1)	Cyanate	Altered electrophoretic pattern		162
Amino acyl ligases (6.1.1.-)	Cyanate	Activity decreased		123
Pyruvate carboxylase (6.4.1.1)	Cyanate	Activity decreased	ε-NH_2 group of lysine	124
Insulin	Cyanate	Loss of biological activity	NH_2 groups of glycine and phenylalanine	125, 126
Glucagon	Carbamyl-P Cyanate	Loss of adipokinetic and hyperglycemic properties	Histidine	127
Oxytocin and oxytocin analogs	Cyanate	Inhibitors or inactivation	OH group of tyrosine	128
Luteinizing hormone	Cyanate	Loss of biological activity		158b
Thyroid-stimulating hormone	Cyanate	Loss of biological activity		158b

Table 3 (*continued*)

Protein	Carbamylating Agent	Effect	Altered Group	Reference
Antidiuretic hormone	Cyanate	Loss of biological activity		158b
Arginine-and lysine-rich histones	Carbamyl-P Cyanate		Lysine	130, 131
Casein	Urea, cyanate		ε-NH_2 group of lysine and α-NH_2 arginine	132
Globin	Urea	Altered electrophoretic mobility	α-NH_2 valine	133
Wheat proteins	Urea	Altered electrophoretic mobility		134
Eye lens α-crystallin	Urea, cyanate	Altered electrophoretic mobility		135
β-Lactoglobulin	Cyanate		—SH groups	90
Glutathione	Cyanate		—SH groups	90
Anti-*p*-azobenzenearsonate antibodies	Cyanate	Loss of antibody sites	ε-NH_2 group of lysine or α-NH_2 group	136
Immunoglobulins	Cyanate	Altered electrophoretic mobility		137
Fibrinogen	Cyanate	Reduction of clottability		150
Hemoglobin	Cyanate Carbamyl phosphate	Altered binding of CO_2, O_2, DPG	α-NH_2 terminal group	26–31, 73–78, 82–88, 28–31, 54, 55

Decomposition of carbamyl phosphate dianion proceeds by C—O bond cleavage, with formation of phosphate dianion and elimination of cyanic acid, which is ionized to yield the stable cyanate ion:

$$H_2N\text{—}\overset{\displaystyle O}{\overset{\|}{C}}\text{—}O\text{—}\underset{\underset{\displaystyle O}{\|}}{\overset{\overset{\displaystyle O^-}{|}}{P}}\text{—}O^- \longrightarrow \begin{cases} HN{=}C{=}O \xrightarrow{H_2O} \begin{cases} CO_2 + NH_3 \\ NCO^- \end{cases} \\ O\text{—}\underset{\underset{\displaystyle O^-}{|}}{\overset{\overset{\displaystyle OH}{|}}{P}}{=}O \end{cases}$$

Since at neutral pH carbamyl phosphate decomposes to cyanate, the possibility that cyanate is an intermediate in carbamylation reactions carried out by carbamyl phosphate has been considered. The intermediate formation of cyanate during enzymatic hydrolysis of carbamyl phosphate (170) and during transcarbamylation to ornithine (171) has been clearly excluded, making even more unlikely the intermediate formation of cyanate during other enzyme reactions involving carbamyl phosphate (32). In contrast, as further discussed below, in all cases tested, carbamyl phosphate seems to act through cyanate in nonenzymatic carbamylation reactions (30, 32, 172).

Carreras et al. (30, 172), by using insulin, histones, and hemoglobin as a model system, tested the possibility that protein carbamylation with carbamyl phosphate proceeds via cyanate. Carbamylation was carried out *in vitro* with cyanate and with carbamyl phosphate, under different conditions of concentration, pH, and temperature. Under all conditions tested, carbamylation with cyanate proceeded more readily than with carbamyl phosphate. Carbamylation of hemoglobin with carbamyl phosphate was tested at low pH, to diminish the decomposition to cyanate, and in the presence of azide, to trap the cyanate formed as carbamylazide (167, 168). Due to the insolubility of hemoglobin, it was not possible to test carbamylation at sufficiently low pH to ensure that all cyanate would be trapped. However, azide decreased significantly carbamylation of both hemoglobins A and S at pH 6. At pH 6 the rate of histone carbamylation with carbamyl phosphate was markedly decreased but was not abolished in the presence of azide. At pH 3.9, in the absence of azide, there was extensive carbamylation which was proportional to the concentration of carbamyl phosphate. In contrast, in the presence of azide, no carbamylation could be detected even at a ratio of carbamyl phosphate to azide of 1:2. Since Mg^{2+} markedly decreases the rate of decomposition of carbamyl phosphate dianion (169), the effect of Mg^{2+} on histone carbamylation with carbamyl phosphate was studied. In the presence of Mg^{2+} a marked inhibition of carbamylation was noted.

On the basis of these experiments, it was concluded (30, 172) that carbamylation of hemoglobin and histones with carbamyl phosphate proceeds through cyanate. In agreement with this conclusion, it has been suggested (100) that carbamyl phosphate is not the primary reactant with glutamic dehydrogenase. Instead, inactivation is due to reaction of the enzyme with cyanate, which is formed by the decomposition of carbamyl phosphate. However, although carbamyl phosphate acts via cyanate in nonenzymatic carbamylation reactions, in the cell the physiological carbamylating agent is carbamyl phosphate, since it is produced in large quantity by mammals whereas cyanate is not (32). Finally, it should be noted that carbamyl phosphate, in addition to participating in carbamylation reactions, may also take part in protein phosphorylation, as illustrated by Grisolia and Hood (32).

Carbamylation with Urea

The formation of carbamylated products during electrophoresis or chromatography in the presence of urea has been observed with several proteins, as summarized in Table 3. In all these cases the carbamylating agent has been shown to be cyanate, present in trace amounts in urea solutions.

The knowledge of the interrelation between urea and cyanate began with the synthesis of urea from ammonium cyanate by Wöhler in 1825. In 1895, Walker and Hambly (173) showed that the reaction was spontaneously reversible in aqueous solutions, and, in 1924, Warner (174) concluded that cyanate was the only intermediate in the hydrolysis of urea to ammonium and carbonate. Dirnhuber and Schutz (175) obtained evidence for the spontaneous formation of cyanate in aqueous solutions of urea and showed that the amount of cyanate formed depends on the concentration of the urea solution, duration of its storage, and temperature. More recently, Hagel et al. (176) by computer calculation correlated data from the literature on the equilibrium between urea and cyanate and studied quantitatively the accumulation of cyanate in urea solutions at different temperature, pH, and ionic strength.

FUNCTIONAL IMPLICATIONS OF PROTEIN CARBAMYLATION

Modification of a protein by a carbamylating agent may result in extensive changes in the properties of the protein, ranging from loss of activity to conformational effects. On the other hand, susceptibility to carbamylation may depend on the presence of ligands and on the conformation as well as the type of protein (32). Some of the enzymes, hormones, and other proteins and peptides that have been shown to be susceptible to cyanate, carbamyl phosphate, and urea are summarized in Table 3.

Carbamylation of Enzymes

Oxidoreductases The first enzyme shown to be inactivated by carbamylation with carbamyl phosphate was the glutamate dehydrogenase (99). At the physiological concentrations of glutamate dehydrogenase and carbamyl phosphate present in liver this reagent rapidly inactivates the dehydrogenase after 1 hour of incubation at 38°C and at pH 7.4. Carbamylation involves the ε-amino group of lysine (32, 99), probably lysine-97, as shown by Veronese et al. (100). These authors have found that inactivation of glutamate dehydrogenase may be attributed to cyanate and, although the ε-amino group of lysine-97 is the major site of carbamylation by cyanate, the α-amino group of alanine-1 and ε-amino group of lysine-85 are also sites of carbamylation. In agreement with the high susceptibility of the enzyme to carbamylation, it has been previously reported that a combination of 0.5 to 1 *M* urea with 1 m*M* NADPH potentiated the cofactor-induced inactivation of glutamate dehydrogenase (101). Chabás and Grisolia (102) have shown that in subcellular fractions of liver and under conditions where there is *de novo* biosynthesis of carbamyl phosphate a loss of glutamate dehydrogenase activity occurs; when ornithine was present in the incubation medium and was converted to citrulline, no inactivation was detected. Inactivation of glutamate dehydrogenase by carbamyl phosphate occurred even in the presence of ATP, an activator of NAD reduction.

Studies on the hexose monophosphate shunt pathway revealed a decrease in the overall activity of the shunt and of glucose-6-phosphate dehydrogenase in erythrocytes on incubation with cyanate (142, 162). Although glucose-6-phosphate dehydrogenase has at least one reactive thiol group related to the functional integrity of the enzyme, irreversible inactivation is consistent with amino, but not thiol, carbamylation of the enzyme.

Incubation of muscle glyceraldehyde-3-phosphate dehydrogenase with carbamyl phosphate (32) results in partial loss of enzymatic activity with parallel formation of carbamyl protein (Table 1). Lack of agreement between the inactivation studies and the extent of carbamylation as judged colorimetrically (141) is due to the fact that the chromogenicity of the carbamyl-amino acids varies a great deal and that inactivation caused by carbamylation depends on many factors, as already mentioned.

The *in vivo* administration of repeated high doses of cyanate in mice (162) produced alterations in the electrophoretic pattern of glyceraldehyde-3-phosphate dehydrogenase from red cells and from other tissues, with appearance of new bands with anodal migration. Similar findings were observed in 6-phosphogluconate dehydrogenase and malate dehydrogenase.

Transferases Inactivation of phosphorylases "a" and "b" by cyanate involves carbamylation of the ε-amino groups of lysine to form homocitrulline (103).

On the other hand, there is no reaction with the sulfhydryl groups of the enzyme. The course of the inactivation follows kinetics typical of a first-order reaction, incorporation of cyanate into phosphorylase "b" being somewhat greater. The rate of inactivation of phosphorylase "a" is affected by several ligands. The activator AMP and glucose-1-P protect against inactivation by cyanate, and further protection is provided when both glucose-1-P and AMP are present together. It is interesting to note that AMP also decreases the rate of inactivation of phosphorylase "b" but glucose-1-P does not, although in combination with AMP it affords further protection (104). The allosteric inhibitors, ATP, ADP, and glucose-6-P, do not affect the rate of inactivation of phosphorylase "b" by cyanate, but they antagonize the protection offered by AMP. The interrelationships among these ligands support the idea of a concerted conformational change between two states of the enzyme and that phosphorylase "b" in these different conformations is inactivated by cyanate at different rates.

Aspartate and alanine aminotransferases (162) show abnormal electrophoretic patterns after either *in vitro* or *in vivo* treatment with cyanate.

Brain proteins are susceptible to carbamylation by carbamyl phosphate (155, 156), and in this regard aminobutyrate aminotransferase (106, 107) is inactivated *in vitro* by both carbamyl phosphate and cyanate. The effect by carbamyl phosphate is less than by cyanate, enzymatic inactivation increasing with time of incubation with carbamyl phosphate.

Chabás and Grisolia (102) studied the susceptibility to carbamyl phosphate of the main enzymes concerned with use of ammonia or related to urea cycle. They found (Table 2) that only liver carbamyl phosphate synthetase (102) was somewhat unstable with carbamyl phosphate, although to a much lesser extent than glutamate dehydrogenase. Under similar experimental conditions no loss in enzymatic activity due to carbamyl phosphate was detectable with the other enzymes. Anderson and coworkers (110) found that glutamine-dependent carbamyl-P synthesis and ATP- and bicarbonate-dependent L-γ-glutamyl hydroxamate hydrolysis activities were inhibited on incubation of carbamyl-P synthetase from *E. coli* with relatively low concentrations of cyanate. The bicarbonate-dependent ATPase activity was stimulated by nearly threefold, and the apparent K_m value for ammonia in the ammonia-dependent carbamyl-P synthesis reaction was decreased. These observations together with the observed similarity of the effects of cyanate and the irreversible inhibitor chloroketone indicated that the inhibitory action of cyanate probably involves an initial binding of the cyanate at the glutamine binding site on the light subunit of the enzyme.

It is known that, in the absence of substrate, phosphoglucomutase can be activated by very short preincubation with Mg^{2+} and a metal-complexing agent such as histidine, imidazole, and cysteine (109). It has been suggested that the process of enzymatic activation involves a change in conformation of the molecule that results in an increased turnover rate of the enzyme. Najjar and coworkers (108) reported that exposure to urea prior to addition of substrate activated muscle phosphoglucomutase such that about 2.5-fold the rate of the native untreated enzyme was attained. Such an increase in activation parallels the increase in the negative rotation of the enzyme, supporting the conclusion that activation of phosphoglucomutase results from a change in its conformation which in some manner involves the active site to augment its catalytic efficiency to a considerable degree. Cyanate has no activating effect but, in conjunction with urea, potentiates the urea-induced activation of phosphoglucomutase; it is uncertain whether cyanate acts by increasing the rate of unfolding of the protein or by stabilizing a particular conformation.

After *in vivo* and *in vitro* treatment with cyanate (162) phosphoglucomutase, phosphoglycerate mutase, and phosphoglycerate kinase from red cells and from other tissues showed an altered electrophoretic pattern. The activity of pyruvate kinase in sickle cells treated with cyanate decreased ($\sim 25\%$) when assayed at either high or low concentrations of phosphoenolpyruvate or in the presence of fructose diphosphate.

Grisolia and Hood (32) pointed out the possible implication of protein carbamylation and of other chemotrophic modifications on protein turnover. They suggested that, on chemotrophic modification, proteins may be rendered either more or less susceptible to proteolytic attack. Thus, protein turnover might be influenced or controlled by this type of effect. Johnson et al. (105) reported the interconversion by carbamylation of three forms of cytoplasmic tyrosine aminotransferase from rat liver and suggested that posttranslational carbamylation of the enzyme might be implicated in the control of its turnover. They found that the treatment of fresh liver homogenates with cyanate promotes the shift from form I (primary product of enzyme synthesis) to forms II and III of tyrosine aminotransferase. Carbamyl phosphate also changed the ratio of the enzyme forms, although this change was difficult to reproduce. Thus, Johnson et al. concluded that forms II and III represent posttranslational modifications of form I, possibly via carbamylation of the amino groups of the enzyme. Interestingly, reaction with cyanate appeared to require a particulate enzyme, since cyanate was not effective on incubation with the purified form I or when the particulate components of the homogenate were removed.

Hydrolases The action of cyanate on human and pig kidney alkaline phosphatases exhibits some interesting features (111). At concentrations of cyanate up to 0.2 M there is an apparently reversible combination with the enzyme, but prolonged treatment with higher concentrations of cyanate (0.6 M) leads to enzymatic inactivation with probable carbamylation of all free amino groups in the enzyme and generation of a new enzyme with decreased V_{max} and increased K_m. The effect of 0.2 *M* cyanate on the enzymatic reaction velocity depends on the substrate concentration. There is inhibition when the substrate concentration is 1.0 m*M* or higher, but, at lower substrate concentrations, cyanate has an activating effect. The pH dependence of the reversible inhibition by 0.2 *M* cyanate suggests that cyanate may react with a thiol group at or near the active site of the enzyme, preventing a conformational change that is believed to be important in the mechanism of action of alkaline phosphatase.

Urea is often tacitly assumed to be a reagent that brings about physical rather than chemical changes in protein molecules. As has been pointed out (90), this assumption is only valid if the urea is completely free of cyanate; thus ribonuclease maintained in 8 *M* urea at 40° C and then dialyzed to remove urea not only lost enzymatic activity but also showed an altered amino acid composition. The amount of lysine concomitantly decreased as the time during which ribonuclease was exposed to urea lengthened. These results suggested that the observed loss of lysine had come about as a result of carbamylation of the ε-amino groups by cyanate present in the urea solution. Further investigations confirmed that cyanate alone would react rapidly with the amino groups of proteins under mild conditions of pH and temperature. Ribonuclease is also unstable to carbamyl phosphate (32). After treatment with carbamyl phosphate, ribonuclease is completely inactivated, and formation of a carbamyl protein, detected colorimetrically, can be observed.

Parallel studies carried out on urea-treated and cyanate-treated lysozyme (112) have shown that changes in activity of the urea-treated enzyme are much like those observed with cyanate, suggesting that carbamylation of the free amino groups by cyanate is responsible for the activity changes in urea-treated lysozyme. In the first stages of carbamylation the resulting loss of free amino groups increases the catalytic activity. Finally, the activity reaches a maximum and then begins to decrease.

Involvement of thiol group will result in reversible inhibition, as in papain (113), whose activity, lost on treatment with cyanate, is slowly recovered when the cyanate enzyme complex is diluted. The thiol group of papain has been shown to be about 3000 times more reactive toward cyanate than is the thiol group of free cysteine, in agreement with previous observations on unusually high reactivity of protein groups in active sites of enzymes.

Cyanate also reacts with chymotrypsinogen A and with α-chymotrypsin, as shown by Hofstee (114–117) and by Shaw et al. (118), respectively. Chymotrypsinogen A has been converted to a carbamylated derivative, chymotrypsinogen X, by unfolding the A-zymogen in 8 *M* urea, heating at 100° C, and refolding the protein by dilution or removal of the urea by dialysis. The cyanate formed on heating the urea carbamylates the terminal α-amino group and most of the 14 ε-amino groups of the zymogen (114). The course of the carbamylation reaction has been followed by isoelectric focusing in polyacrylamide gels, 12 carbamylated protein species being detected. The first product has the same apparent molecular weight as the A-zymogen and appears homogenous on either gel electrofocusing or SDS electrophoresis (117). At pH 8 and low ionic strength, the tryptic activation reactions of the chymotrypsinogen A and its carbamylated derivative are entirely different (116). However, little or no difference is observed at high ionic strength. At low ionic strength the rate of hydrolysis of denaturated serum albumin by activated chymotrypsinogen X is considerably lower than that by the activated A-zymogen, but this difference is also abolished at high ionic strength. Similarly, at low but not at high ionic strength the affinity of a negatively charged substrate for activated X is considerably lower than for the activated A-zymogen. The potential zymogenic capacity of chymotrypsinogen is not decreased by converting the A form to the X form, and little or no effect on the substrate specificity is noted with respect to the relative maximal rates of hydrolysis.

α-Chymotrypsin is inactivated by cyanate more rapidly at pH 6.5 than at pH 8.5 (118). The competitive inhibitor indole propionate protects the enzyme from inactivation at pH 6.5, but the more potent inhibitor indole actually increases the rate of inactivation. By means of finger-prints of peptide digests, evidence has been presented that the inhibition of chymotryptic activity by cyanate could be linked to a carbamylation of the active site, Ser-195. Crystallographic studies (119) have confirmed this and have shown that the carbamyl group is stabilized in a single conformation by a hydrogen-bonding network involving a water molecule, the carbonyl oxygen of the carbamyl group, the peptide carbonyl of Phe-41, and the amide nitrogen of Gly-193. Cyanate reacts differently with the various species of chymotrypsin, and thus with π-chymotrypsin and δ-chymotrypsin the reagent reacts simultaneously at Ser-195 and the amino terminal of Cys-1.

Treatment of trypsin (32) with carbamyl phosphate at pH 7.0 results in formation of carbamyl protein, as detected colorimetrically (Table 1).

As shown by Rimon and Perlmann (120), cyanate reacts both with pepsinogen and pepsin. With pepsinogen, cyanate carbamylates the ε-amino groups of the nine lysine residues present in the NH_2-terminal segment of the polypeptide chain, which is removed on activation to form pepsin. In addition, reaction occurs with four tyrosine residues but not with the α-amino terminal leucine. Pepsinogen, in which three lysines are carbamylated, retains its native conformation and full potential pepsin activity. However, further carbamylation of pepsinogen results in a change in its molecular configuration, which is accompanied by a decreased susceptibility to activation. Pepsin obtained from carbamylpepsinogen is, in its amino acid composition and NH_2-terminal amino acid, indistinguishable from pepsin prepared from the untreated zymogen. However, its specific activity is only 20%. Studies of optical rotatory dispersion and circular dichroism (121) showed that carbamylation of the ε-amino groups of the lysine residues abolishes the electrostatic side-chain interaction between the basic residues and some dicarboxylic acids which stabilize pepsinogen.

In pepsin, cyanate carbamylates six tyrosine residues and the NH_2-terminal isoleucine, although the single lysyl residue of the protein does not react. Carbamylation of pepsin affects its enzymatic activity in a manner similar to that produced by acetylation of the enzyme with acetylimidazole. Hydrolysis of protein substrates (hemoglobin) decreases, whereas the hydrolysis of dipeptides (*N*-acetyl-DL-phenylalanyl diiodotyrosine) increases.

Carbamylation of subtilisin type Novo, another serine protease, by cyanate also involves lysine residues (122), with partial enzyme inactivation. Sedimentation and optical rotatory dispersion studies revealed that both the α-amino groups and the ε-amino groups were freely exposed to react with cyanate, without significant disruption of the native conformation of the molecule and apparently only minor changes in enzymic activity. It was suggested that cyanate could inactivate by reaction with serine at the active center.

Acid phosphatase and adenosine deaminase from red cells and several tissues in mice showed abnormal electrophoretic patterns after both *in vitro* and *in vivo* treatment with cyanate (162).

Lyases, Isomerases, and Ligases Enolase, glucose-6-phosphate isomerase, and triose phosphate isomerase (162) from red blood cells and tissues are affected by treatment with cyanate. New bands in the isozyme patterns appear; most of the normally observed bands are replaced by bands moving toward anode.

In a search for protein "active-site" reagents with selective inhibitory power toward a few aminoacyl ligases, Haines and Zamecknick (123) have found that cyanate induces selective inhibition of the capacity of certain of these enzymes to esterify amino acids to transfer *t*RNAs. When a mixed ligase system was treated with cyanate, the aspartyl, arginyl, leucyl, lysyl, and valyl enzymes lost 64 to 94% of their activities, an average of 1 protein molecule reacting with 17 molecules of cyanate.

Reaction of cyanate, as well as other amino group reagents, with pyruvate carboxylase results in a loss of catalytic activity (124). Modification of the enzyme involves only carbamylation of the ε-amino groups of lysine residues. The allosteric effector acetyl-Co A, which is an absolute requirement for catalytic activity, protects the enzyme against inactivation by these reagents, suggesting that the ε-amino group of lysine may be involved in the enzyme acetyl-Co A interaction.

Inhibition by cyanate involves, in some cases, reaction with the prosthetic group of a metalloenzyme. This has been shown to be the case for carbonic anhydrase (139), which contains one zinc ion per molecule.

Carbamylation of Hormones

As shown by Cole (125), there is a slow transformation of insulin on treatment with urea, due to cyanate, as observed chromatographically. Transformation involves carbamylation of the amino-terminal glycine and phenylalanine. This finding would explain the results of Bischoff and Bakhtiar (126), who observed changes in biological activity, solubility, and electrophoretic properties of insulin treated with urea for prolonged periods at room temperature.

Grande, Grisolia, and Diederich (127) have found that carbamylation of glucagon with either cyanate or carbamyl phosphate resulted in loss of both the glycogenolytic and the adipokinetic effects of the hormone. Only the terminal α-amino group of histidine was modified; the ε-amino group of lysine was not.

Reaction of cyanate with an oxytocin analog yields a carbamylated derivative which exhibited no intrinsic activity on the isolated uterus but inhibited the action of oxytocin (98, 128). It appears from these studies that the tyrosine hydroxyl residue, which was carbamylated, apparently played a fundamental role in binding and was less important for stimulation.

Loss of biological activity has been observed also on *in vitro* treatment of bovine luteinizing hormone, thyroid-stimulating hormone, and antidiuretic hormone with cyanate (157, 158*a*, *b*).

Carbamylation of Other Proteins

Formation of carbamyl derivatives of both lysine-rich and arginine-rich histones has been demonstrated on incubation with added carbamyl phosphate and with a generating system for carbamyl phosphate (130). Lysine-rich histone reacts more readily than arginine-rich histone, yielding homocitrulline (131), but under similar experimental conditions there was negligible carbamylation of albumin. Similar to other proteins, cyanate is also able to carbamylate histones, and, in fact, the reaction proceeds more readily with cyanate than with carbamyl phosphate (30, 130).

In agreement with previous findings (90), formation of carbamylated products in electrophoresis and chromatography in urea solutions has been observed with as distinct proteins as casein (132), globin (133), wheat proteins (134), and eye lens α-crystallin (135). Carbamylation, ascribed to the presence of cyanate in the aqueous solutions of urea, involves in casein (132) both the ε-amino groups of the lysyl residues and the *N*-terminal arginyl amino group, but in globin (133) modification affects only α-amino residues, preferentially valine, leading in any case to altered electrophoretic mobility (134, 135).

Stark et al. (90) demonstrated that when β-lactoglobulin, a typical SH-containing protein, is allowed to react with cyanate, the SH titer diminishes markedly, indicating that carbamylation has taken place.

Studies to establish the antibody-site composition in antibodies directed against a charged group (136) have demonstrated that treatment of anti-*p*-azobenzenearsonate antibodies (anti-Rp) with cyanate until 70 to 80% of free amino groups were carbamylated results in the loss of 20 to 30% of antibody sites. This loss seems to be due to attack on an amino group in the combining region, since the loss of sites can be partially prevented by the presence of the hapten during reaction. The group in the anti-Rp site which is attacked by cyanate seems most likely to be the ε-amino group of lysine or α-amino group, since neither the guanidinium group of free arginine nor the hydroxyl group of tyrosine seems to be attacked by the reagent.

Similarly to other plasma proteins in which carbamylation by cyanate results in altered electrophoretic mobilities (83, 154, 155, 164), migration of human immunoglobulins (137) toward anode increases with increasing concentrations of cyanate until a determined ratio of protein/cyanate is established.

Cyanate affects also the clotting of fibrinogen (150). Incubation of plasma or purified fibrinogen with cyanate progressively reduces the clottability of fibrinogen as the concentration of cyanate increases. The impairment of clottability is partially reversed by dialysis of the cyanate-fibrinogen complex; this suggests involvement of thiol groups, probably from disulfide bridges opened up to provide reactive sites.

Susceptibility of hemoglobin to carbamyl phosphate and cyanate and the clinical implications of carbamylation of hemoglobin on sickle cell disease will be discussed in the next section.

ANTISICKLING ACTIVITY OF CYANATE, CARBAMYL PHOSPHATE, AND UREA

The antisickling property of cyanate was first tested by Cerami and Manning (26, 27) to determine whether cyanate rather than urea was responsible for the antisickling properties of urea (7–13, 17). Cerami and Manning (26) found that when sickle red cells were incubated with cyanate, the cells were irreversibly inhibited from sickling on subsequent deoxygenation. The degree of inhibition of *in vitro* sickling was related to the amount of [^{14}C] cyanate incorporated into acid-precipitable protein, which in turn depended on the concentration of cyanate and the time of incubation. Most of the radioactivity was accounted for by carbamylation of the amino-terminal valine residues of the hemoglobin molecule. Kraus and Kraus (28) reported that carbamyl phosphate was more effective than cyanate in the prevention of sickling *in vitro*. However, Carreras et al. (30, 31) showed that carbamyl phosphate probably acts through cyanate. In recent years, the effects of urea (14–21, 25, 33–38, 58–61, 65, 72), cyanate (29, 39–46, 50–53, 61–70, 79, 81, 83), and carbamyl phosphate (54–57) on red cell sickling have been extensively studied *in vitro* and *in vivo*. The mechanisms responsible for inhibition of sickling by cyanate and carbamyl phosphate are distinctly different from that of urea (60–62, 65).

The basic event responsible for the sickling of the erythrocyte is the intracellular aggregation of hemoglobin S molecules which occurs on deoxygenation. Long-chain polymers form which twist on each other to form the rigid tactoids which distort the red cell membrane and produce the abnormal sickle shape. Aggregation and polymer formation occur only when hemoglobin S is deoxygenated, because of the different conformation of the deoxy- and oxygenated molecular forms. For sickling to occur the proportion of deoxyhemoglobin S in the red cell must exceed a critical level (22, 23, 177–190). Thus, any factor which increases the affinity of hemoglobin S for O_2 will diminish sickling at any given tissue oxygen tension by decreasing the amount of hemoglobin S present in the deoxygenated form. Sickling can also be inhibited by agents which disrupt the intermolecular aggregation process itself or, alternatively, by chemical modifications of the hemoglobin S molecule which prevents the intramolecular aggregation which results in sickling. In addition, sickling can be inhibited by stabilizing the erythrocyte membrane.

Incubation of the sickle erythrocyte with cyanate or carbamyl phosphate results in carbamylation of both hemoglobin (29–31, 66, 69, 78–87) and of the erythrocyte membrane (29–31, 148, 149). Therefore, cyanate and carbamyl phosphate may prevent sickling via any of the following mechanisms: (1) stabilization of the red cell membrane to resist deformation (sickling) induced by intracellular aggregation of deoxy Hb S; (2) carbamylation of hemoglobin with resultant structure-functional alterations: (*a*) increased oxygen affinity, (*b*) prevention of inter- or intramolecular bonding between deoxygenated Hb S molecules, (*c*) other.

Carbamylation of Hemoglobin by Cyanate and Carbamyl Phosphate

Cerami and Manning (26, 27) noted that the inhibition of sickling by *in vitro* treatment of the sickle red cells with [^{14}C] cyanate was directly related to the amount of radioactivity incorporated into hemoglobin S. Deoxygenated sickled erythrocytes also incorporated [^{14}C] cyanate, but the sickling was not reversed. However, when cyanate-treated cells were oxygenated, the normal morphology was restored, and on subsequent deoxygenation these cells no longer sickled. Amino acid analysis of hydrolysates of carbamylated hemoglobin S indicated that carbamylation takes place in the NH_2-terminal valine residues; there was no detectable carbamylation of the lysine and cysteine residues. Carbamylation of both hemoglobin (A, F, and S) in solution and in the intact erythrocyte, with [^{14}C] cyanate and with [^{14}C] carbamyl phosphate, was shown to occur by Grisolia and coworkers (29–31). Several studies

(73–78, 80, 82–88) have been carried out to investigate the specificity and the kinetics of hemoglobin S carbamylation and to evaluate the effects of the hemoglobin ligands on carbamylation to enable more precise selection of conditions under which specific sites on the hemoglobin S molecule are carbamylated.

Lee and Manning (86) reported a detailed kinetic analysis of the relative rates of carbamylation with cyanate of the α- and ε-amino groups of liganded (oxy- and carbonmonoxy- forms) and unliganded (deoxy-form) hemoglobin A and S. They found that, at pH 7.4 and 37° C, the NH_2-terminal residues of hemoglobin S were carbamylated about two orders of magnitude faster than the ε-NH_2 groups of the lysine residues which were carbamylated probably in a random fashion. As reported also by others (78, 82, 83), the NH_2-terminal valine residues of both the α and β chains of oxyhemoglobin were carbamylated to the same extent in liganded hemoglobin. The pseudo-first-order rate constant for the carbamylation of deoxyhemoglobin A was the same as that for deoxyhemoglobin S. Similarly, carbonmonoxyhemoglobins A and S, either in the tetrameric state or as the *p*-hydroxymercuribenzoate derivatives of the individual chains, were carbamylated at the same rates. Both hemoglobin A and hemoglobin S in the deoxy- form were carbamylated about twice as rapidly as in the liganded (oxy- and carbonmonoxy-) forms. In addition, CO_2 inhibited the carbamylation of deoxyhemoglobin (A and S) more than that of carbonmonoxyhemoglobin. Since CO_2 is known to bind preferentially to deoxyhemoglobin and carbamino formation occurs mainly at the NH_2-terminal residues of hemoglobin (73–77), Lee and Manning concluded that isocyanic acid, the reactive form of cyanate, is a structural analog of carbon dioxide (26, 27).

Njikam et al. (87) showed that total carbamylation of hemoglobin in partially deoxygenated whole blood (40 to 50% oxygen saturation) is about $2\frac{1}{2}$ times as extensive as carbamylation in fully oxygenated whole blood and demonstrated that the ratio of carbamylation of the α and β chains isolated from the carbamylated deoxyhemoglobin is not 1 : 1 but 1.7 : 1 in favor of the α-chain. With either A/A or S/S blood, carbamylation of the α-chain in deoxyhemoglobin proceeds about four times faster than it does in oxyhemoglobin. Carbamylation of the β-chain of partially deoxygenated hemoglobin proceeds twice as rapidly as in fully oxygenated hemoglobin.

Jensen et al. (88) also evaluated the influence of oxygenation on the reactivity of cyanate with hemoglobin A and with each chain and noted again preferential carbamylation of the α-chain of deoxyhemoglobin. The α-chains were carbamylated twice as rapidly in deoxy- as compared with oxyhemoglobin in both phosphate and bis-tris buffer (pH 6.0, 37° C). Carbamylation of the β-chains was less dependent on oxygenation of hemoglobin in bis-tris buffer and did not vary at all with oxygenation in phosphate buffer. In bis-tris buffer (pH 6.0) the β-chains of deoxyhemoglobin were carbamylated to the same extent as the corresponding α-chains. In contrast, in phosphate buffer the β-chains of deoxyhemoglobin reacted much more slowly than the corresponding α-chains of oxyhemoglobin. These results showed, in addition to the oxygen linkage of the carbamylation reaction, that inorganic phosphate inhibits the carbamylation of the β-chains. As a possible explanation they suggested that the binding of inorganic phosphate to the NH_2-terminal residues might increase on deoxygenation. Thus, the increased affinity for cyanate subsequent to deoxygenation would be opposed by the increased inhibition by phosphate.

Jensen et al. (88) also studied the effect of pH on the carbamylation of the α and β chains of deoxyhemoglobin. The overall rate of carbamylation which was maximal in the range of pH 6 to 7 dropped sharply at more alkaline pH values. Carbamylation of the α- and β-chains was equal at pH 6. With increasing pH, carbamylation of the β-chains progressively decreased, whereas carbamylation of the α-chain first increased to reach a maximum at $\sim$pH 7.0 and then also declined. At a fixed pH, CO_2 decreased the reaction rate of α and β chain to the same extent. Finally, since the NH_2-terminal groups of the β chains are involved in the binding of organic phosphates (191, 192), Jensen et al. (88) investigated the effect of 2,3-diphosphoglycerate (DPG) and inositol hexaphosphate on the carbamylation of such groups. Both polyphosphates inhibited selectively the carbamylation of the β-chains in both oxy- and deoxyhemoglobin.

In order to establish the specificity of the carbamylation of the globin chains with cyanate, Pacheco et al. (84, 85) carbamylated globin chains from humans (α, βA, βC, and γ), sheep (α, βA, βB, and βC), and rabbit (α and β) with [^{14}C] cyanate. The α-NH_2 terminal amino acid residues (valine, glycine, methionine, or proline) were carbamylated in all these globins. Carbamylation was not limited to this residue, however.

Kraus and Kraus (28) reported that carbamyl phosphate was more effective than cyanate in preventing *in vitro* sickling. Chain separations from [^{14}C] carbamyl phosphate–treated hemoglobin showed (55, 56) that the label appears first in the α-chain and then in the β-chain. At 40 minutes the peptide labeling pattern differed in carbamyl phosphate–treated blood from that of cyanate-treated blood. Competitive studies with *N*-methyl carbamyl phosphate, carbamyl phosphate, or cyanate in various combinations suggested that carbamyl phosphate behaves as an affinity labeling agent, reversibly binding in the DPG binding site of hemoglobin and then reacting covalently to modify specific sites.

To ascertain whether carbamyl phosphate is more effective than cyanate in the carbamylation of hemoglobin, particularly in the intact red cell, Carreras et al. (29–31) carried out comparative experiments with hemoglobins A, F, and S, both in solution and in the intact erythrocytes, under different conditions of concentration, pH, and temperature. Under all conditions tested, carbamylation with cyanate proceeded more readily than with carbamyl phosphate. In addition, at low pH, to diminish decomposition of carbamyl phosphate to cyanate, and in the presence of azide, to trap as carbamylazide the cyanate formed, hemoglobin carbamylation by carbamyl phosphate was significantly decreased. Thus it was concluded that carbamylation with carbamyl phosphate probably proceeds through cyanate.

Carbamylation of hemoglobin occurs *in vivo* also. Hemoglobin was rapidly carbamylated in animals following oral or parenteral administration of cyanate or carbamyl phosphate (155*a*–*c*, 158*a*–*b*, 164). Cerami (62) reported that injection of carbamyl phosphate into a monkey or a mouse did not lead to carbamylation of the hemoglobin, however. He suggested that the carbamyl phosphate is rapidly removed from the animal before it has time to form cyanate. When a single dose of [^{14}C] cyanate (0.5 μmole/g) was injected into a a mouse (157), 7.5% of the total dose reacted specifically with the NH_2-terminal valine of hemoglobin. Similar results were reported by Crist et al. (155), after the injection of a higher dose of [^{14}C] cyanate (1.5 μmole/g) into a rat. The chronic administration of cyanate (157) resulted in progressive carbamylation of the hemoglobin, until approximately 1 mole of cyanate was incorporated per mole of hemoglobin. The specificity of the cyanate for the NH_2-terminal valine of hemoglobin was remarkable; no measureable carbamylation of the ε-amino groups of the lysine residues occurred. The α and β chains were carbamylated to the same extent. Higher values of hemoglobin carbamylation (3 to 3.6 moles of cyanate/mole hemoglobin) have been obtained by the daily injection of cyanate (75 mg/kg/day) into rats for a 3 week period (164).

Oral administration of cyanate to patients with sickle cell disease (42, 45, 46, 87) also resulted in selective carbamylation of the NH_2-terminal valine residues of hemoglobin S. Carbamylation was dose-related and ranged from 0.03 to 0.67 mole of carbamyl group per mole of hemoglobin. Intravenous administration of cyanate (49) in 20 patients (3 to 4 mg/kg/hour) produced a maximum carbamylation of 0.92 carbamyl residue per mole of hemoglobin tetramer within 96 hours.

Effects of Hemoglobin Carbamylation on the Oxygen Affinity

Kilmartin and Rossi-Bernardi (73, 74, 76) found that the affinity of horse hemoglobin for oxygen increased with carbamylation of the α-amino groups of cyanate. Similar results were reported by Roughton (75). Carbamylation of adult, fetal, and sickle cell blood either with cyanate or with carbamyl phosphate increased the oxygen affinity in a degree which was related to the degree of hemoglobin carbamylation (29–32, 78, 79, 82, 83); there are no striking differences in the responses of sickle cell and normal blood (81). The increase in blood oxygen affinity (decrease in P_{50}) is a sensitive indication of the degree of hemoglobin carbamylation.

In vivo studies (62, 147, 158*a*, 159, 164) led to results similar to those from the *in vitro* experiments. The daily administration of cyanate to four animal species (mice, rats, dogs, and monkeys) increased the oxygen affinity of the blood of all. The change in the P_{50} of the blood, although quite dramatic, produced no physiological or pathological sequelae in the cyanate-treated animals (62, 156). The myocardium was able to extract sufficient oxygen from the high oxygen affinity blood to function normally (156*b*). The absence of significant polycythemia in most of the cyanate-treated animals also suggests adequate oxygen delivery by the carbamylated blood without significant tissue hypoxia. However, alternative explanations for the failure to observe marked polycythemia in the cyanate-treated animals have been suggested (164).

Carbamylation of the α-NH_2-terminal groups of hemoglobin inhibits the binding of DPG (80) to the β-chain α-NH_2 group of deoxyhemoglobin (189, 191, 192, 194, 195). DPG decreases the oxygen affinity of hemoglobin both free and in intact erythrocytes. In addition, DPG decreases the binding of CO_2 to deoxyhemoglobin (191, 192). Therefore, the possibility that the increase in oxygen affinity due to hemoglobin carbamylation might be due in part to the interference with DPG binding has to be considered (81). May et al. (83) showed with both normal and sickle cell erythrocytes that the depletion of DPG does not alter the effect of carbamylation on oxygen affinity. The P_{50} response of carbamylated normal and sickle red cells and of hemolysates to DPG was similar to that of noncarbamylated controls (62, 78, 82). Heme-heme interaction and the Bohr response were not affected in cells which had one α-NH_2 group per hemoglobin carbamylated. De Furia et al. (78, 82) concluded that this level of carbamylation does not interfere with the effect of DPG on the oxygen affinity.

Kilmartin et al. (77) further studied the functional properties of carbamylated human hemoglobin under physiological conditions. The effect of DPG on the oxygen affinity of hemoglobin carbamylated specifically at the α-chain α-NH_2 group ($\alpha_2^C\beta_2$) was the same as that of normal hemoglobin. In contrast DPG had a reduced effect on the oxygen affinity of hemoglobin carbamylated in the β-chain α-NH_2 group ($\alpha_2\beta_2^C$ and $\alpha_2^C\beta_2^C$). Carbon dioxide had no effect on the oxygen affinity of fully carbamylated hemoglobin ($\alpha_2^C\beta_2^C$), either in the presence or absence of DPG. The decrease in oxygen affinity produced by CO_2 is much larger in $\alpha_2^C\beta_2$ hemoglobin than in $\alpha_2\beta_2^C$. However, on the addition of DPG, the effect of CO_2 on the oxygen affinity of $\alpha_2^C\beta_2$ is much smaller and similar to that occurring in $\alpha_2\beta_2^C$. The effect of CO_2 on affinity of $\alpha_2\beta_2^C$ is not affected by the presence or absence of DPG. The Hill's constant and the Bohr effect in the carbamylated human hemoglobin derivatives ($\alpha_2^C\beta_2^C$, $\alpha_2^C\beta_2$, $\alpha_2\beta_2^C$) closely resembled the previously described (73, 74, 76) horse hemoglobin derivatives. These studies suggest that at the level of hemoglobin carbamylation which is achieved *in vivo* there is no interference with the effect of DPG on oxygen affinity. As suggested by Cerami et al. (62, 82), the increased oxygen affinity of carbamylated hemoglobin might be the result of a stabilization of the oxy conformation of the hemoglobin molecule. The NH_2-terminal valine residues form salt bridges in the deoxy conformation of hemoglobin (193). Carbamylation of these residues would inhibit the formation of the salt bridges, stabilize the oxy conformation of the hemoglobin, and account for the increased oxygen affinity. The stabilization of the oxy conformation of the hemoglobin molecule also would explain the inhibition of sickling by carbamylation as low as 0.1 carbamyl group per hemoglobin tetramer (26). The carbamylated oxyhemoglobin S molecules would interfere with the aggregation of other noncarbamylated hemoglobin S molecules (82).

Oxygen Affinity–Independent Antisickling Action of Cyanate

The above studies strongly suggest that the inhibition of sickling by carbamylation is due primarily to the increase in oxygen affinity produced by carbamylation (81–83). However, it has been suggested (66, 69) that, in addition, cyanate inhibits sickling directly by a mechanism which is independent of oxygenation. Diederich (81) found that sickling of control and carbamylated erythrocytes was very similar at a given oxyhemoglobin saturation when ~2 moles of cyanate were incorporated per mole of hemoglobin. However, extensive hemoglobin carbamylation (~4 moles of cyanate/mole hemoglobin) did further decrease the sickling propensity. Jensen and coworkers (66, 69) confirmed this observation and reported that the viscosity of carbamylated sickle blood, which at a given hematocrit is dependent on the percentage of sickled cells present, was lower than that of noncarbamylated controls at the same oxygen saturation. Cerami and coworkers (26, 62) claimed that the hemolysates of hemoglobin S incubated with cyanate did not gel on deoxygenation. Charache et al. (63) found that carbamylated hemoglobin was capable of gelation, although a slightly higher concentration of hemoglobin was required than for untreated hemolysates. In agreement with these results, Jensen et al. (66, 69) reported that carbamylation of membrane-free hemoglobin solutions results in an increase in the minimum concentration of deoxy-sickle hemoglobin required for gelation. The concentration of hemoglobin S at the gelling point has been shown to be a sensitive indicator of the sickling propensity. Thus, carbamylation inhibits sickling both by oxygenation-dependent and -independent mechanisms (66, 69, 83). Gelation experiments with membrane-free lysates demonstrated that the oxygen-independent effect of cyanate was not due to the carbamylation of the red cell membrane but rather to the carbamylation of the hemoglobin itself, possibly through carbamylation of amino acid residues of the hemoglobin molecule other than the NH_2-terminal valine residues.

Carbamylation of Erythrocyte Ghosts: Effects on Sickling

The erythrocyte membranes are readily carbamylated either with cyanate or with carbamyl phosphate (30). Decreased osmotic fragility as well as increased oxygen affinity occurs following incubation of erythrocytes with isocyanates (196). The reversibility of sickling is more related to the red cell membrane itself than to the hemoglobin molecule; the formation of irreversibly sickled cells can be prevented by avoiding the accumulation of calcium in the membrane (197). Therefore, the possibility that the antisickling effect of cyanate may be due in part to the carbamylation of the red cell membrane has been considered. Durocher et al. (148) measured the *in vitro* effect of cyanate on the membrane surface charge of red blood cells from humans, rats, and rabbits and reported that carbamylation significantly increased electrophoretic mobility. They (149) also studied the effect of cyanate on the deformability of normal and sickle cell erythrocytes and found that carbamylation improves the decreased deformability of sickle erythrocytes when tested in a microfiltration system. Similarly, carbamylation restored the decreased filterability of normal erythrocytes stored at 4°C for 48 hours. The deformability of normal erythrocytes has been correlated with the maintenance of intracellular ATP; ATP may chelate calcium, preventing the formation of calcium complexes with membrane proteins or lipids (199, 200). However, Durocher et al. (149) reported that cyanate improves deformability of the erythrocyte without significant changes in the cellular ATP. They suggested that carbamylation of hemoglobin may affect the binding of ATP to hemoglobin, thereby increasing the amount of free ATP. In addition, membrane carbamylation may alter membrane binding of hemoglobin and calcium, thereby influencing membrane deformability.

Increased Survival of Carbamylated Sickle Red Cells

To assess whether the antisickling effect of cyanate observed *in vitro* would be retained *in vivo*, Cerami and coworkers (39, 40, 62) compared the survival of untreated erythrocytes with that of cells treated *in vitro* with cyanate in patients with sickle cell anemia. In seven patients the mean 50% survival of ^{51}Cr-labeled sickle erythrocytes was increased from 9.9 to 20.7 days after *in vitro* treatment of the cells with cyanate (2 to 4 moles of cyanate incorporated per mole of hemoglobin tetramer). The increased life span of cyanate-treated cells demonstrated that the antisickling effect of cyanate observed *in vitro* was retained *in vivo* also.

May et al. (83) showed that the prolongation of survival of cyanate-treated sickle red cells was directly related to the increase in the oxygen affinity produced by carbamylation. They suggested that between 25 and 50% of the hemoglobin must be carbamylated to obtain a clinically useful effect.

Alter and coworkers (44) pointed out that the results from Cerami et al. (39, 40) and from May et al. (83) required a cautious interpretation. To obviate the potential effects of cyanate on ^{51}Cr binding to sickle and fetal hemoglobin-containing red cells, Alter et al. (44) studied the survival of sickle reticulocytes containing ^{14}C and ^{3}H-leucine incorporated into the hemoglobin. The labeling system used allowed simultaneous measurement of the survival of two populations of reticulocytes, one of which was subsequently exposed to cyanate. The survival of the carbamylated cells was tripled, although not restored to normal. The enhanced survival of carbamylated reticulocytes suggests that the development of irreversibly sickled cells may be inhibited by cyanate.

Kraus and coworkers (54, 55) reported that *in vitro* treatment of sickle red cells with carbamyl phosphate also improves the erythrocyte survival. Milner and Charache (56, 57) studied the life span of red cells from 20 patients with sickle cell disease after varying degrees of *in vitro* carbamylation with cyanate or with carbamyl phosphate. An increased cell life span proportional to the degree of hemoglobin carbamylation was noted. The greatest prolongation of survival was noted with reticulocyte rich blood; the life span of irreversibly sickled cells was not improved by carbamylation. The degree of carbamylation was not dependent on cell age.

Diederich and coworkers (52, 53) initiated pilot studies of the effects of repeated *in vitro* red cell carbamylation on *in vivo* sickling and on the *in vivo* sickling propensity of cells. At weekly intervals, ~20% of the circulating red cell mass was incubated with cyanate to achieve 1.2 ± 0.4 moles of cyanate incorporation per mole of hemoglobin. Repeated carbamylation decreased the *in vivo* sickling of the erythrocytes, as judged by a progressive decrease in reticulocytosis and in the circulating irreversibly sickled cells and by a general increase in circulating hemoglobin levels. However, no significant decrease in oxygenation-independent sickling propensity could be demonstrated. The only salient factor capable of diminishing *in vivo* sickling was the increased oxygen affinity of whole blood (oxygenation-dependent inhibition of sickling). The number of painful crises during carbamylation was less than during the pretreatment year for the group of patients.

Oral and intravenous administration of cyanate to patients with sickle cell disease (41, 42, 45, 46) also inhibits the *in vivo* sickling of the erythrocytes. Patients receiving cyanate orally over a period of months showed a decrease in the hemolytic rate as evidenced by an increase in hemoglobin (proportional to the degree of carbamylation achieved) and a decrease in bilirubin and plasma lactic dehydrogenase. The frequency of idiopathic painful crises decreased in the cyanate-treated patients as compared with the controls; this decrease was proportional to the degree of hemoglobin carbamylation. The frequency of induced or precipitated crises was unchanged.

Desickling Effect of Urea

In 1954, Ponder and Ponder (201), employing saturated solutions of urea in attempts to solubilize proteins of the red cell membrane, reported the loss of rigidity when sickle erythrocytes were treated with urea. Allison (202), in 1957, observed that the viscosity increment which occurs on deoxygenation of concentrated solutions of hemoglobin S was reduced by 0.1 M urea and inhibited by 1 M urea. Brombert and Jensen (203), in 1965, reported that 1 M urea caused both marked inhibition of *in vitro* sickling and an increase in oxygen affinity of sickle cell blood. Intermediate degrees of both effects were obtained with lower urea concentrations.

Nalbandian and coworkers (7, 8) in 1970 noted inhibition of the gelation of concentrated hemoglobin S hemolysates attributable to 0.5 M urea. They also described urea-induced inhibition and reversal of *in vitro* sickling of intact erythrocytes without apparent hemolysis. 0.2 M urea prevented but 1 M urea reversed sickling of fresh S erythrocyte suspensions containing 2% sodium metabisulfite (11, 23). However, when more physiologic sickling methods were used (oxygen deprivation, carbon dioxide, nitrogen or helium), urea concentrations as low as 10 to 50 mM inhibited sickling (12–14). Barnhart et al. (8, 11, 25) reported optical and electron microscopic evidence suggesting a quantitative correlation between reduction in sickled forms and urea concentration.

The molecular mechanisms proposed (7, 9, 10, 12, 13, 16, 23, 24, 71, 72) for the desickling action of urea were disruption of the implicated hydrophobic bonding critical to sickling and steric hindrance. Urea, because of its high electric dipole moment, may destroy the hydrophobic bonds formed between the deoxygenated hemoglobin S molecules, releasing the tetramers as free and mobile entities no longer rigidly constrained. In addition, because of the concentration of hydrogen groups at one end of its molecule, urea may interact with the hemoglobin S molecule, altering its configuration. All evidence available (7–18, 25–27, 60, 62, 65, 72, 184) negates the thesis that the effectiveness of urea as a desickling agent is due to cyanate, which is in equilibrium with urea in solution (26, 27, 173–175).

A serious problem arises when one attempts to evaluate the reported *in vitro* antisickling effects of urea. Because urea increases oxygen affinity of hemoglobin (203) and because *in vitro* sickling is certainly oxygenation-dependent, it is imperative to examine sickling as a function of oxygen tension and of oxyhemoglobin saturation. Only in this fashion can objectivity be given to the reported antisickling effects of urea.

The effects of urea on *in vitro* sickling phenomenon were examined as a function of oxygen tension by Segal et al. (61). At a urea concentration of 0.2 *M* no inhibition of sickling could be demonstrated. Further, 0.1 *M* urea exerted (*a*) no effect on sickling-induced potassium loss from the erythrocyte, (*b*) a very minimal effect on hypoxia-induced hyperviscosity of sickle erythrocytes, (*c*) no effect on the viscosity and no inhibition of gelation of a 22 g % hemoglobin S solution, and (*d*) no effect on the filterability of 3% suspensions of oxygenated sickle erythrocytes. A urea concentration dependent, reversible increase in oxygen affinity of hemoglobin and of whole blood was noted. Thus, *in vitro* inhibition of sickling by urea requires concentrations exceeding those that can be practically obtained *in vivo*.

Clinical application of urea as an antisickling agent was attempted by Nalbandian and coworkers and by other groups (9, 10, 12, 13, 15–20). Preliminary reports describing improvement in symptoms and in red cell survival generated enthusiasm. Because of conflicting clinical observations concerning the clinical efficacy of intravenous urea in invert sugar in painful crises, a large multicenter double-blind controlled study was undertaken. Urea at two dosage levels was compared with sodium bicarbonate in invert sugars and 10% invert sugar alone. Neither urea nor alkali administration was found to be superior to invert sugar alone in shortening the crisis episodes (36, 37).

EFFECTS OF CARBAMYLATION ON THE METABOLISM AND FUNCTIONS OF THE RED CELLS

The amino terminal valine residues of the hemoglobin molecule have been implicated in several important physiological functions: the formation of stabilizing salt bridges in deoxyhemoglobin (190), the binding of DPG (191, 192, 194), the binding of CO_2 as carbamino groups (73–76), and the alkaline Bohr effect (73–76, 193). Since the carbamylation of the valine residues is irreversible, the treatment of sickle cell blood with carbamylating agents could possibly result in physiologically altered red cells, unable to function properly. In addition, carbamylation could conceivably occur with functional groups of enzymes and other cellular proteins, and alter their physiologic functions.

Cerami and coworkers (62, 78, 82) investigated the effects of cyanate *in vitro* on the metabolism and functions of the sickle red cells, in an attempt to assess the effects of carbamylation at levels necessary to inhibit sickling (~1 mole of cyanate incorporated per mole of hemoglobin tetramer). The most noteworthy effect of the carbamylation was the increase in the oxygen affinity, also reported by others (29–31, 52, 53, 61, 69, 73–80, 83) and already commented. This increase was irreversible and not due to a loss of DPG or ATP. The response of carbamylated erythrocytes to DPG was similar to that of the control cells, showing that this level of carbamylation does not interfere with the effect of DPG on the oxygen affinity. Administration of cyanate to experimental animals (62, 147, 158*a*, 159, 164) also resulted in increasing oxygen affinity. However, no physiological or pathological sequelae were observed.

Kilmartin and Rossi-Bernardi (73, 74, 76) reported that carbamylation of all four of the amino terminal valine residues of horse hemoglobin inhibits the formation of carbamino compounds with carbon dioxide. Cerami and coworkers (62, 78, 82) found, with both normal and sickle cell blood, that the carbamylation of one of the four amino terminal valine residues of hemoglobin slightly diminishes (15 to 20%) the CO_2 capacity of the red cells. However, a compensatory increase in the amount of plasma bicarbonate is produced so that the whole blood CO_2 capacity is not affected. The pH of the blood and the red cell ΔpH does not change significantly after carbamylation.

The protonation of the amino terminal valine residues of the α-chains of the hemoglobin has been shown to be responsible for ~25% of the Bohr effect of isolated horse hemoglobin (73, 74, 76). Cerami and coworkers (62, 78, 82) found that carbamylation of one of the four amino terminal valine residues of hemoglobin does not affect the Bohr effect. There was no demonstrable difference in the response of the oxygen affinity to changes in pH of untreated and carbamylated normal and sickle red cells. Papayannopoulou et al. (163) also found no change in the Bohr effect when blood from primates was treated with cyanate.

Papayannopoulou and coworkers (162) studied by electrophoresis the effect of cyanate on 25 enzymes from the red cells and other tissues. *In vitro* treatment of red cells with cyanate altered the isozyme patterns of all enzymes examined; most of the normally observed zones were replaced by zones with more anodal migration. *In vivo* administration of repeated high doses of cyanate in mice, which resulted in the carbamylation of 0.8 to 1.0 amino terminal valine and 0.12 to 0.14 lysine residues per hemoglobin molecule, produced reversible alterations in the electrophoretic patterns of 15 enzymes in red cells, brain, kidney, liver, and muscle. With low dosage of cyanate minimal alterations were observed, and in survivors of a single LD_{50} dose there were no visible electrophoretic changes. May et al. (62) also found gross changes in the electrophoretic mobility and activities of some red cell enzymes.

De Furia et al. (62, 78, 82) studied the effects of cyanate *in vitro* on red cell metabolism, and found no significant differences in glucose utilization and lactate production in control and carbamylated cells. Similarly, concentrations of glycolytic intermediates in the carbamylated cells were not significantly different from those of control cells except for increased levels of fructose-6-phosphate and 2-phosphoglycerate and decreased levels of pyruvate.

Most of the glycolytic enzymes did not appear to be severely affected after incubation of the red cells with cyanate. However, a light reduction in the activities of phosphofructokinase and pyruvate kinase ($\sim 25\%$) was observed. Glader and Conrad (142) studied the activity of the hexose monophosphate shunt in red cells incubated with cyanate, and reported irreversible inactivation of glucose-6-phosphate dehydrogenase. Kinetic experiments did not reveal any change in the K_m constants for either NADP or G-6-P. In contrast with these studies *in vitro*, the experiments *in vivo* fail to show any change in the pyruvate kinase and glucose-6-phosphate dehydrogenase activities. The activities of both enzymes in erythrocytes from cyanate-treated dogs (158*a*) and rats (160) were in the range observed with control cells.

In the cyanate-treated erythrocytes (142) there was a significant decrease in reduced glutathione that was not associated with an increase in oxidized glutathione. De Furia et al. (82), and Freedman et al. (146) also found a lower concentration of reduced glutathione in carbamylated sickle cells. However, the activities of glutathione reductase and glutathione peroxidase were normal (82). Furthermore the results from De Furia et al. suggested that cyanate did not inhibit the mechanisms available to regenerate the reduced glutathione. Cyanate has been reported to decrease *in vitro* the oxidation-reduction potential of the hemoglobin–methemoglobin system (140). However, methemoglobin was not observed either in the *in vitro* (82) or in the *in vivo* (157) experiments.

The concentrations of ATP, ADP, AMP, and DPG were not modified by *in vitro* carbamylation of sickle red cells (78, 82). DPG-depleted carbamylated cells (62, 78, 82) regenerated DPG and ATP and increased the P_{50} when incubated with phosphate, inosine, and pyruvate, similarly to control cells, showing that they were able to synthesize DPG and ATP. Extracorporeal treatment of blood with cyanate in primates (163), which resulted in hemoglobin carbamylation and increased oxygen affinity, did not significantly change the DPG concentration of red cells. Similarly, the erythrocyte concentrations of ATP and DPG were the same in control and cyanate-treated dogs (158*a*). However, Harkness and coworkers (147, 159) and Alter and coworkers (164) observed a dose-dependent decrease in the DPG concentration in the erythrocytes of rabbits and rats treated with cyanate. In an attempt to investigate the mechanism by which this decrease in DPG occurred, Harkness et al. (147) were unable to demonstrate any *in vitro* or *in vivo* effect of cyanate upon the activities of the enzymes directly involved with DPG metabolism. The activities of both DPG mutase and DPG phosphatase were normal in the erythrocytes from rabbits chronically treated with cyanate. Cyanate-treated cells were able to both synthesize and decompose intracellular DPG at the same rate and to the same extent as the control cells. DPG phosphatase and DPG mutase activities were not affected by the *in vitro* treatment of rabbit erythrocytes with cyanate. Finally, cyanate added directly to the assay system, using hemolysates of erythrocytes from man and rabbit and enzymes purified from human erythrocytes, had no effect upon the activities of these enzymes. Harkness and coworkers suggested that the decrease in the intracellular DPG concentration could result from the inhibition of DPG binding to deoxyhemoglobin by cyanate. A major regulatory mechanism for controlling intracellular DPG is the level of free DPG since DPG mutase is very sensitive to product inhibition. The carbamylation of the α-amino groups of the β-chains of hemoglobin would result in a temporary increase in unbound DPG which would secondarily inhibit the synthesis and result in a new lower equilibrium level of intracellular DPG.

Alter and coworkers (143, 144), and Habib et al. (145) found that cyanate markedly inhibited hemoglobin synthesis *in vitro*. The concentrations of cyanate which inhibited *in vitro* sickling were shown (143, 144) to depress globin synthesis in human reticulocytes and erythroblasts and in rabbit reticulocytes (whole cell or cell-free lysate). Inhibition of globin synthesis was proportional to cyanate concentration (143–145), and both α- and β-chains of hemoglobin S were equally affected. The effects of cyanate on the various stages involved in protein synthesis were studied (143, 144). Transport of radioactive amino acids into the cells was not affected, and free intracellular amino acids were not carbamylated. Aminoacylation of transfer RNA was not inhibited, and the acylated amino acids were not carbamylated. Cyanate induced degradation of polysomes to monosomes, causing patterns that resembled those produced by inhibition of initiation. In addition, cyanate prevented hemin stimulation of initiation in a cell-free lysate. Thus, it was concluded that cyanate inhibits the initiation rather than the termination of the translation process. It was suggested that the impairment of ribosome function might be due to carbamylation of ribosomal proteins, and since part of the inhibition was reversible, it was pointed out that carbamylation could affect sulfhydryl groups in addition to amino groups. Freedman et al. (146) also studied the mechanism of cyanate inhibition of globin synthesis in both human sickle cell and rabbit reticulocytes, and confirmed that the major effect of cyanate appears to be on initiation and/or an early elongation step of translation.

Cerami et al. (158*a*), in experiments to assess the effects of chronic administration of cyanate, found that the hematocrits and hemoglobin concentrations were the same in the cyanate-treated and control monkeys. However, the hemoglobin concentration of some of the cyanate-treated mice was 10–15% higher than in the control animals. Alter and coworkers (164) studied the effects of high doses of cyanate on erythropoiesis *in vivo*. Daily administration of high dose cyanate to rats led to incorporation of 3 to 3.6 moles of cyanate per mole of hemoglobin by the end of a 3 week period. At this time, severe clinical toxicity was observed; treated animals lost weight and the neuromuscular system was markedly affected. In the cyanate-treated rats DPG and P_{50} were decreased, and the mean hemoglobin was higher than in the control animals. However, none of the treated animals had an absolute increase in red cell mass, and the treated animals who were more severely ill did not have elevated hemoglobin concentrations. There was no decline in mean corpuscular hemoglobin concentration in any of the animals and no reduction in red cell iron turnover was observed. Furthermore, *in vitro* protein synthesis by bone marrow from cyanate-treated animals was not decreased and remained sensitive to inhibition by the further addition of cyanate. Thus, it was concluded (164) that cyanate does not inhibit hemoglobin production *in vivo* until signs of toxicity are manifest. The lack of reactive polycythemia secondary to the increased oxygen affinity in the cyanate-treated animals cannot be explained by the inhibitory effect of cyanate on protein synthesis observed *in vitro* (143–145). It might be explained by relative erythropoieitin deficiency due to starvation. Also some shortening of red cell life span due to carbamylation of erythrocyte enzymes might have occurred. However, Cerami et al. (158*a*) reported that survival of red cells from a dog chronically treated with cyanate was normal.

Carbamylation of the erythrocyte ghosts does not appear to affect their functional properties. De Furia et al. (78, 82) found that *in vitro* incubation of erythrocytes with cyanate did not alter the normal concentrations of intracellular Na^+ and K^+, the normal osmotic fragility curves, and the normal patterns of autohemolysis. They concluded that the cation-pumping mechanisms are not affected by carbamylation and that the integrity of the membrane is maintained. *In vitro* incubation of blood with both cyanate and carbamyl phosphate results in extensive carbamylation of plasma proteins (29–31, 83). Similarly, the plasma proteins were carbamylated by *in vivo* administration of cyanate (32, 155*a–c*, 164). Therefore the possible alterations of some functional properties of the plasma has to be considered. Bell and Charache (150) showed that incubation of plasma or purified fibrinogen with cyanate resulted in a striking decrease in clottability. The inhibitory effect, which was proportional to cyanate concentration, was not evident immediately after addition of cyanate, and was partially reversed by dialysis, suggesting the involvement of thiol groups.

PHARMACOLOGY OF CYANATE AND CARBAMYL PHOSPHATE

In the 1940s Schutz and Birch (151–153) carried out a series of classical studies on the pharmacology of cyanate in animals and humans. In the last few years several toxicological studies (154–164) have been carried out in experimental animals to determine the effects of chronic administration of cyanate and carbamyl phosphate. Clinical studies are now being conducted

(41, 42, 45–51) to test the possible use of cyanate as therapeutic agent for sickle cell disease.

Fate and Tissue Distribution of Cyanate in Animals

As it was shown by Crist et al. (155*a*–*d*) and by Cerami et al. (62, 158*a*) administration of cyanate leads to carbamylation of several tissues other than red blood cells. Both [^{14}C] carbamyl phosphate and [^{14}C] cyanate (300 to 375 μmole) injected in rats (250 to 350 g) were extensively incorporated into blood proteins (hemoglobin, red cell membranes and plasma proteins) and tissues, including liver and brain. The major part of the cyanate injected was bound to proteins; 15% of the cyanate was exhaled as $^{14}CO_2$ and 3% was found in urine. The rapid disappearance of free [^{14}C] cyanate was shown to be largely due to protein binding, and the release of $^{14}CO_2$ was found to decrease sharply at 4 hours when cyanate binding was essentially complete.

The values reported by Cerami et al. (62, 158*a*) were somewhat in discrepancy with those of Crist et al. After injection of a single dose (10 μmole) of cyanate to a mouse (20 g), approximately 70 to 75% of the injected dose was broken down to form $^{14}CO_2$ during the first 6 hours, and another 8 to 10% was found in the urine. 7.5% was incorporated in hemoglobin, 3.5% in bones, 2.1% in muscle, and less than 3% in all the other organs.

Cerami et al. (158*a*) showed that the chronic administration of cyanate leads to progressive carbamylation of blood and other tissues. However, the amount of radioactivity in the different organs eventually plateaued after different periods of time, probably reflecting a steady state of proteins being labeled with cyanate equaling the rate of carbamylated proteins being catabolized. When the administration of cyanate was stopped the loss of the radioactivity from the various organs followed different rates. Crist et al. (155*b*) measured the disappearance of carbamylated proteins after extensive carbamylation. At nearly 4 months, when bound radioactivity in blood proteins was negligible, there was extensive residual radioactivity in some tissues (muscle, brain, kidney, liver). Brain proteins retained about 50% of the radioactivity found at 6 hours. No gross or histopathological lesions were found (158*a*) in animals after 12 months administration of cyanate.

Tissue Enzyme Alteration

Papayannopoulou and coworkers (162) studied the tissue isozyme alterations in cyanate-treated mice. The chronic administration of high-dose cyanate produced alterations in the electrophoretic pattern of 15 of the 25 enzymes tested in red cells, brain, kidney, liver, and muscle. These alterations were reversible and disappeared 40 days after cessation of cyanate administration. With low dose of cyanate the alterations were minimal, and no electrophoretic changes were detected in survivors of a single LD_{50} dose. As Papayannopoulou et al. (162) pointed out, the observed change in enzyme electrophoretic pattern does not necessarily imply catalytic dysfunction. In fact, staining activity was present in all the preparations, suggesting that there was no appreciable abnormality in the function of the tested enzymes. Red cell glucose-6-phosphate dehydrogenase has been found to be inactivated by *in vitro* incubation with cyanate (142). However, Gillette et al. (46) found no decrease in the normal activity in patients maintained on oral cyanate therapy for 6 to 12 months. Toskes et al. (160) also tested glucose-6-phosphate dehydrogenase activity in erythrocytes, intestinal mucosa, brain, kidney, liver, and adrenal glands of cyanate-treated rats. Only in the liver was the enzymic activity decreased.

Endocrine Toxicity

In vitro carbamylation abolishes the biological activity of several polypeptide hormones including glucagon, thyroid-stimulating hormone, luteinizing hormone, and antidiuretic hormone (127, 128, 158*b*). However no significant alterations in hormone levels or function were noted in rats receiving oral cyanate chronically (62, 158*b*).

Erythropoietin Gidary et al. (129) examined the effect of cyanate on the erythropoietin and found that exposure of the hormone to 50 m*M* cyanate for 1 hour at 37°C did not inhibit its biological activity. The cyanate-treated erythropoietin preparation was able to stimulate heme synthesis in normal bone marrow cell culture as effectively as native erythropoietin. Polyacrylamide gel electrophoresis suggested that several proteins in the erythropoietin preparation were carbamylated but it was not certain that the active hormone was carbamylated.

Growth Hormone At low dose levels (6 to 32 mg/kg) chronic administration of cyanate to mice did not affect significantly the weight of the animals. Similarly the daily injection of cyanate (25 to 50 mg/kg) to rats during the growing period did not affect adversely the weight gain (158*b*). However, in agreement with results reported by Schutz in 1949 (153), chronic administration of cyanate at high dose (155*a*–*d*, 158*a*–*b*, 164) led to a significant weight loss and eventually to death. The loss of body weight was reversible; even after substantial weight loss, cessation of the cyanate administration allowed complete recovery of the lost weight (158*a*). At autopsy (164) the weight loss appeared to be due largely to loss of body fat. Subcutaneous and peritoneal fat disappeared and all organs decreased in weight.

In vitro incubation of growth hormone (GH) with 100 m*M* cyanate did not diminish the biological activity of the hormone (158*b*). Thus, it is unlikely that the weight loss and failure to grow which was observed in animals was due to *in vivo* carbamylation of GH. Rather the loss of weight and the inhibition of growth observed in cyanate-treated animals resulted from decreased food intake (158*a–b*, 164). Daily injection of 25 to 50 mg/kg of cyanate to mice (158) did not alter food intake, but the injection of a higher dose (100 mg/kg) caused a drastic reduction in the food consumption. This effect was reversible; water intake paralleled food consumption.

Patients with sickle cell disease treated with oral cyanate at a dose greater than 35 mg/kg/day (46) developed anorexia and weight loss. After the dose was lowered, patients lost the anorexia and promptly regained the weight. Occasional anorexia to both solid food and water has been also observed in patients receiving cyanate intravenously (49). The anorectic state persisted a few days beyond the period of infusion. Weight loss was slight (0.9 kg/patient) but consistent.

Thyroid-Stimulating Hormone Thyroid-stimulating hormone (TSH) was inactivated by *in vitro* carbamylation with 10 mM cyanate (158*b*). However when thyroid function was studied in animals treated with cyanate (158*b*) no dysfunction was observed. Similarly, no alterations were found in the serum tri-iodothyronine and thyroxine of patients treated with oral cyanate (46). Serum T_4 and T_3 were followed daily in patients receiving cyanate intravenously (49); no significant differences were observed between the control periods and the infusion.

Glucagon and Insulin Glucagon and insulin have been shown to be carbamylated with cyanate (125–127). Carbamylation of glucagon (127) led to the inactivation of the hormone. In contrast, when insulin was incubated with cyanate (158) no significant loss of the biological activity was observed. However, chronic administration of cyanate seems to alter hepatic glycogen metabolism. Toskes et al. (160) found marked deposition of glycogen in the liver of cyanate-treated rats. Glycogen deposition was more severe in periportal hepatocytes, and electron micrographs showed compression and displacement of smooth and rough endoplasmic reticulum.

Antidiuretic Hormone Birch and Schutz (152), Diker (154) and Alter et al. (164) reported marked diuresis in cyanate-treated rats. Cerami et al. (158*b*) found that cyanate-induced diuresis in rats increased with the doses of injected cyanate (<50 mg/kg), and plateaued at 90 mg/kg of cyanate. Given orally, 100 mg/kg cyanate failed to induce diuresis. Although Birch and Schutz (151) did not observe cyanate-induced diuresis in man, Peterson et al. (49) showed that diuresis was generally increased during the intravenous administration of cyanate to patients with sickle cell disease. Antidiuretic hormone (ADH) is known to stimulate reabsorption of water by the renal tubule. Since cyanate *in vivo* decreases the rate of tubular water reabsorption (158*b*), the cyanate-induced diuresis could result from inactivation of ADH. Cerami et al. (158) found that although ADH carbamylated *in vitro* with cyanate was biologically inactive, it did not act as an inhibitor of native ADH. However, cyanate induced diuresis in the alcohol pretreated (ADH suppressed) rat, suggesting that diuresis is not a result of the *in vivo* carbamylation of ADH but a result of a direct effect of cyanate upon kidney.

Chorionic Gonadotropin, LH, and FSH Cerami and coworkers (62, 158*b*) observed that mice and rats fed a diet containing 1% cyanate did not have normal estrus and instead underwent a constant diestrus, failing to reproduce as long as they were maintained on the diet. Histological examination revealed no lesions in the ovaries or testes of the cyanate-treated animals. To assess the possible inactivation of a gonadotropic hormone by cyanate, carbamylation of human chorionic gonadotropin (HCG), luteinizing hormone (LH), follicle-stimulating hormone (FSH) and prolactin was studied *in vitro*. 100 m*M* cyanate did not affect the activities of HCG and prolactin; FSH lost 30% of its activity, and the activity of LH was nearly abolished. However, indirect evidence was obtained for the fact that LH was not carbamylated *in vivo*: the continued presence of cyanate in the plasma during the LH surge did not prevent ovulation from occurring; pituitary LH activity was not reduced by the administration of cyanate. Furthermore, rats injected with lower doses of cyanate (25 mg/kg/day) gained weight, cycled normally and showed no loss of fertility. Cerami and coworkers (62, 158*b*) concluded that the prolonged anestrus in cyanate-treated animals might simply reflect the reduction in food intake rather than a specific carbamylation of an endocrine secretion. Body water imbalance resulting from the cyanate-induced diuresis might also be involved.

Neuromuscular System Alterations

The neuromuscular system appears to be specifically sensitive to cyanate administration. Treatment of rats with high-dose cyanate (160, 164) led to weakness and/or spasticity, particularly of the hind limbs, although histopathological studies of muscle, peripheral nerve and spinal cord (164) did not provide the basis for the neuromuscular deficit. Cerami et al. (158*a*) also reported a toxic neuromuscular reaction, characterized by muscular tremors and apparent muscular weakness, after 6 months of cyanate administration to dogs.

Patients with sickle cell disease treated with oral cyanate with doses greater than or equal to 35 mg/kg/day developed signs of a peripheral neuropathy (46, 47). Polyneuropathy was confirmed (48) by physical examination, quantitative sensory evaluation, nerve conduction test, electromyography, and sural nerve biopsy. Segmental demyelination was present, but the major lesion appeared to be a neuronal degeneration affecting the distal axons. Nerve conduction studies (48) indicated that the lesion is related to both the peak dose and the total time of cyanate administration, although conduction abnormalities were observed as early as ~50 days after cyanate administration. Distal sensory conduction was more affected than motor conduction. Children appeared to have a higher incidence of subnormal conduction velocities than adults. Improvement occurred with no specific treatment after cessation of cyanate administration.

Central Nervous System Affection

Studies from Birch and Schutz (152), Hildebrand et al. (160), and Alter et al. (164) showed that chronic administration of cyanate at high dose produced marked lethargy, poor hygiene, and decreased activity in rats. Crist and coworkers (155*a*, *b*) reported a dose-related decrease in learning ability in rats chronically treated with cyanate. This retardation persisted when the animals were retested nine weeks after termination of cyanate injections, but when retested at 8 months no retardation could be demonstrated. At this time, only 10% of the originally carbamylated proteins should remain in the brain. Administration of cyanate to animals at doses similar to those proposed for humans (46, 62) did not affect growth and did not produce untoward effects on memory testing. However, motor activity became extensively diminished. Exploratory and grooming activities were reduced in the cyanate treated mice and rats, and the threshold to pentylenetetrazole markedly decreased in the treated animals.

Both oral and intravenous administration of cyanate to patients of sickle cell disease (46, 49) produced occasional drowsiness and tiredness. Other symptoms of subjective toxicity have been occasionally noted: nausea, vomiting, epigastric pain, fullness, abdominal distress, retrosternal distress, frontal headache and burning. All symptoms were minor and were reversed by decreasing the dose or stopping the cyanate infusion.

REFERENCES

1. Pauling, L., Itano, H. A., Singer, S. J., and Wells, I. C. (1949). *Science* **110**, 543–548.
2. Pauling, L. (1971). In R. M. Nalbandian, Ed., *Molecular Aspects of Sickle Cell Hemoglobin: Clinical Applications*, Charles C Thomas Publishing Company, Springfield, Ill, pp. 7–10.
3. Desforges, J. F. (1971). *N. Engl. J. Med.* **284**, 913–915.
4. Freedman, M. L. (1971). *Am. J. Med. Sci.* **261**, 305–308.
5. Editorial, (1971) *Lancet* **2**, 1069–1070.
6. Editorial, (1972) *Lancet* **1**, 671–672.

7. Nalbandian, R. M., Henry, R., Nichols, B., Kessler, D. L., Camp, F. R., Jr., and Vining, K. K., Jr. (1970). *Ann. Int. Med.* **72**, 795.

8. Barnhart, M. I., Lusher, J. M., Henry, R. L., and Nalbandian, R. M. (1970). *Blood* **36**, 837.

9. Nalbandian, R. M., Henry, R. L., Schultz, G., and Avery, N. L., Jr. (1971). *Ann. Int. Med.* **74**, 827.

10. Nalbandian, R. M., Houghton, B. C., Henry, R. L., and Wolf, P. L., (1971). *Fed. Proc. (USA)* **30**, 684A.

11. Barnhart, M. I., Lusher, J. M., Henry, R. L., and Nalbandian, R. M. (1971). *Fed. Proc. (USA)* **30**, 684A.

12. Nalbandian, R. M., Schultz, G., Lusher, J. M., Anderson, J. W., and Henry, R. L. (1971). *Am. J. Med. Sci.* **261**, 309–324.

13. Nalbandian, R. M., Anderson, J. W., Lusher, J. M., Agustsson, A., and Henry, R. L. (1971). *Am. J. Med. Sci.* **261**, 325–334.

14. Nalbandian, R. M. (1971). *N. Engl. J. Med.* **285**, 408.

15. Nalbandian, R. M., Schultz, G., Anderson, J. W., and Lusher, J. M., (1971). *Blood* **38**, 789.

16. Nalbandian, R. M. (1971). In R. M. Nalbandian, Ed., *Molecular Aspects of Sickle Cell Hemoglobin: Clinical Applications*, Charles C Thomas Publishing Company, Springfield, Ill, pp. 128–162.

17. McCurdy, P. R., and Mahmood, L. (1970). *Blood* **36**, 841.

18. McCurdy, P. R., and Mahmood, L. (1971). *N. Engl. J. Med.* **285**, 992–994.

19. Eyre, J. T., Ashcom, R. C., and Black, A. V. (1971). *N. Engl. J. Med.* **285**, 295.

20. Desforges, J. F. (1971). *N. Engl. J. Med.* **285**, 295.

21. Smith R., Sugerman, H., Lubin, B., and Oski, F. (1971). *N. Engl. J. Med.* **285**, 295–296.

22. Murayama, M. (1971). In R. M. Nalbandian, Ed., *Molecular Aspects of Sickle Cell Hemoglobin: Clinical Applications*, Charles C Thomas Publishing Company, Springfield, Ill., pp. 3–19.

23. Nalbandian, R. M. (1971). In R. M. Nalbandian, Ed., *Molecular Aspects of Sickle Cell Hemoglobin: Clinical Applications*, Charles C Thomas Publishing Company, Springfield, Ill, pp. 20–66.

24. Orten, J. M., and Nalbandian, R. M. (1971). In R. M. Nalbandian, *Molecular Aspects of Sickle Cell Hemoglobin: Clinical Applications*, Charles C Thomas Publishing Company, Springfield, Ill., pp. 117–127.

25. Barnhart, M. L. (1971). In R. M. Nalbandian, Ed., *Molecular Aspects of Sickle Cell Hemoglobin: Clinical Applications*, Charles C Thomas Publishing Company, Springfield, Ill., pp. 45–116.

26. Cerami, A., and Manning, J. M. (1971). *Proc. Natl. Acad. Sci. (USA.)* **68**, 1180–1183.

27. Cerami, A., and Manning, J. M. (1971). *Fed. Proc. (USA.)* **30**, 1152A.

28. Kraus, L. M., and Kraus, A. P. (1971). *Biochem. Biophys. Res. Comm.* **44**, 1381–1387.

29. Diederich, D., Carreras, J., Trueworthy, R., Grisolia, S., and Lowman, J. T. (1971). *Blood* **38**, 795.

30. Carreras, J., Diederich, D., and Grisolia, S. (1972). *Eur. J. Biochem.* **27**, 103–108.

31. Carreras, J., Diederich, D., and Grisolia, S. (1971). *XIII Reunion Nacional de la Sociedad Espanola de Ciencias Fisiologicas* (Madrid), *Resumenes.*

32. Grisolia, S., and Hood, W. (1971). In E. Kun, S. Grisolia, Eds. *Biochemical Regulatory Mechanisms in Eukaryotic Cells*, Wiley-Interscience, New York, pp. 137–203.

33. Lubin, B., and Oski, F. A. (1972). *Blood* **40**, 930.

34. Lusher, J. M., and Barnhart, M. I. (1972). *Adv. Exp. Med. Biol.* **28**, 303–323.

35. Nalbandian, R. M. (1973). In M. Murayama, R. M. Nalbandian, Eds., *Sickle Cell Hemoglobin: Molecule to Man*, **00**, 153–187, Little, Brown and Company, Boston.

36. Cooperative Urea Trials Group (1974). *JAMA* **228**, 1120–1124.

37. Cooperative Urea Trials Group (1974). *JAMA* **228**, 1125–1128.

38. Nalbandian, R. M., and Henry, R. L. (1974). *JAMA* **229**, 1285.

39. Gillette, P. N., Manning, J. M., and Cerami, A. (1971). *Proc. Natl. Acad. Sci. (USA.)* **68**, 2791–2793.

40. Gillette, P. N., Manning, J. M., and Cerami, A. (1971). *Blood* **38**, 790.

41. Gillette, P. N., Peterson, Ch. M., Manning, J. M., and Cerami, A. (1972). *J. Clin. Invest.* **51**, 36A.

42. Gillette, P. N., Peterson, C. M., Manning, J. M., and Cerami, A. (1972). *Adv. Exp. Med. Biol.* **28**, 261–278.

43. Ranney, H. M. (1972). *N. Engl. J. Med.* **287**, 98–99.

44. Alter, B. P., Kan, Y. W., and Nathan, D. G. (1972). *Blood* **40**, 733–739.

45. Peterson, Ch. M., Lu, Y. S., Manning, J. M., Gillette, P. N., and Manning, J. M. (1973). *J. Clin. Invest.* **52**, 64A.

46. Gillette, P. N., Peterson, Ch. M., Lu, Y. S., and Cerami, A. (1974). *N. Engl. J. Med.* **290**, 654–660.

47. Charache, S. (1974). *N. Eng. J. Med.* **291**, 212.

48. Peterson, Ch. M., Tsairis, P., Ohnishi, A., Lu, Y. S., Brady, R., Cerami, A., and Dyck, P. J. (1974). *Ann. Int. Med.* **81**, 152–158.

49. Peterson, Ch. M., Lu, Y. S., Herbert, J. T., Cerami, A., and Gillette, P. N., (1974). *J. Pharmacol. Exp. Ther.* **189**, 577–584.

50. Gillette, P. N., Lu, Y. S., and Peterson, Ch. M., (1973). *Prog. Hemat.* **8**, 181–190.

51. Nigen, A. M., Peterson, Ch. M., Gillette, P. N., and Manning, J. M. (1974). *J. Lab. Clin. Med.* **83**, 139–146.

52. Diederich, D., Gill P., Trueworthy, R., and Larsen, W. (1975). In G. Brewer, Ed., *Erythrocyte Structure and Function*, Alan R. Liss, Inc., New York, p 379.

53. Diederich, D., Trueworthy, R., Gill, P., and Larsen, W. (1975), submitted for publication.

54. Kraus, L. M., Rasad, A., Friedman, B. I., Avis, K. E., Allen, C. M., and Kraus, A. P., (1972). *Blood* **40**, 928.

55. Kraus, L. M., Rasad, A., and Kraus, A. P. (1972). *Adv. Exp. Med. Biol.* **28**, 279–296.

56. Milner, P. F., and Charache, S. (1972). *Blood* **40**, 928.

57. Milner, P. F., and Charache, S. (1973). *J. Clin. Invest.* **52**, 3161–3171.

58. Bensinger, T. A., Maisels, M. J., Mahmoos, L., McCurdy, P. R., and Conrad, M. E. (1971). *N. Engl. J. Med.* **285**, 995–997.

59. Scott, R. B. (1971). *N. Engl. J. Med.* **285**, 1025–1026.

60. Nalbandian, R. M. (1972). *N. Engl. J. Med.* **286**, 378–379.

61. Segel, G. B., Feig, S. A., Mentzer, W. C., McCaffrey, R. P., Wells, R., Bunn, H. F., Shoeht, S. B., and Nathan, D. G. (1972). *N. Engl. J. Med.* **287**, 59–64.

62. Cerami, A. (1972). *N. Engl. J. Med.* **287**, 807–812.

63. Charache, S., and Milner, P. F. (1972). *N. Engl. J. Med.* **287**, 1357–1358.

64. Cerami, A. (1972). *N. Engl. J. Med.* **287**, 1358.

65. Nalbandian, R. M., Nichols, B. M., Stehouwer, E. J., and Camp, F. R., Jr., (1972). *Clin. Chem.* **18**, 961–964.

66. Jensen, M. C., Bunn, H. F., Halikas, G. V., and Nathan, D. G. (1972). *Adv. Exp. Med. Biol.* **28**, 297–302.

67. Manning, J. M., Cerami, A., Gillette, P. N., De Furia, F. G., and Miller, D. R. (1972). *Adv. Exp. Med. Biol.* **28**, 253–260.

68. Cerami, A., Manning, J. M., Gillette, P. N., De Furia, F. G., Miller, D. R., Graziano, J. H., and Peterson, C. (1973). *Fed. Proc. (USA.)*, **32**, 1668–1672.

69. Jensen, M., Bunn, H. F., Halikas, G., Kan, Y. W., and Nathan, D. G. (1973). *J. Clin. Invest.* **52**, 2542–2547.

70. Labbe, R. F. (1973). *J. Pharmacol. Sci.* **62**, 1727–1729.

71. Nalbandian, R. M. (1973). In M. Murayama and R. M. Nalbandian, Eds., *Sickle Cell Hemoglobin: Molecule to Man*, Little, Brown and Company, Boston, pp. 67–102.

72. Nalbandian, R. M. (1973). In M. Murayama and R. M. Nalbandian, Eds., *Sickle Cell Hemoglobin: Molecule to Man*, Little, Brown and Company, Boston, pp. 130–152.

73. Kilmartin, J. V., and Rossi-Bernardi, L. (1969). *Nature* **222**, 1243–1246.

74. Kilmartin, J. V., and Rossi-Bernardi, L. (1969). In R. E. Forster, J. T. Edsall, A. B. Otis, and F. J. W. Roughton, Eds. *Carbon Dioxide: Chemical, Biochemical and Physiology Aspects*, U.S. Government Printing Office (NASA no. SP-188), Washington, pp. 73–84.

75. Roughton, F. J. W. (1970). *Biochem. J.* **117**, 801–812.

76. Kilmartin, J. V., and Rossi-Bernardi, L. (1971). *Biochem. J.* **124**, 31–46.

77. Kilmartin, J. V., Fogg, J., Luzzana, M., and Rossi-Bernardi, L. (1973). *J. Biol. Chem.* **248**, 7039–7043.

78. De Furia, F. G., Miller, D. R., Cerami, A., and Manning, J. M., (1971). *Blood* **38**, 795.

79. Charache, S., and Diederich, D. (1971). *N. Engl. J. Med.* **285**, 1147.

80. Caldwell, P. R. B., Nagel, R. L., and Jaffe, E. R. (1971). *Biochem. Biophys. Res. Comm.* **44**, 1504–1509.

81. Diederich, D. (1972). *Biochem. Biophys. Res. Comm.* **46**, 1255–1261.

82. De Furia, F. G., Miller, D. R., Cerami, A., and Manning, J. M. (1972). *J. Clin. Invest.* **51**, 566–574.

83. May, A., Bellingham, A. J., and Huehns, E. R. (1972). *Lancet* **1**, 658–661.

84. Pacheco, J., Melvin, M. N., and Gabuzda, T. G. (1972). *Clin. Res.* **20**, 471A.

85. Pacheco, J., Melvin, M. N., and Gabuzda, T. G. (1972). *Blood* **40**, 929.

86. Lee, Ch. K., and Manning, J. M. (1973). *J. Biol. Chem.* **248**, 5861–5865.

87. Njikam, N., Jones, W. M., Nigen, A. M., Gillette, P. N., Williams, R. C., Jr., and Manning, J. M. (1973). *J. Biol. Chem.* **248**, 8052–8056.

88. Jensen, M., Nathan, D. G., and Bunn, H. F. (1973). *J. Biol. Chem.* **248**, 8057–8063.

89. Vedvick, T. S., Koenig, H. M. and Itano, H. A. (1974). *Proc. Soc. Exp. Biol. Med.* **147**, 255–258.

90. Stark, G. R., Stein, W. H., and Moore, S. (1960). *J. Biol. Chem.* **235**, 3177–3181.

91. Stark, G. R., and Smyth, D. G. (1963). *J. Biol. Chem.* **238**, 214–226.

92. Stark, G. R. (1964). *J. Biol. Chem.* **239**, 1411–1414.

93. Stark, G. R. (1965). *Biochem.* **4**, 588–595.

94. Stark, G. R. (1965). *Biochem.* **4**, 1030–1036.

95. Stark, G. R. (1965). *Biochem.* **4**, 2363–2367.

96. Stark, G. R. (1967). *Methods Enzymol.* **11**, 125–138.

97. Stark, G. R. (1967). *Methods Enzymol.* **11**, 590–594.

98. Smyth, D. G. (1967). *J. Biol. Chem.* **242**, 1579–1591.

99. Grisolia, S. (1968). *Biochem. Biophys. Res. Comm.* **32**, 56–59.

100. Veronese, F. M., Piszkiewicz, D., and Smith, E. L. (1972). *J. Biol. Chem.* **247**, 754–759.

101. Grisolia, S., Fernandez, M., Amelunxen, R., and Quijada, C. L. (1962). *Biochem. J.* **85**, 568–576.

102. Chabás, A., and Grisolia, S. (1972). *FEBS Letters* **21**, 25–28.

103. Huang, Ch., and Madsen, N. B. (1966). *Biochem.* **5**, 116–125.

104. Avramovic, O., and Madsen, N. B. (1968). *J. Biol. Chem.* **243**, 1656–1662.

105. Johnson, R. W., Roberson, L. E., and Kenney, F. T. (1973). *J. Biol. Chem.* **248**, 4521–4527.

106. Alonso, C., Caldes, T., and Gonzalez, P. (1973). *XI Jornadas Bioquimicas Latinas. Salamanca*, p. 160A

107. Caldes, T., Alonso, C., Gonzalez, P., and Santos-Ruiz, A. (1973). *XIV Reunión Nacional de la Sociedad Española de Ciencias Fisiológicas.* (Sevilla) *Resúmenes*, 29.

108. Bocchini, V., Alioto, M. R., and Najjar, V. A. (1967). *Biochem.* **6**, 3242–3249.

109. Ray, W. J., and Peck, E. J., Jr. (1972). In *The Enzymes*, P. D. Boyer, Ed., **6**, Academic, New York, pp. 408–477.

110. Anderson, P. M., Carlson, J. D., Rosenthal, G. A., and Meister, A. (1973). *Biochem. Biophys. Res. Comm.* **55**, 246–252.

111. Carey, M. J., and Butterworth, P. J. (1969). *Biochem. J.* **111**, 745–748.

112. Chang, K. Y. and Carr, Ch. W. (1972). *Biochim. Biophys. Acta* **285**, 377–382.

113. Sluyterman, L. A. (1967). *Biochim. Biophys. Acta* **139**, 439–449.

114. Hofstee, B. H. J. (1967). *Arch. Biochem. Biophys.* **122**, 574–582.

115. Hofstee, B. H. J. (1968). *Arch. Biochem. Biophys.* **125**, 1031–1034.

116. Hofstee, B. H. J. (1968). *J. Biol. Chem.* **243**, 6306–6311.

117. Bobb, D. and Hofstee, B. H. J. (1971). *Anal. Biochem.* **40**, 209–217.

118. Shaw, D. C., Stein, W. H., and Moore, S. (1964) *J. Biol. Chem.* **239**, PC 671–PC 673.

119. Robillard, G. T., Powers, J. C., and Wilcox, Ph. E. (1972). *Biochem.* **11**, 1773–1784.

120. Rimon, S., and Perlmann, G. E. (1968). *J. Biol. Chem.* **243**, 5366–5372.

121. Grizzuti, K., and Perlmann, G. E. (1969). *J. Biol. Chem.* **244**, 1764–1771.

122. Svendsen, I. (1967). *C.R. Trav. Lab. Carlsberg* **36**, 235–246.

123. Haines, J. A., and Zamecnik, P. C. (1967). *Biochim. Biophys. Acta* **146**, 227–238.

124. Keech, D. B., and Farrant, R. K. (1968). *Biochim. Biophys. Acta* **151**, 493–503.

125. Cole, R. D. (1961). *J. Biol. Chem.* **236**, 2670–2671.

126. Bischoff, F., and Bakhtiar, A. K. (1960). *Fed. Proc. (USA.)* **19**, 162A.

127. Grande, F., Grisolia, S., and Diederich, D. (1972). *Proc. Soc. Exp. Biol. Med.* **139**, 855–860.

128. Smyth, D. G. (1970). *Biochim. Biophys. Acta* **200**, 395–603.

129. Gidari, A. S., Cohen, M. H., and Levere, R. D. (1974). *Proc. Soc. Exp. Biol. Med.* **146**, 759–763.

130. Ramponi, G., and Grisolia, S. (1970). *Biochem. Biophys. Res. Comm.* **38**, 1056–1063.

131. Ramponi, G., Leaver, J. L., and Grisolia, S. (1971). *FEBS Letters* **16**, 311–314.

132. Manson, W. (1962). *Biochim. Biophys. Acta* **63**, 515–517.

133. Cejka, J., Vodrazka, Z., and Salak, J. (1968). *Biochim. Biophys. Acta* **154**, 589–591.

134. Cole, E. G., and Mecham, D. K. (1966). *Anal. Biochem.* **14**, 215–222.

135. Gerding, J. J. T., Koppers, A., Hagel, P., and Bloemendal, H. (1971). *Biochim. Biophys. Acta* **243**, 374–379.

136. Chen, C. C., Grossberg, A. L., and Pressman, D. (1962). *Biochem.* **1**, 1025–1030.

137. Weeke, B. (1968). *Scand. J. Clin. Lab. Invest.* **21**, 351–354.

138. Morris, I., and Syrett, P. J. (1963). *Biochim. Biophys. Acta* **77**, 649–650.

139. Binford, J. S., Lindskog, S., and Wadso, I. (1974). *Biochim. Biophys. Acta* **341**, 345–356.

140. Behlke, J., and Scheler, W. (1966). *Acta Biol. Med. Ger.* **12**, 629–643.

141. Hunninghake, D., and Grisolia, S. (1966). *Anal. Biochem.* **16**, 200–205.

142. Glader, B. E., and Conrad, M. E. (1972). *Nature* **237**, 336–338.

143. Alter, B. P., Kan, Y. W., and Nathan, D. G. (1972). *J. Clin. Invest.* **51**, 4*a*.

144. Alter, B. P., Kan, Y. W., and Nathan, D. G. (1974). *Blood* **43**, 57–68.

145. Habib, M., Watson, V., and Schwartz, E. (1973). *Blood* **41**, 635–639.

146. Freedman, M. L., Schiffman, F. J., and Geraghty, M. (1974). *Brit. J. Haematol.* **27**, 303–312.

147. Harkness, D. R., Roth, S., Goldman, P., and Kim, C. Y. (1974). *J. Lab. Clin. Med.* **83**, 577–583.

148. Durocher, J. R., Glader, B. E., and Conrad, M. E. (1973). *Proc. Soc. Exp. Biol. Med.* **144**, 249–251.

149. Durocher, J. R., Glader, B. E., Gaines, L. T., and Conrad, M. E. (1974). *Blood* **43**, 277–280.

150. Bell. W. R., and Charache, S. (1974). *J. Lab. Clin. Med.* **83**, 790–796.

151. Schutz, F. (1946). *J. Physiol. (London)* **105**, 17P–19P.

152. Birch, K. M., and Schutz, F. (1946). *Brit. J. Pharmacol.* **1**, 186–193.

153. Schutz, F. (1949). *Experientia (Basel)* **5**, 133–172.

154. Dicker, S. E. (1950). *Brit. J. Pharmacol Chemotherapy* **5**, 13–20.

155*a*. Crist, R., Grisolia, S., Bettis, C., and Diederich, A. (1972). *Fed. Proc. (USA.)* **31**, 231A.

155*b*. Crist, R. D., Grisolia, S. Bettis, C. J., and Grisolia J. (1973). *Eur. J. Biochem.* **32**, 109–116.

155*c*. Crist, R. D. (1973) Ph.D. Thesis, Department of Biochemistry and Molecular Biology, University of Kansas.

155*d*. Crist, R. D., and Puig, P. (1974). *Physiol. Chem. Physics* **6**, 371–374.

156. Wolk, M., Liebson, P., Beer, N., Cerami, A., and Killip, T. (1972). *J. Clin. Invest.* **51**, 105*a*.

157. Graziano, J. H., Allen, T. A., and Cerami, A. (1972). *Fed. Proc. (USA.)* **31**, 271A.

158*a*. Cerami, A., Allen, T. A., Graziano, J. H., De Furia, F. G., Manning, J. M., and Gillette, P. N. (1973). *J. Pharmacol. Exp. Ther.* **185**, 653–666.

158*b*. Graziano, J. H., Thornton, Y. S., Leong, J. K., and Cerami, A. (1973). *J. Pharmacol. Exp. Ther.* **185**, 667–675.

159. Harkness, D. R., Roth, S., Goldman, P., and Goldberg, M. (1972). *Adv. Exp. Med. Biol.* **28**, 615–630.

160. Toskes, Ph., Hildebrandt, P., Glader, B., Bensinger, T., Rickles, F., and Conrad, M. (1973). *J. Clin. Invest.* **52**, 85*a*.

161. Smith, R. P. (1973). *Proc. Soc. Exp. Biol. Med.* **142**, 1041–1044.

162. Papayannopoulou, T., Stamatoyannopoulos, G., Giblett, E. R., and Anderson, J. (1973). *Life Sci.* **12**, 127–133.

163. Papayannopoulou, T., Finch, C. A., Stamatoyannopoulos, G., and Hlastala, M. P. (1974). *J. Lab. Clin. Med.* **84**, 81–91.

164. Alter, B. P., Kan, Y. W., and Nathan, D. G. (1974). *Blood* **43**, 69–77.

165. Spector, L., Jones, M. E., and Lipmann, F. (1957). *Methods Enzymol.* **3**, 653–655.

166. Jones, M. E., and Lipmann, F. (1960). *Proc. Natl. Acad. Sci. (USA.)* **46**, 1194–1205.

167. Halmann, M., Lapidot, A., and Samuel, D. (1962). *J. Am. Chem. Soc.* **84**, 1944–1957.

168. Allen, Ch. M., Jr., and Jones, M. E. (1964). *Biochem.* **3**, 1238–1247.

169. Allen, Ch. M., Jr., Richelson, E., and Jones, M. E. (1966). In N. O. Kaplan and E. P. Kennedy, Eds., *Current Aspects of Biochemical Energetics*, Academic, New York, pp. 401–412.

170. Diederich, D., Ramponi, G., and Grisolia, S. (1971). *FEBS Letters*, **15**, 30–32.

171. Carreras, J., Chabiás, A., and Grisolia, S. (1971). *Biochim. Biophys. Acta* **250**, 456–457.

172. Carreras, J., Diederich, D., and Grisolia, S. (1971). *XIII Reunión Nacional de la Sociedad Española de Ciencias Fisiológicas*, (Madrid) *Resúmenes.*

173. Walker, J., and Hambly, F. J. (1895). *J. Chem. Soc. (London)* **67**, 746–767.

174. Warner, R. C. (1942). *J. Biol. Chem.* **142**, 705–723.

175. Dirnhuber, P., and Schutz, F. (1948). *Biochem. J.* **42**, 628–632.

176. Hagel, P., Gerding, J. J. T., Fieggen, W., and Bloemendal, H. (1971). *Biochim. Biophys. Acta* **243**, 366–373.

177. Harris, J. W. (1959). *Progr. Hemat.* **2**, 47–109.

178. Harris, J. W., and Kellermeyer, R. W. (1970). In *The Red Cell Production, Metabolism, Destruction: Normal and Abnormal*, Harvard University Press, Cambridge.

179. Murayama, M. (1962). *Nature* **194**, 933–934.

180. Murayama, M. (1964). *Nature* **202**, 258–260.

181. Murayama, M. (1966). *Science* **153**, 145–149.

182. Murayama, M. (1966). *J. Cell. Physiol.* **67**, (Supp. 1) 21–32.

183. Murayama, M. (1967). *Clin. Chem.* **14**, 578–588.

184. Murayama, M. (1973). In M. Murayama and R. M. Nalbandian, Eds., *Sickle Cell Hemoglobin: Molecule to Man*, Little, Brown and Company, Boston, pp. 22–66.

185. Murayama, M. (1973). In M. Murayama and R. M. Nalbandian, Eds., *Sickle Cell Hemoglobin: Molecule to Man*, Little, Brown and Company, Boston, pp. 47–52.

186. Perutz, M. F., and Lehmann, H. (1968). *Nature* **219**, 902–909.

187. Perutz, M. F. (1971). *New Scientist* **50**, 762–765.

188. Magdoff-Fairchild, B., Swerdlow, P. H., and Bertles, J. F. (1972). *Nature* **229**, 217–218.

189. Perutz, M. F. (1970). *Nature* **228**, 726–739.

190. Perutz, M. F. (1972). *Nature* **237**, 495–499.

191. Kilmartin, J. V., and Rossi-Bernardi, L. (1973). *Physiol. Rev.* **53**, 836–890.

192. Benesch, R. E., and Benesch, R. (1974). *Adv. Protein Chem.* **27**, 211–237.

193. Perutz, M. F., Muirhead, H., Mazzarella, L., Crowther, R. A., Greer, J., and Kilmartin, J. V. (1969). *Nature* **222**, 1240–1243.

194. Bunn, H. F., and Briehl, R. W. (1970). *J. Clin. Invest.* **49**, 1088–1095.

195. Arnone, A. (1972). *Nature* **237**, 146–148.

196. Kitajima, M., Sekiguchi, W., and Kondo, A. (1971). *Bull. Chem. Soc. Japan* **44**, 139–143.

197. Jensen, W. N., Bromberg, P. A., and Barefield, K. (1969). *Clin. Res.* **17**, 464A.

198. Jensen, M. C., Shohet, S. B., and Nathan, D. G. (1972). *Clin. Res.* **20**, 491A.
199. Week, R. I., La Celle, P. L., and Merrill, E. W. (1969). *J. Clin. Invest.* **48**, 795–809.
200. Shohet, S. B. (1972). *N. Engl. J. Med.* **286**, 577–583.
201. Ponder, E., and Ponder, R. (1954). *Acta Haemat.* **12**, 282–290.
202. Allison, A. C. (1957). *Biochem. J.* **65**, 212–219.
203. Bromberg, P. A., and Jensen, W. N. (1965) *J. Clin. Invest.* **44**, 1031A.
204. Bruning, W., and Hottzer, A. (1961) *J. Am. Chem. Soc.* **83**, 4865–4866.
205. Whitney, P. L., and Tanford, C. (1962). *J. Biol. Chem.* **237**, 1735–1737.
206. Mukerjee, P., and Ray, A. (1963). *J. Phys. Chem.* **67**, 190–192.

Discussion

Unfortunately, the recording equipment failed during a break preceding the final discussion of the afternoon session. Therefore, the discussion and remarks presented here are incomplete.

Dr. Walser: With regard to the role so-called nonionic diffusion in determining the ammonia gradient between cells and extracellular fluid, I would like to point out that the traditional concept that we teach medical students, namely, that impermeability to the ionized species is a requirement for this process, is in fact incorrect. When a hydrogen ion concentration gradient is present across a membrane, either the hydrogen ion is in electrochemical equilibrium, or it is not, owing to some active transport process. In the former case, the equilibrium ratio for the ammonium ion is the same as that for the hydrogen ion, and it makes no difference whether the membrane is permeable to ammonia or ammonium ions. In the latter case, ammonia will tend to diffuse in and ammonium ions out, partially short-circuiting the pump. The steady-state distribution ratio will depend on the balance of these forces.

Dr. Brosnan: I agree.

Dr. Sainsbury participated also in the discussion on membrane potential of liver. However, her comments are not available.

Dr. Cohen said that it was his impression that glutamine is a precursor of glucose and urea in the liver.

Dr. Lund: This is true at concentrations of 5 to 10 m*M* glutamine.

Dr. Grisolia: Have you considered that acetylglutamate may be required for activation of glutaminase?

Dr. Lund: No. The isoenzyme activated by acetylglutamate is the phosphate-independent glutaminase, and the activation is only of the order of fourfold at 10 m*M* acetylglutamate.

Dr. Chappell made a comment regarding the differences between Dr. Krebs' group and others in relation to glutamine synthesis by isolated liver cells.

Dr. Lund mentioned that joint experiments between the Oxford and Bristol groups should resolve the argument.

Dr. Grisolia: We shall start now the round table discussion on alternate pathways of urea formation and NH_3 use and on future developments.

Discussion

Alternate Pathways of Urea Formation and of NH_3 Use and Future Advances and Predictions

Dr. Cedrangolo: It is well established that ammonia can be converted into urea nitrogen by the liver; the enzymology of the ornithine cycle indicates that one-half of urea nitrogen originates from ammonia via carbamyl-phosphate but the second half may originate either from free ammonia via glutamate dehydrogenase or via transamination through ketoglutarate and oxaloacetate (1). In the present contribution the generally accepted role of free ammonia as an obligatory intermediate between amino nitrogen of proteins and urea will be questioned.

In the past two decades several lines of evidences have been collected in our laboratory which throw doubts on the physiological role of free ammonia along this biochemical pathway (2). Figure 1 shows a simple scheme which summarizes three alternatives related to this problem. Pathway 1 involves ammonia as an obligatory intermediate, pathway 2 implies a facultative role of ammonia, and pathway 3 does not imply any participation of the compound. A number of results which favor pathway 3 will be briefly discussed in connection with evidences reported by other authors.

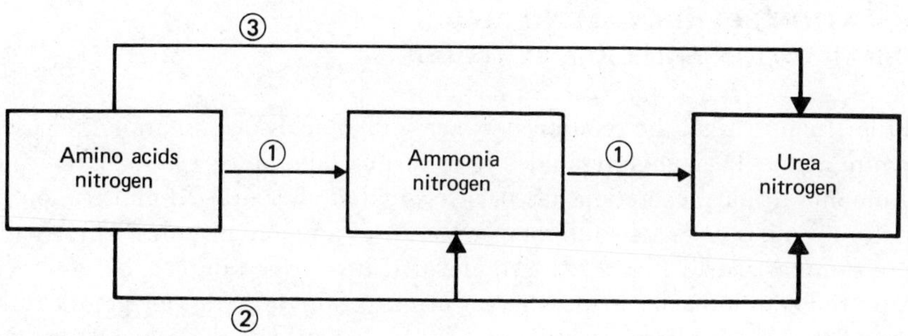

Figure 1. Alternative pathways related to the intermediary role of ammonia in the conversion of protein amino nitrogen into urea.

AMMONIA TOXICITY

Since 1954 (3, 4) we were impressed by the relatively elevated toxicity of ammonia if compared with the amount of urea synthesized by ureotelic organisms, particularly when they are fed with a protein-rich diet. In other words it was difficult to consider such toxic molecule as a "physiological" intermediate in a quantitatively relevant pathway such as urea biosynthesis (4).

In order to verify our hypothesis, the following simple experiment was devised. The urea excreted by rats fed with a protein-rich diet over a given period of time was compared with the amount of urea excreted by controls on a normal diet. The ammonia corresponding in equimolar terms to the difference of urea excreted between the two groups of rats, or even to one-half of that, was slowly administered by intravenous infusion for the same interval (generally 1 hour). The rationale behind the experiment is that ammonia injected should not result in any disturbance, since it is equal to the amount that should be formed within the same interval of time when the animal is fed with a protein-rich meal. On the contrary all animals showed a convulsive symptomatology, typical of ammonia intoxication, followed by death (4).

We are aware of the many hindrances preventing an unequivocal interpretation of this result. In fact, ammonia was injected intravenously, whereas amino acids from dietary proteins reach the liver through the portal system; therefore during ammonia infusion, the brain would be reached before liver detoxication. Furthermore, proteins administered to rats contain several amino acids, such as arginine, which display a significant protective effect toward ammonia intoxication; this effect cannot be exploited in the case of a simple ammonia infusion.

RELATION BETWEEN AMINO ACID DEAMINATION AND UREA SYNTHESIS

From 1956 to 1958 the relation between amino acids deamination and urea synthesis has been investigated *in vivo* in our laboratory (5–7). Urea and ammonia urinary excretion has been evaluated after oral administration of natural amino acids or equimolar amounts of ammonium salts. The results are summarized in Fig. 2 (2). An elevated urea excretion can be observed after the L-amino acids load, whereas urinary ammonia was not significantly modified, except after administration of L-alanine. On the other hand, the intake of ammonium salts resulted in increased ammonia excretion, whereas urea formation was unmodified. Only the administration of ammonium carbonate or acetate resulted in a moderate increase of urea excretion, in any case lower than that observed after administration of L-amino acids. Therefore, under our experimental conditions, ammonium salts were not significantly converted into urea, but the amino nitrogen of L-amino acids was rapidly transformed into urea.

Along the same line of reasoning, another set of experiments has been devised. L- or D-amino acids were orally administered to rats, and urea

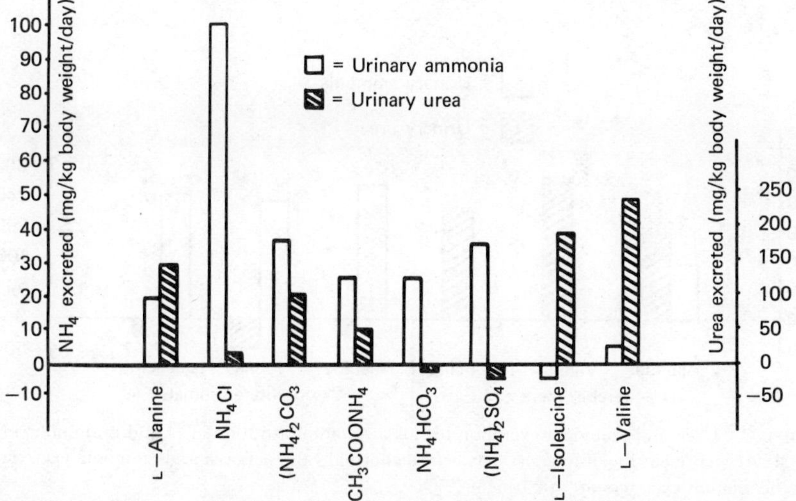

Figure 2. Urea and ammonia excretion in rats after administration of L-amino acids or ammonium salts. The figures represent the differences between the normal values of urea and ammonia excretion and the amount excreted after treatment (2).

excretion was again compared with ammonia excretion (5–7). Since it is well known that D-amino acids are effectively deaminated in rat tissues by D-amino acid oxidase, their administration should affect urea formation more than the corresponding L-isomer, if deamination is an obligatory step for urea formation. The data reported in Fig. 3 show, on the contrary, that administration of D-amino acids results in an increased ammonia excretion, not paralleled by any urea formation, whereas a load with equimolar amount of the corresponding L-isomers causes mainly an increased production of urea (2).

These results suggest the independence of amino acid deamination from the reactions leading to urea biosynthesis, since compounds which generate ammonia, such as D-amino acids, or ammonium salts, are unable to activate urea synthesis, whereas L-amino acids, which are deaminated only at a very low rate, contribute *in vivo* to a larger extent to urea formation.

PHYSIOLOGICAL SIGNIFICANCE OF L-AMINO ACID DEAMINATION

The role of ammonia as intermediate in urea biosynthesis implies its formation from natural amino acids through direct deamination or transdeamination (8, 9). The biological significance and relevance of this process has been evaluated in a series of experiments (10) by following ammonia production from L-amino acids in homogenates from several mammalian tissues; sodium

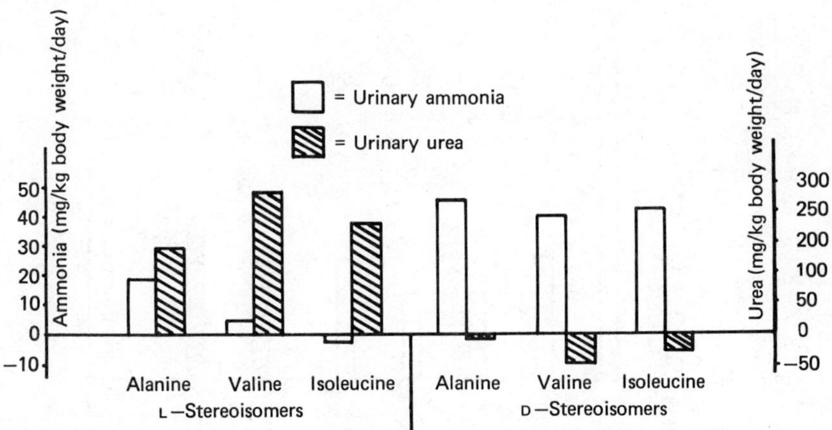

Figure 3. Urea and ammonia excretion in rats after administration of L- and D-amino acids. The figures represent the differences between the normal values of urea and ammonia excretion and the amount excreted after treatment (2).

arsenite was added to the incubation mixtures to minimize keto acid oxidation as well as to prevent ammonia incorporation into glutamine. The concentration of the amino acids employed was two orders of magnitude higher than the physiological ones.

Table 1 shows the deamination of the most common L-amino acids in presence of ox retina and various rat tissue homogenates. Among the amino acids substrates of the "general" L-amino acid oxidase (11) only L-leucine, L-phenylalanine, and, to a lesser extent, L-methionine are significantly deaminated by kidney. In rat liver, muscle, brain, and erythrocytes none of these amino acids is deaminated. It is interesting to note in this respect that histidine, threonine, serine, and cysteine are actively deaminated in rat liver.

The general observation that can be drawn from the data in Table 1 is that deamination is limited only to few amino acids and that it occurs at a very low rate. The results seem to exclude that the "general" L-amino acid oxidase is operative in the tissues investigated. The few amino acids most actively deaminated in liver, that is, histidine, cysteine, threonine, and serine, are substrates of specific nonoxidative enzymes.

In addition, we were unable to observe under our experimental conditions any significant deamination of L-glutamic acid (see Table 1). This result does not support the widely accepted role of glutamate dehydrogenase as the most active enzyme in ammonia production from amino acids in mammalian tissues (8). Furthermore, other experiments performed in our laboratory since 1941 (12) have demonstrated the lack of any enhancing role of α-ketoglutarate on the deamination of several L-amino acids.

Table 1 Ammonia Production from Natural Amino Acids in Various Mammalian Tissues[a]

Amino acids	Ammonia formed (μg/100 mg wet tissue/hour)					
	Kidney	Liver	Skeletal muscle	Brain	Erythrocytes	Retina
Glycine	+0.2	+3.9 ± 0.3	+1.0	+1.2	+0.9	+0.4
Alanine	−0.5	+1.2	−0.1	−0.5	0.0	−1.1
Valine	+1.6	−0.2	−0.1	+2.2	+0.2	+2.4
γ-Aminobutyric acid	−1.4	+3.4	+1.0	+2.1	0.0	−1.0
Leucine	+6.9 ± 2.1	−0.2	−0.1	+1.6	+0.4	−1.1
Isoleucine	+0.3	+0.4	−1.9	+1.9	+0.5	−0.3
Aspartic acid	+1.0	+1.0	0.0	+3.1	+0.3	−1.3
Glutamic acid	+0.7	+0.1	−2.1	−3.4	+0.3	−0.4
Ornithine	−2.3	−4.0 ± 0.3	−1.3	+2.6	+0.2	−5.1 ± 0.6
Lysine	+4.4 ± 0.5	−1.9	+0.3	+1.3	+0.4	−3.2
Arginine	−0.6	−4.4 ± 0.4	−1.4	+2.0	+0.4	−5.9 ± 0.4
Citrulline	+0.5	+4.8 ± 0.9	+0.4	−2.0	+2.6	+5.5 ± 0.4
Serine	+1.2	+30.7 ± 3.2	−0.6	+5.1 ± 0.2	+1.7	−0.7
Threonine	−0.1	+6.5 ± 1.4	−0.5	+1.3	0.0	+7.2 ± 2.4
Cysteine	−0.2	+15.2 ± 0.8	+0.8	+0.8	+3.8 ± 0.7	−1.3
Methionine	+3.5 ± 0.3	−0.7	−1.9	−2.4	+2.1	−0.3
Proline	−0.9	−0.7	+0.8	+1.4	+1.2	+2.9
Histidine	−1.4	+54.9 ± 4.1	0.0	0.0	−0.6	−2.3
Tryptophan	−0.2	−1.6	+0.6	+1.6	+1.3	+2.9
Phenylalanine	+4.3 ± 0.4	−0.5	+0.5	−1.9	−0.6	—

[a] For experimental details see (10).

Figure 4 reports the deamination of L-leucine in liver and kidney homogenates at concentrations ranging from 5 to 150 times the physiological level. Appreciable ammonia formation was observed only in kidney and at high concentrations of substrate, far from the physiological (10). In liver no significant deaminating activity is detectable in the whole range of substrate concentrations. To favor a possible transdeamination, the effect of the addition of α-ketoglutarate has been investigated, at concentrations one-tenth of the substrate (13). No positive effect on the rate of leucine deamination in the two tissues is observable.

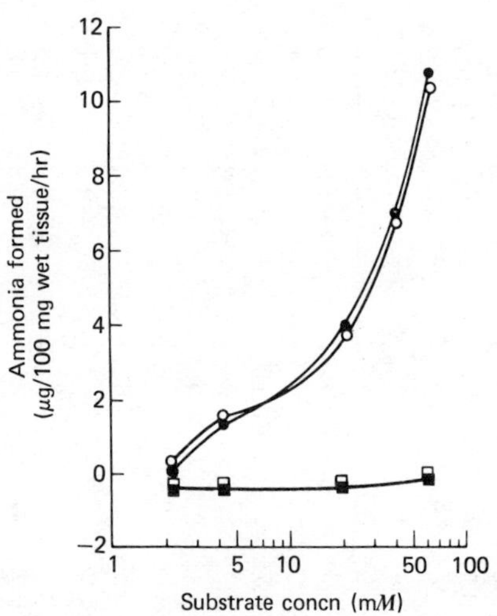

Figure 4. Deamination of L-leucine in rat kidney and in rat liver homogenates in presence or absence of α-ketoglutarate. Circles refer to the ammonia production in kidney in presence (●—●) or absence (○—○) of ketoglutarate; squares refer to ammonia production in liver in presence (■—■) or absence (□—□) of ketoglutarate (10, modified).

It is interesting to mention in this respect that several authors (14–17) have pointed out the role of glutamic dehydrogenase in the reductive amination of ketoglutarate, rather than in the oxidative deamination of glutamate essentially because of the equilibrium constant of the reaction, which favors the synthesis of glutamate, and because of the ratio of concentrations of NADH versus NAD, which is rather high within the mitochondria. Recently Chappell, in an interesting review on the metabolic functions of glutamic dehydrogenase (18), also concludes that the role of the enzyme is in the reductive amination of the keto acid. One of the main observations reported by Chappell (18) is that urea synthesis is lowered when the activity of glutamate dehydrogenase is specifically enhanced by an activator. Since under no circumstances the activation of an enzyme can lead to the inhibition of the pathway in which it is involved, glutamate dehydrogenase seems not to be related with the pathway from amino nitrogen of amino acids to urea. Therefore, all this experimentation indicates that the generally accepted "dogma" of transdeamination as an effective pathway in ammonia formation from L-amino acids has to be revised.

From the results reported in this paragraph we may infer that a common catabolic pattern for amino group detachment from L-amino acids is not operative under physiological conditions. Neither the action of a "general" L-amino acid oxidase nor the transdeamination pathway can be taken as a general pattern for amino group detachment. However, it cannot be excluded that single enzyme activities, specific toward few amino acids, are operative in mammalian tissues.

As far as relationship between deamination and ureogenesis is concerned, the following two facts also seem very pertinent:

1. No large difference in the total extent of ammonia production from several L-aminoacids is observable between elasmobranchs (ureotelic) and teleosts (ammoniotelic) despite their different catabolic end products (19). This means that the detachment of amino groups from aminoacids does not play any significant role regarding the nature of protein nitrogen end products;
2. In liver and also in kidney and brain homogenates, at different incubation times, ammonia production in animals with high protein intake never exceeds that of other two groups of rats (respectively rats fasted and rats fed with normal diet). On the contrary, a decreased NH_3 production is observable in this group of animals if compared to the others (20). These results also seem to suggest a mechanism of urea synthesis without the participation of "free" ammonia.

EFFECT OF PROTEIN INTAKE ON AMINO ACID DEAMINATION

Schimke and coworkers (21) in their classical study on the effect of protein intake in rats have demonstrated that the liver adapts very rapidly to a protein-rich diet by increasing the levels of urea-cycle enzymes. The same kind of adaptation has been reported in primates and men (22).

Following this line, experiments have been recently performed in our laboratory to evaluate the relationship between protein intake and amino acid deamination measured in rat liver homogenates (23, 24). The results are reported in Table 2. The extent of deamination has been compared in two groups of animals: control group (I), fed with normal diet, and group II, fed with high-protein diet. The data in Table 2 show that L-threonine and

L-histidine deamination was actively increased in group II but deamination of all other amino acids tested, including L-glutamate, did not undergo any significant change.

Table 2 Effect of Protein Intake on Ammonia Production from L-Amino Acids in Rat Liver Homogenates[a]

	Group I (controls)	Group II (high protein intake)
Amino acids	Ammonia formed (μmoles/sample/hour)	
Glycine	$+0.43 \pm 0.09$[b]	$+0.53 \pm 0.10$
L-Leucine	$+0.28 \pm 0.08$	$+0.18 \pm 0.09$
L-Arginine	$+0.25 \pm 0.11$	-0.48 ± 0.07
L-Citrulline	$+0.83 \pm 0.20$	$+0.37 \pm 0.10$
L-Threonine	$+0.56 \pm 0.08$	$+5.03 \pm 0.80$
L-Cysteine	$+0.91 \pm 0.08$	$+0.63 \pm 0.07$
L-Histidine	$+1.77 \pm 0.09$	$+13.47 \pm 1.40$
L-Valine	$+0.14 \pm 0.06$	$+0.18 \pm 0.15$
L-Glutamic acid[c]	$+0.27 \pm 0.09$	$+0.26 \pm 0.09$

[a] For experimental details see (23, 24).
[b] Standard deviation.
[c] Unpublished data by F. Cedrangolo, G. Illiano, and L. Servillo.

The reported results suggest that the "general" L-amino acid oxidase, as well as glutamic acid dehydrogenase, is not induced by the increased protein intake and indirectly confirm that these enzymes are not related to the pathway of urea biosynthesis.

Among the amino acids tested, only L-threonine and L-histidine deamination seems to be induced by the increased protein intake. Deamination of threonine in liver homogenates could be catalyzed by threonine dehydrase (E.C. 4.2.1.16), but ammonia can be formed from histidine via urocanic acid pathway. Unpublished data (25) from our laboratory demonstrate that the increased formation of ammonia from L-histidine in rats of group II is paralleled by an increased production of urocanic acid, thus suggesting an induction of histidase (E.C. 4.3.1.3). The physiological significance of ammonia formation from histidine and threonine and the mechanism of induction of the related enzymes are not clear and deserve further investigation.

CONCLUSIONS

In this short review we have summarized several lines of evidence suggesting the independence of the deamination of natural amino acids with respect to urea biosynthesis. The elevated ammonia toxicity, the lack of efficiency of urea biosynthesis under conditions where ammonia formation is artificially promoted or after administration of ammonium salts, the low rate of amino acid deamination *in vitro* under physiological conditions, as well as the lack of induction of the relative enzymes under conditions of adaptative increase of the urea-cycle enzymes, are all evidence which do not support the intermediary role of ammonia along the pathway from amino nitrogen to urea.

ACKNOWLEDGMENT

The author wishes to thank Professors V. Zappia and F. Salvatore for their contribution in the revision of the manuscript.

REFERENCES

1. Ratner, S. (1973). *Advan. Enzymol.* **39**, 1–90.

2. Cedrangolo, F. (1967). *Biochim. Appl.* **14**, 117–150.

3. Cedrangolo, F. (1954). "Giornate Biochimiche Italo-Franco-Elvetiche." Abstracts of the Congress (Ed. C.N.R.) p. 252; Cedrangolo, F., and Piazza, R. (1954). *Bull. Soc. It. Biol. Sperim.* **30**, 1291–1294.

4. Cedrangolo, F., and Piazza, R. (1956). *Enzymol.* **17**, 363–370.

5. Cedrangolo, F., Salvatore, F., and Piazza, R. (1957). *Bull. Soc. Chim. Biol.* **39**, Suppl. III, 135–151.

6. Salvatore, F., and Saccone, C. (1957). *Ital. J. Biochem.* **6**, 103–114.

7. Cedrangolo, F., and Salvatore, F. (1958). *Bull. Soc. Chim. Biol.* **40**, 1849–1857.

8. Braunstein, A. E. (1957). *Adv. Enzymol.* **19**, 335–389.

9. Cedrangolo, F. (1958). *Enzymol.* **19**, 335–374.

10. Salvatore, F., Zappia, V., and Cortese, R. (1966). *Enzmol.* **31**, 113–127.

11. Blanchard, M., Green, D. E., Nocito, V., and Ratner, S. (1944). *J. Biol. Chem.* **155**, 421–440.

12. Cedrangolo, F., and Carandante, G. (1942) *Arch. Sci. Biol.* **28**, 1–12.

13. Still, J. L., Buell, M. V., and Green, D. E. (1950). *Arch. Biochem.* **26**, 413–419.

14. Krebs, H. A., and Bellamy, D. (1960). *Biochem. J.* **75**, 523–529.

15. Borst, P. (1962). *Biochim. Biophys. Acta* **57**, 256–269.

16. Chappell, J. B. (1964). *Biochem. J.* **90**, 237–248.

17. Tager, J. M., Howland, J. L., Slater, E. C., and Snoswell, A. M. (1963). *Biochim. Biophys. Acta* **77**, 266–275.

18. McGivan, J. D., and Chappell, J. B. (1975). *FEBS Lett.* **52**, 1–7.

19. Salvatore, F., Zappia, V., and Costa, C. (1965). *Comp. Biochem. Physiol.* **16**, 303–309; Cutinelli, L., Pietropaolo, G., Venuta, S., Zappia, V., and Salvatore, F. (1972). *Comp. Biochem. Physiol.* **41B**, 905.

20. Cedrangolo, F., Galletti, P., and Federico, A. (1972). *Rend. Acc. Naz. Lincei* **52**, 207–213; see also Cedrangolo, F., Illiano, G., and Cortese, R. (1971). *Rend. Acc. Naz. Lincei* **50**, 37–39.

21. Schimke, R. T. (1962). *J. Biol. Chem.* **237**, 459–468.

22. Nuzum, C. T., and Snodgrass, P. J. (1971). *Science* **172**, 1042–1043.

23. Cedrangolo, F., Zappia, V., Galletti, P., and Oliva, A. (1973). *Abstracts of the Ninth International Congress of Biochemistry*, (Abstract), Stockholm, 1–7 July, p. 80.

24. Cedrangolo, F., Zappia, V., Galletti, P., and Oliva, A. (1974). *Rend. Acc. Naz. Lincei* **56**, 385–388.

25. Cedrangolo, F., Illiano, G., and Servillo, L., unpublished data.

Discussion

Dr. Cedrangolo: At this point it is evident that you would like to know whether our data allow us to suggest a ureogenesis pattern alternative to the Krebs cycle. Unfortunately, my answer is no. Actually, our data are controversial; in a sense they dispute instead of assert. My feeling on the problem is illustrated by the scheme in Fig. 5.

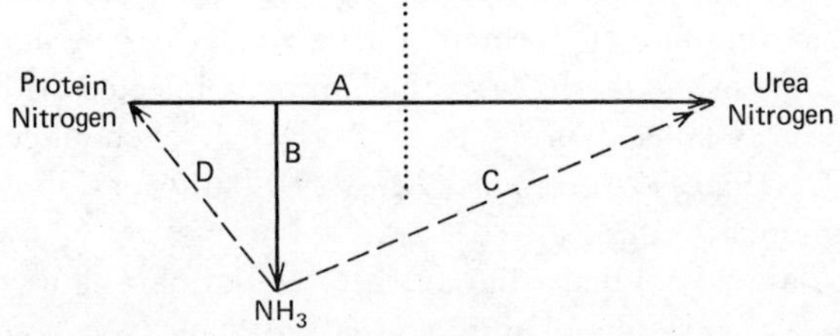

Figure 5. Proposed scheme for the interrelation between protein nitrogen, NH_3, and urea nitrogen.

I think that there is a main pathway (*A*) which converts amino nitrogen to urea without going through free ammonia. At the moment, we are unable to say how this metabolic pathway works. As far as the urea second nitrogen atom is concerned, the Ratner-Pappas reaction could be implicated. But I must add that once in our lab we got results against the exclusive physiological role of this reaction (Cedrangolo, F., Della Pietra, G., Cittadini, D., Papa, S., and De Lorenzo, F.—1962, *Nature*, **195**, 708); a recent report (Bruce Rowe, W., and Miller, L.L—1971, *Proc. Soc. Exp. Biol. Med.*, **138**, 357) supports our view, I believe, showing in perfused rat liver no urea synthesis inhibition by alpha- or betamethyl-aspartate. I must also admit that we couldn't reproduce our results thereafter, but I do not think that they were wrong or inaccurate. As far as the first nitrogen atom of urea is concerned, our data are against general opinion, which considers ammonia its precursor. A hypothetical pathway could consist of the transferring of the L-amino acid amino group to a carrier which could in turn deliver it as ammonia to an accepting CP-synthetase. We suggest that, in this transfer, a kind of "active ammonia" is involved, which, because of its very short half-life, cannot be detected experimentally.

If a biochemical lesion blocks pathway *A*, then the amino nitrogen should appear and be excreted as free ammonia. This point of view, I believe, is based on experimental grounds. For instance, the knowledge that some metabolic disorders (severe degree of hyperammonemia) are due to deficiency of some enzymes of the urea cycle (Cedrangolo, F.—1967, *Bioch. Appl.*, **14**, 117). Related to this problem are results from some of my old experiments (Cedrangolo, F., 1943–45, *Enzymologia*, **11**, 1) showing that deamination and transamination are two distinct and alternative processes; transamination is preferential, and only blocking it through the absence of ketoglutarate allows deamination to take place. Pathway *B* is therefore considered by us as a secondary, usually minor, route which becomes quantitatively important only when pathway *A* is blocked. Ammonia then appears to be at the last step of a secondary pathway and not as an obligatory intermediate of the principal ureogenetic route. We have also indicated another pathway, *D*, which converts ammonia to protein nitrogen. This is in agreement with Schonheimer's experiments showing that animals injected with ^{15}N-labeled ammonia have ^{15}N-labeled proteins (Foster, G. L., Schonheimer, R., and Rittenberg, D.—1939, *J. Biol. Chem.*, **127**, 319). Pathway *C* shows direct conversion of ammonia to urea.

In fact, our data do not deny that at least a minor part of urea derives from free ammonia. The injection of ammonia salts in some cases results in a small increase in urea excretion, however, lower than after L-amino acid administration. I would like to mention also the classical report (Duda, G., and Handler, P.—1958, *J. Biol. Chem.*, **232**, 303) and the experiments carried out in our lab (Balestrieri, C., DeCristofaro, D., and Cittadini, D.—1967, *Life Sci.*, **6**, 337) showing that administration of ^{15}N-labeled ammonia is followed by excretion of ^{15}N-labeled urea. It is possible that the main purpose of pathway *C* is arginine biosynthesis, since it appears to work well only in the presence of ornithine and of aspartate in very high concentrations, at least equivalent to that of ammonia (Salvatore, F. and Bocchini, V.—1961, *Nature*, **191**, 705; Bocchini, V. and Salvatore, F.—*Giorn. Biochim.*, **10**, 493). It is clear, at this point, that the scheme is not conclusive; it points out that the understanding of ureogenesis still requires experimental work. This is usual in science, where the solution of one problem presents us with the riddle of many others. Fermi used to say, "A stone that dropped in the water opens circles over and over again." Thank you!

Dr. Grisolia: I believe that many of the questions which were expected to come up during this Round Table discussion, including some of those raised by the paper and comments of Dr. Cedrangolo, have already been asked during the last few days. It seems apparent that we are making a return, as predicted years ago by many, including myself ("Metabolic Regulation and Enzyme Action," *FEBS Symposium*, **19**, 245, A. Sols and S. Grisolia, Eds. Academic Press, New York, 1970), to the very difficult but pressing problem in biochemistry of translating the events we observe *in vitro* to those that occur *in vivo*; and again Sir Hans's laboratory has been a pioneer in this regard. It is evident that the basic information acquired at the enzyme level in the last 25 years, which has been reviewed in this meeting, has provided a firm basis for the application to disease and disease states, although more work along these lines is needed. I feel that it is, therefore, obvious that the areas of intracellular concentrations and control, in essence nutrition at the molecular level, will receive (and deserve) increased attention both from the basic and practical viewpoints and may lead to a renewed interest in nutrition (Grisolia, S., and Hood, W., in "*Biochemical Regulatory Mechanisms in Eukaryotic Cells*," edited by E. Kun and S. Grisolia, Wiley-Interscience, New York, 1972, 138–202). It is also evident that the control and regulation of pools at different sites within the ureotelic liver cell provides a unique model but that as such it exemplifies the caution needed to refrain from generalizing to other systems; for example, this may be a unique tissue in which the PCO_2 is lower than anywhere else. I would now like to open the meeting for general discussion. Are there any questions?

Dr. Cohen: Dr. Crist, your extended discussion of the use and misuse of cyanate in the treatment of sickle cell anemia under the title "Use and Misuse of Urea-Cycle Intermediates in Medicine" suggests that you are resurrecting Wohler's concept of urea biosynthesis. I doubt that you intended that cyanate be considered an intermediate in urea biosynthesis. If so, what was the purpose of your presentation on the therapeutic use of cyanate in a symposium dealing with urea biosynthesis?

Dr. Grisolia: I should like to answer this question. The title of the Symposium includes Physiopathology. Urea, as such, has been used extensively, particularly by surgeons, in efforts to control brain edema, and then its use in sickle cell anemia led to the "rediscovery", mostly by clinicians, that cyanate and urea are in equilibrium and to the extensive clinical use of cyanate.

I should not need to remind this audience, and certainly not Dr. Cohen, who, as you may remember, has been very vocal in this issue, about the moral obligations of biochemists, particularly in relation to clinical guidance and participation in clinical work. Unfortunately, and I hope Dr. Cohen is aware, as well as many other scientists, this has not been the case seemingly for the use of urea and particularly cyanate clinically. There is evidence of untoward clinical manifestations which could have been prevented by more cautious and extensive animal experimentation.

I should like to make a few more brief comments to clear up this and related issues, for we have so little time left.

Of course Sir Hans tested and excluded cyanate as a possible intermediate in the urea cycle long ago, as we have ourselves excluded it as an intermediate with ornithine transcarbamylase (Carreras, J., Chabás, A., and Grisolia, S., *Biochim. Biophys. Acta*, **250**, 456–457, 1971) and CPase (Diederich, D., Ramponi, G., and Grisolia, S., *FEBS Letters*, **15**, 30–32, 1971). Carbamyl glutamic acid has really been used for quite some time in Spain, and recently there has been some interest in its use in the United States, due particularly to experiments in Cohen's laboratory by Paik (Kim, S., Paik, W. K., and Cohen, P. P., *Proc. Nat. Acad. Sci.*, **69**, 12, 1974). The clinical use of carbamyl glutamate, as well as arginine, was first illustrated by Brown and others (Brown, R., Manning, R., Delp, M., and Grisolia, S., *The Lancet*, **1**, 591, 1958) over 15 years ago, and it is my belief that, based on this report, many laboratories in Spain and in other European areas have developed commercial preparations. Other questions or comments?

Dr. Tatibana: I would like to ask a question to Dr. Crist. I understand that your therapy with cyanate on patients with sickle cell anemia could increase the life span of the erythrocytes as assayed with the use of DIPF (diisopropylphosphofluoridate, ^{32}P-labeled). Could you please tell me whether your values for the life span represent the real longevity of the cells with normal values being 120 days or the values show the half-time of disappearance of the label, due to DIPF bound to the cells, from the blood stream. I would like to know the extent of the therapeutic effect as represented by the improved life span of erythrocytes in comparison with that of normal cells.

Dr. Crist: Thank you, Dr. Tatibana, for your question concerning red cell survival. The survivals were done with the diisopropylphosphate labeling, using two isotopes with untreated and subsequently with cyanate treated cells. The mean survival time was increased with carbamylation but never approached 60 days, half of normal survival time.

Closing Remarks

Dr. Grisolia: And I now come to the most pleasant part of this interesting meeting, in which we have strengthened and made new scientific and personal bonds.

I first want to thank Sir Hans for having consented to come and for having worked so hard during these days, 9 to 10 hours of lectures and discussions every day, not counting, I am sure, the many hours he has worked in solitude. He has been again a source of inspiration for all of us.

I would like to thank all of you individually, for I am indebted to many of you for past kindnesses and now even more for being willing to take the time to come and be here with us to honor Sir Hans. However, because of the time limitations, I will thank individually only a few of you.

I would single out Phil Cohen for having accepted me in his laboratory, as his first postdoctoral, I believe, at a time when he was a very young man and I was scarcely more than a boy. Sometimes when I am disturbed about something which may have happened in the laboratory, my wife, a graduate of Phil's at the time, reminds me of an anecdote. We had been talking about rearranging a very large laboratory, so that one evening I assembled most of the graduate students in physiological chemistry and commenced to "remodel" the laboratory according to my ideas. We ripped out several benches (none of the students, and certainly not I, had no idea that they were anchored to the wall by butterfly bolts which, when pulled out by brute force, as we did, left horrible holes in the wall). By four to five in the morning we were all tired and went home. I suppose it was fortunate, for when I came back very late the next day, all that Phil said was, "Santiago, before you do that again, let's talk it over first!" And I must mention Sarah Ratner for her charming attitude toward me, always, but particularly during my first year in the United States in Ochoa's laboratory; she was already a most outstanding and well-known scientist, but she had the graciousness to treat me as an equal, even if I was at most a fledgling biochemist. I would, in addition, like to thank my students, for they are the ones who, in reality, have taught and helped me the most. I would, therefore, like to single out those who are here in the audience, that is Drs. Carreras, Chabás, Crist, de la Morena, Feijoo, Gonzalez, Guerri, Kennedy, Puig, Raijman, Ramponi, and Rubio. Also, I would like to compliment a very gifted group of young people who have contributed greatly to the efficiency and success of this meeting; these are Miss Báguena, Mr. Viña, my son Jim, Ms. Viqueira Niel, and Miss Garcia-Conde. Last, I would like to thank our coordinator for this meeting, Ms. Judi Mendelson, who has worked not only very hard and diligently but with alertness, initiative, and inspiration to solve the many problems as they arose.

H. A. Krebs: I have not yet had an opportunity to express my thanks to Dr. Grisolia, to all participants, and to the sponsors of this symposium, and so I must ask you to allow me to impose on your time and patience.

It was Santiago Grisolia—alone—who conceived the idea of honoring me through this event. It was Santiago Grisolia who secured the necessary financial backing from the sponsors listed in the Agenda Paper. It was he who undertook the responsibility for the whole organization, involving an enormous amount of correspondence and quite a lot of commuting between Valencia and Kansas City. We all appreciate his efforts, crowned by undoubted success. I know that everybody enjoyed the scientific as well as the social aspects of the symposium. I myself, of course, appreciate especially the spirit behind this event, the idea to do something that would give me pleasure. It has given me indeed an enormous amount of pleasure. This was a truly memorable occasion which, because of its unique character, will always remain vivid in the minds of the participants—not only mine.

My sense of gratitude also extends to the participants who have traveled from far afield. Their contributions maintained a high standard of scientific content, and besides they were exceedingly kind to me personally. I appreciated especially that they prepared a copy of their papers which will soon be in my hands. There are many points which I would like to study thoroughly, and I am very pleased that I need not wait for the appearance of the printed version.

Finally my expression of gratitude must go to our host: to Professor Báguena, and the University of Valencia. On the occasion of the beautiful, splendid, dignified, and touching ceremony I had an opportunity of thanking the university for the honor bestowed on me. But Professor Báguena and the university have done much more, by acting as our host, providing facilities, entertaining us generously and beautifully at meals and at a splendid concert where we enjoyed modern music in an ancient setting.

If at this stage of the proceedings you find me somewhat talkative, please bear in mind that after being lauded for three days I am in a special frame of mind, described to me sometime ago by a visiting speaker after I had introduced him to the audience praising his achievements. Somewhat flustered, he began his lecture, "I never knew what a marvelous fellow I am, but now that you have told me, I can hardly wait to listen to myself." Forgive me if I am in a somewhat similar frame of mind and have talked longer than I ought.

Dr. Báguena: The day before yesterday, I began the formal ceremonies saying: "Excelentisimos y Magnificos Señores Rectores, Excelentisimas e Ilustrisimas Autoridades, etc." Now, after three days of being together, I say:

"Dear Friends: Thank you very much again for being here and Sir Hans for your kind words. I am an internist and, of course, couldn't discuss properly your work. In a few days you will leave Valencia, but you will remain in my heart and in that of this university. I hope and I wish to see you again. In the name of His Highness, the Prince of Spain, the Symposium is adjourned."

Index